Secure Smart Embedded Devices, Platforms and Applications

Konstantinos Markantonakis
Keith Mayes
Editors

Secure Smart Embedded Devices, Platforms and Applications

Foreword by Fred Piper

 Springer

Editors
Konstantinos Markantonakis
Keith Mayes
Information Security Group
Smart Card Centre
Royal Holloway
University of London
Egham, Surrey
UK

ISBN 978-1-4939-5192-5   ISBN 978-1-4614-7915-4   (eBook)
DOI 10.1007/978-1-4614-7915-4
Springer New York Heidelberg Dordrecht London

Printed on acid-free paper

Springer is part of Springer Science+Business Media (www.springer.com)

I would like to dedicate this book to the memory of my father, Georgios Markantonakis. Thank you dad!
Konstantinos Markantonakis

I would like to dedicate this book to my family and friends, and to people who succeed despite disadvantage
Keith Mayes

Foreword

This is the second book to be co-edited by Keith Mayes and Konstantinos Markantonakis. The first, *Smartcards, Tokens, Security and Applications* was published in 2008 and this volume is a natural 'companion' of that earlier publication and greatly expands on the range of content. Both are the result of experiences gained in managing the Smart Card Centre (SCC) at Royal Holloway, University of London.

The SCC, which was founded 10 years ago by Vodafone and Giesecke & Devrient, and has since been supported by numerous other companies, teaches a specialist module to students studying for the M.Sc. in Information Security. That module, which has the same title as their first book, focuses strongly on the relevant technical, practical and security issues.

Just as with the earlier book, the editors have produced an informative volume that is easy to read and its wide range of topics, which includes RFID, NFC, Mobile communications and wireless sensor nodes, (to list only a few), will appeal to a much wider audience than the Masters students at whom it is primarily aimed. This wider audience is likely to extend to researchers and experts form industry and governments. The two editors are both active researchers and their enthusiasm for research adds extra interest to a fascinating area.

It is clear that technology has advanced enormously over the years, although the fundamentals of Information Security may not have changed very much. Whether we are dealing with pencil and paper or advanced super computers, the motivation for fraud or security attacks and many of the reasons why vulnerabilities exist still have very human origins. Perhaps the biggest change is that people have transformed from being occasional users of technology to being dependent upon it and its underlying security properties. We also have generations that have grown up with the computer, mobile phone and Internet connectivity as essentials for life and they consume services and share personal data with carefree enthusiasm, whereas older heads might worry about how the technology works, who controls the system and data etc. In response to this it is certainly possible to focus on the security of something "big" like the Internet, but is very important to remember that much of what we rely on to keep us and our data and activities secure, is a collection of increasingly complex smaller devices. For example, the mobile phone is really a concept and what we actually buy and use is an electronic assembly with processors, memories, security modules, displays, batteries, speakers etc. We could almost describe a car in the same way as it only functions correctly because a large number of embedded electronic modules, processors, sensors and communications links work as they should. Therefore if we are to fully understand the threats to modern systems and services, and then to help protect against them, we should keep abreast of developments in embedded systems. A textbook on secure smart embedded devices, platforms and applications would therefore seem a welcome addition to the bookshelf.

Fred Piper
Founder, Director of the Information Security Group
Royal Holloway, University of London

Preface

As we progress into the twenty-first century it seems that the pace of technological advance shows no sign of slowing. We are in fact becoming increasingly dependent on technology in our normal day-to-day lives, which means that we are critically reliant on the security of systems and services that are built upon this technology. In exploring this issue within a textbook, one could consider the high-level design aspects or concentrate more on the nuts and bolts of security systems. This book focuses mainly on the latter approach, as the editors and authors felt there was no introductory overview that covered a sufficient breadth of available technology and related issues. Generally speaking, a complex system is made up of smaller components such as devices, processors, security modules, memories etc. and knowing which of these can be trusted (and to what extent) to resist attacks and misuse, is critical to the security of the complete system. For example, a very sophisticated and expensive car might be reliant on a tiny embedded device (chip) in the engine management system, for it to start and for protection against theft. It is hoped that this book will help to clarify the role of embedded devices, their capabilities, and how best to exploit them in secure system designs.

Structure of the Book

The book consists of 24 chapters organised in four sections. Part I introduces some typical embedded devices and hardware, before some more generic information on security issues is provided in Part II. The Part III (which is the largest section) considers a wide range of application aspects and considerations. Part IV is provided for readers who are interested in application development for embedded devices. The chapters are written as self-contained texts, from a range of expert authors and can be read individually or in the book order. The chapters are briefly introduced below.

Part I: Chapter 1 provides an overview of smart cards and (RFID), their security capabilities and attack resistance, and their widespread use within a range of security sensitive applications. Chapter 2 then introduces Digital Signal Processor devices which are widely used in modern devices, such as mobile phones. Chapter 3 relates the historical development of microprocessor and microntroller chips and goes onto cover the specialist design of secure embedded microcontrollers. Chapter 4 introduces a specific type of secure controller, the Trusted Platform Module (TPM) and its mobile equivalent, that are intended to ensure (amongst other things) the safe boot up of a computing platform, so it is a reliable platform on which to load applications. Chapter 5 considers the Very Large Scale Integration (VLSI) approach to the design of electronic hardware and the potential for security attacks and associated countermeasures.

Part II: Chapter 6 provides a general recap on information security best practices. Although we are focussing on embedded devices we must not forget that without a secure theoretical design the implementation security will be fundamentally flawed. Chapter 7 illustrates how a theoretically sound security design can be undermined by a poor implementation that lacks attack resistance. The chosen attack target is the smart card; however the principles are applicable to most embedded security devices. Chapter 8 considers the Graphics Processing Unit (GPU), a processing platform that is often overlooked for its security capabilities. It can be used as a cryptographic processor; however it is also a target for malware and general misuse. Chapter 9 focuses on the FPGA, which has been exploited both to protect and to attack security systems. The discussion also extends to the protection of valuable Intellectual Property loaded into FPGAs used in commercial systems.

Part III: Chapter 10 considers a range of options for providing mobile communications security controllers. It begins with the conventional Subscriber Identity Module (SIM) and the associated personalisation, management and usage processes, but goes on to consider other possibilities, including software SIMs and TPMs. The action taken by a mobile device depends not just on the security controller, but the validity of the data that it receives, which increasingly can include a representation of physical location. Chapter 11 discusses practical approaches to location estimation, highlighting the possible security vulnerabilities. Car Satellite Navigation systems are just one obvious example of this; however as discussed in Chap. 12 motor vehicles are packed with processing technology that has important safety and security aspects. By contrast, payment card systems tend not to have such emphasis on safety, but they are required to safeguard significant financial transactions. The potential to undermine the payment terminals is discussed in Chap. 13 with reference to published attacks. Another technology where the misuse may have both safety and security implications is the (WSN) which is described in Chap. 14. For example, if a sensor value is modified, replaced or blocked the resulting effect could be serious and/or costly if the system was used for say telemedicine or metering. In fact a number of sensing and terminal solutions are proposed around mobile devices and this seems to be expanding with the arrival of (NFC) Technology. Chapter 15 considers NFC and its security in detail, and how the phone (or laptop, PDA, tablet) may emulate an RFID, or act as an RFID reader, or communicate with other NFC phones over a close proximity link. Although NFC includes a Security Element (SE) some aspects of the functionality are reliant on the phone platform security, which has vulnerabilities similar to conventional PCs. To clarify this problem, Chap. 16 provides a recap on BIOS and Rootkit infections on computing platforms. Specialist computing/server equipment can get around this problem to some extent by the use of security hardened peripheral devices for sensitive processing. These are commonly known as Hardware Security Modules (HSM), and are discussed in Chap. 17. Such devices are normally required to be formally security evaluated and the Common Criteria approach to this is outlined in Chap. 18. In Chap. 19 there is a description of Physically Uncloneable Functions (PUFs) that have generated significant academic interest and then in Chap. 20 there is an overview of SCADA systems security that has generated significant industry concerns.

Part IV: Chapter 21 provides an overview of the PIC family of microcontrollers that are intended for general-purpose non tamper-resistant implementations; however they are often used as clone platforms, as well as for research experiments. More secure implementations are commonly implemented on Java Card platforms and the programming aspects are introduced in Chap. 22. Java has also been a preferred approach for mobile phone platforms and this approach plus important APIs are described in Chap. 23. Finally, for readers interested in experimenting with Wireless Sensor Nodes, some practical guidance on available platforms is presented in Chap. 24.

The ISG Smart Card Centre   Keith Mayes
Royal Holloway, University of London   Konstantinos Markantonakis
www.scc.rhul.ac.uk; www.isg.rhul.ac.uk

ISG Smart Card Centre—Members Message

The (SCC) was established more than 10 years ago at Royal Holloway, University of London. The primary objective was to create a World-Wide Centre of Excellence for training and research in the field of Smart Cards, applications and related technologies. Over the years this has expanded into RFID, NFC, mobile devices and general embedded/implementation system security. Following the success of its first textbook in 2008 (Mayes and Markantonakis (eds), *Smart Cards, Tokens, security and Applications*, Springer) it was felt that this new book was now needed to cover more aspects of Secure Embedded Devices.

The SCC is part of the World renowned Information Security Group (ISG) that is one of the oldest and largest such groups and is one of the UKs Cyber Security Academic Centres of Excellence, with alumni of over 2,000 M.Sc./Ph.D. postgraduates. The SCC in common with ISG principles is very strongly engaged with industry, focussing on responsible research into real world projects of significant impact, and actively engaging industry experts into postgraduate training, research and publication.

As representing the range of supporting industrial members, we are pleased to be associated with the work and publications of the SCC.

Orange Labs (UK)
Transport for London
UK Cards Association
ITSO

Acknowledgments

This book would not have been possible without the help and support of a number of organisations and individuals. Firstly we would like to thank Orange Labs (UK), Transport for London, The UK Cards Association, ITSO and Royal Holloway, University of London for their tremendous support of the ISG Smart Card Centre. We owe an enormous debt of gratitude to all chapter authors and reviewers for their expert contributions and patient co-operation. We would also like to extend our thanks to Fred Piper for writing the foreword. Last, but certainly not least, we must thank Raja Naem Akram for his tremendous efforts in helping to bring this book to print and to Sheila Cobourne for proof reading on an epic scale.

Contents

Contributors

Raja Naeem Akram received B.Sc. from University of Punjab, Pakistan in 2002. After graduating, he studied for M.Sc. Computer Science from University of Agriculture, Pakistan. In 2004, after completing the Master's course in Distinction he worked as IT/Computer Science teacher at Government M.C. High School, Samanabad. He completed his M.Sc. Information Security from Royal Holloway, University of London (RHUL) with Distinction in 2007. In 2012 he completed his Ph.D. under the supervision of Dr. Konstantinos Markantonakis. Worked as Senior Research Fellow on the RatTrap project, designing fraud detection techniques in on-line affiliate marketing at Edinburgh Napier University. Currently, working as a Research Fellow at the Cyber Security Lab, Department of Computer Science, University of Waikato, New Zealand. His research mainly focuses on user centric security and privacy models in different computing environments.

Dr. Jérémie Albert received his M.Sc. in Computer Science from the University of Bordeaux in 2007 (with honors). Next, he pursued a Ph.D. in Computer Science in this same University under the supervision of Prof. Serge Chaumette. During his Ph.D. he designeda process calculus suitable for the modeling of Highly Mobile Ad Hoc Networks. After his graduation (with highest honors) in 2010, he was an Assistant Professor at the Polytechnic Institute of Bordeaux. Since 2011, he is a Senior Solutions Architectat Ezakus, Bruges, France. His current research interests are related to distributed computing, large datasets processing, semantic computing, machine learning and game theory.

Dr. Lionel Barrére is the Director of Research and Development team at H5 Audits. He previously graduated from the University of Bordeaux, France, and then received his Ph.D. in2009 under the supervision of S. Chaumette for his work on services over military MANets (Mobile Ad hoc Networks), work that was funded by the DGA (French Army). At the end of 2008, he joined the H5 Audits company to create its R&D Department. His research areasare oriented towards the development of passive network monitoring tools such as network probes.

Tony Boswell began working in IT security as a security evaluator in one of the original UK government Evaluation Facilities in 1987. Since then he has worked on a wide range of secure system developments and evaluations (including the

ITSEC E6 certificationsof the Mondex purse and the MULTOS smart card operating system) in the government and commercial domains. Much of Tonys recent
work has been on integrated circuit security projects and on server and application-
level virtualisation. Tony has been involved in UK and international interpretation
of evaluation requirements for smart cards and payment terminals since 1995, and
continues to contribute to multi-national technical community work on interpretation and maintenance of Common Criteria evaluation requirements, as well as
assisting hardware and software developers to take their products through Common Criteria evaluations. He is currently a Principal Consultant and CLEF
Technical Manager at SiVenture (www.siventure.com).

Serge Chaumette is a Professor in Computer Science at the University Bordeaux
1, Leader of the Mobility, Ubiquity, and Security research group at Laboratoire
Bordelais de Recherche en Informatique. He has been using the Java technology
for distributed programming since its early beginning. Being concerned by security
issues he participated in thedesign of tools to help in the process of evaluating Java
Cards (and applications) within government funded industrial projects. He then
naturally moved to the domain of mobile systems and networks and he is collaborating with the French Army in the area of MANets; his group is designing
peer to peer applications to support battlefield/emergency situation management
(fleets of mobile terminals, robots, or drones). He is an expert by the European
Union for Framework Program 7 (FP7) and non FP programs, expert by the ANR
(French National Research Agency), and expert by the AERES (French Research
and Higher Education Evaluation Agency). He is a member of IEEE, of IEEE
Portable Information Devices (PID) group, of Situation Management SubCom-
mittee of the Communications and Operations Technical Committee of the IEEE
Communication Society, of IFIP groups 8.8 Smart Cards and 11.2 Pervasive
Systems Security.

François Durvaux is a Ph.D. Student at Universit Catholique de Louvain. He
received the Electro-mechanical Engineering Science degree from UCL in 2010
with his master thesis under the supervision of Pr. Jean-Didier Legat. He joined the
Pr. Legat's team in October2010 to work as a researcher in the field of digital
nanoelectronics. In January 2011 he started a Ph.D. thesis in cryptography under
the supervision of Pr. Franois-Xavier Standaert at UCL. His researches are currently focused on cryptographic hardware design, side-channel analysis, and
intellectual property protection.

Dr. Igor Nai Fovino Igor is the Head of the Research Division of the Global
Cyber Security Center. Igor has deep knowledge in the fields of ICT Security of
Industrial Critical Infrastructure, Energy and Smart Grids, Risk Assessment, IDS,
Cryptography. He is author of more than 60 scientific papers published on inter-
national journals, books and conference proceedings; moreover, he serves as
reviewer for several international journals in the ICT security field. In May 2010
he received the IEEE HSI 2010 best paper award in the area of SCADA Systems.
He is also an expert in European Policies (mainly in CIIP field). From 2012 he is

member of the European Commission Experts Working Group on the security of ICS and Smart Grids. During his career Igor worked as Contractual Researcher at the University of Milano in the field of privacy preserving datamining and computer security and as Contractual Professor of Operating Systems at the University of Insubria. From 2005 to 2011 he served as Scientific Officer at the Joint Research Centre of the European Commission, providing scientific support to the EU Policies related to the EPCIIP program. Since 2007 he is member of the IFIP Working Group on Critical Infrastructure Protection.

Lishoy Francis is a Security Researcher with background in Computer Science and Engineering, and has specialised in Information Security. In 2002, he graduated with a B.E. in Computer Science and Engineering from Visvesvaraya Technological University (VTU), Belgaum, India. In 2004, he graduated with a M.Sc. degree in Information Security from Information Security Group (ISG) at RHUL. He is currently in his final stages of a Ph.D. degree in Information Security at RHUL. He has extensive practical experience in security testing and product prototyping of smart card, mobile, location, contactless, RFID and proximity technologies. He is an acknowledged and published expert on NFC security. He started his career by working as a Software Engineer in Wipro Fluid Power LTD (Wipro Group), Bangalore, India; and more recently he worked as an Expert Consultant at Crisp Telecom UK LTD. He is currently employed as a Senior Research Engineer at France Telecom R&D UK LTD (Orange, UK) where he is enterprising excellence and innovation in information security.

Dr. Mick Ganley is an independent security consultant who has worked in the industry for 25 years. He specialises in the security of card payment systems, cryptography and key management, hardware security modules and security management. His current client list includes a number of the worlds largest multinational corporations. Until recently, he provided consultancy services to the prestigious ISG at RHUL, and was on the editorial board of the Information Security Technical Report, published jointly by the ISG and Elsevier. In previous lives he was an academic mathematician and Head of Security Analysis for the security division of Racal (now Thales).

Univ.-Prof. DI Dr. Thomas Grechenig is a Senior Architect in large IT systems and nation-wide IT-infrastructures. He is a Professor for Industrial Software Engineering at the Vienna University of Technology. He and his teams have planned, designed, and built several large scale NFC-solutions in payment, mobile keys, mobile ticketing, railway and public transport applications. In science and research the focus of interest goes towards (a) enhancing the stability and fine-tuning of the NFC-mass-concept in all its critical sectors (usability, security, IT-infrastructure, performance, integration and interoperability) (b) redefining every day use cases and interactions for the consumers via NFC in a way that preserves the users' old experiences while providing "the new ubiquitous feeling of simple touch interaction" in a natural form (c) this leads into a science aiming at re-understanding daily processes like payments, locking doors, showing tickets, or personal identification in its newly adapted "Gestalt" in fusion of the old meta-phors, new facilities as well as its privacy and security needs. From a more abstract scientific point of view Thomas Grechenig promotes NFC being one out of 5 major technology enablers towards a "vitalized" environment on all three relevant operative levels: (1) "desktop"/individual (2) buildings/groups (3) urban and regional/social.

Dr. Gerhard Hancke (B.E., M.E., Ph.D., CSCIP, SMIEEE) received a Bachelor and Masters of Engineering degrees in Computer Engineering from the University of Pretoria (South Africa) in 2002 and 2003, and a Ph.D. in Computer Science for the Security group at the University of Cambridge's Computer Laboratory in 2008. He joined the ISG in 2007 as a post-doc, working within the ISG Smart Card Centre managing the RFID/Contactless research track and RF/Hardware labora-tory. In 2011, he was appointed as a Fellow within the ISG. His main interests are smart hardware tokens and their applications, mobile systems and pervasive computing.

Graham Hili B.Sc. I.T. Hons (Malta), M.Sc. (Royal Holloway) began his career with the Vodafone Group (Malta) where he was in charge of the security and the availability of the mobile value added systems (SMS, MMS, WAP). After this he moved to a consultancy position with Orange Group in The Netherlands. His current fields of specialisation include smart card security and the development of digital identity and digital security in IT systems and virtual worlds.

DI Christian Kantner graduated in Communications Engineering from the Vienna University of Technology in 1997 Christian Kantner joined Ascom Business Systems in Switzerland. He was responsible for the design and imple-mentation of data and fax protocols for the Thuraya satellite phone system. Before he joined Mobilkom Austria in 2003 he was working as Freelancer for Ascom (Solothurn), Philips Semiconductors (Zurich) and Hughes Network Systems (San Diego, CA) in the field of GSM and Data Protocols. He taught Real Time Oper-ating Systems for several years at the University of Applied Science in Upper Austria. In 2003 he joined Mobilkom Austria's TechLab, where he was respon-sible for analyzing new technologies for mobile phone operators, focusing mainly on mobile terminal technology. He has been investigating NFC technology since 2004 and got involved in NFC Forum activities in 2005. He is coeditor of the NFC technical guidelines white paper published by the GSMA. Christian Kantner joined Mobilkom Austria's m-commerce team as Product Manager for NFC in 2007. He was in the leading team for key NFC projects at Mobilkom Austria. From 2010 to 2012 Christian Kantner was head of IT and Services at Mobilkom Austrias daughter paybox Bank. Paybox Bank operates ad hoc mobile payment services for customers of A1, T-mobile Austria and Orange Austria. Christian Kantner has dedicated his career to wireless communications. Starting with 36,000 km satellite links and arriving at 3 cm NFC transactions. He has deep understanding about technological aspects as well as market insight. Christian Kantner is now driving the payment innovation roadmap in A1 Telekom Austria (former Mobilkom Austria).

Stéphanie Kerckhof is a Ph.D. student at Universit Catholique de Louvain. She received the Electro-mechanical Engineering Science degree from UCL in 2007 with her master thesis under the supervision of Pr. Jean-Didier Legat. She was a hardware developer for two years at intoPIX, Louvain-la-Neuve, Belgium. In April 2010, she started a Ph.D. thesis in Cryptography under the supervision of Pr. Franois-Xavier Standaert at UCL. Her researches are currently focused on cryptographic hardware design, side-channel analysis, and intellectual property protection.

Mario Kirschbaum received the B.Sc., M.Sc., and Ph.D. degrees in Telematics from Graz University of Technology in Austria, in 2005, 2007, and 2011, respectively. He is currently working as a member of the Secure Entities for Smart Environments (SEnSE) group of the Institute for Applied Information Processing and Communications (IAIK), Graz University of Technology, Austria. His research interests include implementation attacks, development and investigation of countermeasures, and the implementation of cryptographic hardware modules.

Dr. Gerald Madlmayr is an IT and Telecommunication Architect based in Vienna. In his daily work he is confronted with technology strategy for mobile network operators, software system and IT Integration in banking and payment systems as well as customer focused mobile technologies and devices. Besides that, he is Lecturer at the Vienna University of Technology at the Research Group for Industrial Software. There his research is focused on mobile technology in society as well as energy and environmental topics. Before that he worked as a Research Associate at the Research Center Hagenberg. There his work was focused on NFC/RFID based applications as well as security and privacy in such systems. He is an authority on NFC technology and applications, actively participating in the standardization of NFC. Within the scope of this job one of the most sophisticated NFC trials was launched in 2006. Previously Gerald Madlmayr was working as a visiting Researcher in Princeton/New Jersey at Siemens Corporate Research (SCR) dealing with the design and implementation of CSCW Systems. Before that he was part of the innovations department of Siemens mobile in Munich. There he also wrote this diploma thesis with the focus on image processing on mobile devices. Gerald Madlmayr holds a Diploma in Media Technology from the University of Applied Sciences of Hagenberg and a Ph.D. in Computer Science from the University in Linz.

Konstantinos Markantonakis B.Sc. (Lancaster University), M.Sc., MBA, Ph.D. (London) received his B.Sc. (Hons) in Computer Science from Lancaster University in 1995, his M.Sc. in Information Security in 1996, his Ph.D. in 2000 and his MBA in International Management in 2005 from RHUL. He is currently a Reader (Associate Professor) in the ISG. His main research interests include smart card security and applications, secure cryptographic protocol design, Public Key

Infrastructures (PKI) and key management, embedded system security, mobile phone operating systems/platform security, NFC/RFID security, grouping proofs, electronic voting protocols. Since completing his Ph.D., he has worked as an independent consultant in a number of information security and smart card related projects. He has worked as a Multi-application Smart Card Manager in VISA International EU, responsible for multi-application smart card technology for southern Europe. More recently, he was working as a Senior Information Security Consultant for Steer Davies Gleave, responsible for advising transport operators and financial institutions on the use of smart card technology. He is also a member of the IFIP Working Group 8.8 on Smart Cards. He has published more than 90 papers in international conferences and journals. He continues to act as a consultant on a variety of topics including smart card security, key management, information security protocols, mobile devices, smart card migration program planning/project management for financial institutions, transport operators and technology integrators.

Stathis Mavrovouniotis was born in Athens, Greece on June 27th, 1981. Stathis attended the Athens University of Economics and Business (AUEB) and graduated in 2004 with a degree in Business Administration. Following his graduation from AUEB, Stathis attended the RHUL and received two M.Sc. degrees, in Business Information Systems (2005) and in Information Security (2006). After serving his military service back in Greece, he was offered the job of IT Security Analyst in First Data Greece International, having the main responsibilities of key management, compliance, audit preparation and Incident Investigation/Report as well as Implementation of security related tools. Stathis soon became the IT Security Manager for SE Europe, Middle East and Africa in First Data International, focusing in implementing the information security policy and addressing it with procedures and guidelines, maintaining compliance with payment schemes, PCI DSS and ISO 27001, running IT Security related audits and gap analysis, security planning, risk assessments and implementation of security awareness programs. He has been also assist in consulting and assessments around key management in different First Data sites. He has been so far qualified with the following certifications: CISM, SSCP, ISO 27001:LA, PCI ISA, CTGA and is member of ISC2, ISACA and active member of the local OWASP chapter.

Keith Mayes is the Director of the Information Security Group-Smart Card Centre (ISG-SCC) (www.scc.rhul.ac.uk) at RHUL. He is also the Founder and Managing Director of the consulting company Crisp Telecom Limited (www.crisptele.com). He is currently a non-executive independent Director of AIMs listed GMO ltd., a provider of mobile services in China and a Director of IWICS Europe Limited, a 4G mesh radio network company. Dr. Mayes has a Bachelor of Science degree in Electronic Engineering and a Ph.D. in Digital Image Processing from the University of Bath. He is a Chartered Engineer and Member of the Institute of Engineering and Technology. He is also a Member of the Licensing Executives Society and a Founder Associate Member of the Institute of

Information Security Professionals. During a long and varied industry career he has worked for Philips, Honeywell Aerospace & Defence, Racal Research and finally for the Vodafone Group as the Global SIM Manager responsible for SIM card strategy and harmonisation. Aside from his current research and teaching focus on smart cards, RFIDs and security, he has maintained an active interest in mobile communications, hardware and software development, Intellectual Property and radio relay trials.

Mehari G. Msgna received a Bachelor of Engineering degree in Electronics and Communication Engineering from Mekelle Institute of Technology (Ethiopia) in 2007. In 2009 he received a Masters of Science degree in Information Security from RHUL and he started his Ph.D. with the ISG in 2011 at the same institution. His research interests are virtual machines for embedded devices, smart cards/tokens security, biometrics and side channel analysis.

Jan Pelzl Since 1994, Dr. Pelzl works in the area of IT-security. In 1997, he received the certificate as telecommunication technician from the company Bosch Telecom. Since 1999, Dr. Pelzl is working in the area of embedded IT-security. He successfully accomplished many national and international projects and released numerous related publications at renowned international conferences and in journals. As a researcher, Jan Pelzl investigated practical aspects of elliptic-curve-based cryptography and cryptanalysis. Dr. Pelzl is teaching data security and introduction to cryptography for industry courses, e.g. for TV-Akademie Rheinland, gits AG and Ruhr-University of Bochum. From March to August 2007, Dr. Pelzl was Chief Technology Officer (CTO) of ESCRYPT GmbH. Since September 2007, Dr. Pelzl is Managing Director of ESCRYPT GmbH.

Thomas Plos received the B.Sc. and M.Sc. degrees in Telematics from Graz University of Technology (TU Graz) in 2004 and 2007, respectively. In 2011 he received the Ph.D. degree in Computer Science from TU Graz. His research interests include digital VLSI design with a focus on low power and low-area circuit design, information security, RFID technology, and implementation attacks such as side-channel analysis and fault analysis. Currently, he is a post-doctoral researcher at the Institute for IAIK at TU Graz.

Konstantinos Rantos is an Assistant Professor at the Industrial Informatics Department of the Technological Educational Institute of Kavala. He received his Diploma in Computer Engineering and Informatics from the University of Patras, Greece, and both his M.Sc. and Ph.D. in Information Security (sponsored by Marie Curie Research and Training Grant) from RHUL. He has extensive project involvement and substantial (more than 15 years) private- and public-sector experience in the area of information security which he gained while holding positions in both sectors as well as in academia. His scientific interests lie in the areas of public-key infrastructures, embedded systems, e-government services, authentication systems, smart cards, electronic payment systems and security awareness. He is a reviewer to a number of conferences and scientific journals and authored many articles and papers.

Serendra Reddy has earned his B.Sc. in Engineering from the University of Natal and a Masters in Engineering from the University of Pretoria. He is currently completing his Ph.D. in Engineering at the University of Cape Town, South Africa. His doctoral research is involved in the investigation and development of methods for autonomous three dimensional conversion of two dimensional monocular images. Between 1999 and 2005 he worked and consulted for Siemens Telecommunications, having been involved in projects locally and on-location across Africa, the Middle East and Europe. In 2007 he joined the academic staff of the Department of Electronic Engineering at the University of Pretoria, where he lectured on Digital Systems and Digital Signal Processing, and was involved in the Intelligent Systems research group. In 2011 he joined the academic staff of the Department of Electronic Engineering at the Durban University of Technology, where he currently lectures on Radio Engineering. He serves as the Chair of Communications and is a founding member of the Intelligent Systems research group. His research interests include artificial intelligence, machine learning, computer vision, pattern recognition, embedded systems and robotics.

Francesco Regazzoni is a post-doctoral researcher at ALaRI Institute of University of Lugano (Lugano, Switzerland). He received his Master of Science degree from Politecnico di Milano (Italy) and his Ph.D. degree from University of Lugano (Switzerland). He has been a post-doctoral researcher at the Crypto Group of the Universit Catholique de Louvain (Louvain-la-Neuve, Belgium) and has been a visiting researcher at several institutions, including NEC Labs America (Princeton, NJ, USA), Ruhr-University of Bochum (Bochum, Germany), and EPFL (Lausanne, Switzerland). His research interests are mainly focused on embedded systems security, covering in particular side channel attacks, cryptographic hardware, and electronic design automation for security.

Prof. Dr.-Ing. Ahmad-Reza Sadeghi is the Head of the System Security Lab at the Center for Advanced Security Research Darmstadt (CASED), Technische Universitt Darmstadt and the Scientific Director of the Fraunhofer Institute for Secure Information Systems (SIT), Darmstadt, Germany. Since January 2012 he is the Director of the Intel-TU Darmstadt Security Institute for Mobile and Embedded Systems in Darmstadt, Germany. He received his Ph.D. in Computer Science from the University of Saarland in Saarbrcken, Germany. Prior to academia, he worked in research and development of telecommunications enterprises, amongst others Ericson Telecommunications. He has been leading and involved in a variety of national and international research and development projects on the design and implementation of trustworthy computing platforms and Trusted Computing, security hardware, Physically Unclonable Functions (PUFs), cryptographic privacy-protecting systems, and cryptographic compilers (in particular for secure computation). He has been continuously contributing to the IT security research community and serving as general or program chair as well as program committee member of many conferences and workshops in information security and privacy, Trusted Computing and applied cryptography. He is on the Editorial Board of the ACM Transactions on Information and System Security.

Damien Sauveron is Assistant Professor at the XLIM (UMR 6172 University of Limoges/CNRS—France) Laboratory since 09/2004. Damien Sauveron worked during three years for the ITSEF of SERMA Technologies on the Java Card security. During his thesis that he carried out in the Distributed Systems and Objects team of the LaBRI he was one of the main developers of a Java Card emulator, he introduced the concept of pre-persistance in Java Card and he highlighted a new category of attacks on the open multi-application smart cards. From 01/02/2006 to 10/08/2006, he was an invited researcher at the ISG-SCC of the RHUL. He is member of the IFIP WG 8.8 Smart Cards, member of the IFIP WG 11.2 Small System Security and member of IEEE.

Dipl. Ing. Steffen Schulz received his Diploma degree in Information Security from Ruhr-University Bochum, Germany. He works as Research Assistant at the System Security Lab at Ruhr-University Bochum and at the CASED, Technische Universitt Darmstadt, Germany. He currently pursues his Ph.D. in Trusted Infrastructures and Trust Management in a joint cooperation between the System Security Lab in Bochum and Macquarie University Sydney, Australia. Steffen Schulz was involved in several national and international research projects, where he participated in the design and development of trustworthy operating systems, trust establishment in resource-constrained environments and trusted virtualization infrastructures (TVDs). Furthermore, he has worked on different aspects of network security and covert channels, with several publications in international conferences.

Peter Schwabe is a Post-Doctoral Researcher at the Research Center for Information Technology Innovation of Academia Sinica, Taiwan. He graduated from RWTH Aachen University in Computer Science in 2006 and received a Ph.D. from the Faculty of Mathematics and Computer Science of Eindhoven University of Technology in 2011. His research area is the optimization of cryptographic and cryptanalytic algorithms in software. The target architectures of this software range from high-end desktop and server CPUs through parallel architectures such as the Cell Broadband Engine and graphics processing units to embedded processors such as ARM and AVR. He has published articles at several international conferences on fast software for a variety of cryptographic primitives including AES, hash functions, elliptic-curve cryptography, and cryptographic pairings. He has also published articles on fast cryptanalysis, in particular attacks on the discrete-logarithm problem.

Chris Shire has a background in security technologies and semiconductor hardware. He joined Infineon (then Siemens) in 1998 in the Chipcard and Security business line, with many years experience in the industry. His current focus of activity is on projects in the government and finance sectors. He is active on several advisory committees helping to set standards for the UK, and support new security solutions. Chris is an active member of the IET, Intellect, UK Smart Card Club and has been a guest lecturer for several years on the RHUL M.Sc. course on Smart Card Security. He has written several articles on security technology and contributed to textbooks on the subject.

Francois-Xavier Standaert received the Electrical Engineering degree and Ph.D. degree from the Universite Catholique de Louvain, respectively in June 2001 and June 2004. In 2004–2005, he was a Fulbright visiting researcher at Columbia University, Department of Computer Science, Network Security Lab and at the MIT Media Lab, Center for Bits and Atoms. In March 2006, he was a founding member of IntoPix s.a. From 2005 to 2008, he was a post-doctoral researcher of the UCL Crypto Group and a regular visitor of the two aforementioned laboratories. Since September 2008, he is Associate Researcher of the Belgian Fund for Scientific Research (F.R.S.-FNRS) and Professor at the UCL Institute of Information and Communication Technologies, Electronics and Applied Mathematics (ICTEAM). In June 2011, he has been awarded a Starting Independent Research Grant by the European Research Council. His research interests include digital electronics, FPGAs and cryptographic hardware, low power implementations for constrained environments (RFIDs, sensor networks, ...), the design and cryptanalysis of symmetric cryptographic primitives, physical security issues in general and side-channel analysis in particular.

Michael Tunstall has been involved in the research and development on the implementation of cryptographic algorithms on embedded platforms for close to nine years. He was originally employed by Gemplus (now called Gemalto after a merger with Axalto) to develop authentication algorithms for GSM SIM cards. After several years working for Gemplus Michael changed roles within the team to focus on research into attacks and countermeasures that could be applied to smart cards. He was involved in evaluating Gemplus' products to determine whether a suitable level of security had been achieved. The research conducted while Michael was at Gemplus enabled him to start a Ph.D. At RHUL resulting in his thesis entitled "Secure Cryptographic Algorithm Implementation on Embedded Platforms". Michael is currently employed at University College Cork as a postdoctorate researcher, and is currently funded by an Enterprise Ireland grant to develop side-channel countermeasures for FPGA implementations of AES and elliptic curve cryptographic algorithms.

Dipl. Ing. Christian Wachsmann received his Diploma degree in Information Security from Ruhr-University Bochum, Germany. He worked as a Research Assistant at the System Security Lab at the Horst Grtz Institute for IT Security (HGI) at Ruhr-University Bochum. He is currently employed as a Research Assistant at the System Security Lab at the CASED, Technische Universitt Darmstadt, Germany and pursues his Ph.D. on privacy-protecting protocols for mobile and resource constrained embedded devices, in particular RFIDs and smartphones. His work focuses on the development, design and formal modeling of cryptographic primitives and protocols based on physical security features, in particular PUFs. He was and is involved in a variety of national and international research projects and has been continuously contributing to IT security research with several publications at international conferences.

Colin Walter has spent the last 25 years concentrating on the practical implementation of cryptography, partly in industry and partly in academia. He helped design one of the first RSA chips for Plessey-Crypto in 1989. He published the first fully systolic array for modular exponentiation in 1993 and this is now widely used in SSL accelerator chips. In the late 90s he did some consultancy for Multos to understand and reduce side channel leakage from public key cryptography on smart cards. This led to some important work on the implementation of Montgomery modular multiplication and some improved algorithms for exponentiation. He joined the ISG at Royal Holloway in 2009 after 8 years working on product development as head of cryptography at a well-known certificate authority. For many years he was on the steering committee of the IACR CHES workshops, and was programme chair and local organiser for two of these. He is a senior member of the IEEE.

Marko Wolf Dr.-Ing. Marko Wolf is a senior IT security expert and branch manager of ESCRYPT GmbH in Munich. Marko is primarily active in the area of automotive data security and privacy protection for various industry customers in Europe, Asia, and the US as well as for different national and international government authorities and standardization bodies. Marko studied Electrical Engineering and Computer Engineering at the University of Bochum (Germany) and at Purdue University (USA). After receiving his M.Sc. in 2003, he started his Ph.D. in the area of Trusted Computing and vehicular IT security at the Chair for Embedded Security hold by Prof. Dr. Christof Paar. Wolf completed his Ph.D. in 2008 with the first comprehensive work about vehicular IT security engineering. He is editor/author of the books *Embedded Security in Cars* (Springer, 2006) and *Security Engineering for Vehicular IT Systems* (Vieweg+Teubner, 2009), program chair of the international Embedded Security in Cars (escar) workshop series, and has published over 30 articles in the area of embedded IT security and privacy.

Thomas Wollinger Dr. Wollinger has worked in the area of data security and embedded security since 1997. He implemented and led several projects, for instance, at secunet AG. Dr. Wollinger has published numerous articles at international conferences and in relevant journals in the area of security. Dr. Wollinger frequently gives invited talks and teaches data security courses (e.g. at Motorola Labs Paris, gits AG, and TV Academy Rhineland). He obtained his B.S. from the University of Dieburg and obtained his Master of Science at the Worcester Polytechnic Institute, USA. In June 2003, he obtained his Ph.D. with honors from the University of Bochum. Dr. Wollinger worked as Chief Sales Officer (CSO) at ESCRYPT from 2005 to 2007. Dr. Wollinger established the technical sales and marketing structure of the company. He was involved in all acquisitions regarding ESCRYPT projects. Since 2007, Dr. Wollinger is Managing Director of ES-CRYPT GmbH.

Part I
Embedded Devices

Chapter 1
An Introduction to Smart Cards and RFIDs

Keith Mayes and Konstantinos Markantonakis

Abstract Security systems often include specialised modules that are used to build the foundations of attack-resistant security. One of the most common modules has been the smart card; however, there are often misconceptions about the definition of the smart card and related technologies, such as Radio Frequency Identification (RFID), as well as the requirement and justification for using them in the first place. These misconceptions are fuelled by the ever evolving nature of applications, security technology, personal devices and the growing threats that they must deal with. There is also a question of whether smart cards/RFIDs should really be in a book about embedded security, but we will see that the "embedded" aspect is growing ever stronger especially with developments in the mobile phone area. This chapter will consider a range of smart cards and RFIDs, and associated applications. It will also briefly cover the traditional manufacture, personalisation and management aspects, illustrating how they are challenged by new mobile developments.

1.1 Introduction

Book chapters describing smart cards and Radio Frequency Identity (RFID) often put a lot of focus on early history. This is all very interesting, but not really the primary focus for a new book. We will direct our discussions towards the electronic "chips" that are utilised in smart cards and RFIDs and which are now also being incorporated into other electronic equipment such as mobile phones. However, for

K. Mayes (✉) · K. Markantonakis
Information Security Group, Smart Card Centre, Royal Holloway, University of London, London, UK
e-mail: keith.mayes@rhul.ac.uk

K. Markantonakis
e-mail: k.markantonakis@rhul.ac.uk

K. Markantonakis and K. Mayes (eds.), *Secure Smart Embedded Devices, Platforms and Applications*, DOI: 10.1007/978-1-4614-7915-4_1,
© Springer Science+Business Media New York 2014

readers that would feel disappointed without a little history, here are a few critical developments.

- 1940: A generally held view is that the earliest RFID system appeared during WWII and was based on RADAR transmissions to/from aircraft. It was called Identification Friend or Foe (IFF), although strictly speaking it could only identify friends that still had functioning equipment. In one mode, a radar pulse striking an aircraft would trigger a "friendly" coded response, i.e. an identity transmitted by radio means.
- 1968: This is probably the earliest appearance of what might one day be regarded as a smart card. At the time it was referred to as the automated chip card; and the invention was attributed to Helmut Grttrup and Jrgen Dethloff. The associated patent was granted much later in 1982 to Giesecke and Devrient.
- 1974: The first memory card appeared and was attributed to Roland Moreno.
- 1977: The first microprocessor card was attributed to Michel Ugon from Honeywell Bull.
- 1978: Honeywell Bull also patented the first Self-Programmable On-chip Micro-computer.
- 1991: The first GSM [1] mobile SIM [2] card was manufactured by Giesecke and Devrient.
- 1996: The first EMV [3] card specifications were issued by Mastercard and Visa.
- 1996: The first Java Cards [4] were introduced by Schlumberger.
- 1997: The Octopus [5] smart card travel ticket was launched in Hong Kong.
- 2003: The Oyster [6] transport card was launched by Transport for London.
- 2004+: The introduction of European e-passports in accordance with International Civil Aviation Organisation (ICAO) [7].
- 2006: Nokia launched the 6131 NFC phone.

Note that throughout this chapter we will use the term smart card to indicate both smart cards and RFIDs, unless there is a need to differentiate between them. The meaning of the other terms and smart card types mentioned in the previous list will become clearer as we move through the chapter, although it is worth emphasising from the outset that whenever such a technology has been introduced it has been subject to attack. Even back in WWII an enemy would generate "fake" friendly radar signals to trick aircraft into responding with information and location. More than 70 years on this approach has similarities with fake reader attacks on modern RFIDs.

The study of smart cards has a very broad scope in which we find a wide range of devices with diverse functional and attack-resistant capabilities. However, it would not be a good start to our discussions if we did not explain why these devices are at all necessary, so we will begin by extracting some requirements from relevant applications.

1.2 Application Requirements

The fact that smart cards exist in their billions might be grounds to waive analysis of requirements as they must surely exist in overwhelming strength. This would be rather dangerous as we should satisfy ourselves that smart cards were actually needed for these applications or that those deployed are actually fit-for-purpose. We will base our brief analysis around a few well-known applications of smart cards:

- Mobile Communications.
- Banking Cards.
- Satellite TV.
- Passports/Identity Cards.
- Transport Tickets.
- Product tagging.

The reason for putting mobile communications at the top of the list is due to its dominance of the smart card market, as illustrated by Fig. 1.1.

The total size of the market is immense with 6.5 billion units shipped in 2011 rising to a predicted 8 billion by 2014. Note that Fig. 1.1 does not include an entry for the RFIDs used in tagging, which is expected to reach 3 billion units by 2014.

1.2.1 Mobile Communications

Mobile communications uses a smart card which is typically referred to as a Subscriber Identity Module (SIM) [2]; although these days it is strictly speaking a Universal Integrated Circuit Card (UICC) with a SIM application (and/or UMTS [8] variant USIM) hosted on it. It came about initially in GSM standards because the early analogue systems had poor security protection implemented in the phone, which led to call eavesdropping and account cloning. The fundamental SIM requirements were as follows:

Fig. 1.1 Smart card market by application in 2011 (*source* Infineon)

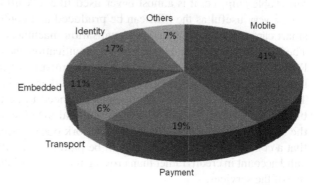

Fig. 1.2 SIM example (*source* Giesecke and Devrient)

- A portable identity/security element for easy transfer between phones.
- An attack-resistant chip.
- Algorithms/protocols for authentication and cipher key generation.
- Protected storage for unique identity and diversified cryptographic secret keys.
- Internationally standardised solutions.
- High security with supplier attested evaluation.

It is important to realise that the SIM (in common with many other smart cards) is a personalised device. The industry way of working was to have separate supply chains for phone manufacturer and SIM card manufacture and personalisation. An example SIM card is shown in Fig. 1.2.

The example shows three supported SIM sizes. The full-card (ID-1) format, the more common and smaller "plug-in" and the even smaller "third" form factor.

The first requirement in the above list is not as significant as it once was. Originally phones were very expensive and the SIM was a full size card moved between say your portable handset and your car kit. Today, the SIM is usually in a plug-in format and may be pre-installed in a purchased mobile phone, in which case the user may not even be aware of it. The SIM is a bit of an oddity in that it is the most widely used smart card (2 billion+), but is used more like an embedded, yet removable chip, i.e. it is almost never used like a "card". However, the card form is still very useful as the SIM can be produced and configured using conventional smart card manufacturing and personalisation machinery and associated processes. The SIM supports the mobile network authentication and cipher key generation (see Ref. [9] for GSM/UMTS description and comparison) in which the SIM and the back office Authentication Centre are the trusted end points in the security protocol. This means that the mobile phone does not need to be trusted and there is direct back-office control means to disable individual SIM Identities (IDs) from accessing the network. Historically, the mobile network operators have been more concerned that a communication service/call will be paid for (the SIM ID is associated with a valid account in credit) rather than proving that the legitimate phone owner is making use of the services.

1.2.2 Banking Cards

Banks make use of many Automatic Teller Machines (ATM), Point of Sale terminals (POS) and credit/debit cards to secure financial transactions. Their core requirements include those listed below:

- A portable identity/security card for use at standardised ATMs and POSs.
- An attack-resistant chip.
- Algorithms/protocols for authentication, ciphering.
- Protected storage for unique identity and diversified cryptographic secret keys.
- Support for user identity checking via Personal Identification Numbers (PIN codes).
- Card body authenticity check mechanisms (for manual inspection).
- Mag-stripe support for legacy systems.
- Internationally standardised solutions.
- High security requirements with independent evaluation.

An example banking card is shown in Fig. 1.3.

The example shown is a rather special card as not only is it a conventional EMV [3] contact bank card, but it is also a contactless card (RFID) that permits travel on the London Oyster [6] card system. Clearly, the bank card is much more card-like than the SIM and the body is part of the overall security solution, especially for human checks and fall-back situations. Because the transfer of significant amounts of money is involved it is usually a requirement to determine that the legitimate owner is using the card and hence the support for PIN codes. This is in contrast to the approach taken with SIM cards.

Transactions of significant value are still online, so effectively managed by a back-office. The POS/ATMs are intended to be controlled and trustworthy (as compared to the mobile phone) and so can form part of the overall security solution. An interesting aspect is that the banking industry has a history of trading security against cost and accessibility. A first example of this is the continued support of the mag-stripe for

Fig. 1.3 OnePulse bank card (*source* Barclaycard)

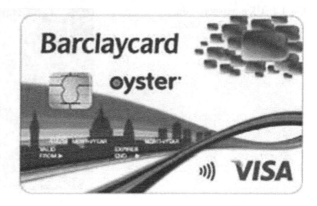

locations where the chip card cannot be used and indeed the long use of the mag-stripe before chip and PIN was introduced, despite the fact that the security was extremely weak. A more current example is the use of touch & pay contactless transactions in which there is an offline transaction and the PIN is not used. Moving from a two-factor (card and PIN) to a single factor authentication reduces security, however, the potential losses are intended to be contained by caps on the number of transactions and associated value. The business thinking is that the easy/simple customer experience will generate more transactions, and fees will be captured that outweigh any money lost to fraud.

1.2.3 Passports

Passports have existed for many time without smart card functionality and because of their relatively long lifetime, legacy and chip-enabled passports exist side by side. Therefore the body of the passport is very important, and some requirements are listed below:

- A portable identity/security card for use at border control.
- An attack-resistant chip.
- Algorithms/protocols for authentication, ciphering.
- Protected storage for unique identity and diversified cryptographic secret keys.
- Support for user identity checking (PIN codes).
- Advanced passport body authenticity check mechanisms (for manual inspection).
- Optical machine readable support for legacy systems.
- Internationally standardised solutions—ICAO [7] compliant.
- High security requirements and independent evaluation.

The picture in Fig. 1.4 shows the symbol used to recognise an e-passport, i.e. one containing the special RFID.

Fig. 1.4 UK e-passport show-ing logo (*source* www.direct. gov.uk)

The use of the chip provides an additional anti-counterfeit measure as well as traveller convenience for automated checks (compared to presented machine readable printed strips). Improved security is possible as the chip may contain secret credentials to support protocols that are not based on printed information on the passport.

1.2.4 Satellite Pay-TV

Satellite TV companies broadcast valuable media content (e.g. TV programmes, films, sports, etc.) and use smart cards within Set Top Boxes (STB) as part of their conditional access systems. Their core smart card requirements are as follows:

- A replaceable identity/security card for use in the satellite TV company's STBs.
- An attack-resistant chip.
- Algorithms/protocols for ciphering and privilege /access control.
- Protected storage for unique identity and diversified cryptographic secret keys.
- Support for user identity checking (PIN codes).
- Usually proprietary/non-standardised solutions.
- High security requirements—proprietary.

An example of a typical Pay-TV card is shown in Fig. 1.5.

Satellite TV differs from most other smart card applications in that it is a broadcast system so transmissions can be potentially received by anyone and also that there is often no return channel for the protocol. The Satellite TV companies are quite secretive and are suspected of having non-standardised proprietary defensive measures within their conditional access solutions, i.e. security by obscurity. These facts coupled with the value/desirability of the protected contact have led to a great

Fig. 1.5 Skytv card (*source* www.skytv.co.nz)

deal of attacker activity and so the requirements for security countermeasures are high. Given that the STB is under the company control it would be reasonable to suggest an alternative strategy in which the full conditional access security solution is implemented in the STB, thus avoiding the need for smart cards. The reasons for not doing this tend to be economic. The Satellite TV industry recognises that its security and possibly account details may need to be updated over time and so the companies that use smart cards have decided that it is simpler and cheaper to personalise and issue them than replace STBs.

1.2.5 Transport Ticketing

Transport service providers are increasingly turning to smart cards as electronic tickets. Their core requirements are summarised below:

- A portable identity/security card for use at their station gates/buses.
- A fast transaction.
- An attack-resistant chip.
- Algorithms/protocols for authentication, ciphering.
- Protected storage for unique identity and diversified cryptographic secret keys.
- Protected wallet/ticket functionality/storage.
- Moderate security usually supplier attested evaluation.

Figure 1.6 shows two popular examples of contactless/RFID travel cards. The Oyster [6] card is the most well known in the UK and has been very successful since its introduction in 2003. However, the Octopus [5] card from Hong Kong was introduced much earlier (1997) and is now being used for a range of purchases in addition to travel.

Smart card tickets are often used alongside legacy tickets, and are popular with customers for their ease of use and avoiding the need to queue for tickets. They also help with fraud control at gated stations and reduced cash handling as well as

Fig. 1.6 Oyster and Octopus travel cards (*source* Transport for London)

supporting statistical journey analysis and optimisation. Transport tickets have been attacked in a public manner (notably the MIFARE Classic [10] -based cards) and in response the security of the solutions has been improving, albeit driven more by reputational issues than actual measured losses from fraud.

1.2.6 Product Tagging

Product tagging and logistics is a growth area for smart card devices, although in this field they are most often exclusively referred to as RFIDs. There is in fact a wide range of devices to consider from extremely simple IDs to high-end smart cards with similar capabilities to SIMs. Core requirements include:

- Storage/memory including at least an ID—preferably protect by a security protocol.
- A fast transaction.
- Algorithms/protocols for authentication, integrity—optional.
- Low to moderate security usually supplier attested evaluation.

Product tagging was originally driven mainly by convenience, and the choice of tags by cost; however, tag sophistication and security is growing as manufacturing costs reduce.

The examples in Fig. 1.7 illustrate the diversity of tag types. The left-hand side image shows a typical self-adhesive tag which in this case is being used for medicine identification, by being stuck onto the container. The tag on the right can be used for the identification of pets and is inserted under the skin of the animal, yet can still

Fig. 1.7 Medicine and animal tags

Table 1.1 Comparison of application requirements for smart cards

	Speed	Security protocols	Storage amount	Portability	Low cost	Standards	Security evaluation
Mobile	M	H	H	L	M	H	L
Banking	M	H	M	H	M	H	H
Passport	M	H	H	H	L	H	H
Satellite	M	H	M	L	M	L	L
Transport	H	M	M	H	M	M	L
Tagging	H	L	L	M	H	L	L

be accessed by an external reader device. The diversity is possible because of the contactless interface and so the form factor is far less restricted than a contact smart card, and RFID tags can be made smaller, physically robust and reliable.

1.2.7 Comparing Requirements

The above discussions are summarised in a subjective manner within Table 1.1. Each characteristic is rated by importance as High Medium or Low (H:M:L).

We can explain some of the differences in the table, using the mobile SIM as a reference.

- In terms of speed, the SIM may have complex functionality and so needs to be reasonably swift; however, it is within a powered device and most transaction times do not inconvenience the user. By contrast, a transport card (such as an Oyster card) has to be extremely fast to maximise safe throughput at station gates during busy times.
- A SIM will support secure protocols, however, the impact of such protocols has more significance for bank cards and passports where considerable sums of money or proof of identity may be at risk.
- SIMs tend to have the highest storage of any mass market smart cards, whereas a simple RFID tag might just have a few bits of memory to hold a fixed ID.
- The SIM is not very card-like and today is not really portable, but rather transferable between devices, and in this respect is not unlike the cards used in Satellite pay-TV systems. Bank cards, passports and transport tickets rely far more on portability.
- Because they are produced in huge volumes, SIM cards are not expensive compared to their high level of functionality, however, cost is perhaps the biggest issue for RFID tagging of products where just a few pennies are available for tag purchase. Passports are perhaps the least cost-sensitive, especially as in the UK the citizen is required to pay a significant amount to obtain a passport.
- Standards are well developed in the mobile, banking and passport applications, whereas proprietary solutions are still common in satellite pay-TV and tagging.

The transport industry is also still dominated by proprietary solutions, although there are some moves towards a more interoperable approach.
• Formal security evaluation (e.g. common criteria) has historically been important to banking and passport applications. In principle, the mobile industry could also insist on formally evaluated SIMs, although cost and process delay issues have prevented this in the past.

Having established some requirements related to the real-world applications of smart cards, the next step is to consider the available devices that might satisfy those requirements, and this is discussed in the following section.

1.3 Contact and Contactless Smart Cards/RFIDs

Smart card products exist to satisfy the full range of application requirements mentioned in the preceding section. Before considering the product categories we need to first cover some basic characteristics and differences of contact and contactless smart cards, passive RFIDs and active RFIDs.

1.3.1 Cards with Contacts

Referring back to Fig. 1.2 we can see some electrical contacts behind which sits the chip. In normal use the card is inserted into a reader that makes electrical connection to the chip via these contacts. The pin-out for the contacts has changed a little over the years and Fig. 1.8 shows the traditional definitions.

VCC is the power input and GND is the ground (0V) return. CLK provides the chip with a clock signal (it does not have an internal clock), I/O is used for input/output and RST is there to reset the chip. Vpp is a throw-back to old EPROM technology when a voltage higher than VCC was need to actually write to the chip memory. These days this pin is used in SIM cards for the Single Wire protocol (SWP) which enables a SIM hosted Security Element (SE) to communicate with the phone's Near Field Communication (NFC) modem. Furthermore, in the most modern SIMs, the contacts marked RFU are now used for the USB connection which is much faster

Fig. 1.8 Smart card contacts

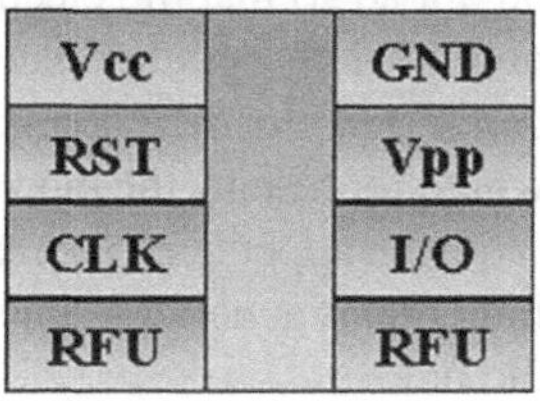

Vcc	GND
RST	Vpp
CLK	I/O
RFU	RFU

Fig. 1.9 Laminated construction of a contactless smart
card (source Giesecke and
Devrient)

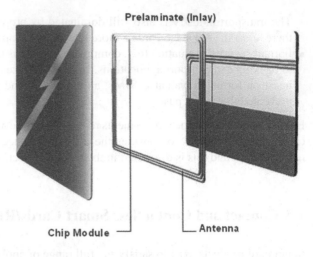

Prelaminate (Inlay)

Chip Module ———   —— Antenna

than the traditional I/O PIN interface. More information on general contact cards can
be found in Ref. [11], and [2] is a good starting point for SIM information.

1.3.2 Contactless Smart Cards/RFIDS

Clearly a contactless smart card/RFID requires a different method for powering and
communication. This is typically achieved by connecting the chip to an antenna as
shown in Fig. 1.9. Although we have shown a card example, the great flexibility
of RFIDs is that they can be made in all manner of shapes and sizes, provided
that the antenna design and size is sufficient to provide powering and support chip
communication. The basic operation is described in the following text, although the
interested reader is referred to Ref. [12] for a more detailed description.

The reader device has no contacts, but instead creates an electromagnetic field.
When the card is placed in this field, the antenna gathers energy from it (rather like
a transformer) in order to power the chip. The field is modulated by the reader in
a controlled manner so that the card can detect a clock signal and the information
transmissions/requests from the reader. The card communicates back to the reader
by modulating the field amplitude. This is basically achieved by the chip switching
on a load so that the electromagnetic field strength momentarily shrinks lower than
normal. This is rather like the way a battery voltage will drop when you switch on a
connected electrical load. This can be seen more clearly with reference to Fig. 1.10,
which represents the electromagnetic field for the reader and card transmissions.

In the upper trace the reader is able to exert strong control over the electromagnetic
field that it generates, whereas in the lower trace the card can only weakly modulate
the field.

Fig. 1.10 RFID Reader
and card signal modulation
examples

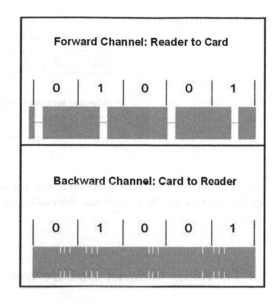

Active RFIDS: What we have just briefly described is a passive RFID which is by far the most common type in use. However, there are also active RFIDs that have their own batteries and transmitters, with perhaps the most common being used for remote locking in vehicles. Whilst active RFIDs are useful devices, their cost, size and maintenance aspects means that they are used far less than passive RFIDs, so they will not be considered further within this chapter. Instead we will focus on the family of passive devices that support standard Application protocol Data Unit (APDU) communication.

1.3.3 APDU Communication

Whether we are using a smart card with contacts or a contactless device, the relevant reader needs some logical way to communicate with the chip, once we have got beyond the physical interfaces (wires or RF). This is achieved via a simple command-response protocol in which the reader issues command messages and expects appropriate responses from the card. The commands are structured into APDUs and an example of command and response formats is shown in Fig. 1.11.

The CLA represents the "class byte" which is a static value for a given type of application. INS indicates the instruction/command and is what the reader wants the card to do, which could be to read a memory location or perhaps run an algorithm. P1 and P2 are parameters relevant to the particular INS and P3 is a data length indicator. P3 can be used to indicate the length of data that is supplied with the command, or the length of data field expected in the response from the card.

Fig. 1.11 APDU format

APDUS

Message format: Reader to Card

CLA	INS	P1	P2	P3	DATA

Message format: Card to Reader

DATA	SW1	SW2

The card response will always provide the status words SW1 and SW2 indicating the outcome of the request and depending on the INS some data may also be returned.

1.4 The Range of Smart Card Devices

Smart card products are not simply split into contact and contactless, and RFIDs are not just passive and active. There is in fact a very wide range of products with differing capabilities and costs, designed to satisfy the variety of application requirements. The product range for passive devices is depicted in Fig. 1.12 and the generic types are described below.

1.4.1 Simple ID Tag/Card

The simplest devices offer convenience and fast machine readability, but not much more. They are usually called tags or RFIDs and almost never smart cards. They contain a very small amount of memory to represent an ID which is transmitted when the device interacts with a reader. In tags that are used as barcode replacements the ID need not be unique, but perhaps represents a product type. Tags can also have unique IDs in which case the only security feature is if the ID field is read-only. There is no security protocol other than reporting the ID to the reader and so it is

Fig. 1.12 The range of smart card/RFID devices

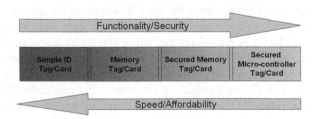

very easy for an attacker to discover the ID and program it into an emulator device or a similar ID/Tag that permits control of the ID field. In the latter case it could be a legitimate tag in its pre-configured state, i.e. before personalised with its ID.

Anyone with some interest in information security might wonder why such tags (which are usually proprietary) are used when they have almost no security protection and there are better devices available. Cost is not far from the answer, but you also avoid the trouble of key management (as there are none!). This is not to say that the use of such tags is inappropriate, it depends on the application. For tagging your groceries it could be fine, although using it as a building access control card would be very worrying. It also depends on what you are replacing. A system based on paper vouchers and human inspection could be revolutionised by the introduction of automated inspection of machine readable tags and whilst this could introduce potential opportunity for fraud, it might be far less than for the legacy system it replaces.

1.4.2 Memory Tag/Card

Some applications require more data to be stored on the tag than just the ID and so need more capable devices, which we will refer to as memory tags. Usually they have a unique read-only ID (like the ID/Tag) plus a small open memory with read/write access, although in principle the whole memory could be read/write. In common with the simplest ID tags they are usually proprietary and there is no low-level security protocol and so it is possible for anyone to read the memory contents and re-use them in an emulator or clone platform. The tags are quite fast and speed is a function of the memory size, or at least how much data are read or written during a transaction. The tag system operator can add some security measures at the application level if required. There is not too much that can be down about tag authentication, but the data integrity and authenticity could be protected by an associated stored Message Authentication Code (MAC) or digital signature and/or the data privacy could be maintained by encryption. If the tag data is effectively interpreted by an online secure server then this is reasonably secure and manageable, but otherwise it means that keys will need to be distributed to reader devices that must also run algorithms, which makes them and the associated Key Management System (KMS) processes critical parts of the system. If you are going to this trouble it might be better to opt for a Secured Memory Tag/Card.

1.4.3 Secured Memory Tag/Card

These tags/cards have key-based cryptographic protocols to control access to memory contents. They are perhaps the first devices in the card family that deserver the "smart" description. Typically, the tag and reader will mutually authenticate before allowing access to memory; usually with data transfers encrypted under session keys. Some cards divide the memory up into smaller partitions that have different keys, so multi-application support is possible with different keys assigned to various application providers. There are other devices that divide the memory into a hierarchical file structure using the security protocol to establish access rights and privileges. One of the most maligned products in this class is the MIFARE Classic [10], yet it has also been one of the most popular and successful products. Although its security has been comprehensively compromised by widely published attacks (sec Chap. 6), it at least has some security measures to attack, which cannot be said for the simpler Tag/Card products, and so for very simple low risk applications it may still be useful. It would be better to opt for the newer MIFARE Plus [13] which takes the same basic product approach, but uses the AES [14] algorithm. For the file-based secured memory tag/card there is the DESFIRE EV1 [15], which despite the name can also use AES. The products are fast and although proprietary, the more modern types have undergone independent security evaluation.

1.4.4 Secured Microcontroller ID/Tag

At the head of the family we find the most sophisticated products based on secured microcontroller chips. In card form this is what we most definitely refer to as a "smart" card, although it is really the chip that matters and this could be used within many other form factors and assemblies. These devices can be used for just secure data storage, but more importantly for hosting secure functionality and especially advanced security and transaction protocols. A modern device would typically have be a Java Card [4] and include GlobalPlatform [16] support for management. The most advanced products include cryptographic co-processors for common symmetric and asymmetric cryptographic functionality such as encryption, verification and signing. These products tend to conform to international standards and industry guidelines and are often security evaluated either in an internationally recognised manner (e.g. common criteria [17]) or via private lab tests. The downside of such devices is that they tend to be slower and more expensive than the alternatives; and usually more complex to develop and manage.

Note: It is of fundamental importance to understand that the Secured Memory and Secured Microcontroller products can only be used with confidence in target applications because they are based on tamper-resistant hardware, incorporating physical countermeasures against all known practical attacks. Attacks on security devices are covered in detail elsewhere within this book, but given the importance it is worthwhile just briefly recapping on these capabilities.

1.5 The Importance of Providing Attack/Tamper-Resistance

When we talk of attack/tamper-resistance we are referring to attacks that can be performed on the implementation of sensitive applications, algorithms and protocols, etc., of a particular device. This is not to be confused with the logical design of the solution, e.g. algorithm and keysize choices, which are covered by best-practice considerations (see Sect. 6.8). For an assembly, tampering may be removing the lid, rewiring the connections, probing chips, etc., and with smart cards we are concerned with similar things at the chip level. There are a lot of tools from the manufacturing world that facilitate this and there is evidence of physical attacks and reverse engineering. An attacker may seek to physically inspect the chip design, probe memories and buses and make changes to low-level hardware. To hinder this the special smart card chips have shields which can be simple fixed barriers or current carrying meshes, they scramble the design layout so it is hard to access areas of interest and the low-level encrypt buses and memories. The designers also add environmental sensors for detecting light, temperature and voltage. Light is an indicator that the chip is outside of its package and so a reason to render the device inoperable. Temperature and voltage extremes may be associated with fault attacks whereby the attacker seeks to disrupt normal operation for a security advantage. The chip will also include measures to disguise/break the linkage between chip current and operation performed to prevent exploitation of side-channel leakage attacks. At chip level these measures usually include power smoothing, noise addition and variable processing delays. The very intrusive physical attacks are usually only attempted for reverse engineering, whereas the fault and side-channel attacks could be justified against individual cards and need not destroy the test target.

Just looking at a chip it is impossible to appreciate its level of tamper-resistance and physical attack resistance and given the sensitive nature of the protection measures they will not be detailed within a data sheet. It is therefore very important for someone seeking a smart card product to be able to gain assurance of the implementation security. Fortunately there are well-known means to do this. The Common Criteria [17] framework is a means to achieve an internationally recognised level of evaluated security on a range of products including smart cards. The levels start at EAL1 and rise to EAL7, with smart cards commonly evaluated to EAL4+, where the + means that some higher level features are included. Where such an evaluation certificate is not available the next best thing is a report from a credible expert lab stating that the products resisted all known attack strategies during the period of the tests.

To finally hammer home the point, many common smart card applications would be flawed and indeed pointless without the property of attack/tamper-resistance. Fortunately, these principles have been well understood in the smart card world and products, specifications and processes have emerged to successfully provide the necessary safeguards. However, these measures are now being stressed by the emergence of new technologies, which challenge the way that we implement smart card (and particularly RFID) functionality.

1.6 Mobile and NFC

Conventional smart cards show little sign of disappearing from the market and more and more are used each year. However, there is a disruptive technology that might change this in the longer term and which is already challenging assumptions and processes related to security modules. The technology in question is Near Field Communication (NFC) [18]. This is discussed in detail within Sect. 14.5.2.3, so here it will suffice to say that it gives a mobile phone the ability to act as a smart card reader, to emulate a smart card or to communicate with another phone in peer-to-peer mode. In our discussions here we will just focus on card emulation and the challenges for evaluated tamper-resistance and associated security processes.

A good example for this is shown in Fig. 1.13. This represents a collaboration between Orange and Barclaycard to offer the first NFC wallet (Quicktap [19]) in the UK, whereby the mobile phone may be used in place of a contactless bank card. Clearly, as financial transactions are involved there will be interest from attackers and so it is important to emulate the bank card in a secure manner. In NFC the smart card is emulated by something called a Security Element (SE). In the early NFC phones this was provided in the form of a chip embedded within the phone hardware. As the chip

Fig. 1.13 Quick-tap application (Orange/Barclaycard)

was a high-end smart card device (e.g. SmartMX [20]), then from a physical point of view there should be no problem with it, although there are serious challenges for personalisation, ownership and management that we will defer until later. Other options include the SE as an integrated part of the SIM and a plug-in on a memory card port, which again should be capable of satisfying physical protection requirements. More of a concern is the Soft-SE approach whereby the SE emulation is running in the phone CPU. The history of mobile phone security is very poor and the complexity and fast moving nature of modern smart phone hardware and software development is unlikely to see the problem resolved overnight. There are efforts to improve mobile phone security, however, even if an enterprising company comes up with a physically (and not just logically) secure solution there is the problem of convincing application and service providers that this is the case and bearing in mind that the general market will not be restricted to products from one supplier.

Whichever solution is used there are big challenges for the associated processes and trust management. Typically, a smart card has an Issuer who owns, personalise, issues and securely manages the device during its useful life. The processes for this are very well established and proven. The disadvantage is that you end up with a lot of smart cards in your wallet, although some might argue that this provides some diversity; if you lose one card you still have others. An NFC phone could in principle replace all your cards which can be very convenient, but potentially disastrous if it fails to function or gets lost or stolen. Exactly what types of smart cards might be displaced by NFC is not yet known although low value applications such as metro travel or touch & pay purchases seem reasonable candidates.

The first NFC payment services in the UK are likely to be constrained and proprietary. The Quicktap [19] is available from Orange and uses Barclaycard for the financial aspects. Whilst other products are expected, it is doubtful that they will be compatible beyond the card to reader interface. It therefore seems very likely that the configuration management of SEs will be a major issue, especially as customers often change, phone, SIM, credit cards and sometimes banks. This will challenge the lifecycle processes used for conventional smart cards, which are briefly described in the following section.

1.7 Conventional Smart Card Lifecycle Management Processes

The typical stages in the preparation, issuance and management of a smart card or RFID are shown in Fig. 1.14. The overall process is normally triggered by the Issuer (who provides cards to end users) placing an order with the Manufacturer (sometimes called the Smart Card Vendor) for a batch of cards according to a specification (profile) and using Issuer input data (input file). The Issuer will eventually receive the smart cards (or perhaps they are shipped direct to end-users) plus the response file containing data, keys and PINS to securely manage the smart cards.

Chip Manufacture is handled by the chip fabrication plant (FAB) and for a masked Read Only Memory (ROM) style device would include the Operating System (OS)

Fig. 1.14 Typical smart card lifecycle stages

from the particular smart card vendor. Completion is getting the chip into a usable state which may mean patching the OS using some of the chip's non-volatile memory. Initialisation is the process of getting all the standard data and functionality loaded onto the card. Here the word "standard" means for the particular Issuer's product and other Issuers may have very different requirements. Personalisation is configuring a smart card for a particular account, or if known, a particular end-user. The Issue stage represents the ways and means to get the card into the hands of the end-user. In-use Management includes updates to keep the smart card functioning in an optimum manner, including data and functional updates, which have to be applied in a security protected manner. End of Life can be a tangible update that disables the card, the removal of its corresponding back office data/functionality or simply a soft ending, e.g. that the card is probably not in use any more.

The Completion through to Issue stages are often all handled by the smart card vendor. Basically, the Issuer requests smart cards to a certain specification (profile) from the vendor and also provides an input file for personalising them to either generic unique accounts or particular end-users. In the latter case, the vendor may also pack/send the cards to end-users via the mail. Thereafter the smart cards are deemed In-use and any updates are handled by the Issuer using credentials received from the vendor in the form of a response file. The response file is highly sensitive as it contains all the security credentials, IDs, keys, PINs, etc., for the issued cards and gives the holder the ability to manage the smart card contents and functionality. This includes the capability to disable the card at end of life should this be necessary and indeed the means to create a clone of the smart card.

The time span of the lifecycle is quite different depending on the application. A banking card might be expected to have a lifecycle of about five years after which it will be expired and no longer work. A mobile communications SIM card might be used for less than a year, but then again some continue 10+ years as there is no fixed expiry limit. Chips in passports would be expected to last for 10 years. If we think ahead to chips within electronic assemblies such as phones and even cars we again have a wide variation. Mobile smart phones fall rapidly out of fashion, whereas a chip in an automobile subsystem might have to survive 15+ years.

Returning to NFC SEs the above situation is not much changed if we have a SIM-based SE or indeed a memory card-based version. However, the embedded chip is markedly different. The chip should go through Completion before it ends up in the phone hardware. If we buy the phone as an unlocked device then it has in a way been issued to us, but without the Initialisation and Personalisation stages and (at the moment) no real certainty over where the response data/credentials are and indeed who now has the privileges and responsibilities of the Issuer. If we get the phone locked to a network, the SE might have been through Initialisation for the network, but Personalisation probably requires some subsequent effort/set-up. Bearing in mind that Initialisation and Personalisation are conventionally carried out in highly secured physical environments there is quite a challenge to replicate the same level of security in the field. There is of course a queue of companies lining up to meet the challenge with the ambition of becoming the Trusted Service Manager (TSM) who would be in charge of the remote security management of SEs. However, whilst agreement on technical issues is eventually likely, agreement on who should be the TSM seems far more elusive and indeed a battle ground for conflicting business interests.

1.8 Conclusion

In this chapter we have briefly introduced smart cards and some of the major applications that make use of them. There is no absolute right or wrong choice for a smart card device as it depends on the requirements of the particular application. The most successful application in terms of number of standards-compliant devices has been mobile communications, which has used SIMs with advanced functionality, yet paid less attention to formal security evaluations than other standardised solutions such as bank cards or passports. Furthermore, the SIM today is far more like an embedded security module than a conventional smart card and there are suggestions that it should simply be replaced by a chip in the phone, although this raises all kinds of issues related to personalisation, ownership and control. In some respects the SIM is similar to the cards used in proprietary satellite Pay-TV security systems as all they are manufactured and distributed in the card form, but once installed they are used like embedded modules. The bank card is most obviously still a conventional and portable smart card, making full use of personalised card body features as well as the chip security. Historically, the less sophisticated and usually proprietary devices are found in tagging and transport systems. For tags the capabilities are normally restricted due to very tight cost constraints, whereas for transport it is the speed of operation that is critical.

The diversity of application requirements has led to a wide range of available products that offer varying degrees of functionality and security for a given cost. For many applications the tamper resistance of the chip is of vital importance and a wide range of attacks against the implementation should be resisted, including physical tampering, side-channel and fault attacks. For the cryptographic algorithm and protocol design aspects it is highly advisable to make a selection based on best-practices of information security (see Sect. 5.9), however cost and legacy compatibility often mean that a compromise has to be made and indeed much of information security is about working with imperfect solutions. It is important to appreciate that the less secure products can be quite suitable for some applications, although one should always check that the security of a device has not been compromised, otherwise the wrong design decisions will be made. A good example is the MIFARE Classic, as it was originally offered as a small key secured memory card, although today it is best considered as a basic memory card; so extra security protection may need to be added at the application layer.

A lot of industry experience has been gained from the manufacture, initialisation, personalisation and management of smart cards, however, new technological developments may challenge the conventional way of working. In particular, an NFC phone may emulate several smart card devices via the SE. Whilst the hardware SEs are based around smart card chips and should offer attack resistance, the configuration, personalisation, management and ownership of the phone-embedded and memory module SE options may be quite different from that of the SIM card, or indeed any other issued smart card. The security challenges and added complexity has led some parties to suggest the use of software SEs hosted by the phone processor. This is a worrying development and history risks repeating itself if too much trust is placed in mobile phone software security, without proper consideration. What may eventually emerge is a hybrid solution using mobile phone software underpinned by specialist hardware features, either provided by separate chips or possibly included within the phone processor itself.

As a final remark, there appears little sign that the billions of smart cards and RFIDs produced each year will reduce and in fact they are expected to rise significantly. NFC will not make much difference in the short term, especially while companies are squabbling over roles and standards options, however it might result in an acceptable security solution for the remote management of SEs. If this is the case then the same solution might be used to logically justify the use of embedded SIMs, although this would no doubt be resisted by mobile network operators.

References

1. Mouly, M., Pautet, M.B.: The GSM System for Mobile Communications, Cell & Sys. Correspondence, 1992.
2. ETSI, 3GPP TS 11.11 V8.14.0 Specification of the Subscriber Identity Module -Mobile Equipment (SIM - ME) interface, (2007–06)
3. EMV Books 1–4, Version 4.3, November 2011. www.emvco.com
4. Java Card Platform Specifications V3.04, Oracle, 2011. www.oracle.com/technetwork/java/javacard/overview/index.html
5. Octopus, www.octopus.com.hk/home/en/index.html
6. Transport for London, Oyster Card, www.tfl.gov.uk/oyster
7. International Civil Aviation Organisation (ICAO) Doc 9303, www.icao.int
8. Friedhelm Hillebrand: GSM & UMTS - The Creation of Global Mobile Communication Wiley, 2002, ISBN: 978-0-470-84322-2.
9. Mayes and Markantonakis: Smart Cards, Tokens, Security and Applications, Springer 2008, Chapter 4, p 85–112.
10. Philips Semiconductors (NXP), MIFARE Standard Card IC MF1 IC S50 Functional Specification, revision 4.0 1998.
11. International Organisation for Standardisation, ISO.IEC 7816 1–4, 1999.

12. International Organisation for Standardisation, ISO.IEC 14443 Identification cards - Contact-less integrated circuit cards - Proximity cards, 2000.
13. NXP, MIFARE Plus data sheet MF1SPLUSx0y1, February 2011. www.nxp.com/documents/short_data_sheet/MF1SPLUSX0Y1_SDS.pdf
14. Federal Information processing Standards, Advanced Encryption Standard (AES), FIPS publication 197. http://csrc.nist.gov/publications/fips/fips197/fips-197.pdf
15. NXP, MF3ICx21_4_81 MIFARE DESFire EV1 contactless multi-application IC short data sheet, revision 3.1, December 2010
16. GlobalPlatform, GlobalPlatform Card Specification 2006 www.globalplatform.org
17. Common Criteria V3.1, 2009, www.commoncriteriaportal.org
18. NFC Forum, NFC Forum technical Specifications, www.nfc-forum.org/specs/spec_list
19. Orange UK, Quicktap, www.shop.orange.co.uk/mobile-phones/contactless
20. NXP, SmartMX Platform features, Revision 1.0 Short form Specification, 2004

Chapter 2
Embedded DSP Devices

Serendra Reddy

Abstract As a consequence of the rapid surge in digital signal processing (DSP) technologies, DSP components and their specific algorithms continue to find uses in broad application areas, including the embedded systems arena. Embedded systems generally refer to systems that include dedicated hardware and computationally specific software. When several fundamental components of an embedded system are integrated onto a single silicon substrate it is referred to as a system-on-chip (SoC). These embedded systems, including SoCs, can either stand-alone or seen as a subsystem of a much larger and/or complex system. However, these systems are not without constraints, and constantly need to adapt to the drawbacks associated with limited hardware, restricted computational power and fewer resources. Recently, there has also been an increased interest in the use of field-programmable gate arrays (FPGAs) and application-specific instruction-set processors (ASIPs) within embedded DSP devices. This can be seen as a trade-off between size, speed and flexibility, with the latter being the driving force. Embedded DSP devices have proliferated through society so much so that we have become virtually oblivious to their impact. Among the countless applications of embedded systems, some products that require a DSP component include our mobile phones, digital radios, digital televisions, digital satellite set-top boxes, DVD players, MP3 players, heart-rate monitors, GPS navigation devices and automotive control systems. This chapter gives a brief introduction into the theory of DSP, followed by a more detailed examination of the architectures, implementations, security and applications within real-time embedded systems.

S. Reddy (✉)
Department of Electronic Engineering, Durban University of Technology,
KwaZulu-Natal, South Africa
e-mail: serenr@gmail.com

K. Markantonakis and K. Mayes (eds.), *Secure Smart Embedded Devices,*
Platforms and Applications, DOI: 10.1007/978-1-4614-7915-4_2,
© Springer Science+Business Media New York 2014

2.1 Overview

A signal can be described as information within a form of detectable energy that is generated by a physical occurrence, like changes in electromagnetic radiation or air pressure. In order to investigate these signals this energy is first converted into a continuous electrical signal using, for example, a photosensor or microphone, as in the case of light or sound, respectively. These continuous electrical signals are commonly termed analog signals and the variations in these signals are represented by voltage values that are theoretically infinite, both in amplitude and precision. A digital signal is then the discretisation of an analog signal i.e. the representation of the continuous signal by a discrete (non-continuous) set of quantised values. This is achieved by taking samples (measuring the amplitude/voltage) of this continuous signal at successive non-zero time intervals i.e. a snapshot of the relative space-time. This process of conversion from an analog signal to a digital signal is called an analog-to-digital (A/D) conversion.

Digital signals can be represented in multiple dimensions, like one-dimensional, in the case of sound, and two-dimensional (2D), in the case of images. Although photons hitting a photosensor (like a CCD array in a digital camera) arrive at the speed of light, the image (or photograph) is merely a representation of the individual voltage levels on all the sensors at a single instance in time, arranged and stored in a 2D matrix. Digital video can then be extrapolated as a collection of 2D matrices captured sequentially, like at 0.033 s intervals in the case of a standard 30 frames per second movie, hence the term motion- or moving-picture.

If not clearly defined, the acronym DSP is often ambiguous, as it can describe the specific hardware/software processes used in handling digital signals, as well as a hardware processor. In this chapter, DSP refers to digital signal processing (DSP), which is the generic term applied to the hardware/software processing of digital signals and data by digital electronic devices. Digital electronic systems can range from super-computers, desktop computers and laptops, to tablet computers and smartphones, to small DSP specific systems and SoCs, including application-specific integrated circuits (ASICs), application-specific instruction set processors (ASIPs), field-programmable gate arrays (FPGAs), general-purpose DSP processors (GP DSPs) and general-purpose microprocessors (GPPs). The aim of the processing is to analyse the information content of the cached or stored signal data and sometimes modifying the signal depending on the desired output. This can range from, among others, noise reduction to data enhancement to data compression to pattern recognition, and to whether the system should operate in real-time. Real-time systems can be defined as those systems that respond in a timely manner to external actions or triggers [26]. In DSP, this implies that an output is produced from the current set of data before the next set of input data is collected and/or available to process. Once the digital signal has been processed and possibly enhanced (or altered) by the DSP system, it might be required to be sent back out into the real world, as in the case of music being outputted from a digital amplifier or equaliser. This process is, fundamentally, the reverse of an A/D conversion, where discrete digital data is transformed or converted into a continuous analog signal, and is, therefore, called a digital-to-analog (D/A) conversion.

An embedded system is a combination of computer hardware and software, and perhaps additional mechanical or other parts, designed to perform a dedicated function [13]. Unlike a personal computer which serves a general-purpose, an embedded system is generally designed to serve a specific purpose, and is usually limited in size, cost, power consumption, processor speed, memory and hardware functionality. An embedded system can also be seen as being a part of larger system, containing possibly a collection of smaller embedded subsystems, including those with DSP functionality, each responsible for a separate task, like handling decompression and decoding of audio files on a MP3 player or the capture, compression and memory card storage of images on a digital camera. An embedded processor is a specialised processor, like a GP DSP or ASIC, designed to meet the requirements of a specific application, i.e. the functionality is often limited or tailored to just the intended purpose, e.g. with perhaps low power consumption and low heat dissipation, and restricted clock speed [26]. In order to handle real-world analog signals, including speech, images, video and music, an embedded processor must interface with external hardware such as input/output (I/O) devices like coders/decoders, memories and displays.

An embedded DSP device is typically a combination of one or more individual pieces of hardware (or subsystems) integrated into a single stand-alone system performing a specialised function, that requires a DSP specific hardware processor, like an ASIP and/or a FPGA, that uses DSP specific software algorithms and techniques to process and/or transform real-time input signals into a desired output.

2.2 Digital Signal Processing

Almost all information in the physical world is represented in the form of an analog signal, and as a result the processing of these analog signals represents a fundamental component within the field of electrical engineering. This subfield came to be known as analog signal processing (ASP); and entails the use of analog hardware in the manipulation of these signals. However, ASP had its shortcomings in the form of complicated electronic circuitry, inflexibility, varying accuracy and inconsistent reproducibility. In addition, sophisticated applications, like speech and image processing, were not suitable to ASP technologies. These challenges needed to be addressed and were eventually solved by the advent of digital systems and DSP.

Although the mathematics of DSP algorithms had been in existence for many years, it was only the emergence of the GPP and GP DSP in the 1970s that marked the turning point in digital systems. The original systems were primarily fixed-point machines [39, 45]. The mid- to late 1970s saw the introduction of floating-point machines and together with supporting memory devices gave rise to the era of DSP, which by 1980s included multi-processor systems with massively parallel architectures, allowing for efficient execution of the fast Fourier transform (FFT)[1] and vector-based processing schemes. However, non-conventional schemes, like adaptive and high-resolution signal processing remained a bottleneck [18, 49], until recently. Advances in super-scalar and massively parallel processor technologies have seen the GP DSP go from being able to perform several hundred million multiply accumulates (MACs) per second (or about 21 ns per MAC) in 2000 to around 5 billion MACs per second (or about 3 ns per MAC) in recent years [22, 27].

2.2.1 The DSP Processor

The core purpose of a DSP processor (or GP DSP) is to perform signal processing, and almost every single DSP application is based on efficient mathematical implementations of one or more of algorithms [15, 36, 37] shown in the following table:

Fourier transforms are used for representing signals in the frequency domain; convolution can be used to perform filtering in the spatial domain; correlation is used to detect similarities in signals, like in the case of Radar; finite impulse response (FIR) and infinite impulse response (IIR) filters can be used in noise attenuation and other frequency selective processes; 2D Fourier transforms are used for image processing in the frequency domain, and discrete cosine transforms are used in image compression, like JPEG.

The DSP processor was thus optimally designed around the ability to efficiently execute the above algorithms. This was done by exploiting the inherent similarities between the algorithms, like the summation and multiplication operations. The combination of the summation, which can be described as a "for" loop in software, together with the multiplication operations, results in the accumulation of a large number of multiplied elements. It was, therefore, logical to develop a processor that was able to resourcefully accommodate the operands of multiplication and accumulation. In addition, there are intrinsic structures in these algorithms which allow for non-dependent parts to be operated on separately. In the end, the common factors observed in the digital signal algorithms allowed for the tailoring of a DSP specific processor that could achieve tremendous execution savings.

The GP DSP differs mainly from the GPP in its memory architecture, internal architecture and instruction set. The memory architecture of the GPP is based on the Von Neumann single memory model, whereas the GP DSP is based on the Harvard dual memory model (Refer to Fig. 2.1), which is designed for parallel access to the program and data memory, allowing for the fetching of multiple data and/or instructions at the same time. An advancement to the Harvard architecture is referred to as the super Harvard architecture and includes an instruction cache and dedicated I/O controller (with DMA) [42]. The internal architecture contains several multipliers and accumulators, in order to optimally perform fixed-point and floating-point

---

[1] The discrete Fourier transform (DFT) is the method of translating any sequence of discrete values into its frequency domain equivalent, by representing the signal as a composition of sine and cosine waves. The FFT is the more efficient method of generating a DFT.

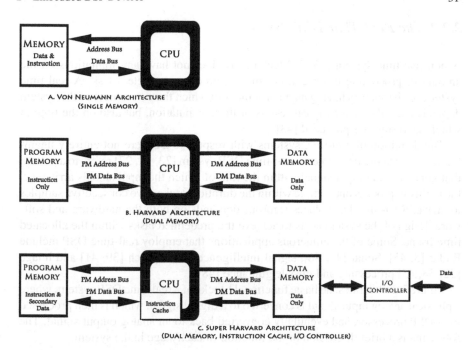

Fig. 2.1 Microprocessor architecture

multiply-accumulate (MAC) operations extremely fast, which is a necessary requirement in order to efficiently perform the DSP algorithms described in Table 2.1. The specialised instruction set was designed to maximise the use of hardware, minimise memory space and increase efficiency. In addition, it incorporated measures to alleviate some of the problems associated with the limitations of working in the digital arena, like quantisation errors, round-off errors, finite wordlength effects and overflow. All this, in the end, means that a GP DSP performs signal processing much more proficiently than a GPP, which is optimised for non-signal processing centric applications.

Table 2.1 Common DSP algorithms

Discrete fourier transform (DFT)	$X(k) = \sum_{n=0}^{N-1} x(n) e^{-j2\pi kn/N}$
Convolution	$y(n) = \sum_{k=0}^{M-1} h(k) x(n-k)$
Correlation	$r_{xy}(n) = \sum_{k=0}^{M-1} x(k) y(n-k)$
Finite impulse response (FIR) filter	$y(n) = \sum_{k=0}^{M} b_k x(n-k)$
Infinite impulse response (IIR) filter	$y(n) = -\sum_{k=1}^{N} a_k y(n-k) + \sum_{k=0}^{M} b_k x(n-k)$
2-D DFT	$F(u,v) = \sum_{x=0}^{M-1} \sum_{y=0}^{N-1} f(x,y) e^{-j2\pi(ux/M + vy/N)}$
2-D Discrete cosine transform (DCT)	$F(u,v) = \sum_{x=0}^{M-1} \sum_{y=0}^{N-1} f(x,y) cos(\frac{\pi(2x+1)u}{2M}) cos(\frac{\pi(2y+1)v}{2N})$

2.2.2 The Real-Time DSP System

A non-real-time system [23, 24] technically does not have any limitations on the amount of processing or execution time required to complete a task. A real-time system can be seen as having an environment in which the correctness of the system depends not only on the logical results of the computation, but also on the time in which the results are produced [43].

The definition of real-time systems with respect to DSP are not entirely unambiguous, hence in this chapter, a real-time DSP system [23, 24, 27] refers to a system that generates an output signal within the rate of which the input signals arrive, i.e. the ability to process one sample within the duration it takes two consecutive samples to arrive. Real-time DSP places rigorous demands on both the hardware and software design of the system so as to achieve the predefined tasks within the allocated time frame. Some of the numerous applications that employ real-time DSP include Radar [5, 45], Sonar [45, 46], signal intelligence [48], speech [39, 41] and image [22, 38, 45] processing and missile guidance [48].

Figure 2.2 illustrates the basic functional blocks of a real-time DSP system, where a physical analog input signal is converted to a digital signal, which is then processed by a DSP processor, and eventually converted back to an analog output signal. The following is a brief discussion of the function of each stage in the system:

- Input signal. An electronic sensor (microphone, photosensor, etc.) will first convert the variances of the analog signal (temperature, pressure, sound, light, etc.) into an electrical signal.
- Analog signal pre-processing. The aim of this circuitry is to perform some pre-processing on the incoming electrical signal. This can include amplification, voltage regulation, anti-aliasing filtering and possibly other analog enhancements that could benefit the A/D conversion.

Fig. 2.2 Functional blocks of a real-time DSP system

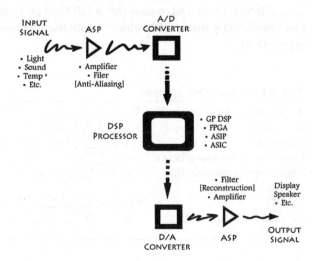

- Analog-to-digital convertor (ADC). The proverb, "a chain is only as strong as its weakest link", when applied to a real-time system, is applicable to the ADC, owing to the fact that the sampling rate defines the digital filter's working frequency. There are three components that comprise the ADC: sample-and-hold (S/H), quantisation and coding. In order to obtain the digital representation of the electrical amplitude, the electrical signal is first sampled, which is the process of tracking and tapping into the fluctuating electrical signal; this tapped value is then held constant for a single A/D conversion cycle, in order to be read and translated (or digitised). N.B. For a complete and unambiguous reconstitution of a continuous analog signal by a digital process the electrical signal must be sampled at a frequency that is at least greater than twice the highest frequency of the respective signal. This has been defined by the sampling theorem [37]. Quantisation is the process of mapping an electrical value of theoretically infinite precision to a value of finite precision. This is dependent on the quantisation step size, which is the minimum significant value allowed i.e. the value to which the discrete time signal must be rounded up or down to. A system that has, for example, only one significant digit after the decimal point will result in the values 0.4862969 and 2.348672 to be rounded to 0.5 and 2.3, respectively, with a quantisation error of 0.0137031 and −0.048672, respectively. The final phase is the coding of the quantized values into binary, which is, like the quantisation stage, also dependent on the significance of the system. In order to maximise efficiency, the resolution, or step-size, of both the quantiser and encoder are jointly optimised. The errors due to quantization and wordlength effects can be modelled to determine the effective accuracy and throughput of the ADC.
- DSP processor. This represents the hardware and/or software responsible for the analysis and/or modification of the digital bits or signals by DSP methods; a process which largely involves the application of at least one of the algorithms given in Table 2.1 (Sect. 2.2.1).
- Digital-to-analog convertor (DAC). The transformation from the digital world to the real world is done at this stage. A voltage is generated at the output, proportional to an electrical signal, corresponding to a binary word input to the DAC. In order to go from a series of discrete values to a continuous signal requires a type of interpolation. Although there are several methods of interpolation [37], most DACs are zero-order-hold. In this system a constant voltage value is outputted until another sample value is received; the result is a staircase effect. In order to create a continuous smooth analog output signal, the output from the DAC is processed through a post-filter.
- Analog signal post-processing. This represents the smoothing and anti-imaging filter. The final reconstruction of the analog signal is done by smoothing off the corners of the staircase signal generated by the interpolation or zero-order-hold process at the output of the DAC. In addition, the interpolation could cause image (high frequency) components being passed above the folding frequency, resulting in replications (or images) at multiples of the sampling frequency (sometimes referred to as post-aliasing [37]); this could generate possible degradation of the output signal due to overloading of certain components of the external circuitry, and thus are subsequently filtered and removed at this stage.

- Output signal. The final converted, modified and/or improved analog signal is processed at this stage, usually by some human-compatible real-world device, like a speaker, display or computer.

Real-time systems are occasionally influenced by response time constraints, like unanticipated delays or latencies. The effect of these inconsistencies on the overall system is vitally important, consequently creating two types of real-time systems viz. soft and hard [33]. Soft type implies that in the event of a missed deadline there is a degradation of performance, but no system failure; hard type subsequently implies a hard deadline, which if not met, results in system failure e.g. the navigation system of an aircraft.

Real-world embedded DSP systems are usually qualified as hard real-time systems. Compared to desktop programming, embedded DSP real-time systems must be able to respond predictably and reliably to all external events, meaning that the DSP code must not only execute as predicted, but execute on time, without little or no exception.

2.2.3 The FPGA in DSP

In the past, whenever there was a need to meet extreme performance requirements, and a programmable device could not handle the load demand, or there was a solution requiring ultra-low power ultra-small silicon size, an ASIC was the only alternative for an embedded DSP system [27]. ASICs are produced by directly mapping algorithms to an integrated circuit; this, however, provides extremely limited flexibility, and virtually no room for changes once fabricated. In addition, as DSP applications become vastly more complicated, the mapping of the DSP functions to circuits become vastly more difficult.

As a result, over the past decade, the DSP market has seen a drastic increase in the use of alternate processing systems, most especially the FPGA [12]. The FPGA is seen as a compromise between the rigid ASIC and the programmable GP DSP. Although both the ASIC and GP DSP have their own merits, the attractiveness of the FPGA comes in its flexibility, or configurable hardware, which has shown to have a cost/performance of between 10 and 25 times that of a typical high performance GP DSP [11, 47].

Although GP DSPs have conquered the world over, the requirements of certain newer applications tended to exceed their processing capabilities. The FPGA was able to meet these demands, owing to its ability to reconfigure the logic, thus allowing for the design of computationally effective structures, in the form of special purpose functional units that can perform limited tasks very efficiently [34]. The massive computational resources, as well as the ability to configure highly parallel architectures, surpassed the throughput of even the high-end GP DSPs. Due to the nature of FPGAs, field upgrades of the configuration file are rare; whereas, GP DSP code and product software patches and updates are widespread. Since a FPGA is hardware-programmed, by manipulating logic at the gate level, it is possible to construct DSP-oriented processors in parallel that can efficiently and simultaneously solve a desired DSP task.

The downside to the FPGA is the effort of complexity required to map DSP applications, including ADCs (although significantly higher frequency ADCs can be achieved with FPGAs), DMA controllers, bus interfaces, etc.; requiring architecture implementations at the register-transfer-level (RTL)^2 usually via some hardware description language (HDL)^3 like very-high-speed integrated circuits HDL (VHDL). This complication is, however, being addressed with steadfast advancements in FPGA tools (including simulators) and libraries. Another issue concerning the FPGA, is working with floating-point cores, which can consume a significant amount of area within the FPGA, especially when parallel processing is considered; one alternative is to settle for using integer-based coefficients, or to possibly pipeline the floating-point operations, which will reduce the required gate area, but at the expense of the response times. A final issue that is worth noting is the issue of "excess baggage"; the FPGA essentially implements a function by turning on or off different logic gates; the problem being that once an application is programmed into the FPGA there can be a lot of redundant gates taking up unnecessary space. These aforementioned issues are relative to each project and need to be considered in determining the tradeoff between performance and size, cost and development times.

2.2.4 The ASIP in DSP

Another avenue of interest that has recently emerged in the embedded systems sector is the use of embedded soft-cores. Embedded soft-cores can be seen as ASICs with application-specific parameterisable components, and are thus referred to as application-specific instruction-set processors (ASIPs) [32]. Although the instruction set of an ASIP is designed around specific application requirements that tailor the processor for these applications, it is also a programmable machine with a degree of flexibility, allowing it to run various software programs [27]. The former allows for a product that can be high performing while low in power consumption, and high volume manufacturing can be used to lower costs. ASIPs can even come with their own development environment, debug tools, simulators and compilers, with the option of adding peripherals for communications, I/O, memory control, etc.

FPGA-oriented-ASIPs [25] are also possible, whereby a "soft-core" processor can be configured and downloaded onto a FPGA and used just like any other embedded processor, including a GP DSP. A DSP ASIP mapped on an FPGA will consist of

---

^{2} The RTL is the level above the transistor or logic gate level that translates the circuits described by the HDL into their equivalent sequential (usually consisting of registers comprising a number of D-type flip-flops) and combinational logic structures.

^{3} A HDL, which is implemented at a level above the RTL, is a method of using text-based expressions to represent an algorithm that describes the behaviour of a digital circuit.

an instruction set optimised for a class of DSP applications; the reduced instruction-set-like architecture can take a form of a parallel process-unit array that can realise high processing speeds and high levels of parallelism [51], thus combining the programming capabilities of a GP DSP and the high throughput of an ASIC. Any type of microprocessor can be implemented on an FPGA; however, in recent times the vendors have provided soft-core processors specifically designed for FPGAs. These soft-cores include instruction sets, arithmetic-logic units, register files, and other features specifically optimised for the FPGA resources, and dually serve as preventative measures against inefficient and/or incorrect use of FPGA resources. It has been shown that the performance overhead between general circuit implementations on FPGAs versus ASICs is far superior when comparing overheads between such soft-core processor implementation on FPGAs versus ASICs [40]. A key value of an ASIP on a FPGA is that they are composed of smaller building blocks that can be reconfigured on the fly to implement more than one high-level function [25]. Example relevant to DSPs would be FIR filters and Fast Fourier Transform (FFT) blocks. Since these two high-level algorithms share many common sub-blocks, the ASIP can be easily altered to implement the FIR, instead of the FFT, in hardware, by changing the interconnect between these subblocks.

The customisable ability of the ASIP is not without challenges, as currently, each application that is compiled must be simulated and run on all possible configurations of the ASIP, which can be exponentially exhausting owing to the increase in configurations with the number of parameter values [17]. This can result in an increase in the time-to-market and subsequent cost, prompting a possible decrease in the number of customisable parameters or opting for a general-purpose chip solution. However, there are methods to improve these drawbacks, like having customisable development, simulation and test-bench environments, which aid in optimising the values of these architectural parameters. Although soft-core processing is still in its relative infancy, their applications on SoC platforms make them hugely attractive within the embedded systems market [17], thus creating a consistent drive to adapt and advance this technology.

2.3 Embedded DSP Systems

Both a general-purpose computer and an embedded computer (or system) can simply be seen as devices created to store, retrieve and process data; with the fundamental distinction lying in their application, size, power, cost and predictable speed. A general-purpose computer is typically a multi-tasking system that can respond to the requirements of multiple applications, such as playing music or accessing web pages or generating spreadsheets. An embedded computer might only perform one specific task, like the removing of noise from an audio stream. In addition, unlike desktop computers, embedded devices do not necessarily have a user interface; alternatively they might have a basic user interface, like a simple button with a LED, to a midrange user interface, such as an alpha-numeric liquid crystal display (LCD) menu system, all the way up to a complex graphical user interface (GUI), like a touchscreen icon-based menu system on a mobile phone.

With the rapid evolution and requirements of modern dedicated embedded circuitry, the modern embedded computer is seen as a viable replacement to application-specific electronics, with the fundamental advantage being that the former can be used to define functionality using software and/or hardware, as opposed to the inflexible dedicated hardware of the ASIC [6]. The embedded system can subsequently be partitioned into the hardware component, which is responsible satisfying the performance requirements of the application, and the software component, which provides for bulk of the functionality and flexibility within the system [33].

In the case where there is a need for the handling of time critical tasks by an embedded system, a DSP component may be required. Given the high demands of current embedded devices, like smartphones, it has become an unavoidable necessity for most modern embedded systems to include a DSP subsystem.

A DSP subsystem can also be fabricated within an embedded GPP system together with additional components, like a direct memory access (DMA) controller, a programmable interrupt controller (PIC), a programmable general-purpose timer and even Ethernet interfaces. These self-contained platforms are referred to as system-on-chip (SoC) processors, and due to the reduction in the need for external peripheral devices the overall size and cost of a product can be optimised.

Embedded systems that incorporate DSP technologies include three-dimensional high-definition televisions (3D HDTVs), digital cameras, media players, digital voice recorders, fingerprint scanners, unmanned aerial vehicles (UAVs), software defined radio (SDR) and innumerable others.

2.3.1 The Embedded DSP Architecture

The architecture of an embedded DSP system is modelled on the architecture of a generic embedded system. Both comprise four basic units, including processor, memory, general-purpose I/O and bus subsystems, with the embedded DSP system containing a DSP processor and a few other I/O peripherals.

Figure 2.3 illustrates the basic architecture of an embedded DSP system. The following is a brief discussion of each subsystem:

- Embedded processor unit (EPU). This is essentially the brains of the operation, and its key responsibility is in the processing of instructions and data. The EPU can include a master processor (or several master processors) and can either have one or several associated slave processors (or none at all). A master processor can be classed as a stand-alone microprocessor or a microcontroller, with the latter representing a subsystem where one or more microprocessors are integrated together with other memory and I/O components [30]. Furthermore, the embedded DSP processor can be an add-on processor which accompanies a GPP within the processor subsystem, or a stand-alone processor representing the entire EPU. Depending on the application, the co-processor scheme can be advantageous, in the sense

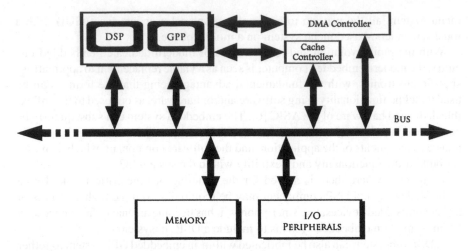

Fig. 2.3 Embedded DSP architecture

that the GPP can handle the miscellaneous and/or mundane tasks, as for example, hardware management and connection protocols, or interacting with a keypad or display, while the DSP processor focuses on the computationally intensive real-time demands. In addition, cache and DMA controllers (discussed later) can also be included as part of the EPU. These are, in essence, slave processors that function primarily to improve the overall performance and efficiency of the processor cores, and, therefore, when included, define the EPU as a microcontroller.

- Memory. This subsystem supports data and program storage for the EPU, and represents multiple levels of memories that range hierarchically from the local and/or cache memories within the processor cores, to on-chip memory (as in the case of SoC), to the extrinsic or off-chip main memory. These three memory architectures are called Level 1 (L1), Level 2 (L2) and Level 3 (L3) memories respectively, and vary incrementally in size, and decrementally in performance [21]. L1 memory is literally closest to the core on the silicon die and is fabricated for maximising interoperable speeds. Within the GP DSP this L1 memory is configured using the Harvard architecture scheme, where the instruction and data segments are split.
- Cache controller. Cache is essentially a small, fast advanced memory that holds duplicates of some of the data and/or instructions stored in main memory. The cache controller takes on a mediator role between the embedded processor and main memory. It works by simultaneously sending memory requests to both the cache and main memory. If the location requested is in the cache, it aborts the main memory request, while forwarding the location's content from the cache to the embedded processor [50]. Cache is a valuable L1 memory which is costly in terms of silicon size, and thus is limited in its capacity. The main aim of the cache is to maximise the instances of successfully finding what it needs in the cache as an alternative to having to wait to retrieve it from the larger and slower main memory [21]. DSP processors typically contain smaller and simpler caches than GPPs.

It has been shown that once the cache reaches a certain size there is a saturation of the performance benefits. This typically occurs between 256 KB and 512 KB [33]. DSP processors frequently use caches to free memory bandwidth rather than to minimise memory delays.

- Input/Output Peripherals. These interfaces are in charge of interacting with functional units that are connected to the outside world, with the primary responsibility of bringing data into, and getting data out of, the embedded system. They are principally managed by a slave processor known as the I/O controller which interfaces with the communication interface and I/O buses in order to facilitate data communication between the EPU and I/O device [29]. There are multiple roles performed by these I/O devices, which can be extremely diverse in their range, and consist of simple circuit components, like LEDs, up to other complex embedded systems. These I/O peripherals can be broadly categorised as follows.

  - Human-machine I/O: Keypad/keyboard, touchscreen, mouse, etc.
  - Graphical and audio I/O: Camera, microphone, speaker, displays, etc.
  - Real-time I/O: ADCs and DACs.
  - Communication and networking I/O: IEEE 802.11 g/n, Ethernet, etc.
  - Data transfer I/O: USB 2.0, IEEE 1394 (Firewire), etc.
  - Subsystem controls: Timers, counters, low-speed serial interface, etc.
  - Storage I/O: Optical disk, Flash, SDRAM/DDR, etc.
  - Debugging I/O: JTAG, parallel and serial ports, etc.

- Bus. The bus is the mechanism responsible for the interconnection and exchange of data, address and control signals between all subsystems. It does not just physically represent a collection of related wires connected between functional units, but also conceptually represents a set of protocols that are used to allow for communication between the EPU and memory, EPU and I/O, and memory and I/O. The embedded bus network can be split into three types of busses via system, backplane and I/O buses [29]. The system bus interconnects the EPU with the cache and the main external memory. The backplane bus represents a single bus that connects all three subsystems. The I/O bus is an extension of the system bus and handles communication between the I/O peripherals and the system bus, including interrupt requests. The performance of a bus is measured in bandwidth, which is dependent on the design, protocols, number of bus lines and the interconnect permutations of the respective bus.

- DMA controller. DMA is a bus operation that provides for the movement of data between subsystems without the need to interfere with the more important tasks being performed by the processor, thereby improving overall system performance. The DMA controller essentially oversees the DMA operation. This is done by first requesting control of the bus from the main processor followed by the transfer of data to and from IO and memory, or ports and memory, or internal and external memories [33]. By off-loading memory transfers from the processor it allows for efficient parallel processing. Without DMA the main processor is consumed for the entire time required for the bus transfer, and since the buses often function at a reduced clock speed compared to the main processor, this can significantly reduce overall efficiency of the embedded system [50]. In a DSP, for example, when data comes onto the serial port from the A/D converter, rather than interrupting the DSP processor to transfer the data to memory (and subsequently consume a large amount of processing cycles), this can be handled by the DMA controller; the DSP processor can then focus on the execution of the algorithms while the DMA controller is handling the movement of data [33]. This feature of overlapping operations, via DMA, is extremely common in real-time embedded DSP systems.

2.3.2 The Embedded DSP Processor and RISC

At the lowest depth of any processor is ultimately a collection (albeit millions) of transistor-based hardware gates. These gates are connected into groups of combinational and sequential logic circuits that perform the instructions of a higher level application. These logic circuits are in the end the essence of the machine, and operate using binary language. This binary or native language is known as machine language. The problem encountered is that application layer instructions are formulated at a non-binary level, and therefore a process of translation between this level and the lowest level hardware is required. This hierarchical process contains several intermediary stages with the most crucial sublevel being the Instruction set architecture (ISA).

The functionality of any application is determined by a collection of interdependent instructions. These instructions can be constructed by a myriad of high-level languages like C or FORTRAN. ISA is the common platform that provides for the interpretation and execution of high-level instructions independent of the high level language employed [10]. This is achieved by first having all high-level language instructions compiled into a single universal language. This universal language is known as assembly language and uses mnemonics to express instructions. These assembly language mnemonics are then finally encoded into the binary machine language.

At the ISA level there exists two design categories complex instruction set computing (CISC) and reduced instruction set computing (RISC). CISC as the name suggests is the system of employing complex or complicated instructions, like those used when working directly with multiple array elements. RISC, on the other hand, is the system of employing reduced or simple instructions, as in the case of just adding two integers.

An early downside of complex instructions was that there was a need for complex and expensive complementary hardware to carry out these instructions. To work around this problem, a micro-programmed computer was introduced [10]. This came in the form of a small run-time interpreter, which was located between the ISA level and the hardware, which converted complex instructions into simple ones. This, therefore, eliminated the need for complex hardware, and meant that complex instructions could be executed on simple hardware.

An inherent shortcoming to RISC systems was that they were first constrained to a set of simple instructions, and second these limited operands were restricted to the processor's internal registry, as opposed to the main memory. CISC, on the other hand, does not have such restrictions and allows for an increased volume of complex instructions, which can be placed in the main memory.

With the staggering increase in functionality and features demanded of our modern embedded DSP systems, comes the need for higher processing capabilities, increased number of chips per system, and associated power constraints. Within the embedded systems environment, each of these issues comes with a whole host of related problems, ranging from processor size, to access to a compatible, sufficient and sustainable power supply, and heat dissipation. Heat dissipation is not only an aesthetic concern, but also of vital importance to both processor performance and lifespan. In addition to these issues, a problem that has been encountered is in the advancement of battery technologies, which have not kept abreast with the increase in consumer expectations for more functional and feature driven smart embedded devices.

As a consequence of the aforementioned problems, the hardware for these specialised high-demanding embedded DSP systems had to adapt almost independently; as a result a separate evolutionary branch emerged in conjunction with their generic computer counterparts, and prompted the rise of RISC.

Although CISC has some inherent advantages, the complexity of the instruction set requires additional processing, hardware and memory, prompting an increase in physical size and power needs, which consequently is inadequate with regard to embedded systems. Desktop and server computing usually do not suffer so severely from power requirement issues and size constraints, and these systems have the advantage of large cooling units, including heat sinks and fans, and air-conditioned environments.

RISC systems, having instructions that are simple in nature, can be executed directly on the simple hardware, thus eliminating the need for any additional middleware, and as it turns out, the confinement of the register-based operands not only results in a simplified processor design, but additionally creates improved application performance. This, in the end, means that RISC is ideally suited for embedded systems.

Applications, like video encoding, that require high computational complexity as well as data bandwidth, can be designed using just a DSP core; however, higher performance would entail either an increase in the clock frequency or the use of multiple DSP processors. The disadvantage of these alternatives is increase in silicon size, which subsequently results in the rise in cost and power demand [28]. A better approach would be to have a SoC design that combines a DSP processor with a RISC processor [14]. This kind of design can split tasks thus providing video-specific hardware acceleration necessary for optimal encoding and decoding. The comparative trade-offs between possible SoC architectures that can be employed in video encoding are shown in Table 2.2 [28].

Currently, the most popular RISC processor architecture employed in embedded DSP systems is the Advanced RISC Machine (ARM) [4, 7]. SoC platforms for these high-performing embedded DSP devices can consist of multi-core ARM processors,

Table 2.2 SoC architecture

Solution	Programm-ability	Performance	Power	Cost (Area)	Development time (Reuse)
AISC	Low	High	Low	Low	High
FPGA	High	Medium	High	High	Medium
Multi-Processor + Multi-Core	High	High	High	High	Medium
DSP	High	Medium	Low to Medium	Medium	Low
DSP + Co-Processor	High	High	Medium	Low to Medium	Medium

together with multi-core DSP chips, and various other interfaces. These ARM/RISC processors feature a highly specialised and optimised architecture, with extremely low power consumption [3]. There is, currently, a whole subdivision dedicated to DSP processors using the ARM architecture. A possible configuration for the above video encoding problem could be to use an ARM 32-bit RISC processor which could be extended into an efficient co-processor scheme that offers a standard ARM processor integrated with a DSP-oriented data path and an associated DSP instruction set. This provides for a small low cost embedded DSP chip that is optimised for performance while providing low power consumption [9].

2.3.3 Embedded DSP and Security

The security issues that affect large computer systems also apply to the embedded systems environment. However, the latter comes with added complexity, due not only to the limited hardware and software capabilities, but also because of the diverse environments which these devices are deployed.

Security concerns are especially crucial in embedded DSP systems that are involved in safety critical hard real-time functions, like automotive breaking systems, and in protection of private and confidential information using encryption, as needed for cellular voice and data transmissions.

The malicious exploitation of sensitive data as well as catastrophic system failure (owing to internal or external influences) can signify severe security risks apparent in using an embedded device. Among the numerous possible security threats inherent in embedded systems [44] there are four types that can be considered crucial when considering an embedded DSP device:

- Physical/Environmental. Considered when there exists a potential vulnerability in the manipulation of the physical components of the device, thereby either causing the device to malfunction or fail completely. This can also include complete device destruction.

- Internal. Involves the breakdown of the hardware and/or software of the embedded system, as in the case of a processor overheating or an erroneous/corrupt software routine, thereby causing erratic/incorrect operation and/or complete failure of the device. Owing to the often unpredictable nature of hardware malfunction it is common practice, especially in large-scale production, to perform some form of system/load testing prior to full-scale production. The more reputable embedded systems designers may also follow a form of best practices and software lifecycle management in order to improve the success of their development projects.

- External. Involves hacking, hijacking and/or purposely corrupting the embedded device, usually, but not always, by some form of software tampering. Off-the-shelf PC security products, like anti-virus and firewall, are designed to be general and often very flexible. However, application dependent embedded devices are unique and specific, and therefore a universal solution is implausible. A system cannot be guaranteed secure even when using cryptography, as employing the most cryptographically astute algorithm is almost rendered futile should someone be allowed to access the information, directly from the source, prior to encryption [19]. Another type of security violation can involve a denial-of-service attack, whereby, as in the case of a mobile phone, the signal can be "jammed" so as to prevent the transmission or reception of calls.

- Information/Data. Involves the access, interception and/or decryption of stored and/or transmitted information. As highlighted in the previous point, software-based encryption usually has some kind of security infirmity due to the high risk environment inherent when managing the certificates/keys; an alternative might be to employ the use of an embedded DSP device that incorporates an on-board encryption module which can store and encrypt sensitive information [19]. As in the case of cellular communications, the GSM speech service is relatively secure, up to the point of entry of the network provider; however, this can be construed as a potential security vulnerability, as the cellular network provider, and not the subscriber, is controlling the encryption/decryption. A malicious intrusion of the network provider's system could therefore render all subscribers susceptible to privacy violations. A possible solution will be for the subscriber to incorporate some form of embedded DSP device that can provide personalised encryption prior to the speech entering the handset, thereby providing exclusive protection [20]. Not that DSPs generally do not have the attack resistant capabilities found in security modules (such as smart card chips).

However, with the role of the embedded DSP device in many vulnerable applications, especially hard real-time safety-critical systems, it is vitally important that a thorough risk versus reward analysis be undertaken when pondering the repercussions of a security failure.

Table 2.3 The mobile phone evolution

Mobile phone: features/applications before 2000	Mobile phone: features/applications between 2000–2005	
	• Camera	• GPRS
	• MMS	• 360 KB RAM
• Voice	• Push-Email	• Colour touchscreen
• Keypad	• Monochrome touchscreen	• 128 MB Memory
• Basic monochrome display	• QWERTY keyboard	• Video camera
	• 64 MB memory	• Colour LCD display
• Short message service (SMS)	• FM radio	• Extended battery life
	• Infra-red	•Basic games
• Monochrome LCD display	• Calendar	• 3G
• WAP	• Video playback	• Size (Thinner)
• MP3 Playback	• Bluetooth	• HTML browsing
	• Size (Shorter)	• 1 Mega-pixel camera
	• Colour screen	• Windows mobile

Mobile phone: features/applications between 2006–2010

• Dual-processor	• GPS navigation and google maps	• GPU
• High-resolution LCD screen	• Voice-over-IP	• High-resolution graphics gaming
• Multi-format document viewing	• Downloadable applications (Apps)	• High-definition video capture
• Multi-touch sensor screen	• Document/office editing suites	• 1.2 GHz ARM processor
• Instant messaging	• Programmable open-source O/S	• 1 GB RAM
• Instant messaging	• Resistive touch screen	• 3D autostereoscopic displays
• Accelerometer	• 8 Mega-pixel camera	• 3D stereoscopic capture
• 5 Mega-pixel camera	• Smile and blink detection	
• 8 GB memory		

2.3.4 Embedded DSP and the Mobile Phone

The worldwide ascension of the mobile phone has been met with significant evolutionary changes to the capabilities of the device; having transformed from a wireless gadget that made and received phone calls, into a handheld computer integrated with a mobile phone (known as a smartphone) containing myriads of features and seemingly limitless capabilities (Refer to Table 2.3).

In the original mobile phone the DSP processor was connected to the audio and RF interfaces and responsible primarily for modulation, demodulation, decoding, encoding, encryption, filtering and noise reduction. The GPP was responsible for hosting the operating system and controlling the RF modem, user interface/keypad and few other control functions. The roles of these processors had to adapt significantly to

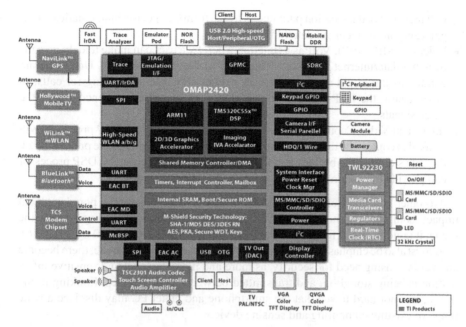

Fig. 2.4 Smartphone SoC processor (courtesy of Texas instruments)

meet the new demands of the modern mobile phone. As a consequence, the chip man-ufacturers had to find innovative ways to develop multi-integrated SoC processors for products that were smaller, lighter, more energy efficient and richer in features.

These processors, like the smartphone SoC processor shown in Fig. 2.4 [31], were designed for complete solutions in one embedded system. The embedded DSP system in Fig. 2.4 has a co-processor EPU that includes a RISC/ARM11 quad-core processor and a DSP multi-core processor. These process the basic telephony functions but also handle a multitude of additional peripherals, including GPS navigation, 3G and WiFi network connectivity, high resolution touch screens, and cameras, as well as support security. The EPU also interfaces with an image accelerator for image and video processing and video decoding, and a 3D graphics accelerator for gaming and other 3D applications.

In such an embedded system the DSP processors or processor cores have to perform several functions, a few of which include:

- Security: A DSP multi-core processor can be integrated in a SoC to perform the function of a security module that contains a cryptographic engine responsible for the logical protection of the voice, data and internetworking systems, as well as providing support for the defense against malicious software and firewall acceler-ation [1, 2, 48].
- GPS Navigation: In the GPS link there will be a DSP-based multi-channel satellite receiver that handles the signals received from the satellites. The DSP processor

will then obtain the location parameters by performing specific mathematical computations on the received data [35].

- Voice-over-IP (VoIP): VoIP is a function whereby a user can make telephone calls over the data/internet network as opposed to the GSM network. Under certain conditions this could result in the user paying very little or nothing for calls. In this scenario the DSP will perform the low bit rate coding and encoding of the voice signals.
- Image and Video Capture: During image and video capture, a DSP processor will be used to control the display prior to capture and then perform image processing on the data stored in the buffer memory after capture. In addition, the DSP processor can provide for features like face, smile and blink detection.

No other consumer electronics device in history has had such a popular global impact, both socially and culturally, as the mobile phone. Since emerging onto the scene, it rapidly spread from a just handful of countries to around 200 by 2010 [16].

It must also be emphasised that with the evolution of the smartphone, there become an ever increasing need for security, as more and more personal and sensitive information is being stored on, and transmitted between, these devices. Looking to the future, it is not hard to see that the smartphone and tablet PC may displace a wide range of existing computing and sensing devices.

2.4 Discussion

DSP is pivotal in systems that involve speech, vision, high-fidelity audio, modulation-demodulation, image compression and compositing, beamforming, echo cancellation, spectral estimation and real-time processing. These types of DSP subsystems have broad ranging applications within the embedded devices market and are incorporated extensively into both the military and commercial sectors.

With the demand for more complicated, secure, smaller, faster and cheaper real-time embedded systems, there will be a need to develop newer more advanced DSP techniques and technologies. However, it is also important to note that as the number of computationally expensive processes increase, so, too, does the power consumption. A collaborative effort between improved RISC processor designs and more energy efficient power sources is necessary in order for these systems to continue to evolve.

References

1. ADSP-2141 SafeNet™ DSP Security System on Chip. Analog Devices. http://www. analog.com/static/imported-files/product_highlights/ADSP2141_Brief.pdf (2000). Accessed 15 September 2011
2. ADSP-2141L Data Sheet. Analog Devices. http://www.analog.com/static/imported-files/data_ sheets/ADSP-2141L.pdf (2000). Accessed 15 September 2011

3. ARM Architecture Reference Manual, 3rd edn. ARM Limited. http://www.altera.com/literature/third-party/archives/ddi0100e_arm_arm.pdf (2000). Accessed 15 July 2011
4. ARM Holdings profits up on tablet and smartphone sales. BBC News. http://www.bbc.co.uk/news/business-13207150 (2011). Accessed 15 July 2011
5. Blackman, S.S.: Multiple-Target Tracking with Radar Applications. Artech House Inc., Norwood, MA (1986)
6. Catsoulis, J.: Designing Embedded Hardware. O'Reilly (2005)
7. Clarke, P.: ARM reports sales, profits up in Q2. EE Times, News and Analysis. http://www.eetimes.com/electronics-news/4204942/ARM-reports-sales-profits-up (2010). Accessed 15 July 2011
8. Cong, J., Fan, Y., Han, G., Zhang, Z.: Application-Specific Instruction Generation for Configurable Processor Architectures. In: Proc. ACM International Symposium on Field-Programmable Gate Arrays, pp. 183–189 (2004). doi: 10.1.1.122.9578
9. DSP and SIMD. ARM Limited. http://www.arm.com/products/processors/technologies/dsp-simd.php. Accessed 15 July 2011
10. Dundamudi, S.P.: Guide to RISC Processors for Programmers and Engineers. Springer, United States (2004)
11. FPGAs for High-Performance DSP Applications. Altera Corporation. http://www.altera.com/literature/wp/wp_dsp_comp.pdf (2005). Accessed 11 July 2011
12. FPGAs Provide Reconfigurable DSP Solutions. Altera Corporation. http://www.altera.com/literature/wp/wp_dsp_fpga.pdf (2001). Accessed 11 July 2011
13. Ganssle, J., Barr, M.: Embedded Systems Dictionary. CMP Books (2003)
14. Goldston, J., Bhattacharya, R.: Reaping the Benefits of SoC processors for Video Applications. Texas Instruments. http://focus.ti.com.cn/cn/lit/wp/spry096/spry096.pdf (2007). Accessed 16 July 2011
15. Gonzalez, R.C., Woods, R.E.: Digital Image Processing, 3rd edn. Pearson Prentice Hall, New Jersey (2008)
16. GSM Roaming and Coverage Maps. Mobile World Live. http://www.mobileworldlive.com/maps/ (2010). Accessed 14 July 2011
17. Gupta, V.K., Vinod, T., Gupta, K.: Compiler directed Customization of ASIP Cores. In: Proc. of the 10th Int'l Symp. on Hardware/Software, Codesign, pp. 97–102 (2002). doi: 10.1.1.16.8455
18. Heath, S.: Embedded Systems Design, 2nd edn. Newnes (2003)
19. Hu, j., Hoang X. D., Khalil, I.: An embedded DSP hardware encryption module for secure e-commerce transactions. In: Security and Communication Networks 4(8), 902–909 (2011). doi: 10.1002/sec.221
20. Islam, S., Ajmal, F.: Developing and implementing encryption algorithm for addressing GSM security issues. In: International Conference on Emerging Technologies, pp. 358–361 (2009). doi: 10.1109/ICET.2009.5353146
21. Katz, D.J., Gentile, R.: Memory Systems. In: Ganssle, J. (ed) Embedded Hardware, pp. 183–238. Newnes (2008)
22. Katz, D.J.: Embedded Media Processing (Embedded Technology). Newnes (2005)
23. Krishna, C.M., Shin, K. G.: Real-Time Systems. McGraw-Hill (1997)
24. Laplante, P.A.: Real-Time Systems Design and Analysis, 3rd edn. Wiley-IEEE Press (2004)
25. Lau, D., Blackburn, J., Seely, J.A.: The Use of Hardware Acceleration in SDR Wave-forms. Altera Corporation. http://www.altera.com/literature/cp/cp_sdr_hardware_acceleration.pdf (2005). Accessed 12 July 2011
26. Li, Q., Yao, C.: Real-Time Concepts for Embedded Systems. CMP Books (2003)
27. Liu, D.: Embedded DSP Processor Design: Application Specific Instruction Set Processors (Systems on Silicon). Morgan Kaufmann (2008)
28. Mody, M.: Video encoding, SoC development, and TI's DSP architecture. Texas Instruments. http://www.eetimes.com/design/signal-processing-dsp/4013053/Video-encoding-SoC-development-and-TI-s-DSP-architecture (2006). Accessed 15 July 2011
29. Noergaard, T.: Embedded Board Buses and I/O. In: Ganssle, J. (ed) Embedded Hard-ware, pp. 137–182. Newnes (2008)

30. Noergaard, T.: Embedded Processors. In: Ganssle, J. (ed) Embedded Hardware, pp. 63–136. Newnes (2008)
31. OMAP™ 2 Architecture: OMAP2420 Processor. Texas Instruments. http://focus.ti.com/pdfs/wtbu/TI_omap2420.pdf (2005). Accessed 17 July 2011
32. Oraioglu, A., Veidenbaum, A.: Guest Editors Introduction: Application Specific Microprocessors. IEEE Design & Test of Computers 20(1), 6–7 (2003). doi: 10.1109/MDT.2003.1173046
33. Oshana, R.: DSP Software Development Techniques for Embedded and Real-Time Systems. Newnes (2006)
34. Parker, M.: FPGA vs. DSP Design Reliability and Maintenance. Altera Corporation. http://www.altera.com/literature/wp/wp-01023.pdf (2007). Accessed 11 July 2011
35. Prasad, K.V.K.K.: Embedded/ Real-Time Systems: Concepts, Design and Programming. Dreamtech Press, New Delhi (2009)
36. Pratt, W.K.: Digital Image Processing: PIKS Scientific Inside, 4th edn. Wiley-Interscience, New Jersey (2007)
37. Proakis, J.G., Manolakis, D.G.: Digital Signal Processing: Principles, Algorithms and Applications, 4th edn. Prentice Hall (2007)
38. Quereshi, S.: Embedded Image Processing on the TMS320C6000™ DSP: Examples in Code Composer Studio™ and MATLAB. Springer, New York (2005)
39. Rabiner, L.R.: Digital Processing of Speech Signals. Prentice Hall (1978)
40. Sheldon, D., Kumar, R., Lysecky, R., Vahid, F., Tullsen, D.: Application-Specific Customization of Parameterized FPGA Soft-Core Processors. In: International Conference on, Computer-Aided Design (2007). doi: 10.1.1.76.3659
41. Sinha, P.: Speech Processing in Embedded Systems. Springer, New York (2010)
42. Smith, S.W.: The Scientist and Engineer's Guide to Digital Signal Processing. California Technical Pub. (1997)
43. Stankovic, J.: Misconceptions About Real-Time Computing: A Serious Problem for Next-Generation Systems. IEEE Computer 21(10), 10–19 (1988). doi:10.1109/2.7053
44. Stapko, T.: Practical Embedded Security: Building Secure Resource-Constrained Systems. Newnes (2007)
45. Stergiopoulos, S.: Advanced signal processing handbook: Theory and implementation for radar, sonar, and medical imaging real-time systems. CRC Press LLC (2001)
46. Stergiopoulos, S.: Implementation of adaptive and synthetic-aperture processing schemes in integrated active-passive sonar systems. Proc. IEEE. 86(2), 358–396 (1998). doi:10.1109/5.659491
47. The Evolving Role of FPGAs in DSP Applications. BDTI. http://www.bdti.com/MyBDTI/pubs/fpga_article.pdf (2007). Accessed 11 July 2011
48. Tretter, S.A.: Communication System Design Using DSP Algorithms: With Laboratory Experiments for the TMS320C6701 and TMS320C6711. Springer (2003)
49. Windrow, B., Stearns, S.D.: Adaptive Signal Processing. Prentice-Hall, Englewood Cliffs, NJ (1985)
50. Wolf, W.: Computers as Components: Principles of Embedded Computing System Design, 2nd ed. Morgan Kaufmann (2008)
51. Zhang, L., Li, S., Yin, Z., Zhao, W.: A Research on an ASIP Processing Element Architecture Suitable for FPGA Implementation. In: International Conference on Computer Science and, Software Engineering, 3, pp. 441–445, 2008. doi: 10.1109/CSSE.2008.580

Chapter 3
Microprocessors and Microcontrollers Security

Chris Shire

Abstract This chapter will consider the chip architectures used in embedded security; how they have evolved over the past three decades, the current designs, and the future trends. The chapter will consider the evolution of the microcontroller Central Processing Units (CPU) cores such as the 8051, 6805. It will look at the wide range of innovative and reduced instruction set designs, including popular off-the-shelf microcontroller designs, microprocessors, and digital signal processors. It will also consider other reduced instruction set designs, with reference to known attacks and options for protection. It will look at the vulnerability of functions within the chips such as memories and interfaces, and possible enhancements. Further security measures for different memory types will be reviewed. Enhanced security concepts using defensive designs, anti-tampering measures, and other hardware protection are discussed.

3.1 Microcontrollers and Microprocessors Security Needs

From an abstract perspective there is little difference in the function of a microcontroller and microprocessor, and in embedded applications the implementation becomes blurred as to the outside world the computing device in the system is often literally a "black box". This chapter will use the euphemism of "Embedded CPU" to cover all the options in design and integration, unless discussing a specific nuance of a design. This is because the designers or test engineers of the system are the only people likely to appreciate the difference. There are several misconceptions about the security of embedded systems. First that attacking the Embedded CPU is difficult, because it is often deep inside a complete assembly. Second there is little value in the embedded software or intellectual property. Finally, that people lack the motivation

C. Shire (✉)
Infineon Technologies UK, Bristol, United Kingdom
e-mail: chris.shire@infineon.com

K. Markantonakis and K. Mayes (eds.), *Secure Smart Embedded Devices,*
Platforms and Applications, DOI: 10.1007/978-1-4614-7915-4_3,
© Springer Science+Business Media New York 2014

to attack an embedded control system. Attacking a single smart meter and turning off the lights in one house seems trivial, it is only when this attack could be scaled up to a city may people worry. As a result many people consider an embedded control system is secure, because it has never been attacked. Justifying embedded security is often a major issue. In addition by definition to embed an item is "to fix something firmly in a surface or object"[1] so that it is an integral part of a larger system. It is therefore perceived that as an Embedded CPU cannot easily be physically removed without damaging the system so it is not open to abuse and must be secure. While from a physical point of view this may often be the case, it does not imply the device is secure electronically.

The first issue to clearly determine is why the Embedded CPU might be attacked, as this will likely determine the method of attack. When a thief steals intellectual property, e.g. software or chip design, then destruction of the system may be of little consequence. Alternatively the attacker may want to deny the use of the system to others for a period of time so as to gain profit directly or indirectly. The physical location of the system will have a bearing on the method of attack. If a system can be accessed remotely and even better covertly, then the potential for attack is greater. A system with strong physical security which is difficult or expensive to breach, such as the cashbox of an arcade gaming machine might be attacked via their test port. Not properly understanding the complete system's security mechanisms and potential attacker's methods can render any Embedded CPU security useless. Of course the more money spent on a security mechanism is in theory the better but there may be consequences. One example is an intrinsically valuable object, such as a royal seal stamp, could be stored in a very secure vault, but if fast frequent access is required such security might impede the signing of official documents. There has to be a balance of risk versus performance.

The detailed methods of attack are described in another chapter, but for an Embedded CPU in an enclosed system the physical connections to the outside world are often the weakest links especially those used for manufacturing tests, e.g. a JTAG (the industry standard Joint Test Action Group serial interface on many CPUs), remote programming, and peripheral connections, e.g. USB, Ethernet, etc. From a physical point of view the housing of the system can include anti-tamper mechanisms, conformal coatings, or epoxy encapsulation. It may even have some countermeasures such as one-way screws or include some "security by obscurity" such as the deletion of product identifiers on the Embedded CPU chip packages. Beyond the physical assembly the electronic architecture of the Embedded CPU and its associated components should be considered in any security analysis. To understand the strengths and weakness of an Embedded CPU it is worth considering the historical development of common architectures.

---

[1] Definition from Macmillan Dictionary http://www.macmillandictionary.com/thesaurus/british/embed#embed_4

3.2 Historical Development

Embedded CPUs can be broken into two broad categories: microprocessors with various peripheral components and microcontrollers which have most of its memory and peripherals on chip, thereby reducing cost and size and often designed for dedicated applications. In contrast to the personal computer and server markets, a large number of basic CPU architectures are used today; these are Von Neumann as well as various degrees of Harvard structures, Reduced Instruction Set Computer (RISC) as well as non-RISC and Very Long Instruction Word (VLIW); word lengths vary from 4 bit to 64 bits and beyond, mainly in Digital Signal Processors (DSPs) although the most typical remain 8/16 bit. Most architectures have a large number of different variants and formats, the most popular of which are manufactured by several different semiconductor companies.

Some common architectures are or were denoted by the following code numbers: 65816, 65C02, 68HCxx, 68K, 78K, 8051, 80251, ARM, C167, ColdFire, COP8, H8, MIPS, MSP430, PIC, PowerPC, SHARC, SPARC, ST6-ST32, TriCore, V850, x86, Z8-Z8000. Information on any of these architectures can be found from their entry in an Internet search engine. It can be seen that this multitude of different designs may present an obstacle to any attacker by simple obfuscation of what CPU is used in an embedded system. However, the underlying format of any design and therefore its weaknesses can be traced back to the basics of logic designs.

Since Integrated Circuit (IC) logic designs started to include arithmetic processing units with software programming, the potential for misuse either accidently or deliberately has been an issue. The earliest CPU's in the 1970's were made from off-the-shelf components such as Bit Slice Processors (BSP). BSP's used arithmetic logic units (ALUs) that typically came in 4 bit increments. These could be assembled together to make larger word lengths (8 bit, 16 bit, etc.) and were controlled using Programmable Read Only Memories (PROM). From these early beginnings developed the one chip microcontroller solutions found in every type of electronic product today, to the multi-core processors powering laptops, servers, and games machines. A direct descendant of the BSP is the Digital Signal Processor (DSP).

In 1971 Intel released its first real microprocessor, the 4004 and with it, the era of microcomputers began [1]. Microcontrollers which included some form of on chip memory first appeared in 1974 with the Texas Instruments TMS1000 with 1K Byte of masked ROM and 64 × 4 bits of RAM for use originally in calculators[2]. In 1976, Intel introduced one of its earliest microcontrollers, the 8748 and 8048. They were used extensively for computer keyboards, or programmed to perform certain data conversion operations. Other Embedded CPUs were programmed with specific stand-alone functions such as a calculator. From the 8048 came the 8051 and later the 16 bit version the 80251. These designs were extensively licensed to over 20 semiconductor suppliers both as embedded cores and later synthesisable software cores. To date it is estimated that over 5 billion embedded designs have been based on the 8051 and its derivatives, mostly in smart cards.

---

[2] http://www.ti.com/corp/docs/company/history/timeline/semicon/1970/docs/74tms_1000.htm

From the start two issues regarding the reliability and security of systems became evident, one internal, the other external. The internal problems were found either when initially testing devices to ensure correct operation as construction faults within multi-chip systems or faults in the PROM could lead to misbehaviour. These faults could appear later in the field, due to the semiconductor process impurities or because of the environment. Excess vibration, voltage, or electrostatic charge, either applied by accident or on purpose, might damage the devices. The other threat was to the intellectual property of the design. Clones from state-run semiconductor companies in the USSR and other eastern bloc countries appeared within a few years[3]. The need for clones was driven by the block on export of high technology by the USA and the need to produce computers for military and commercial applications. The only electrical protection at this time was often in the form of external devices that would protect the Embedded CPU from electrical damage. Hardware protection consisted of strong epoxy encapsulations to deter intruders and to ensure the component assembly stayed together when vibrated.

From the beginning there was also seen a need to protect the software in the ROM of a Embedded CPU as it represented the results of expensive software development and sometimes key intellectual property. The masked ROM of the TMS1000 protected the code of the developer by making the command to read out the ROM nominally inactive. These devices were used in simple games machines and so became, at least as a hobby, the target of attack to allow people to sell cloned or modified games. Various articles still exist (on illicit hacking websites) on how this ROM might be read out using various hardware test functions.

3.3 The Microprocessor

The microprocessor is the portion of a computer system that carries out the instructions of a computer program. This term has been in use in the computer industry at least since the early 1960s. The form, design and implementation of microprocessors have changed dramatically since the earliest examples, but their fundamental operation remains much the same.

Early microprocessors were custom designed from logic circuits as a part of a larger computer. However this has given way to the development of standard mass-produced microprocessors. This trend generally began in the era of discrete transistor mainframes and minicomputers and has accelerated with the popularity of the Integrated Circuit (IC) and the underlying technology trend sometimes known as Moore's Law [2]. Semiconductor technology has allowed increasingly complex CPUs to be designed and manufactured to tolerances in the order of a few tens of nanometers.

Early single chip CPUs supported 8 bit data and address buses such as the Intel 8080 or 16 bit as was TI's TMS 9900. There are extensive references to the history of CPU developments [3], but with regard to embedded systems there are a few

---

[3] http://www.cpucollection.ca/Russian_and_ussr.htm

significant steps which have driven this technology. Western Design Center Inc. introduced the Complementary Metal Oxide Semiconductor (CMOS) 65816 a 16 bit upgrade of the WDC CMOS 65C02 in 1984. This was the core of the Apple II and later the Super Nintendo Entertainment System, making it one of the first and most popular 16 bit embedded designs of all time. Intel followed a different path, and upgraded their 4 bit 4004 designs to the 8 bit 8080t and eventually the 16 bit Intel 8086, the first member of the x 86 families. Their derivatives were used in a number of embedded systems. Intel introduced the 8086 as a cost effective way of porting software from 8080 code. The 8088, a version of the 8086 that used an external 8 bit data bus, was the microprocessor in the first IBM PC, the model 5150, but also used for several early embedded applications. Following up their 8086 and 8088, Intel released the 16 bit 80186, 80286. The 8086 and 80186 had a crude method of memory segmentation, while the 80286 introduced a full-featured segmented memory management unit (MMU) which could be used to protect access to some software. The Intel family became the target of various clone manufacturers, both legitimate licensees and reverse engineered chip suppliers. This lead to several litigious incidents around chip design and intellectual property theft [4]. However it also lead to the acceptance that chip design needed further security to, if not stop, at least deter such issues in the future.

3.3.1 32 Bit Microprocessor Designs

16 bit designs had only been on the market briefly when 32 bit implementations started to appear. The most significant of the 32 bit designs is the Motorola MC68000, introduced in 1979. The 68 K, as it was widely known, had 32 bit registers, but used 16 bit internal data paths and a 16 bit external data bus to reduce pin count, and supported only 24 bit addresses. Motorola described it as a 16 bit processor, though it clearly has a 32 bit architecture. The combination of high performance, a large 16 MByte memory space, and low cost made it the most popular CPU design of its class. The Apple Macintosh designs made use of the 68000, as did a host of other designs in the mid-1980s and its derivatives such as the 68020 ever since. Other large companies designed the 68020 and its derivatives (such the 68EC020 with reduced 24 bit addressing) into embedded equipment such as laser printers Today's ColdFire [20] processor cores are derivatives of the respected 68020. The 68000 had several clones again mostly legitimate.

From 1985 to 2003, the 32 bit Intel x86 architectures became increasingly dominant in desktop, laptop and server markets and these microprocessors became faster and more capable. In 1994 Intel introduced its Smart Die™ program for its 386, 486, and Pentium products so it could supply CPU cores and peripherals for multi-chip modules, for use in embedded computers such as hand held terminals and communication equipment.

3.3.2 64 Bit Microprocessor Designs

While 64 bit microprocessor designs have been in use in several markets since the early 1990s, the early 2000s saw the introduction of 64 bit microprocessors targeted at the PC market. One example is the PowerPC a RISC architecture created by the 1991 Apple–IBM–Motorola alliance [5], also known as AIM. Derivations of this design are now found in high end embedded systems in the network communications and automotive engine management units. Several derivatives have been developed as cores for embedding in a Field Programmable Gate Array (FPGA) by various suppliers such as Altera, LSI logic, Lattice and Xilinx and as cores for various Apple products. There are a multitude of Linux-related operating systems developed with this platform for embedded applications.

Overall in the past decade or more as world trade became more open and the semiconductor technology became more complex the benefit from cloning complex CPU's became less economic. The so called "Grey" market for unofficial sales moved to remarking or repackaging lower specification original devices as high spec units. This practice continues to this day. In the past 10 years, CPU designers have now started to include hardware security functions, ranging from serial numbers to dedicated encryption engines to ensure that users can verify on-line the source of the device. This technique is effectively two factor authentication. The first step is to use the serial number or credential, which can include a check sum, to verify that it is a valid formatted serial number or credential. This will not stop copied hardware serial numbers. The second step is for a hash function of data using this identity to be verified by a remote trusted third party. This then ensures that clones cannot be active in a population of connected devices. It does not solve the problem of cloning for embedded devices with little or no remote connectivity. In these circumstances the Embedded CPU's architecture has to have further security protection integrated at the start and to be continuously active when operational. This may include hidden unique properties programmed into each individual Embedded CPU that cannot be cloned, or a public/private key pair generated on chip that can be challenged to test for authenticity of the device.

3.3.3 RISCs and ARM

A microprocessor is a general purpose product that may have many commands both logical and arithmetic to allow easy programming. Several specialised processing derivatives have followed from this basic concept. A digital signal processor (DSP) has limited instructions but often with a very large data bus to allow for high data throughput. Graphics processing units similarly in the past had limited or no general programming facilities. However, ever since the first "general purpose" CPUs were developed there has been a demand for high performance, dedicated functionality, low cost designs; so-called RISC Machines. By implementing fewer instructions, the chip designer is able to dedicate some of the precious silicon real-estate for performance enhancing features. In addition the benefits of RISC design simplicity are a smaller chip, smaller pin count, and low power consumption. Among some of the typical features of a RISC processor are a Harvard architecture which allows simultaneous access to all the memory by having separate buses for instructions and data. The overlapping of some operations increases processing performance. Probably the most popular RISC family today is the Acorn RISC Machine (ARM) architecture which first appeared in 1985. It has since come to dominate the 32 bit embedded systems processor space due to its efficiency, the low cost licensing model, and its wide selection of system development tools. Many mobile phones include an ARM processor, as do a wide variety of other embedded products. There are microcontroller-oriented ARM cores without virtual memory support, as well as multi-core processors with virtual memory. It has been estimated that by 2011 that over 25 billion ARM cores will have been shipped [6], the vast majority into embedded systems. ARM has been licensed to over 60 commercial companies, including nearly all major IC manufacturers, and several other institutions. Only a few vendors are licensed to modify the ARM cores. This approach has lead to common design and layout, without security features making it an easy target for attack. So derivatives with security functions were requested by the smart card industry. In 1999, ARM announced it was looking at derivatives incorporating security features, called SecureCore, the first public implementation was with Samsung in 2001 with the SC100 core [21]. This core besides being a fully synthesisable design, offered randomised layout options, secure debugging, controlled development to stop reverse synthesis, plus memory protection features and anti-Differential Power Analysis (DPA) functions. From this has come the Cortex-M series of microcontrollers including the special secure core M3 (SC300) range and the ARM company has allowed further modification by at least one vendor. The SC300 has become the preferred architecture for most of the major smart card IC vendors, with various different extra security features. It provides a relatively common platform for the major software developers, i.e. the smart card vendors. This allows at least some software portability from one IC vendor to another, which has been a major hurdle to the industry in the past. In addition second sources silicon suppliers for high volume smart card designs can be provided quickly and less expensively. These derivatives can be considered for assessment to Common Criteria, with certification to EAL5 High and potentially up to EAL6 High.

   In 2003, ARM announced the introductions of security extensions for its microprocessor range, marketed as TrustZone® [7] first for its ARM1176 and later found in the Cortex A5 to A15 CPU range. It provides a low cost alternative to adding an additional dedicated security core to a System-on-Chip, by providing two virtual processors backed by hardware-based access control, including a 33rd bit to identify secure commands. It offers a combination of MMU and caches uses tables to determine whether a particular level of memory exists in "Secure Mode" or "Non-secure Mode". This enables the application core to switch between two states, referred to as "worlds" (to reduce confusion with other names for domains), in order to prevent information from leaking from the more trusted domain to the less trusted domain. This domain switch is generally orthogonal to all other capabilities of the processor, thus each domain can operate independently of the other while using the same core. Memory and peripherals are then made aware of the operating domain of the core and may use this to provide access control to secrets and code on the device. Over 20 companies have taken licenses for the TrustZone® extensions, and variants have been used in applications as varied as MP3 players, smart-phones and payment terminals. The implementation the TrustZone® extensions are specific to each design, and some may include other hardware security features. However, it would be difficult to apply any sort of formal security certification on a device which includes both secure and insecure functions on the same chip.

3.4 Security Design of Embedded CPU Architectures

Today Embedded CPUs are at the heart of a huge range of commercial and industrial equipment [22], including domestic appliances such as microwaves, DVD players and televisions. They are used in cars for engine-control and service functions, in medical instruments, and in many other areas. The widespread availability of Embedded CPUs is a measure of their flexibility and cost when compared to a dedicated hardware function. Usually, they have a high level of input and output (I/O) device options including serial interfaces, e.g. SPI Serial Peripheral Bus (SPI), Control Area Network (CAN), Universal Serial Bus (USB), general use I/O and other interfaces. A microcontroller may minimise the number of external devices used in the system by integrating much of the external interfacing to analog signals as many of them have built-in analog-to-digital (ADC) and digital-to-analog converters (DAC), comparators and pulse width modulators (PWM).

Early Embedded CPUs had 4 bit or 8 bit internal data buses, while modern microcontrollers have 16 bit or 32 bit data buses to access external memory. Obviously, the wider the data bus, the more difficult it is to micro-probe it and reverse engineer. While the complexity, size, construction and general form of Embedded CPUs have changed drastically over the past 40 years. It is notable that the basic design and function has not changed much at all. Most modern Embedded CPUs can be described as von Neumann stored-program machines, but a few do have a Harvard architecture. The first architecture uses separate memory for instructions and data, while the latter uses a single memory structure to hold both instructions and data. From the security point of view, the Harvard architecture should offer better protection against micro-probing attacks. When attacking a von Neumann architecture, an attacker could interrupt the CPU so that it will no longer execute branch instructions or fetch instructions. The contents of the memory can be revealed by micro-probing the data bus and storing the signals. The same attack when applied to a Harvard microcontroller might reveal only the program code, whereas the data memory, which usually contains passwords and decryption keys, may not be available. If a RISC design is too simple with few instructions, it might be easier to reverse engineer it. A RISC with a complex instruction set may leave more distinctive power traces making identification of each instruction through power analysis easier.

Some Embedded CPUs derivatives have core designs with speed enhancement features. One is an instruction pipeline. Each instruction is divided into some simple subinstructions, which are executed by the CPU in step, with a pipeline controller watching the process. Hence, the Embedded CPU will not be immediately execute the code, instead it executes two or more instructions simultaneously. This makes the power analysis more difficult, because two or more instructions can contribute to the power trace. Some secure Embedded CPUs may have one or more slave crypto-coprocessors [8]. These support encryption and related processing of either various symmetric algorithms such as AES, 3DES or asymmetric algorithms such as RSA or ECC. These firmware coded coprocessors typically provide hardware acceleration of a range of functions such as multiple XORs, Galois Field multiplication/addition or modulo multiplication while storing the intermediate results in a dedicated memory of the crypto-coprocessor (SRAM). Such devices have numerous protection features that prevent unauthorised analysis of their data, or reverse engineering. A crypto-coprocessor block may include functions such as a random number generator in order to house them in the same protective environment as the encryption function. If the Embedded CPU is to be certified such random number generators have to use bit sources with a high level of entropy, for example based on two independent free running oscillators. Another common feature in many Embedded CPUs is a cache memory that stores instructions and data that requires frequent and fast access. For example, if the Embedded CPU executes a loop, then the instructions will be fetched from the cache memory rather than from the external memory thus saving time. This makes micro-probing attacks harder, as some data will not appear on the external data bus.

The demand for hardware security began with embedded systems for consumer products. Thirty years ago there was almost no protection against cloning of such devices except legal and economic forces. Often Application-Specific Integrated Circuits (ASICs) were widely used. Such ASICs were simple state machines replacing discrete logic components, thus reducing the size of the assembly and at the same time protecting against competitors with less integrated designs incurring more cost and larger solutions. These ASICs did not carry much security. Their functionality could be determined by a simple analysis of the signals using an oscilloscope or doing an exhaustive pattern analysis of their inputs and outputs. For example as the consumer demand for a clock in every room grew, digital clock ICs were heavily cloned. From the late 1970s, microcontrollers offered a very good replacement for ASIC-based designs. They not only had internal memory and useful interfaces such as LCD drivers, but some sort of security protection against unauthorised access to the internal memory contents. Unfortunately, early microcontroller's semiconductor technology did not offer non-volatile storage for large programmes or variable data, so this had to be stored in a separate chip outside the microcontroller thus allowing the attacker to access them. Games machines had ROMs made with low-cost mask technology allowing easy reverse engineering their contents. Replication of the design could involve using a microcontroller with EPROMs, which although expensive, it was economically viable if the games machine was very popular. This trend continues today as even recently news of attacks on dongles used for "software protection" in the consumer games market has been published [9].

The next step in security design was to place an Electrically Erasable Programmable Read Only Memory (EEPROM) data storage chip next to the microcontroller inside the same plastic package or on the same die. To attack such a chip is not easy; a professional attacker would have to take apart the sample and micro-probe the chip. Such methods require equipment that cannot be afforded by a "hobbyist" attacker, and so their only hope was to exploit a software bug to get access to the data. Even today most microcontroller EEPROMs do not have any special hardware security protection, with the exception of the obscurity of the programming algorithm. In some cases the ROM read-back function is disguised, or replaced with a verify-only function. The verify-only approach can be very powerful if implemented properly, as it is in most microcontrollers used for smart cards.

Microcontrollers with on-chip program memory today often have one or more security fuses that control access to the information stored in on-chip memory. These fuses can be implemented in software or in hardware [10, 11]. Software implementation means that a password is stored in the memory or a certain memory location is assigned as a security fuse. The earliest implementation was for the fuse to be in the logic for the read-back function of the programming interface. The drawback of this design is that the size of the fuse makes it easy to locate and perform an invasive attack. For example, the state of the fuse could be reconfigured by connecting the fuse logic output directly to the supply or ground line. Another well-known example of such attacks is erasing the security fuse under a UV light. The next concept in designs was to make the security fuse part of the memory access circuit, so that any external access to the data is disabled if the fuse is set, usually the fuse is located very close to the main memory or even shares some control lines with it. If the fuse shares the same technology as the main memory array it makes it harder to locate and reset directly. One solution, used in the Motorola MC68HC705C9A microcontroller, was to place fuse cells bit-lines mixed in between the main memory cells. However, other noninvasive attacks are possible, because a fuse cell was often a one-time programmable location in the memory so the fuse may operate differently from the normal memory. As a result a combination of signals could be found under which, thus allowing the access to the information stored in the on-chip memory. Noninvasive attacks could be automated reducing time and effort. Alternatively, the attacker may try using glitch attacks to confound the security check subroutine, or using power analysis to see whether a password guess is correct or even partially correct. This is useful if the fuse is in a separate memory cell to the main memory array. For example, this was the case for early Microchip Peripheral Interface Controller (PIC) and early Atmel AVR (this is an Atmel brand not an acronym) microcontrollers. In both cases, the fuses could be easily found and disabled by one or another method. The simplest way is to check the state of the fuse on power-up, on reset, or on entering the programming mode. The state of the fuse might be changed for a short time by power glitch or laser pulse. Storing the fuse state in a register may not help, because the fuse state is checked only once and the register could be changed by fault injection. The PIC16 × 84 became popular in many hobbyist applications because it uses a simple serial programming algorithm. It also used an EPROM memory, which was easy to erase. It also has a 64 byte EEPROM for storage of user data. The PIC16 × 84 was easily tweaked to allow hackers to read its protected contents, simple disassembly software could then reproduce the source assembly files. Microchip corrected this by introducing the PIC16F84 (and later the PIC16F84A) and discontinuing the PIC16C84.

Fuses are more secure when located within the same memory array but with separate control and signal lines. For example, fuse and main memory cells can touch each other with bit-lines, as in the Zilog Z86E33 microcontroller; or with word-lines, as in the STMicroelectronics ST62T60 microcontroller. Even if the fuses could be erased with electromagnetic radiation it is likely the main memory area would be damaged trying to erase them. At the same time, semi-invasive methods may work on some Embedded CPUs if the fuses have a separate control circuit that could be attacked without affecting the main memory. Apart from different implementations, the security fuse can be monitored in different ways. It is preferable to ensure the fuses are checked each time there is a data access. It may be more secure if the fuse state is monitored in real time and any change affects memory access. In this case, any attacker will have to disable permanently the fuse to access the information. A further improvement is the Anti-fuse [12]; this is a different kind of One-Time-Programmable (OTP) memory that uses programmable interconnection links between metal wires inside the chip. As these links are extremely small, ($\sim$100 nm wide) it is virtually impossible to identify their state and that gives an extremely high security level to the devices based on this technology.

In some early Embedded CPU architectures there were various undocumented features in their command sets, e.g. Z80, 8085, 8048 besides the occasionally obscured ROM read out command. These commands were available, depending of the particular vendor, to offer commercial advantage to special customers. They could provide test routines or to preserve compatibility with other members of the family, e.g. the 8085 with the 8086. These features varied by licensee, but it was common practice in the early developments. One apocryphal command said to exist in some CPUs was the HCF command—the Halt and Catch Fire command, [13], it offered either a hazard to hobbyist programmers or a target for hackers. The 68000 HCF command is believed to be used as a memory checker during production, as it halts the processor and reads through all memory locations as fast as possible and can only be stopped by resetting the system. Certainly some exotic commands did exist on other devices, some of which were discovered by the use of various disassembler tools that had been developed to regenerate source code. Even in later designs such as the Intel Pentium and some of its derivatives there was the so-called F00F command or a bug. This instructed the CPU exception handler to stop servicing interrupts. As a result, any Embedded CPU must be reset. This so- called "Bricking" command of an Embedded CPU can be considered a serious security flaw and can be the target of various "denial of service" attacks. Newer designs include non-executable bits to make some address space secure. ARM has instigated this feature in its TrustZone(r) concept to provide such a secure execution environment for some code.

To make invasive attacks more difficult the Embedded CPUs designs have used a final layer metal or poly-silicon mesh for some time [14]. Logic paths in this mesh are monitored for interruptions and short circuits, and cause reset or zeroing of the EEPROM memory if triggered. In addition, sensors for light, voltage, frequency and temperature maybe included to test for invasive attacks. Normally, such protection is not used in ordinary Embedded CPUs because it increases the design cost. Such sensors could be also triggered unintentionally in their environment such as in automotive applications. Ordinary microcontrollers sometimes have been seen with a fake top layer mesh, whilst this may not stop the determined reverse engineering attempt it is an effective hurdle for simple optical analysis and basic micro-probing attacks. In secure Embedded CPUs such meshes have incorporated various tamper detection mechanisms and sensors. The logic lines in the mesh are polled to check for timing interference, indicating a short circuit and the data on the lines may have randomly changing encrypted data, thus discouraging false signal injection. However, not all meshes are perfect and flaws make micro-probing attacks possible. Some semi-invasive attacks are still possible if the mesh has gaps between the wires and light/radiation or a micro probe can pass through the gap into to the active areas of the circuit. Some user programmable Embedded CPU designs have a non-standard programming interface, allowing a one-time programmable option, effectively a WORM function. In some recent Embedded CPUs further protection against micro-probing attacks is provided by bus and memory encryption. This means even if the chip is stripped down to its active layers without having access to the key materiel the sensitive information cannot be retrieved. This protection process often prevents invasive and semi-invasive attacks. In the past noninvasive attacks could still be possible if the CPU used unencrypted data. The data reaching the CPU could then be vulnerable. An example was the encrypted data stored in external program memory of the old Dallas Semiconductor DS5002FP encryption engine. Weaknesses were found in the data encryption method used in this CPU that lead to a relatively low cost attack published several years ago [23]. In a standard Embedded CPU like the PIC microcontroller, an attacker can easily trace the data bus coming from the memory to the CPU. To reduce further the chance of a micro-probe attack, various non-standard circuit layout processes have been used in secure Embedded CPUs. The standard circuit blocks used in a CPU such as the register file, ALU, instruction decoder, have been laid out in a pseudo-randomly way. This approach is sometimes called 'glue logic layout' and it is widely used in ASICs. Glue logic makes it very difficult to monitor the CPU information physically. Semi-invasive attacks will be difficult due to random layouts of blocks. Of course given time the probing can be automated to test every possible point and then cross-analyse the results. This approach takes a long time and may not be successful. It may be easier to attack a memory block or its control circuit as they cannot be implemented with a glue logic structure and are often physically separate.

Semiconductor process changes to facilitate faster processing and smaller lower cost production has become more expensive with each new generation of geometric feature reduction. This progress is also increasing the costs to the attackers. Ten years ago it was possible to use a laser cutter and a simple probing station to get access to any point on the chip surface. Even today, second hand semiconductor production test equipment can be acquired quite cheaply. The original owners may have used these tools to repair devices that might have errors in the top layer metal mask, which might include the ROM of the user. These tools then provide potential for attackers, with deep pockets, to attack some chips. However, with the latest multi-metal layer semiconductor ICs, with silicon geometries measured in tens of nanometres, most potential attackers are excluded and new attack methods must be found. For example, the structure of the old Microchip PIC16F877 microcontroller was easily observable and could be reverse engineered under a microscope. The second metal layer and poly-silicon layer can still be determined even when buried under the top metal layer. This is because each mask layer in the semiconductor fabrication process follows the shape of the underlying layer. An observer may determine not only the top layer logic functions but also shapes of circuits in structure of the deeper layers. With newer technologies, for example in the Microchip PIC16F877A microcontroller, each layer is smoothed using both chemical etching, and mechanical polishing before the application of the next layer. In this way, the top metal layer does not indicate the features of the deeper layers. The result is that for an attacker to identify the circuit functions they carefully have to etch the chip layer by layer. Currently, many circuit functions are spread across several layers, the result is a three dimensional jig-saw with no big picture.

3.4.1 Security of Embedded CPU Memory

An Embedded CPU operates according to the program located in its memory. There are many different memory types and most of them are used inside microcontrollers. The majority of recent Embedded CPUs are made with CMOS technology. Embedded CPUs often have different memories on the same die. Developers can then use the appropriate memory technology for each different data functions, Static Random Access Memory (SRAM) for cache, Read Only Memory (ROM) for programs, and EEPROM for user variable data, or program updates. This has led to attackers trying to identify the memory cells and the control circuits as a focus of different sensitive data. The EEPROM is likely to hold the user credentials, the SRAM an individual session key. The ROM may hold the communication or encryption algorithms and of course the designers IP for all sorts of software processes. From the traditional security point of view, an Embedded CPU with ROM has less risk than one with EPROM memory, and in turn better than one with EEPROM or Flash memory as the number of possible attack vectors are limited. Most external memory devices are not designed with security in mind. For example, serial EEP-ROMs can be read in-circuit, usually via the SPI or inter-IC (I2C) bus. It is also difficult to securely and totally erase data from RAM and non-volatile memory.

Early Embedded CPUs incorporated a masked ROM or relied on external Ultraviolet EPROM for program storage and SRAM for data storage. Masked ROM is still used where large-quantity production and low cost are required.

Such microcontrollers may not be marked with their part number on the package and have only a manufacturer's logo and a ROM version number. Masked ROM offers very good performance, but cannot be reprogrammed or updated. Normally, the ROM of a standard Embedded CPU does not allow any form of external access. There are few examples where a ROM is the last layer metal mask as it is intended to be modified during production as a way of personalising the devices. In standard CMOS masked ROM the data is stored as a NOR function, this allows active layer programming; the logic state is encoded by the presence or absence of link to a transistor. Information from this type of memory is observable under an optical microscope. This type of feature was offered in Dallas Semiconductor/Maxim iButton products [24] for serialisation and the information is programmed by cutting memory bits with a laser cutter. This memory allows an attacker with sophisticated tools to change the memory contents on a nominally secure product. For semiconductor geometries smaller than 0.5 microns, further processing might be required to remove the top metal layers, which may deter observation.

With the advent of microcontrollers with integrated UV EPROM, a reprogrammable single chip embedded design became possible. In fact, there were usually two versions—one for prototyping, in ceramic packages with a quartz window allowing write and erasure of the program and another in standard plastic packages for mass production allowing a single One-Time Programming (OTP). The early devices had some disadvantages as they required high voltages for programming, which might not be available on the circuit board so in-circuit programming was not possible. This was in effect a security measure as data could only be written only one byte or word at a time, so taking a long time to program a whole chip. Some plastic packages were not 100 % UV opaque allowing OTP devices to be erased, but the time for an erase operation is around 20–30 minutes under a very intensive UV light source so it was unlikely to be attacked without some careful planning. However, attacks on devices using photographic flashgun have been known for some time [15].

George Perlegos at Intel developed Electrically Erasable PROM (EEPROM) memory in the late 1970s. The first products were discrete memory devices and it offered a great advantage over the EPROM by allowing full electrical control over both write and erase operations. Due to high manufacturing cost and complexity, it was not widely embedded in single chip Embedded CPUs until the early 1990s. Even today, many embedded system may have a small serial bus EEPROM on board to store configuration settings or transaction logs. The more recent Embedded CPUs have relied on EEPROM, which has several advantages over the UV EPROM: it can be reprogrammed electronically, in- circuit, up to hundreds of thousands of times; the high voltages are usually generated by on-chip voltage charge-pump circuits; and programming is much faster.

A further improvement of the EEPROM memory, called Flash EEPROM, is becoming the main memory storage for modern Embedded CPUs. It offers much faster programming, it can be reprogrammed in blocks saving a lot of time, and this can be repeated thousands of times. Most of the modern microcontrollers with Flash memory offer internal memory programming, thus allowing field code upgrades without expensive programming tools. Flash memory also has high density offering

3–5 times more storage capacity than the same area of EEPROM. The downside of this memory type is that it can only be erased in blocks, which are relatively large. That puts some strain on embedded software design where program updates are required. Some microcontrollers offer an alternative solution to this problem, having both new memory cells with a combined Flash and EEPROM type behaviour. Flash EEPROM has many different layouts and structures; every IC manufacturer normally has its own design process. The structure is made up of a floating gate memory with either a NOR or a NAND structure. From the security point of view, all floating-gate memories offer very good protection against invasive attacks, because of the very small electrical charge used during programming, which is buried deeply inside the memory cell, so it cannot be detected directly.

Another memory type uses a ferroelectric function to store the data. So called FRAM, has been promoted as an alternative for EEPROM and Flash memories. FRAM has a very fast write cycle and does not require internal high voltage generators, so could also replace some of the functions as SRAM used in an Embedded CPU. FRAM has a two-transistor cell with nonlinear capacitors, which are polarised depending on the applied electric field; the cell will keep its state even when unpowered. FRAM has a disadvantage in that the Read operation destroys the contents of the cell so that a refresh Write is required. However, FRAM offers very good security because its logic state cannot be detected either optically or with probes. Micro-probing of the memory data bus is of course still possible, unless the information is encrypted. However, current FRAM has a limited number of read/write cycles, the cell size is 3–5 times larger than a Flash cell and its fabrication technology is more complex, so there are very few areas where FRAM-based memories are used.

Attacks on the regular layout areas of memory on a chip have forced chip designers to introduce additional protection. For example, modern secure Embedded CPUs may have a default setting of a one-time bootstrap software loader located in the Flash memory that overwrites itself during initialisation. This eliminates any possible access to the information, unless disabled by the system designer with a password. The password can be stored at a certain address location in non-volatile memory. For example, in the Texas Instruments MSP430F112 microcontroller, the read-back operates only with the correct 32 bytes password. Although such protection seems to be more effective than previous offerings, it is open to low-cost noninvasive attacks such as timing attacks and power analysis. If the security code is sampled from the memory during power-up or reset, it could present an attacker the chance to identify the password. This could be done using a combination of brute force attack and power glitches, or by trying to force the checking circuit to get the wrong state of the memory.

Another hardware security issue for all types of memories is data remanence. Remnants of stored data may exist and be retrievable from devices long after nominally being erased and with the power removed, which could be useful to obtain program code, temporary data, crypto keys, etc. In many modern Embedded CPUs, a monitoring circuit is usually implemented, causing a reset of the hardware programming interface or preventing any write/erase operations below or above certain voltages, frequencies etc. Some system designers have assumed that the erased data will disappear. In reality, some traces of the data may be left behind. Even in SRAM, after power removal, have shown examples of data remanence, as when frozen some SRAM cells retain information for hours [16]. To retrieve the trace of the data is not easy, but for example during the chip erase, operation if the security fuse was deactivated, the memory may be accessed normally. Then each transistor inside the memory array has to be checked by micro-probing the internal memory bus. In general, SRAM memory offers a very good level of protection by placing sensors into the circuit to avoid low-temperature attacks.

3.4.2 Security of Embedded CPU Interfaces

Almost every modern piece of assembly equipment in a factory with an electronic control unit is connected to a network. These networks carry information that controls production flow, transfers manufacturing data, and provide remote equipment management. It may be possible to diagnose and repair many failures, if the unit equipment is connected to a network. Thus avoiding expensive on-site service calls, and reducing production down time. It may be required to provide secure access to the control system in a number of situations:

- To interrogate an industrial Embedded CPU for data, even when the manufacturing machine is switched off.
- Reboot a controller station remotely.
- Ensure an operator panel is safeguarded with latest health and safety policies, without halting operations or operator intervention.

All of these scenarios and more represent possible threats. An embedded system designer must consider that attackers will attempt to access the Embedded CPU and consider the following points of attack:

- Software programming interface
- Hardware programming interface
- Third party unverified protocols
- Read-back functions
- Hardware security fuses
- Software security fuses
- Discrete memory separate from the on chip memory
- Shared memory control lines
- Shared bit-lines
- Password locations
- Verification checks at power-up
- Permanent real time monitoring

Some Embedded CPU manufacturers intentionally leave a side channel access to the code for testing or programming purposes after fabrication. Normally, the information on these test protocols is kept secret by the manufacturers. The programming interface allows writing, verifying, reading and erasing of data in on-chip memory. It could be implemented either in hardware such as a JTAG state machine, in a proprietary interface, or in software (e.g. Mask ROM or Flash bootloader). Before initial programming and a fuse has been set, some microcontrollers offer a software controlled boot loader for in-system programming. Others offer a fast hardware interface for mass production programming. For example, an Embedded CPU may have in-circuit serial programming via a synchronous interface (e.g. SPI, JTAG), fast industrial parallel programming, and a software boot loader via an asynchronous serial interface (e.g. USB). The JTAG (IEEE 1149.1) interface maybe an Achilles' heel of the system. JATG can provide a direct interface to the internal registers of Embedded CPU and so has become a common attack vector. A JTAG interface to USB test harness can be bought or self-assembled with a few low cost commonly available components, allowing automated attack routines to set easily set up. Removing JTAG functionality from a device is difficult. System designers usually disguise links, cut traces, or blow fuses. However, a determined attacker can easily repair most of them. Such test lines are used in smart card ICs only during the initial wafer manufacture. These lines are routed into the sawing corridors of the die during chip layout. These lines are then destroyed during chip separation. This technique when used with the combination of fuses make micro-probing for the lines useless.

The In-Circuit-Emulator (ICE) is a commonly adopted tool as a software program debugging technique. The GUI interface debugging software can help a legitimate user to debug easily. If freely available, these tools may reduce the time taken to attack an embedded CPU design. One solution for embedded system designers that need to protect their embedded software, from competitors and counterfeiters, is to use a secure Embedded CPU as an in-system software authentication device. To protect embedded software from cloning a challenge is sent at random intervals from the secure Embedded CPU. The response to the secure Embedded CPU is then compared to the expected response. By providing a large number of challenges and placing those in unique areas, the source code can be relatively well protected. This makes it extremely difficult for anyone to reverse engineer the source code. This added difficulty will make it more cost effective for attackers to develop an entirely new system rather than modify the existing source code.

3.5 Advanced Chip Design

Advanced embedded designs may use more than a single chip CPU, and as part of an ASIC or other VLSI design. Synthesisable logical blocks such as DSPs and RISCs and CPUs have been available for nearly 20 years. Their first development in the early 1990s, was to reduce the system cost by using the minimum functions needed for an application and to improve the system size or performance or security. Implementation tools are often optimized for a specific FPGA family or ASIC library and for a given range of clock frequencies. When designing a synthesised Embedded CPU, a few important aspects must be taken into account; area usage, performance in terms of throughput, and added value. To protect what maybe a non-secure hardware platform various techniques have been employed to protect the design from an illicit observer, such as introduction of random or spurious logic blocks but these may impact performance or power consumption. Synthesis tools have led to multi-core designs for embedded applications incorporating two or more CPU's with DSP functions. Safety and security features are sometimes included such as error correction on the memory, parity checking on some interfaces and interrupt registers, redundancy checking functions, and advanced memory lock protection. In addition as previously described the software in an Embedded CPU-based system may be protected often by a mixture of encryption and fuse protection, against unauthorised attacks. Although this will provide a barrier to any reading of the memory optically or by micro-probing the data bus, this data normally has to be decrypted somewhere, often in or by the main CPU and stored in a SRAM cache pipeline ready for operation. The focus of the attacker may then try to detect any plaintext on the data bus close to the CPU, or better still the key to decipher the ROM. It may be possible by stopping the clock and literally freezing the circuit to read the contents of the SRAM with impunity. As a further precaution, Embedded CPU chip manufacturers offer an enhanced verify-only approach. In this case, a hash value of the content of memory is compared to a secured value and a single-bit response in the form of pass/fail sent back. The verification process can take place both in hardware or in software. It may be impossible to verify the whole memory in one go, so the process is split into blocks with their size limited by the available SRAM buffer or hardware register. The result of the verification is either to stop the Embedded CPU on detection of the first incorrect memory block, or to flag the status in a register. Of course, as described ever more sensors can be included to test for invasive attacks and tighter geometry meshes added to deflect probing or reverse engineering.

A radical departure to this concept has been developed by Infineon in the past few years, given the name "Integrity Guard™" [17] it consists of three features; error detection, full data encryption, and a new type of mesh shield. The concept is focused on not, as in earlier generations, to add more sensors and protective devices to the periphery of the Embedded CPU but to concentrate on protecting the data at all times, so no plaintext is ever used. The concept includes a mesh that uses a new shielding concept combined with intelligent secure wiring. The electrical signal lincs inside the chips are rated concerning their relevance, and on the basis of this classification they are automatically routed and checked. An intelligent shielding algorithm checks the chip's layers, providing the final so-called "Active I^2-shield". The error detection is based on a microcontroller with a dual CPU allowing error detection in real time, even while processing. Both CPUs deliver their operational results independently from each other. A comparator detects whether an operation was performed the same, or if an erroneous operation was made. In the case of an error, an alarm is issued. Even the cache is an active part of the error detection, which is essential, as cache-based attacks will become a major threat for embedded security in the near future. In addition the concept applies full encryption over the complete core and memories, leaving no plain text on the chip. The dual CPUs utilise full hardware encrypted operation, with different secret keys used in each of the CPUs. All memories are completely encrypted: for the memory buses and blocks of RAM, ROM, EEPROM, and FLASH, a strong block-cipher hardware encryption engine has been utilised. Data is enciphered from the memory encryption system to the encrypted CPU without exposing plaintext. Peripheral buses are protected using dynamically changing keys, and some peripherals work in encrypted modes. For example, the new crypto coprocessor "SCP" (a Symmetric Crypto Processor for Triple-DES and AES) utilises internal, dynamic encryption — just like the encrypted CPUs. This prevents the presence of plaintext inside important parts of the chip. This type of enhanced digital security is required as advanced attacks are developed, such as micro coil-based localised Differential Electromagnetic Analysis (DEMA). This concept is being applied to both traditional 16 bit Embedded CPU architectures and to modified ARM designs. As with all innovation it can be expected that other Embedded CPUs will also start to incorporate such concepts, but the cost of such developments and the cost of such technology has to be balanced against the threat risks.

3.6 Conclusion

The impact of embedded devices is huge. Overall, it is usually estimated that for every desktop computer chip sold, 100 microcontrollers are sold for embedded systems. Techniques for creating secure high-reliability embedded systems have focused historically on safety-critical markets, e.g. the aerospace, medical, and automotive industries. In these sectors system failures can have fatal consequences. These applications remain important, but embedded microprocessors and microcontrollers now also have an enormous impact in much broader areas of product development, such as consumer applications as diverse as simple washing machines and as sophisticated as games consoles. Manufacturers need to maximise the reliability, and the security, of the key components in embedded systems in order to reduce the cost of warranty repairs, minimise product recalls and ensure continued business. In summary, it is clear that Embedded CPUs have become more sophisticated and more secure in the past 40 years. A typical modern average house may contain 10–20 devices with an Embedded CPU, mostly independent of each other. The average citizen may carry 5–10 portable devices with Embedded CPUs as smart cards or in smart phones etc. The average mid-range car may have over 50 Embedded CPUs with over 50% networked together. A factory employing 500 operators may have over a 1,000 networked devices. While most of these Embedded CPUs may have little access to the outside world, and little information of worth, the few such used for payment or access rights may present a target for theft either physical or virtual. However, as the issues of the "Semantic Web" [18] become reality there will be need to increase security as threats from "denial of service" or malware attacks on the user or the network provider will become more attractive. With the increasing complexity of software and the possibility to provide remote updates there will be need for remote authentication, and integrity management of an Embedded CPU's status. It is likely that self-checking of systems will have to move past parity checks or error correction and look at frequent hardware and software image verification. The need for embedded hypervisor Embedded CPU elements will increase. This idea has been seen in the Trusted Platform Modules (TPM) for notebook computers, and the various secure elements in mobile phones, engine management systems, gaming consoles and smart meters. If the current trend [18] continues then by 2020 the current level of machine-to-machine communications could have more than quadrupled. There is little doubt that as the value of these communications increases, so will be the need for embedded security.

References

1. Ed Glynnis Thompson Kaye "Intel: Innovator of the information revolution", Published by Intel Communications Dept 1984 [Online Available] http://www.intel.com/Assets/PDF/General/15yrs.pdf
2. Gordon Moore: "Cramming More Components Onto Integrated Circuits" Electronics Magazine, 1965, [Online Available] http://download.intel.com/museum/Moores_Law/Articles-Press_Releases/Gordon_Moore_1965_Article.pdf
3. Mike MALONE, "The Microprocessor - A Biography", Springer-Verlag 1995, 0–387-94342-0, [Online Available] http://www.computerhistory.org/
4. An early example :Article 10 of 31, Article ID: 8901130503, Published on February 16, 1989, San Jose Mercury News (CA), "Start-Up used Stolen Trade Secrets, Intel Charges".
5. Charles R Mooore & Russell Stamphill: "The Making of the PowerPC", Association of Computing Machinery Communications of the ACM, June 1994, 37(6), [Online Available] http://zmoore.net/CACM%20PPC%20Alliance.pdf
6. Mark Hachman, ARM: "We'll Own over Half of the Mobile PC Market by 2015": PC Magazine May 31st 2011 [Online Available] http://www.pcmag.com/article2/0,2817,2386209,00.asp
7. ARM PLc; "Building a Secure System using TrustZone Technology", ref PRD29-GENC-009492C, [Online Available] http://infocenter.arm.com/help/topic/com.arm.doc.prd29-genc-009492c/PRD29-GENC-009492C_trustzone_security_whitepaper.pdf
8. Helena Handschuh: "Smart Card Crypto-Coprocessors for Public-Key Cryptography" 2000 The Pennsylvania State University [Online Available] http://citeseer.ist.psu.edu/viewdoc/summary?doi=10.1.1.1.9288
9. Ronald Huizer, "Don't Give Credit: Hacking Arcade Machines", Immunity Inc, May 2011 [Online Available] http://www.immunitysec.com/infiltrate/presentations/Arcade_Attacks.pdf
10. Smart card handbook By Wolfgang Rankl, Wolfgang Effing Smart card Handbook 2003 Wiley, ISBN 0-470-85668-8 page 535.
11. US patent: 5083293, "Prevention of alteration of data stored in secure integrated circuit chip memory", Gilberg, Robert C. (San Diego, CA), Moroney, Paul (Cardiff-By-The-Sea, CA), Shumate, William A. (San Diego, CA), January 1992.
12. Inventors Cutter, Douglas J. (Boise, ID), Beigel, Kurt D. (Boise, ID), Ong, Adrian E. (Santa Clara, CA), Ho, Fan (Boise, ID), Mullarkey, Patrick J. (Meridian, ID), Luong, Dien S. (Boise, ID), Debenham, Brett (Meridian, ID), Pierce, Kim M. (Meridian, ID) United States Patent 5631862 "Self current limiting antifuse circuit", Micron Technology, 20 May 1997.
13. HCF was first mentioned in conjunction with the Motorola 68000, in BYTE Magazine, Vol 2, December 1977.

14. Inventors Gilberg, Robert C. (San Diego, CA), Knowles, Richard M. (San Diego, CA), Moroney, Paul (Cardiff-by-the-Sea, CA), Shumate, William A. (San Diego, CA) US patent:4933898, Secure integrated circuit chip with conductive shield, 1990.
15. Hinds, D.J. Stokoe, J.C.D. British Telecom Research Laboratories, Ipswich, UK, IET Electronics Letters: June 20 1985 Volume: 21 Issue: 13, On page(s): 553–554 ISSN: 0013–5194.
16. Peter Gutmann. "Secure Deletion of Data from Magnetic and Solid-State Memory", 6th USENIX Security Symposium Proceedings, San Jose, California, July 22–25, 1996.
17. Peter Laakmann, Markus Janke "A New Security Concept For The Next Decade" Secure Magazine Issue 2 2006. www.infineon.com/integrityguard.
18. Tim Berners-Lee, James Hendler and Ora Lassila: "The Semantic Web" Scientific American Magazine (May 17, 2001). [Online Available] http://www.scientificamerican.com/article.cfm?id=the-semantic-web
19. Harbor Research: 2010–2014 M2M & Smart Systems Forecast Report 2010 M2M & Smart Systems Report Brochure.
20. Motorola Press release High Performance Embedded Systems Division today announced the ColdFire(TM) MCF5200D Developer's Chip at the Embedded Systems Conference, September 12–15, 1995 in San Jose, CA. [Online Available] http://www.thefreelibrary.com/MOTOROLA+ANNOUNCES+COLDFIRE+MCF5200D+DEVELOPER'S+CHIP-a017376831
21. Samsung to use ARM core with security accelerator in 32-bit smart card chips 10/09/2001 [Online Available] http://www.eetimes.com/electronics-news/4105054/Samsung-to-use-ARM-core-with-security-accelerator-in-32-bit-smart-card-chips
22. Designing Embedded Hardware, 2nd Edition By John Catsoulis Publisher: O'Reilly Media Released: May 2005 Print ISBN:978-0-596-00755-3I ISBN 10:0-596-00755-8 Ebook ISBN:978-0-596-55662-4I ISBN 10:0-596-55662-4.
23. Pirate War Causes Shut-Off of Battery Cards article by David Lawson Scrambling News. 1996 [Online Available] http://www.mycal.net/Group42/hack/tv/sat/scnews/news0306.htm
24. Dallas/Maxim datasheet DS1427 17.02.1998 [Online Available] http://datasheets.maximintegrated.com/en/ds/DS1427.pdf

Chapter 4
An Introduction to the Trusted Platform Module and Mobile Trusted Module

Raja Naeem Akram, Konstantinos Markantonakis and Keith Mayes

Abstract The trusted platform module (TPM) is a tamper-resistant component that provides roots of trust in secure computing and remote attestation frameworks. In this chapter, we briefly discuss the TPM architecture, operations and services. The discussion is then extended to the mobile trusted module (MTM)—to contrast and compare different approaches to implement a trusted platform architecture. This illustrates the vital role the ecosystem of a computing platform plays in the architectural design decisions regarding the root of trust in a trusted platforms.

Keywords Trusted Computing · TPM · MTM · Mobile phones · Tablets

4.1 Introduction

The concept of a trusted platform is based on the existence of a trusted and reliable component that provides evidence of the state of a given system. How this evidence is interpreted is dependent on the requesting entity. Trust in this context can be defined as an expectation that the state of a system is expected and so secure. This definition requires a trusted and reliable entity called the trusted platform module (TPM) to provide the trustworthy evidence regarding the state of a system. Therefore, a TPM is a reporting agent (witness) not an evaluator or enforcer of the security policies. It provides a root of trust on which an inquisitor relies for the validation of the current state of a system.

R. N. Akram (✉)
Department of Computer Science,
University of Waikato, Hamilton, New Zealand
e-mail: rnakram@waikato.ac.nz

K. Markantonakis · K. Mayes
Information Security Group, Smart Card Centre, Royal Holloway,
University of London, London, UK
e-mail: k.markantonakis@rhul.ac.uk

K. Mayes
e-mail: keith.mayes@rhul.ac.uk

K. Markantonakis and K. Mayes (eds.), *Secure Smart Embedded Devices,*
Platforms and Applications, DOI: 10.1007/978-1-4614-7915-4_4,
© Springer Science+Business Media New York 2014

The TPM specifications are maintained and developed by an international standards group called the Trusted Computing Group[1] (TCG). Today, TCG not only publishes the TPM specifications, but also those for the mobile trusted module (MTM), Trusted Multi-tenant Infrastructure and trusted network connect (TNC). With emerging technologies, service architectures and computing platforms, TCG is adapting to the challenges they present. In this chapter, we described the TPM and MTM architectures along with the influence exercised by the ecosystem of the targeted computing platform on their design. Finally, we discuss the future challenges facing the TCG in the shape of restricted and technology specific initiatives proposed by individual organisations.

4.2 The Trusted Platform Module

In this section, we open the discussion with a brief description of the trusted platform framework that is then extended to the TPM architecture.

4.2.1 Trusted Platform Framework

The basic framework for the trusted platform is to have a root of trust (preferably in hardware) that must be trustworthy, if an entity has to measure the trustworthiness of a system. The root of trust in the TCG specifications [1, 2] is a collection of—a root of trust for measurement (RTM), a root of trust for storage (RTS) and a root of trust for reporting (RTR). The RTM is an independent computing platform that has a minimum set of instructions, which are considered to be trusted for measuring the integrity matrix[2] of a system. On a typical desktop computer, the RTM will be part of the basic input output system (BIOS) and in this scenario, it is referred as core root of trust for measurement (CRTM). Where the RTS and RTR are based on an independent, self-sufficient and reliable computing component that has a predefined set of instructions to provide the platform authentication[3] and attestation[4] functionality; such a component is referred to as a TPM.

---

[1] TCG: It is a non-profit industry standard organisation that "develop, define and promote vendor neutral specifications for trusted computing". Web site: http://www.trustedcomputinggroup.org/.

[2] Integrity Matrix: To provide integrity assurance of a platform component, a TPM generates the hash of individual subcomponents, this individual measurement is referred to as an integrity measurement. Whereas, integrity matrix is the condensed value of the integrity measurements that represent the overall state of the respective platform component (Sect. 4.3.4.1).

[3] Platform Authentication: It provides the proof the platform's identity and this identity may or may not be associated with the respective user. A TPM can have unlimited number of platform identities that are usually generated by the TPM itself (discussed further in Sect. 4.3.3).

[4] Platform Attestation: It provides the proof that a platform can be trusted by providing the cryptographically signed integrity matrix of the respective platform (further discussed in Sect. 4.3.4.4).

A platform can be considered to be a trusted platform if it has a TPM and support-ing architecture for the "trusted building block" (TBB). The TBB includes CRTM, physical connection between a CRTM and the motherboard (of the platform), con-nection between a TPM and the motherboard, and functionality to detect the physical presence. The physical presence implies the direct interaction of a user with the plat-form, which is traditionally based on a secret credential that in theory is only known to the respective user. By verifying the credentials, the platform assumes that the platform owner is physically present. Credentials can be a simple password, USB dongle or a smart card, etc. The TPM specification does not specify any implemen-tation technique regarding the physical presence check. The specification defines the physical presence as a signal from the platform to the TPM that indicates the user has manipulated the hardware of the platform (e.g. typing on the keyboard) [1]. Figure 4.1, illustrates the trusted platform framework.

The trust boundary is a collection of the TBB and roots of trust (e.g. RTM, RTS and RTR). A TPM extends the trust from roots of trust through transitive or inductive trust. A transitive trust is a process that enables a root of trust to provide a trustworthy-description (e.g. hash generation) of a second function (e.g. software). Therefore, the requesting entity can then verify whether it can trust the second function based on the integrity measurement provided by the respective TPM. The rationale behind transitive trust is that if an entity trusts the TPM of a platform, it will also trust its measurements.

4.2.2 Basic Architecture

In this section, we discuss the basic TPM architecture and its different components that are shown in Fig. 4.2 and discussed subsequently.

Fig. 4.1 Trusted platform framework

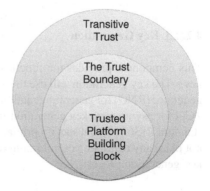

Transitive
Trust

The Trust
Boundary

Trusted
Platform
Building
Block

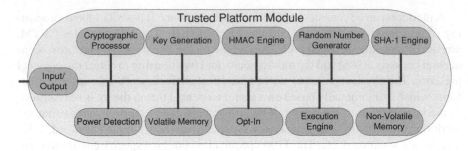

Fig. 4.2 Generic architecture of trusted platform module [1]

4.2.2.1 Input and Output

The input/output (I/O) component of a TPM provides protocol encoding/decoding for data transfer over the communication bus. Furthermore, the I/O enforces the access policies stated by the Opt-In (Sect. 4.2.2.8) or other TPM components. The structure and design of the I/O are not specified by the TPM specifications, but left to the discretion of the platform implementers.

4.2.2.2 Cryptographic Processor

The cryptographic processor is a dedicated hardware for cryptographic operations that supports: (a) asymmetric encryption and decryption (including signature algorithms), (b) asymmetric key generation, (c) Hashing (e.g. SHA-1) and (d) Random Number Generation. A TPM might implement symmetric encryption, but TCG specification states that such algorithms should only be used internally by a TPM and not exposed to general users.

4.2.2.3 Key Generation

This component provides the functionality to generate asymmetric and symmetric keys. The key generation will use the algorithm implemented by the random number generator (RNG) for generating random sequences. The requirements imposed by the TCG for asymmetric key generation include mandatory compliance with the IEEE standard P1363 [3] that specify primality tests. Whereas, the TCG specification does not place any restriction or minimum requirements regarding the performance of the key generation process.

4.2.2.4 HMAC Engine

The TCG specification requires that the HMAC implementation in a TPM should conform to the RFC 2104 [4]. A TPM uses HMAC engine to ascertain the validity of the Authentication Data (AuthData) and prove that the request received by the TPM is not modified during transmission. Each object in a TPM that does not allow public access has an associated AuthData of 160-bits.

4.2.2.5 Random Number Generator

Trusted platform module use the RNG for random-nonce and key generation along with providing randomness for digital signatures. The TCG is open regarding the implementation of the RNG and leave the design decision to the individual vendors.

4.2.2.6 SHA-1 Engine

The TCG specifies the SHA-1 hash algorithm [5] to support integrity measurement generation. The hash algorithm is available for public access, but a TPM is not regarded as a cryptographic accelerator; therefore, there are no performance restrictions on a TPM-based SHA-1 engine.

4.2.2.7 Power Detection

The TPM specification requires that a TPM should be notified of any power state changes, which is obviously the responsibility of the host platform. On such an event, a TPM might perform some tasks based on the pre-defined security and reliability policy. For example, a TPM might restrict certain commands while the system is in a particular power consumption state (e.g. power-on self-test, hibernate, and sleep, etc.).

4.2.2.8 Opt-In

The Opt-In component manages different TPM states that include: on/off, enabled/disabled and activated/deactivated. It also maintains and enforces the policies associated with individual states. These policies describe the requirements for authorisation of a TPM user, or/and how to ascertain the physical presence. The TPM states are discussed in detail in Sect. 4.3.2.

4.2.2.9 Execution Engine

The TPM commands (instructions) [6] are executed by the execution engine in a secure and reliable manner. The execution engine is an on-chip (within the boundary of a TPM) processor that provides execution isolation.

4.2.2.10 Non-volatile and Volatile Memory

The non-volatile memory is used to store persistent data items that relate to a TPM's identity and associated state. Whereas, volatile memory is used to store temporary data items including keys during the signing or decryption operations. Persistent data items that are moved to the volatile memory location to facilitate the associated operations (i.e. keys for signing or decryption operations), are stored back to non-volatile memory (if there are any changes to them) at the end of the associated operation.

4.3 TPM Operations

In this section, we discuss the main operations of a TPM including generation and use of a endorsement keys (EKs) and attestation identity keys (AIKs). Later, we will dive into the integrity measurement and reporting mechanisms. Finally, we detail the delegation process for moving a user's credentials (cryptographic keys, and related data associated with a user) from one TPM to another.

4.3.1 TPM Endorsement Key

During the TPM manufacturing process, the TPM Entity (TPME) will initiate the generation of a TPM endorsement key (EK). The EK is a 2048-bit RSA key pair that is bonded to the respective TPM, and it is certified by the TPME [1]. Typically, TPM manufacturers take the role of the TPME. The issued certificate also validates the association of the EK with the respective TPM. The EK remains the same for the entire lifetime of a TPM and due to security and privacy reasons, the EK is not sanctioned to generate signatures. However, if a TPM implements (optional) the "revoke trust" [1] functionality then the TPM can revoke the old EK and generate a new EK (if required).

The public portion of the EK, which is termed as PUBEK by the TPM specification does not have any security or privacy issues associate with its exposure, assuming there is no association with other information [1]. A PUBEK can be considered a platform identifier that is associated with the TPM hardware, and typically it is not associated with personal information. Nevertheless, a PUBEK can be considered personally identifiable information (PII) if it is associated with other PIIs or part of the personal information in some way. The PII enables an external entity to uniquely identify a relation between a platform and its user(s). Therefore, for privacy reasons use of any such information that is related to the PII should be under the TPM owner's control.

For example, an attestation service (e.g. Certification Authority) might include personal information related to a user in the AIK credentials that may also contain the PUBEK; making AIK-PUBEK as PII. In such case, the PUBEK becomes privacy sensitive information from the respective user's point of view. To provide privacy to the TPM users, the TPM specification requires that a TPM should allow the respective owner to specify whether to include/disclose PUBEK along with the AIK. During the AIK credential generation (Sect. 4.3.3), the user should be notified what personal information is being included. Therefore, a PUBEK on its own is not privacy sensitive information until the respective user permits it to be associated with an AIK.

4.3.2 TPM Ownership

A TPM can have up to eight operational modes shown in Table 4.1 that are formed by combining the three states discussed in Sect. 4.2.2.8.

The S1 is the fully operational state of a TPM, and the S8 is the most restrictive state where all functions of a TPM are off except the state change operations. The TPM has to be under the ownership of an entity to perform all of the designated operations. The ownership acquisition process in the TPM includes the generation

Table 4.1 TPM operational states

Operational mode	Description
S1 Enabled–Active–Owned	The TPM is enabled with all supported features available and under the ownership of an entity
S2 Disabled–Active–Owned	This state is similar to the S1 except all TPM operations are restricted with exception of the reporting TPM capabilities and Platform Configuration Registers (PCRs) update
S3 Enabled–Inactive–Owned	The TPM is enabled, but disables all TPM operations except the once that changes the TPM's operational state (e.g. ownership change or activating the TPM)
S4 Disabled–Inactive–Owned	The TPM is disabled and inactive; however, it is still in the ownership of an entity
S5 Enabled–Active–Unowned	Similar to state S1 without any owner
S6 Disabled–Active–Unowned	Similar to state S2 without any owner
S7 Enabled–Inactive–Unowned	Similar to state S3 with no owner
S8 Disabled–Inactive–Unowned	Similar to state S4 but not under any entity's control

of RTS, insertion of the shared secret (e.g. AuthData) and setting TPM policies. The proof of ownership of the TPM constitutes the knowledge of the shared secret by an entity.

The TPM ownership process generates a new storage root key (SRK) that is associated with the particular user, and it acts as RTS. For each new user, the TPM will generate a new SRK and the new user will not inherit any objects from previous users. If the new user wants to inherit objects from other users then data migration is requested that is discussed in Sect. 4.3.5. A TPM use SRKs to secure the data that is stored on the external storage; an SRK provides confidentiality and integrity to a given data—providing secure storage.

The new AuthData is inserted into a TPM using the AuthData insertion protocol (ADIP). To change the AuthData, an entity can execute either AuthData change protocol (ADCP) or asymmetric authorisation change protocol (AACP) [1]. During the lifetime of a TPM, an entity can provide the proof of ownership to the TPM by either using the protocols: object-independent authorisation protocol (OIAP), object-specific authorisation protocol (OSAP) and delegated-specific authorisation protocol (DSAP) [1]. The basic design principle of these protocols is to provide ownership, command and parameter authentication along with protection against replay and man-in-the-middle attacks. In this chapter we avoid detailing these protocols for the sake of brevity; however, interested readers can consult TCG specification regarding these protocols [1, 7].

4.3.3 Attestation Identity Keys

An AIK is a 2048-bit RSA key, which is generated on the request of the TPM owner and stored in non-volatile storage outside the TPM; it is encrypted and integrity protected by the respective SRK. The usage of an AIK is under the control of the user who requested its generation. The AIK is used to generate signatures on data that is only generated by the TPM internally (e.g. integrity measurements, keys, and TPM status information, etc.). To provide assurance that the AIK is genuine and associated with a trustworthy TPM, the TCG has defined two possible solutions. One based on the privacy certification authority (privacy CA) and second on direct anonymous attestation (DAA).

The AIK credentials are generated by an external entity, by the privacy CA. The TPM will disclose the AIK (public part) with or without EK to the respective privacy CA. The CA will attest the AIK and its association with the endorsement, platform and/or conformance credentials.

In this solution, the privacy CA has to take part in every transaction between an entity (verifier) that requests the TPM to provide platform authentication. This imposes high availability restrictions on the privacy CA along with a possible issue of a CA colluding with the verifier, which can create privacy concerns. Furthermore, if a malicious user is able to obtain privacy CA (historic) transaction records in which a TPM has participated, the malicious user will be able to uniquely identity the respective TPM in future transactions. However, on a positive point, the privacy CA- based architecture can effectively detect rogue TPMs, which have been compromised by a malicious user.

Another possible approach is DAA [8]. The DAA protocol enables a user to prove to a verifier that she has a trustworthy (certified) TPM, and the verifier cannot even tell whether it has communicated with the user before. The DAA protocol is based on three entities: the TPM, DAA issuer and verifier. The DAA protocol consists of two steps, first joining in which a TPM and DAA issuer executes a two-way protocol. At the successful conclusion of the protocol. The DAA issues the AIK credential to the TPM. In second step, the TPM uses the DAA credentials to prove to the verifier the trustworthiness of the TPM—using zero knowledge proof. The DAA scheme provides the same level of privacy along with the capability of detecting the rogue TPMs without the availability requirement of the privacy CA. Furthermore, the DAA scheme also effectively prevents the problem in which a malicious user can identify a TPM based on previous communications.

4.3.4 Measurement and Reporting Operations

Before we discuss the integrity measurement process and how it is reported to the local and remote entities, we should first discuss the platform configuration register (PCR) that plays a crucial role in the trusted measurement and reporting framework.

4.3.4.1 Platform Configuration Register

A PCR is a 160-bit (20 bytes) data register that stores the result of an integrity measurement, which is a generated hash of a given component (e.g. BIOS, operating system or an application). Therefore, a group of PCRs form the integrity matrix. The process of extending PCR values is as: $PCR_i = Hash(PCR'_i || X)$. Where i is the PCR index, PCR'_i represents the old value stored at index i, and X is the byte string that updates the PCR value. The "||" indicates the concatenation of two data elements in the given order. The starting value of all PCRs is set to zero.

The TPM specification requires a minimum of 16 PCRs, numbering from PCR-0 to PCR-15. The first eight PCRs are reserved for TPM use, and this set is termed as hardware PCRs as they store integrity measurement relating to the BIOS, hardware configuration and ROM code (see Table 4.2). The remaining registers can be utilised by the operating system (OS). A TPM manufacturer can opt for more PCRs as desired and the decision is left to the respective manufacturer. The TPM specification requires the use of a 32-bit index that indicates the maximum number of PCRs a TPM can have; however, it also specifies that a TPM should not have PCRs with indices 230 or higher. The indices 230+ are reserved by the TPM specification for future use. This means that a TPM manufacturer can only implement 214 optional PCRs with indices 16-229.

Table 4.2 Platform configuration registers

PCR	Description
PCR-0	Stores the integrity measure of the BIOS
PCR-1	Integrity measurement of the motherboard configuration
PCR-2	ROM code (third party BIOS code)
PCR-3	ROM configuration (BIOS configuration)
PCR-4	Initial program loader (IPL) code
PCR-5	IPL configuration
PCR-6	Platform state data (sleep or hibernate)
PCR-7	TPM reserves this PCR for the manufacturer use
PCR-8 to PCR15	Not assigned by the TPM specification; however, in user of the operating system and installed applications

The PCR values should be in the shielded location (inside TPM) and they are stored in a volatile memory. On every system boot, a TPM will calculate new values for individual PCRs and these values are then used to provide secure boot and platform state attestation. A TPM does not make any decision regarding the validity of the software whose hash is generated. Therefore, how the hash will be interpreted is left to the discretion of the requesting entity.

The PCR values can be used to seal data, which will only be decrypted if the PCR values matches on every boot. If a malicious user tries to boot a different OS or is able to compromise the OS (i.e. installed a rootkit or backdoor to an OS), the PCR values will be different and the data will not decrypt properly. However, the TPM will not detect any changes to the OS once it is booted up or detect any malicious activity due to exploitation of an existing bug in the OS or software. The TPM can only detect any changes to the existing state of the OS or software before they are initiated (i.e. begin execution).

The attestation mechanism discussed in Sect. 4.3.4.4, keeps track of all PCRs associated with the software states. When an entity requests the attestation, it will sign the PCR value with the sealed private key. If there is a change in any of the PCR values during the system boot, the signature will not verify. This will indicate to the requester that the system is not in the same state as before; whether it is still secure is a decision that the requesting entity has to make.

4.3.4.2 Boot Process

When a user boots up her computer, the first component to power up is the system BIOS. On a trusted platform, the boot sequence is initiated by the Core BIOS (i.e. CRTM) that first measures its own integrity. This measurement is stored in PCR-0, and later it is extended to include the integrity measurement of the rest of the BIOS. The Core BIOS then measures the motherboard configuration setting, and this value is stored in PCR-1. After these measurements, the Core BIOS will load the rest of the code of the BIOS. The BIOS will subsequently measure the integrity of the ROM firmware and ROM firmware configuration, storing them in PCR-2 and PCR-3, respectively. At this stage, the TBB is established and CRTM will proceed with integrity measurement and loading of the OS.

The CRTM measures the integrity of the "OS Loader Code" (also termed as Initial Program Loader: IPL) and stores the measurement in the PCR. The designated PCR index is left to the discretion of the OS. Subsequently, it will execute the "OS Loader Code" and on its successful execution, the TPM will measure the integrity of the "OS Code". After measurement is taken and stored, the "OS Code" executes. Finally, the respective software that initiates its execution will first be subjected to an integrity measurement, and values will be stored in a PCR then sanctioned to execute. This process is shown in Fig. 4.3, that illustrates the execution flow and integrity measurement storage.

By creating the daisy chain of integrity measurements, a TPM provides a trusted and reliable view of the current state of the system. Any software, whether part of an OS or an application has an integrity measurement stored in a PCR at a particular index. If the value satisfies the requirement of the software or requesting entity, then it can ascertain the trustworthiness of the system. As discussed before, a TPM does not make any decision, it only measures, stores, and reports integrity measurement

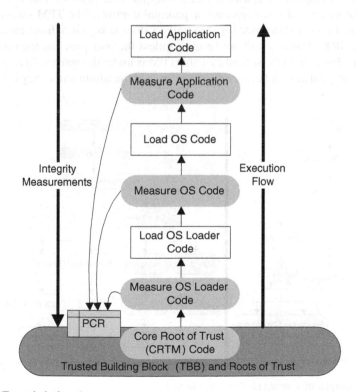

Fig. 4.3 Trusted platform boot sequence

in a secure and reliable manner. When a TPM reports an integrity measurement, it is recommended that it should generate the signature on the value - to avoid replay and man-in-the-middle attacks [1].

The secure boot process relies on the BIOS of the system to provide the CRTM. However, we note that this is not the only method to perform a secure boot using the TPM. The intel trusted execution technology (Intel TXT) [9] allows CRTM to be moved from the BIOS to the CPU.

4.3.4.3 Secure Storage

The internal non-volatile memory of a TPM is not large enough to store all data related to a system and its users. Therefore, the TPM specification allows storage of data on general-purpose non-volatile memory in a secure manner. This involves encrypting data before communicating out of a TPM (chip) including the cryptographic key material, except for the RTS. The RTS is actually the SRK that is discussed in Sect. 4.3.2.

In Fig. 4.4, two of the possible key hierarchies on a trusted platform are shown and each child node is sealed (encrypted) under the parent key. Figure 4.4, illustrates scenarios of a single-user platform, and a multiple user platform that is under an administrative control (i.e. corporate or parental control). The TPM owner in the first case is a user, and the user key (basically a storage key) is a direct child of the respective SRK. This key will not be used unless the user provides the associated AuthData. However, in the second case the TPM is under the ownership of a system administrator, and all user keys are direct children of the administrator key in Fig. 4.4.

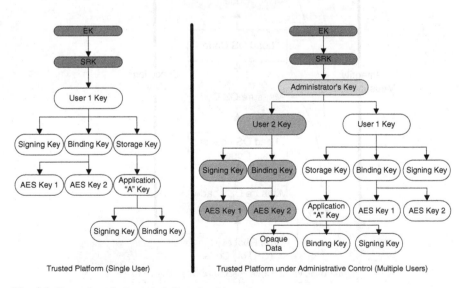

Fig. 4.4 Examples of a trusted platform key hierarchy

Next to the user's key, (depending upon the privileges associated with each user) there can be three types of keys: a signing key (i.e. AIK), binding key and storage key. The signing key is used during the attestation process to provide the requester a signed integrity measurement. The binding key is used to store symmetric keys, and it is not used for signing any data. The process of binding and sealing is different in the TPM specification. In binding, data is encrypted with a key, and before it is used the data is required to be decrypted. Where in the sealing, not only the data is encrypted, but it is associated with a PCR state, and it will only be decrypted if the PCR value is the same as at the time of encryption.

Finally, the storage key gives the privilege to a user to load keys (or generate them in the TPM). As in Fig. 4.4, "User 1" loads the key material for an application "A". The opaque data can be any data related to the application that it is requested to be encrypted. A point to note is that only the first two keys in the hierarchy (EK and SRK) are stored on the TPM. All other keys and opaque data are stored on general-purpose storage encrypted by their parent keys in the hierarchy (Fig. 4.4); thus, providing the secure storage.

In the second scenario, an administrator takes the ownership of the TPM and then creates two users. Each user has her own storage key, and they are restricted by the policies set by the administrator. For example, in Fig. 4.4 "User 2" cannot load keys for other entities or applications as the administrator did not sanction her to have a storage key. Where "User 1" is allowed by the administrator to load keys for the application "A".

4.3.4.4 Attestation

This is the process of providing proof that the respective TPM is trustworthy to report the integrity matrix to the requesting verifier. The process, involves the generation of a signature using the respective AIK on the (associated/requested) PCR values. The signature proves to a verifier the validity of the integrity measurement stored in PCRs. The choice of the AIK and the PCR index is dependent on the verifier, platform (OS), and application.

4.3.5 Migration Model

The TPM specification provides a flexible architecture to support trusted platforms. One feature is the possibility that keys and related data can be migrated either for backup or transfer purposes. From the migration point of view, there is no particular difference between the backup and transfer of TPM keys. The backup process can be considered as key archiving, in which TPM keys are stored on a protected and safe server. The key material can be restored on to the respective TPM in case a system restore is required after the system crashes (failure). The transfer mechanism comes to play when data related to a user is migrated from one system to another; such a

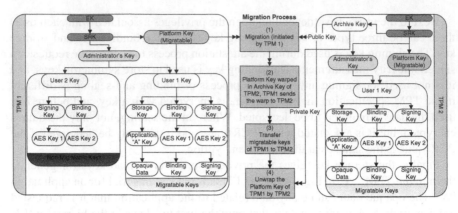

Fig. 4.5 Migration process between two TPMs

scenario may be possible in a corporate environment where employees move around and at each location they might have different (computer) systems.

The pre-requisite of the migration architecture is that all keying material that is required to be migrated should be flagged as migratable. As shown in Fig. 4.5, the TPM 1 has two users and keys belonging to User 1 are not migratable. However, key material related to User 2 is flagged as migratable by the TPM owner (in this case the system administrator). At the destination, whether it is an archiving server or a TPM–there should be an Archive Key. In Fig. 4.5, we have chosen TPM 2 as the destination that has an Archive Key as a child to the SRK. The migration process initiated by the user (or system administrator) is shown in Fig. 4.5 with control flow depicted as double arrowheads and described below:

1. The TPM 1 initiates the migration process by requesting the public part ($PK_{ArchiveKey}$) of the TPM 2's archive key. The TPM 2 provides the public part of the archive key. The archive key is a RSA key of length 2048-bits.
2. The TPM 1 will warp (encrypt) the migratable Platform Key using the $PK_{ArchiveKey}$ of TPM 2. The encrypted key is then transferred to the TPM 2.
3. All key material that is encrypted by the Platform Key of the TPM 1 is transferred to the general-purpose memory accessible to the TPM 2.
4. The TPM 2 will decrypt the wrapped Platform Key and make it a child of the SRK.

During the backup process, the archiving server can take the role of the TPM 2 and step four will be deferred until the backup is required to be restored onto the same or new TPM.

4.4 The Mobile Trusted Module

The growth of the mobile computing platforms has encouraged TCG to propose the MTM. In this section, we briefly discuss the MTM architecture and operations along with how it differs from the TPM.

4.4.1 Basic Architecture and Operations

The ecosystems of mobile computing platforms (e.g mobile phone, tablet, and PDA) are fundamentally different to traditional platforms. Therefore, the architecture of the MTM has only some features from the TPM specification, and introduces new features to support its target environment. The main changes introduced in the MTM that make it different than the TPM specification are stated below:

1. The MTM is required not only to perform integrity measurement during the mobile-device bootup sequence, but also enforce the security policy that aborts the systems from initiating securely if it does not meet the trusted (approved) state transition.
2. The MTM does not have to be in hardware, it is considered as a functionality, which can be implemented by the mobile manufacturers as an add-on to their existing architectures. It is not mandatory to have MTM in hardware, but it is profoundly recommended.
3. Some commands that are mandatory for the TPM are optional for the MTM (e.g. DAA).
4. The MTM specification supports parallel instances of MTM, associated with different stakeholders.

The MTM specification [10] proposes an abstraction architecture enabling multiple instances of the MTM supporting different stakeholders as shown in Fig. 4.6. Each instance of the abstract MTM is referred to as an engine, where each of the engines are under the control of a stakeholder including mobile manufacturer (Device Engine), Mobile Network Operator (Cellular Engine), Application Provider (Application Engine), and User (User Engine); as illustrated in Fig. 4.6. A point to highlight is that each engine is an abstraction of a trusted services associated with a single stakeholder. Therefore, on a mobile platform there can be a single hardware that supports the MTM functionality, which is being accessed by different engines.

Each abstract engine on a mobile platform supports: (1) provision to implement trusted and non-trusted services (normal services) associated with a stakeholder, (2) self-test to ascertain the trustworthiness of its own state, (3) storage of EK (which is optional in MTM) and/or AIKs and (4) key migration.

We can further dissect each abstract engine as components of different services as shown in Fig. 4.7. The non-trusted services in an engine cannot access the trusted resources directly. They have to use the APIs implemented by the trusted services.

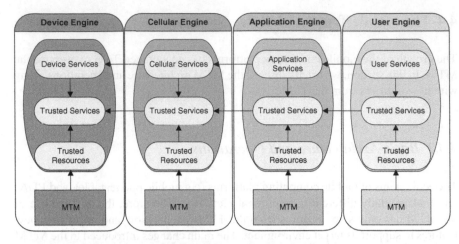

Fig. 4.6 Possible (generic) architecture of mobile trusted platform

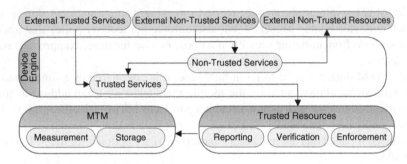

Fig. 4.7 Generic architecture of an abstract engine

The trusted resources, including reporting, verification, and enforcement are new concepts that are introduced in the MTM specifications. The MTM measurement and storage services shown in Fig. 4.7 are similar to the TPMs discussed in previous sections.

The MTM specification defines two variants of the MTM profile depending upon who is the owner of a particular instant of an MTM. They are referred as mobile remote ownership trusted module (MRTM) and mobile local ownership trusted module (MLTM). The MRTM supports the remote ownership, which is either held by the mobile manufacturer or mobile network operator, where MLTM supports the user ownership.

The roots of trust in the MTM include the ones we discussed in TPM including RTS, RTM and RTR; however, the MTM introduces two new roots of trust that are root of trust for verification (RTV) and root of trust for enforcement (RTE). During the MTM operations on a trusted mobile platform, we can logically group different roots of trust; like RTM and RTV together to perform an efficient measure-verify-extend operation, illustrated in Fig. 4.8. Similarly, RTS and RTR can be

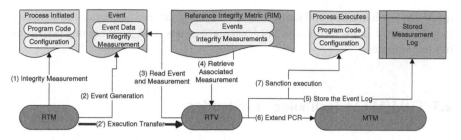

Fig. 4.8 MTM measurement and verification process

grouped together to provide secure storage and trustworthiness of the mobile platform.

The MTM operations as shown in Fig. 4.8 begin when a process starts execution, and they are listed as below:

1. The RTM will perform the integrity measurement of the initiated process.
2. In the second step, the RTM will register an event that includes the event data (application/process identifier) and associated integrity measurement. The RTM then transfers the execution to the RTV.
3. The RTV will read the event registered by the RTM.
4. The RTV will then search the event details from the reference integrity metric (RIM). The RIM includes the (trusted) reference integrity values associated with individual events; populated by the engine owner. The RTM will measure the integrity value of the event and RTV will perform the comparison with associated (trusted) reference integrity value stored in the RIM. If the measured integrity value from RTM does not match, the MTM will terminates the execution or disable the process. If the verification is successful then it will proceed with step 5 and 6 along with sanctioning the execution (step 7). This operation makes the MTM different than the TPM, as the later does not make any decision regarding the trustworthiness of the application or process.
5. The RTV will register the event in the measurement logs. These logs give the order in which the measurements were made to generate the final (present) value of the associated PCR.
6. The RTV will extend the associated PCR value that is stored in the MTM.
7. If verification is successful, it will sanction the execution of the process.

The RIMs are issued as part the RIM certificates (RIM_Certs) by an authentic and authorised entity referred as RIM_Auth. A RIM_Auth can be an external entity or the engine owner itself. Each RIM-Cert is verified by the TPM verification keys, which is a hierarchical structure of keys, set up by the engine owner. The engine owner referred to as the root verification authority (RVA), has a public key that is pre-configured to the engine. The public key of the RVA, also referred as RVA identifier (RVAI), should be integrity protected as any successful modification to the RVAI can enable a malicious user to load RIM_Certs. The RVA acts as a root CA in the MTM key hierarchy, and it might sign the RIM_Cert itself, delegate the authority to a third party, or even authorise third parties to issues RIM_Cert signing privileges to their affiliates.

4.5 TPM/MTM Technology Contenders

In the mobile platforms market, new technologies are introduced that claim to offer secure platform services making them competitors to the MTM (and to some extend also to the TPM). In this section, we discuss some of these new proposals then compare them with the MTM/TPM.

4.5.1 ARM TrustZone

The ARM TrustZone provides an architecture for a trusted execution platform, which has its application in mobile platforms. The underlying concept is providing two virtual processor with hardware level segregation and access control [11, 12]. This enables the ARM TrustZone to define two execution environments termed as: Secure world and Normal world. The Secure world executes the security and privacy sensitive components of applications, where as normal execution can be taken place in the Normal world. The ARM processor manages the switch between the two worlds that also include the switch in the contents of the memory and peripherals (i.e. they are notified of the switch between the worlds). The ARM TrustZone is implemented as security extensions to the ARM processors (e.g. ARM1176JZ(F)-S, Cortes-A8 and Cortex-A9 MPCore, etc.) [12].

4.5.2 M-Shield

The Texas Instrument has designed the M-Shield, a secure execution environment for the mobile phone market [13]. The M-Shield is a stand-alone secure chip (unlike ARM TrustZone), and it provides a secure execution and limited non-volatile memory.

4.5.3 GlobalPlatform Device

The GlobalPlatform device (GPD) is the GlobalPlatform's initiative for trusted execution environment (TEE) [14, 15] for mobile phones, set-top boxes, utility meters and payphones, etc. The GlobalPlatform defines a specification for interoperable secure hardware, which is based on the experience they had with the smart card industry. It does not define any particular hardware, which can be based on either a typical secure element or any of the previously discussed tamper-resistant devices. The underlying ownership of the device still predominately resides with the GPD issuing authority, which is similar to the GlobalPlatform's specification for smart card industry [16].

4.5.4 Trusted Personal Devices

The term trusted personal devices (TPD) was coined by the Integrated secure platform for interactive Trusted Personal Devices (InspireD) project. The aim of the project was to develop a next generation of smart cards to meet the challenges of privacy, trust and security form emerging technologies like mobile devices and pervasive environments [17].

The TPD gives the management privileges to the user but not the "ownership" of the platform. The platform is still control by a centralised authority (i.e. card issuer) and the user gets the privilege of using the device or not. They cannot request installation or deletion of an application. The architecture of the TPD is similar to the smart card, with an exception of different form factors that include SIM card, secure digital (SD) card and universal serial bus (USB) memory stick, etc [17].

4.5.5 Secure Element

Secure elements (i.e. smart cards) are tamper-resistant and reliable devices that are used as security tokens. Traditionally and unlike the TPM, secure elements are under the control of a centralised authority and users do not have any authority except for the decision to use it. However, there are proposals that enable a user to gain ownership of her secure element and request installation/deletion of any application she is entitled to [18, 19]. During the discussion in next section, we assume a secure element that is under the user's control similar to the MTM.

4.5.6 Comparative Analysis of TPM/MTM Technology Contenders

In this section, we list some security and reliability requirements that a typical mobile computing platform might require. Later in Table 4.3, we compare the MTM and other proposals discussed in previous sections, against these requirements.

1. Execution protection: Defined commands related to security and privacy sensitive processing are executed in a secure and reliable environment.

Table 4.3 Comparison of different candidate tamper-resistant devices for a mobile platform

Criteria		MTM	ARM TZ	M–Shield	GPD	TPD	SE
1.	Execution protection	Yes	Yes	Yes	Yes	Yes	Yes
2.	Storage protection (volatile)	-Yes	-Yes	Yes	Yes	Yes	Yes
3.	Storage protection (non-volatile)	-Yes	-Yes	Yes	Yes	Yes	Yes
4.	Tamper-resistant	Yes	No	Yes	Yes	Yes	Yes
5.	Tamper-evident	Yes	No	Yes	Yes	Yes	Yes
6.	Scalability	Yes	Yes	Yes	No	No	Yes
7.	Interoperable architecture	No	NA	NA	Yes	Yes	Yes
8.	Dynamic relation	Yes	NA	No	No	No	Yes
9.	User ownership	Yes	NA	NA	No	No	Yes
10.	Administrative architecture	Yes	Yes*	Yes*	No	No	Yes
11.	Open design	-Yes	No	No	-Yes	-Yes	Yes
12.	Extendible design	No	No	No	Yes*	Yes *	Yes
13.	Secure execution platform	No	-Yes	Yes	Yes	Yes	Yes
14.	Independent security evaluation	Yes*	No	No	No	-Yes	Yes

2. Storage protection (volatile): The device has a volatile memory on the chip intended for the secure storage of temporary data and code related to the executing application.

3. Storage protection (non-volatile): The device provides non-volatile storage on the chip, that is intended to be secure.

4. Tamper-resistant: The device provides tamper-resistant protection that is based on hardware techniques (for example, physical, side-channel and fault attacks).

5. Tamper-evident: The device has the capability to detect potential tampering with the hardware and respond in a pre-defined manner.

6. Scalability: The architecture of the device is scalable so that it can provide services to any application or application provider.

7. Interoperable architecture: The architecture deals with the idea that the candidate device can be interoperable with different computing devices (i.e. mobile phones, tablets and personal computers, etc.).

8. Dynamic relation: A third party can establish a direct relationship based on the security and reliability of the device. The dynamic relation requires that an application provider can trust a device without requiring it to be part of a syndicated scheme (i.e. one adopted by Apple App Store, etc.) and vice versa.

9. User ownership: The device is in the control of its user and she can install, delete and execute any application she desires.

10. Administrative architecture: The device also provides for administrative controls as might be required in a corporate network or in the case of parental control. This option is to accommodate different deployment scenarios.

11. Open design: The design should not be proprietary; it should be in the public domain.

12. Extendible design: The design should allow third parties to design and deploy their proprietary credentials algorithms or trust architectures on-board the secure hardware.
13. Secure execution platform: The device allows the execution of an (arbitrary) application code (from third parties) in a secure and reliable manner as long as it complies with the device's security and operational requirements.
14. Independent security evaluation: As part of the design, the device is subjected to a third party (e.g. Common Criteria [20]) security and reliability analysis.

In the Table 4.3, "Yes" indicates that that the device fully supports the requirement, "-Yes" means that the device generally supports the requirement but there are instances where it does not (e.g. (U)SIM are not required to be independently evaluated whereas in the case of EMV cards it is mandatory), "Yes*" means that the device could support the requirement with adequate design. The notation "No" means not supported, and "NA" means that the given criterion is not applicable.

4.5.7 What Lies Ahead?

The TCG has been on the forefront of trusted computing frameworks, and it has proposed specifications that cater to specialised needs of individual computing devices (e.g. TPM and MTM). However, as the services accessible to different computing devices are converging, the trusted computing framework must evolve.

The trusted computing frameworks that are based on specialised hardware and a software architecture and provide a trusted execution and data storage have already been proposed by several parties. Examples of such proposals are ARM TrustZone, M-Shield and GPD and as can be seen from Table 4.3 they satisfy the requirements to a greater or lesser extent. These technologies are challenging the traditional view of trusted hardware that has been championed by the TCG. As shown in Table 4.3, smart card (or secure element) technology has the opportunity to become a unified trusted device to provide security, privacy and trusted services. In response, the proposal User Centric Tamper Resistant Devices [19] provide the ground work for giving the control of the device to its respective user. Whether these proposals will become viable or not is still to be ascertained; nevertheless, they are providing an alternative to the TCG specification.

4.6 Conclusion

In this chapter, we briefly discussed the TCG framework, its architecture and operations. We discussed the key hierarchy, integrity measurement and secure storage architecture proposed by the TCG. Later, we dived into the details of the MTM that is targeting mobile platforms. We discussed the differences between the TPM and

MTM to give an appreciation that different platforms have different ecosystems that play a crucial role in the design of trusted platforms. Finally, we compared the MTM with other proposals for trusted mobile platforms. These proposals are challenging the TPM architecture, and each of them takes a different view on how to provide trusted mobile platforms.

Acknowledgments The authors want to thank the reviewers for their constructive comments which were helpful to improve this chapter.

References

1. TPM Main: Part 1 Design Principles, Online, Trusted Computing Group (TCG) Specification 1.2, Rev. 116, March 2011.
2. ISO/IEC 11889–1: Information Technology - Trusted Platform Module - Part 1: Overview, Online, International Organization for Standardization (ISO) Standard 11 889–1, May 2009.
3. Standard Specifications for Public Key Cryptography, Online, Institute for Electrical and Electronics Engineers (IEEE) Standard 1363–2000, January 2000.
4. H. Krawczyk, M. Bellare, and R. Canetti, HMAC: Keyed-Hashing for Message Authentication, Online, Network Working Group Requst for Comments 2104, February 1997.
5. *FIPS 180–2: Secure Hash Standard (SHS)*, National Institute of Standards and Technology (NIST) Std., 2002.
6. TPM Main: Part 3 Commands, Online, Trusted Computing Group (TCG) Specification 1.2, Rev. 116, March 2011.
7. *ISO/IEC 11889–2: Information technology - Trusted Platform Module - Part 2: Design principles*, International Organization for Standardization (ISO) Std., May 2009.
8. E. Brickell, J. Camenisch, L. Chen, "Direct anonymous attestation", in *Proceedings of the 11th ACM conference on Computer and communications security*, ser. CCS '04. New York, NY, USA: ACM, 2004, pp. 132–145. [Online]. Available: http://doi.acm.org/10.1145/1030083.1030103
9. "Intel Trusted Execution Technology (Intel TXT)", Intel Corporation, Software Development Guide 315168–008, March 2011. [Online]. Available: http://download.intel.com/technology/security/downloads/315168.pdf
10. TCG Mobile Trusted Module Specification, Online, Trusted Computing Group (TCG) Specification 1.0, Rev. 6, June 2008.
11. P. Wilson, A. Frey, T. Mihm, D. Kershaw, and T. Alves, "Implementing Embedded Security on Dual-Virtual-CPU Systems", *IEEE Design and Test of Computers*, vol. 24, pp. 582–591, 2007.
12. , "ARM Security Technology: Building a Secure System using TrustZone Technology", ARM, White Paper PRD29-GENC-009492C, 2009.
13. —, "M-Shield Mobile Security Technology: Making Wireless Secure", Texas Instruments, Whilte Paper, February 2008.
14. GlobalPlatform Device Technology: Device Application Security Management - Concepts and Description Document Specification, Online, GlobalPlatform Specification, April 2008.
15. , "GlobalPlatform Device: GPD/STIP Specification Overview", GlobalPlatform, Specification Version 2.3, August 2007.
16. *GlobalPlatform: GlobalPlatform Card Specification, Version 2.2,*, GlobalPlatform Std., March 2006.
17. F. C. Bormann, L. Manteau, A. Linke, J. C. Pailles, and J. D. van, "Concept for Trusted Personal Devices in a Mobile and Networked Environment", in *15th IST Mobile & Wireless Communications Summit*, June 2006.

18. R. N. Akram, K. Markantonakis, and K. Mayes, "A Paradigm Shift in Smart Card Ownership Model", in *Proceedings of the 2010 International Conference on Computational Science and Its Applications (ICCSA 2010)*, B. O. Apduhan, O. Gervasi, A. Iglesias, D. Taniar, and M. Gavrilova, Eds. Fukuoka, Japan: IEEE Computer Society, March 2010, pp. 191–200.
19. —, "User Centric Security Model for Tamper-Resistant Devices", in *8th IEEE International Conference on e-Business Engineering (ICEBE 2011)*, J. Li and J.-Y. Chung, Eds. Beijing, China: IEEE Computer Science, October 2011.
20. *Common Criteria for Information Technology Security Evaluation, Part 1: Introduction and General Model, Part 2: Security Functional Requirements, Part 3: Security Assurance Requirements*, Common Criteria Std. Version 3.1, August 2006.

Chapter 5
Hardware and VLSI Designs

Mario Kirschbaum and Thomas Plos

Abstract Efficient and secure hardware implementations have become a very popular topic during the last decades. In this chapter, we discuss the fundamental design approaches to successfully implement integrated circuits (ICs) as well as testing methods and optimization techniques to achieve an adequate solution for various application scenarios. A major topic handled in this chapter is security in the context of hardware implementations. We elaborate on the characteristics of modern CMOS circuits with regard to side-channel attacks and we discuss possible countermeasure approaches against such attacks. Furthermore, we describe a comprehensive practical example of combining cryptographic instruction set extensions with hardware countermeasures on a modern 32-bit processor platform. In the last section of this chapter, we argue about the assets and drawbacks of implementing test structures in digital circuits with regard to unintentionally opening security holes as well as about intentionally introducing malicious hardware structures, also called hardware Trojans.

Keywords VLSI design cycle · Design space · Advanced encryption standard (AES) · Secure hardware design · Side-channel analysis · Power-analysis attacks · Masking · Hiding · Dual-rail precharge · Instruction-set extensions · Design for test · Hardware Trojans

M. Kirschbaum (✉) · T. Plos
Institute for Applied Information Processing and Communications,
Graz University of Technology, Inffeldgasse 16a, 8010 Graz, Austria
e-mail: Mario.Kirschbaum@iaik.tugraz.at

T. Plos
e-mail: Thomas.Plos@iaik.tugraz.at

K. Markantonakis and K. Mayes (eds.), *Secure Smart Embedded Devices,*
Platforms and Applications, DOI: 10.1007/978-1-4614-7915-4_5,
© Springer Science+Business Media New York 2014

5.1 Introduction and Motivation

During the last decades, integrated hardware circuits have become more popular and are an integral part of our daily life. Hardware circuits are not only found in personal computers and laptops, but also in cars, domestic appliances, and any kind of communication and multimedia device. Continuous migration to smaller process technologies has allowed us to dramatically increase the transistor count and consequently also the functionality of hardware circuits. In order to handle the increasing complexity of hardware circuits, a whole new branch of industry has been built up: Very large scale integration (VLSI) design.

Mainly two different approaches can be followed when designing complex hardware circuits: a general-purpose approach or a special-purpose approach. The general-purpose approaches based on hardware circuits like microprocessors that provide a fixed set of functionality. Customization of the microprocessor for a concerning application is done through program development which provides high flexibility. The special-purpose approach on the other hand, involves design of a dedicated hardware circuit for a more specific application. This hardware-based concept is less flexible and causes longer development times, but allows optimizing a design toward a certain goal, for example: low area, low power consumption, low energy consumption, or high throughput. Reducing the hardware overhead is desirable for cost-sensitive high-volume products that aim for minimum chip area. Achieving low power consumption or low energy consumption is important for passively-powered devices (e.g., RFID tags) and battery-operated devices, respectively. Especially when integrating complex and resource-intensive operations into a device, like for example cryptographic operations in a security-related application, fulfilling the design requirements is often only achievable if dedicated hardware circuits are used.

In the following sections, we discuss some principles of hardware and VLSI design. We start with describing the fundamental VLSI design cycle which is the basis of every integrated circuit (IC). Hardware designers may choose between different design perspectives at various abstraction levels encountered during the design cycle. Starting with the system specification, a designer defines the submodules, uses hardware description languages (HDLs), applies tools for standard-cell mapping, and finally comes to the geometric layout of the design.

During the design cycle, repeated testing and simulation of the design at different abstraction levels is inevitable in order to obtain a flawless and well functioning circuit. In case errors occur, only a small step has to be made back in the design cycle instead of returning to square one. Another important topic we will discuss is the extensive design space which is at the designer's disposal during the whole design cycle. A designer has almost indefinite possibilities to reach a design goal, which is, e.g., high performance, low area, or low energy consumption. By means of several practical examples, we illustrate the impact of decisions made during the design cycle on the outcome of a hardware design.

In the next part, we concentrate on *security* in hardware and VLSI design, which is an important topic. The use of security-related devices is steadily increasing, e.g., web servers protected with SSL, encrypted hard-disc drives, wireless car keys, or radio-frequency identification (RFID) tags, to name but a few well-known examples. Mathematically secure cryptographic algorithms become highly vulnerable to various types of attacks when implemented in hardware due to the strong relation between the data processed within a device and the power consumption of conventional complementary metal-oxide semiconductor (CMOS) circuits. In order to address this issue, researchers began to develop countermeasures to protect sensitive hardware devices. We have a detailed look at the fundamental characteristics of CMOS circuits that enable side-channel attacks in the first place and we elaborate on possible approaches for implementing countermeasures against such attacks. Similar to the importance of repeatedly simulating a design during the design cycle to obtain a reliably-working circuit, we point out the possibility of verifying the effectiveness of countermeasures by means of simulations.

Combining the topics of efficient hardware implementations, cryptographic algorithms, and security, we contrast different approaches for implementing cryptographic algorithms on modern embedded devices. As we will see, pure software as well as pure hardware solutions have both drawbacks. A more efficient solution in case of embedded systems is the usage of instruction-set extensions (ISEs). Furthermore, we discuss the implementation of countermeasures against side-channel attacks in the presence of ISEs and we elaborate on a practical example of a modern 32-bit processor platform.

The testability of a device during the design cycle as well as after manufacturing of an IC is very important to prevent distribution of malfunctioning parts. Unfortunately, the integration of test structures may counteract the efforts to protect a device from various attacks, since test structures could be used by an adversary to break security-enabled devices. We describe different testing approaches and discuss their relevance and possible impact on secure implementations. Finally, we discuss the topic of hardware Trojans. Contrary to test structures implemented in ICs that may unintentionally cause a vulnerability of the device, hardware Trojans are malicious structures intentionally implemented by an adversary with the goal, for example, to possibly bypass one or more security features of an IC.

5.2 VLSI Design Cycle

Today's VLSI designers are faced with two challenges, increasing circuit complexity and shorter design cycles. Following Moore's Law, the transistor count of hardware circuits doubles about every 18 months. This prediction has been formulated more than 40 years ago and is still adhered to by the semiconductor industry by migrating towards smaller and smaller process technologies. Most recent microprocessors, for example, have already reached a transistor count of one billion and more. Circuit complexity grows faster than the productivity of designers and the increase in efficiency of electronic design automation (EDA) tools. This has opened a so-called "design gap" over the years. In order to close this gap, design of VLSI circuits has

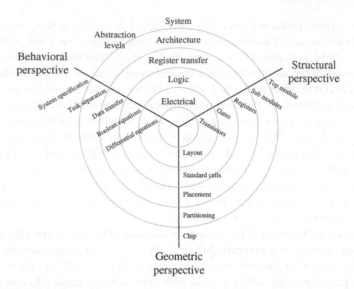

Fig. 5.1 Y-diagram according to Gajski and Kuhn showing the different design perspectives and abstraction levels of hardware circuits

been brought to a higher abstraction level. Designing a circuit at a higher abstraction level also addresses the requirement of shorter design times to improve cost effectiveness.

A good overview of the different abstraction levels and design perspectives of hardware circuits is provided by the Y-diagram illustrated in Fig. 5.1. The Y-diagram has been introduced by Gajski and Kuhn [6] and has its name from the three axes that are arranged in a y-shape. Each axis relates to a different design perspective. The three design perspectives are behavioural perspective, structural perspective, and geometric perspective. Behavioural perspective focuses on the functionality of a circuit, whereas structural perspective describes the interconnection of different blocks within it. Geometric perspective deals with the arrangement of the components, including the final layout of a circuit. Concentric circles indicate the various abstraction levels, which are system level, architectural level, register-transfer level, logic level, and electrical level. Starting from highest abstraction level at the outermost circle, the various development steps of a hardware circuit are passed through when moving toward the center of the diagram, marking the final outcome of the design (i.e., layout of the circuit).

When moving toward the center of the diagram to reach the design goal, the level of detail increases continuously. Different perspectives can be used for entering lower abstraction levels and changing between perspectives is possible as well. Behavioral perspective is the most-suitable domain for describing hardware designs with high complexity. Consequently, behavioral perspective is used for starting the design process. The first step is creating a software model that implements the specification of the system and that allows exploring different algorithm variants. This first software model also eases communication among design teams and enables concurrent development of hardware and software components (important to shorten overall development time). Next step is finding an appropriate architecture that is reflected by a cycle-accurate high-level model. When the architecture is fixed, HDLs like VHDL and Verilog are deployed to transfer the high-level model into a register-transfer level representation. The combined use of HDLs and EDA tools for circuit synthesis allows an automated transformation from behavioral perspective to structural perspective. The outcome of this step is a netlist that contains a circuit representation with logic gates, flip flops, and the appropriate wire connections.

The following steps after netlist creation relate to the so-called back-end design where the structural perspective is left and the geometric domain is entered. Automated tools are again applied to deduce a standard-cell representation and the layout of the design. During back-end design, various verification techniques are utilized to ensure proper operation and manufacturability of the circuit. Verification techniques comprise for example, design-rule checks, electrical-rule checks, layout-versus-schematic checks, timing verification, and simulation of power consumption. With the layout of the circuit, the final design step (tape out) is reached and data can be sent to a semiconductor manufacturer.

Following this top-down approach gives a good understanding of the involved steps of state-of-the-art VLSI design. Implementing a circuit within behavioral perspective through HDLs and deploying automated tools for further processing eases not only the handling of circuit complexity, but also brings also more flexibility. A circuit in HDL representation can be easily mapped to different process technologies and targets by using circuit-synthesis tools. This significantly shortens the time required for migrating a design to a new process technology and allows also first-level tests on field-programmable gate array (FPGA) prototypes.

Continuously testing the functionality of a design within all abstraction levels is an important aspect of modern VLSI design. Required test data is typically derived from the high-level model and repeatedly used for tests on lower abstraction levels. When a test fails, designers can immediately step back and fix the problem. This allows detection of issues as early as possible, following the first-time-right concept to launch products on time.

When building hardware circuits that contain security-relevant components, functional tests alone are no longer enough. Additional considerations have to be taken into account like evaluating the resistance of the implementation against side-channel analysis (SCA) and fault analysis. Such evaluation tests are mainly conducted after chip production on first prototype samples, but also during design phase. Power-simulation results of the circuit can be used to deduce first information about side-channel resistance of a design. Other examples are side channel and fault attacks on FPGA prototypes that contain a synthesized version of the design.

5.3 Design Space of Hardware Circuits

Hardware circuits can be designed toward different optimization goals, depending on the targeted application. Typical optimization goals are high throughput, low area, low power consumption, and low energy consumption. Optimization can be conducted on different abstraction levels. However, the higher the abstraction level, the larger is the impact of the optimization techniques and the lower the required effort. Optimizing a design at system level or at architectural level is therefore more promising than optimizing it for example on logical level. Various metrics are used to quantify the effectiveness and the influence of a certain optimization measure. Widely-used metrics includes chip area, throughput, execution time, maximum clock frequency, latency, and average power consumption.

Optimization at system level typically involves finding more suitable protocols or looking for alternative algorithms that lead to the same result but provide advantageous behavior in terms of computation time or resource usage. A good example is the representation of the substitution box (S-box) used in the AES. The S-box is a non-linear operation that is applied on a single byte of data. Hence, the result of the S-box operation can be precomputed for all possible 2^8 input values and stored in a look-up table. This will result in an area requirement of more than $1,000$ GEs when implementing the look-up table with standard cells. However, the S-box operation can also be realized by calculating the multiplicative inverse in the finite field $GF(2^8)$ followed by an affine transformation (see [15] for more details). Using combinatorial logic to calculate the S-box operation in that way, leads to an area requirement of 300 GEs. This is less than a third of the value required by the look-up table approach. Achieving such an area saving through optimization at lower abstraction levels is hardly possible.

Architecture is another abstraction level that has significant potential to optimize a design toward a certain direction. Well-known optimization techniques at architectural level are functional decomposition, pipelining, and parallel computation [8]. Functional decomposition aims at breaking a complex function into smaller subfunctions that can be computed sequentially. This method is most effective when the subfunctions compute similar operations that allow reusing of a single hardware unit that decreases the overall chip area. Execution time remains roughly the same, since the shorter critical path allows a higher maximum clock frequency, which compensates for the increased number of required clock cycles.

Pipelining is another effective optimization method at architectural level. The data path of a function is cut into smaller parts (ideally of equal length) by inserting storage elements called pipeline registers. This shortens the critical path and leads to a higher maximum clock frequency. For computing the result of one data item, as many clock cycles are required as there are pipeline stages. However, once the whole pipeline is filled, the result of a data item is computed with every clock cycle. It is important to note that this works only if there are no recursive data dependencies, since they would prevent the pipeline from getting filled. Pipelining is very efficient because a marginal increase of chip area that is introduced by adding pipeline registers, results in a significant computational speed-up.

Computing operations in parallel is the opposite of functional decomposition. Instead of reusing components to reduce chip area, additional hardware modules are introduced to lower computation time. Trading chip area for speed is some kind of brute-force approach and is used if other measures like pipelining are not applicable (e.g., if low latency is required). In contrast to pipelining, the critical path of a design is not shortened and therefore increasing the clock frequency is not possible. Chip-area requirements increase significantly and relate to the degree of parallelism.

An overview of the impact of all three optimization techniques within the design space is given in Fig. 5.2. Functional decomposition and pipelining are efficient approaches to decrease chip area and execution time of a design, respectively. Both techniques significantly lower the area-time product. Parallel execution of operations increases chip area to lower execution time, by keeping the area-time product roughly constant.

The following examples illustrate the effects on hardware implementations of the advanced encryption standard (AES) when focusing on different design goals and implementing different optimization techniques. AES is a symmetric block cipher and has been standardized by the National Institute of Standards and Technology (NIST) in 2000 [15]. Let us first have a look at the low-power AES implementation of Feldhofer et al. [4]. The design goals of this AES implementation have been low area and low power in order to apply AES in highly resource-limited devices like RFID tags. The AES module supports encryption and decryption including the key schedule and is based on an 8-bit architecture. In order to reduce the area to a minimum the design contains only one S-Box instance (combinational, one pipeline stage) and one multiplier for the MixColumns operation. The usage of only one S-Box instance corresponds to functional decomposition, i.e., one S-Box instance is used several times during one cryptographic computation. The pipeline stage within the S-Box implementation helps to shorten the critical path of the design. One encryption/decryption can be done in roughly 1, 000 cycles. Strictly following low area and low-power guidelines, the developers produced the AES module in a

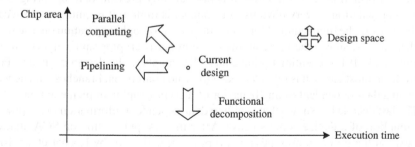

Fig. 5.2 Impact of *functional decomposition*, *pipelining*, and *parallel computation* on *design space*

0.35 μm CMOS technology and were able to achieve a very low area requirement of 3, 400 GEs and an extremely low power consumption of 3.0 μA when operating the AES module at 100 kHz and a supply voltage of 1.5 V. In this configuration, the module achieves a throughput of approximately 12.5 kbps.

Contrary to the low-area implementation of Feldhofer et al., Mangard et al. [12] proposed a high-performance hardware implementation of the AES based on a 32-bit architecture. The AES module follows the parallel computation approach and contains 16 S-Box instances and 16 multipliers implementing the MixColumns operation. The AES implementation has an area requirement of 16 kGEs, needs 34 cycles per encryption/decryption, and reaches a maximum clock frequency of 64 MHz and a throughput of 241 Mbps produced in a 0.6 μm CMOS technology. There also exist some extreme-performance implementations of AES, deploying a 128-bit architecture and highly optimized implementations of the AES operations, resulting in larger implementations (beyond 20 kGEs) and significantly higher throughput ($\geq$ 1 Gbps).

These examples show that a designer has almost indefinite possibilities to exploit the design space in many different directions. Often a designer decides to go in more than one direction at once: e.g., low area and low power, high throughput, and low area. In the end, a designer is mostly forced to accept compromises between throughput, area, and power consumption in order to achieve an adequate hardware solution suitable for the particular application.

5.4 Secure Hardware Design

The use of security-related devices has been steadily increasing during the last few years. Besides meeting appropriate design goals in terms of throughput, chip area, and power consumption, security goals started to play a major role in hardware design. Various attacks on hardware circuits in the past have pushed the emergence of a completely new research field.

Additionally to the intended output, e.g., the result of a cryptographic computation, each physical device also emits various other information, the so-called side-channel information. This information is permanently present, before, during, and after a computation. A very obvious side channel is timing. It is quite easy to accurately measure the time required for executing a cryptographic operation on a device. Kocher has first shown the vulnerability of asymmetric cryptographic algorithms to timing attacks [9]. Preventing timing attacks lies manly at the designer's hands. For example, in most cases, it is relatively easy to avoid conditional branches, and hence, to avoid a data-dependent timing behavior of a cryptographic implementation.

The last decade has shown that avoiding data-dependent information in the power consumption of a device is not so easy. After the first publication on SCA attacks that exploit the power consumption of cryptographic devices by Kocher et al. [10], the security of hardware designs against power analysis (PA) attacks became a major research topic in the fields of cryptography and hardware implementations. As it turned out that almost any cryptographic device implemented in CMOS technology is highly vulnerable to PA attacks, designers of algorithms as well as hardware developers began to think about possible solutions to overcome the inherent relationship between the processed data within a CMOS circuit and its total instantaneous power consumption.

5.4.1 Power Consumption of CMOS Gates

The total power consumption of CMOS gates is the sum of the *static power consumption* and the *dynamic power consumption*. The static power consumption is caused by a small leakage current that is flowing through the metal-oxide semiconductor (MOS) transistors that are turned off. An actual example is given in [27]: the leakage current of a MOS transistor in a 100 nm process is typically in the nA range. In most applications, the static power consumption of CMOS circuits is neglected, except for low-power applications. In such applications, special low-leakage process technologies come into play which are able to significantly reduce the static power consumption. From a security perspective, the static power consumption of CMOS circuits can be neglected, as the leakage current only shows an extremely low data dependency. Dynamic power consumption, on the contrary, is significantly higher than static power consumption, and even more important, it shows a strong dependency to the data processed by the CMOS circuit.

In the following, the fundamental characteristics of CMOS circuits, which enable the execution of power-analysis attacks in the first place, are described by means of a conventional CMOS inverter. The schematic of a CMOS inverter is depicted in Fig. 5.3 (left), it basically consists of a pMOS transistor p and an nMOS transistor n. The output line q of the inverter is naturally afflicted with a vast number of parasitic capacitances. In a simplified model, we can assume two significant-pooled parasitic capacitances (indicated as C_{L1} and C_{L2} in Fig. 5.3). Depending on the state transition of the CMOS inverter one of the capacitances is charged. The events in case input

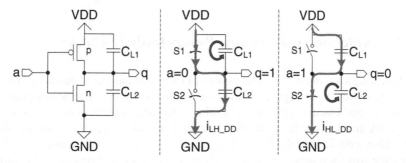

Fig. 5.3 Depiction of the power consumption of a CMOS inverter: the schematic of the inverter (*left plot*), the equivalent circuit in case input $a : 1 \rightarrow 0$ (*middle plot*), the equivalent circuit in case input $a : 0 \rightarrow 1$ (*right plot*)

a switches from 1 to 0 (i.e., $q : 0 \rightarrow 1$) are illustrated in the equivalent circuit in Fig. 5.3 (middle). The pMOS and nMOS transistors are represented by switches $S1$ (closed) and $S2$ (opened), respectively. Assuming that C_{L1} is charged from the previous state and C_{L2} is discharged, the following charging processes occur: C_{L1} is discharged internally via $S1$ and C_{L2} is charged via i_{LH_DD}, i.e., the CMOS inverter consumes power to charge C_{L2}. In case input a switches from 0 to 1 (i.e., $q : 1 \rightarrow 0$), very similar charging processes occur in the circuit (Fig. 5.3, right): $S1$ is opened, $S2$ is closed, C_{L2} is discharged internally via $S2$, and C_{L1} is charged via i_{HL_DD}. The CMOS inverter consumes power to charge C_{L1}.

We can summarize the events happening in the CMOS inverter in the following way: if the input of the inverter does not change (i.e., $a : 0 \rightarrow 0$ or $a : 1 \rightarrow 1$), the nMOS and pMOS transistors keep their state, and none of the output capacitances C_{L1}, C_{L2} needs to be charged. Neglecting the static power consumption, we can state that the power consumption of a CMOS inverter is *zero* in case the input signal remains in its state. On the other hand, if the input of the inverter changes its state (i.e., $a : 0 \rightarrow 1$ or $a : 1 \rightarrow 0$) the nMOS and pMOS transistors change their conductivity and the output capacitances C_{L1}, C_{L2} are charged/discharged accordingly. We see a change of the input value causes a significant amount of power consumption. Furthermore, we can also imagine the following: assuming we know the initial state of the inverter and we record the power consumption of our small CMOS circuit, we are able to determine the actual state of the circuit at any time by simply considering the power spikes we see on our record. That is exactly the reason why power-analysis attacks pose a serious threat to CMOS circuits. An attacker can quite easily figure out what is happening in a circuit by analysing the power consumption.

As the power consumption is closely related to the electromagnetic (EM) emanation of a device, EM analysis attacks can also reveal very small data dependencies within a device. More specifically, EM-analysis attacks and usually even more powerful than power-analysis attacks, because with the appropriate equipment, the measurement of the EM emanation of a device can be limited to a very small portion of the circuitry. This way, the signal-to-noise ratio (SNR) of the measurement can be significantly increased.

5.4.2 Countermeasures Against Power-Analysis Attacks

The following section gives a broad outline of ideas that have been developed during the last few decades to impede PA attacks. We introduce the basic approaches and indicate the main assets and drawbacks. The underlying concept of countermeasures against PA attacks is to break the dependency between the processed intermediate data within a device and the device's instantaneous power consumption. Basically, there exist two approaches: masking countermeasures and hiding countermeasures. Figure 5.4 depicts the points where the two approaches apply. Both approaches can be utilized at the architecture level (implemented in software and/or hardware) as

Fig. 5.4 Depiction of the two basic countermeasure approaches: masking and hiding

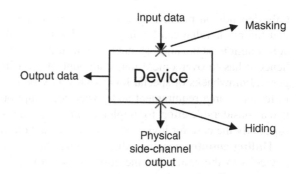

well as at the cell level (purely implemented in hardware). Usually, a combination of both approaches is worthwhile to achieve an adequate level of security.

Masking countermeasures conceal the processed data within the device by a random mask, i.e., the input data is altered before any operations are executed. Hence, the device's power consumption only depends on the masked data. After performing the critical operations within the device, the mask has to be removed again. Some additional effort is associated when implementing a masking countermeasure, in many cases the algorithm's operations need to be adapted in order to process the masked data correctly, e.g., when masking an AES S-Box operation [3].

Special logic styles belong to the strongest methods to prevent PA attacks. The secure logic-style front is highly competitive. Many approaches have been presented during the last decade, although for most one or more flaws have been discovered by the research community. Let us have a closer look at one example for such a special logic style implementing a masking technique: the masked dual-rail precharge logic (MDPL), proposed by Popp et al. [16], Popp and Mangard [17]. The MDPL style is based on the dual-rail precharge (DRP) principle which prevents the occurrence of glitches by representing each signal with complementary electrical wires in the circuit. It is well known that glitches, also called *dynamic hazards*, do occur in CMOS circuits and that they have a significant effect on the power consumption [19]. Considering their effect on the power consumption, it appears obvious that glitches also play a major role in the context of countermeasures against PA attacks. It has been shown several times that glitches may have a negative influence on the effectiveness of countermeasures [13, 21]. Preventing glitches is achieved by strictly applying only monotonic logic functions and by introducing a precharge phase[1] and an evaluation phase[2] in each clock cycle. In a DRP circuit, only one of the two complementary wires of each signal is HIGH, depending on the value of the signal, the other wire remains LOW. Additionally, each cell within an MDPL circuit unexceptionally processes masked signals, with the result that the power consumption

---

[1] In the precharge phase, every signal (both complementary wires) within a digital circuit is charged to the precharge value, which is in most cases logic '0'.

[2] Similar to a standard clock cycle in a conventional CMOS circuit, the combinational blocks start to evaluate according to their input signals.

only depends on masked values. A not yet eliminated flaw in the MDPL style is that the mask value can be discovered due to significant differences in the power consumption, as the mask signal is connected to every MDPL cell in the circuit and hence it has to overcome a major amount of parasitic capacitances [20, 25]. The general drawbacks of special logic styles may be manifold: a significant overhead in terms of area requirement and power consumption, a decrease of performance, and a considerable effort for implementing the logic style in the first place, which is especially the case for logic styles that are based on full-custom cells.

Hiding countermeasures directly alter the side-channel characteristics of the device with the result that the correlation between the processed data within the device and the device's power consumption is weakened or even completely dissolved. This way, an attacker is not able to draw any conclusions from the power consumption about the intermediate data processed in the device. There are basically two approaches to implement hiding countermeasures: randomizing the power consumption of a device (also called *hiding in time*) and equalizing the power consumption of a device (also called *hiding in amplitude*). Unfortunately, both approaches are almost impossible to be *perfectly* implemented in practice.

An example for hiding in time is the random insertion of additional operations during the execution of a cryptographic algorithm. The additional operations do not contribute anything to the actual cryptographic computation and therefore they are called *dummy operations*. The random insertion of dummy operations increases the runtime of the whole computation with the result that the power consumption of the critical operations on actual data is randomized in time. This approach simply adds noise to power measurements and thus complicates a side-channel attack. The random insertion of dummy operations has to be implemented with great care: the dummy operations must not be distinguishable from the actual operations, otherwise the execution times of dummy operations can be detected and filtered. Furthermore, once the countermeasure is activated, the number of dummy operations that are executed has to remain constant for every cryptographic computation. Otherwise the implementation is vulnerable to timing attacks. As the runtime of the computation correlates with the number of dummy operations inserted, another drawback is that the countermeasure has to be adapted to every cryptographic algorithm for which it is implemented.

A very simple example for a countermeasure following the approach of hiding in amplitude is the application of noise generators within a device. The obvious drawbacks of noise generators are the requirement of area and power, without contributing anything to the actual functionality of the circuit. A further example for hiding in amplitude is a heavily parallelized design of an algorithm. However, this approach strongly depends on the implemented algorithm, as data dependencies limit the possible degree of parallelization.

More-advanced techniques are based for example on the previously mentioned DRP logic style. Pure DRP styles, which do not implement a masking technique, try to keep the power consumption constant and thus independent of the processed data. This is achieved by balancing each pair of complementary wires, i.e., the designer tries to adjust the complementary wires in a way that their electrical characteristics (resistance, inductance, and capacitance) perfectly match. Ideally, a circuit containing perfectly balanced wires would consume a constant amount of power, and hence, it would be secure against PA attacks. The irrefutable flaw of this approach is that an exact balancing of wires is almost impossible to achieve in practice. Even if the most powerful EDA tools are used, the smallest variations in the chip-fabrication process would cause differences in the electrical characteristics of the complementary wires once more.

5.4.3 Verification of Countermeasures by Means of Simulations

As mentioned in Sect. 5.2, simulations play a major role during the design phase to verify the functionality of the IC. When implementing countermeasures, it is also highly desirable to be able to verify the efficiency of the implemented protection techniques. Various simulation techniques can be used to perform a detailed investigation of the implemented countermeasures without the need of actually fabricating an IC. The following section describes to what extent simulations at different levels can be used to estimate the impact of countermeasures.

One of the main advantages of simulations is the possibility to detect errors in a design before a chip goes into production. This is also the case for detecting flaws in countermeasures implemented in a secure hardware design, before developing a costly prototype chip. Simulations also offer the possibility to simply narrow down the simulated parts of a digital circuit and hence to easily detect and improve faulty submodules or vulnerable parts of an implemented countermeasure. For investigating countermeasures, two simulation levels are interesting: transistor-level simulations and logic-level simulations.

Transistor-level simulations based on SPICE [18] models of transistors represent a highly accurate yet very time consuming way to verify the correct functionality of digital circuits and countermeasures. SPICE simulations may include detailed parasitic information about each element in a circuit and about each wiring in a placed and routed chip design. This results in power-estimation results that are highly comparable with power measurements on an actual chip. Hence, transistor-level power simulations are very suitable to perform power-analysis attacks and to investigate the effectiveness of countermeasures. The main drawback of transistor-level simulations is the complexity and the associated expenditure of time due to solving countless algebraic equations based on nonlinear transistor models. If a designer does not have a powerful computer cluster at hand, the transistor-level simulation of a medium-sized design consisting of approximately 2 million transistors may easily take several hours for a few-hundred clock cycles. Considering that hundreds of power simulations are potentially required to perform a meaningful PA attack, transistor-level simulations become a rather impractical.

Logic-level simulations (also called gate-level simulations) in their simplest form have the advantage of operating at a significantly higher level (i.e., not including any low-level circuit information) compared to transistor-level simulations. This implies a significant speed up of performed simulations, but also a decrease in simulation accuracy. Furthermore, a conventional logic-level simulation is not able to provide something similar to a power consumption trace, it is merely possible to obtain logic-level transitions of each signal within a digital circuit. As described in Sect. 5.4.1, the state transition of CMOS gates is directly related to the dynamic power consumption. Hence, it is possible to derive a simulated power trace from the logic-level transitions obtained from the simulations. A common technique is called *toggle counting* or *transition counting*. At each point in the simulation time where a signal transition $(0 \rightarrow 1$ or $1 \rightarrow 0)$ occurs, the power-consumption value for this specific point in time is increased by 1. Constant signals $(0 \rightarrow 0$ or $1 \rightarrow 1)$ do not contribute anything to the power consumption. This way a designer is able to obtain power consumption traces in a fraction of the time needed to perform transistor level simulations.

There are some limitations in case of power traces derived from basic logic-level simulations. First, there is no timing information at all included in the simulations: some simulators work with unit delay (i.e., all logic gates have the same constant propagation delay) or zero delay (i.e., all transitions occurring in a specific time are summed up to one point in time, usually the clock event). Second, all signal transitions consume the same amount of power, which is highly unrealistic compared to an actual digital circuit. The accuracy of power-consumption traces derived from logic-level simulations can be substantially increased if back-annotated delay information is included in the simulations. This approach has minor effect on the performance of logic-level simulations but greatly increases the accuracy of signal-delay information. A second measure to increase the accuracy of toggle-count power traces is to randomly weight the signal transitions or to include parasitic information when processing the signal transitions. The latter approach would result in a time-consuming preprocessing step to build an appropriate transition-weight database.

Although we have seen that various simulation techniques may be used to verify the efficiency of implemented countermeasures, unforeseen effects may cover actual side-channel leakages during simulation.

5.5 Instruction-Set Extensions

Efficiently implementing cryptographic algorithms on embedded devices is highly challenging due to the limited resources (energy, clock frequency, and memory). A widely deployed processor for embedded devices is for example the LEON CPU core [5], which is a SPARC V8-compliant processor. The LEON core has a 32-bit architecture and follows the Reduced Instruction Set Computing (RISC) concept. When implementing a cryptographic algorithm on such an embedded system, a designer has mainly two options: selecting a software approach or a choosing a hardware approach. The software approach uses only the existing instructions of the processor and requires no additional hardware. This concept provides maximum flexibility, but is costly in terms of code size and allows achieving only a moderate computation speed. A hardware solution on the other hand requires the integration

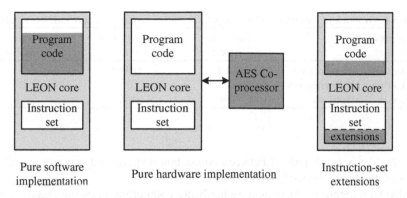

Fig. 5.5 Design alternatives for implementing AES on a LEON core. *Dark-gray* colored areas contribute to the AES implementation

of a dedicated coprocessor that is optimized for a special algorithm. Relying on an optimized hardware module allows very short execution times, which comes at cost of additional chip area and loss of flexibility. Another aspect that has to be considered is the communication overhead between embedded processor and coprocessor. As reported in the work of Hodjat and Verbauwhede [7], much more time is typically spent for the communication between processor and coprocessor, than for the actual computation of the algorithm within the coprocessor. This overhead dramatically lowers the performance gain of a coprocessor approach.

ISEs are techniques that combine the advantages from both pure software implementations and pure hardware implementations. ISEs provide the flexibility of a software solution together with the high computation speed of a dedicated hardware circuit. Moreover, there is no communication overhead between processor and coprocessor. A schematic overview of the three different design approaches is depicted in Fig. 5.5. ISEs provide a processor with additional instructions that are optimized for a certain purpose, like the execution of a cryptographic algorithm. The additional instructions require extra hardware circuits and can be used in a program as any other instruction. Hardware costs of the ISEs are much lower than those of a corresponding coprocessor.

A concept for ISEs on a LEON core has been presented by Tillich and Großschädl [22]. These ISEs aim for improving the computation speed of the AES algorithm on the embedded processor. The proposed ISEs allow computing AES within 196 clock cycles on the LEON core using only 896 bytes of code. The additional hardware costs introduced by the ISEs are estimated by the authors at 3 kGEs. For comparison, a pure software implementation of the AES algorithm on the LEON core takes 1,637 clock cycles and requires 2168 bytes of code. A pure hardware implementation on the other hand, as presented for example by Mangard et al. [12] (cp. Sect. 5.3), requires only 34 clock cycles for computing AES, but leads to additional hardware costs of 16 kGEs. Table 5.1 summarizes the performance numbers of the three design approaches and clarifies that ISEs are a highly efficient approach

Table 5.1 Overview of the different design approaches with corresponding performance numbers

Design approach	Code size [Bytes]	Execution time [Cycles]	Hardware costs [kGEs]
Pure software implementation	2168	1637	–
Pure hardware implementation	–	34	16
Instruction-set extensions	896	196	3

that provides a good tradeoff between computation speed and resource usage (in terms of code size and hardware overhead).

Also with regard to implementing hardware countermeasures (i.e., secure logic styles) against side-channel attacks, cryptographic ISEs have a significant advantage over dedicated coprocessors. In the case of cryptographic coprocessors, critical data is running countless times from one submodule to another and vice versa. For example, an AES coprocessor the input data is first XORed with the secret key. As the key represents the critical data in an AES computation, the following operations process critical data. In the next step, the critical data runs through the S-Box module and a probably directly-integrated ShiftRows module, followed by the MixColumns module back to the XOR operation with the next RoundKey. We see, in order to secure an AES coprocessor we need to implement countermeasures in many submodules, as all submodules directly process critical data. Securing a cryptographic coprocessor by means of hardware countermeasures usually results in implementing the whole coprocessor in a costly secure logic style, which entails a significant increase in terms of area requirements and power consumption.

In case of cryptographic ISEs implemented on a processor platform, operations that are actually transforming critical data are confined to the functional units of the processor. This circumstance enables us to implement only the functional units in a costly secure logic style and to implement a much cheaper countermeasure to the rest of the chip. The next section introduces a concept for developing a secure processor with ISEs and hardware countermeasures.

5.6 A 32-Bit Processor with ISEs and SCA Countermeasures

In the following we investigate a practical example of how ISEs can be combined with countermeasures against side-channel attacks on a modern processor platform. We discuss the main features of a comprehensive concept proposed by Tillich and Großschädl [23], Tillich et al. [24] for implementing AES ISEs and hardware countermeasures on a 32-bit SPARC V8-compliant processor.

In this concept the majority of the 32-bit processor remains unmodified. All critical operations are executed within a single hardware module which acts like a conventional function unit, the so-called *secure zone*. Only the secure zone is implemented in a secure logic style and contains some additional hardware blocks, the rest of the processor remains untouched and is implemented in standard CMOS logic. A protected functional unit within the secure zone provides a set of custom instructions that can be used for a flexible implementation of different cryptographic algorithms. In fact, all operations that may potentially become a target of side-channel attacks have to be unconditionally executed by the protected functional unit. Hence, a software developer still has to proceed with great care during the process of software development in order to avoid unintentionally implemented security leaks.

Within the secure zone, all operations are protected by a secure logic style, outside of the secure zone all data is strictly masked. Function operands entering the secure zone are unmasked, critical operations are performed on the unmasked data, and before leaving the secure zone the data is masked again with a freshly generated mask. This way it is ensured that transformations on critical data as well as the masks do not leak any side-channel information because of the secure logic style, and the masked data running outside of the secure zone may not leak any useful side-channel information because of the masking technique and the anonymity of the masks.

All mask-handling modules are also part of the secure zone, i.e., they are protected by the secure logic style, and the masks themselves must not leave the secure zone in their original form. The secure zone contains a mask generator and a mask storage for generating, storing, and retrieving masks. The retrieval of masks corresponding to input operands and the storage of a mask corresponding to a result can be linked to the addresses of the operands and the result, respectively.

We now want to take a look at the implementation costs of the practical example proposed by Tillich et al. [24] based on a SPARC V8-compliant LEON3 processor [1]. A full version of the secure zone has an area requirement of approximately 22 kGEs. In the following, we go through some theoretic numerical examples, calculating the total area overhead when implementing the secure zone in different secure logic styles.

We assume that a typical LEON3 processor implementation containing a debug support unit, RAMs, and caches requires approximately 580 kGEs. In total, a LEON3 processor equipped with a secure zone implemented in unprotected CMOS logic requires 602 kGEs. A pure DRP logic style like DWDDL increases the area requirement approximately by a factor of about 12 [31]. In our example the secure zone implemented in DWDDL would require approximately 264 kGEs, the whole LEON3 processor would thus result in 844 kGEs. The total area overhead would increase by a factor of only 1.4 compared to the LEON3 implementation in unprotected CMOS logic. In case of a coprocessor that would have to be completely implemented in DWDDL we would encounter the full area overhead factor of 12. In case of iMDPL [16], which causes an area overhead of a factor of 18, the overall area for our LEON3 processor would be 976 kGEs. This would result in an area overhead factor of approximately 1.7.

These numerical examples illustrate the following: although secure logic styles may result in a significant drawback in terms of area requirements as well as power consumption, they may be implemented much more economically if secure logic styles are combined with advanced concepts like the secure-zone approach.

5.7 Testability and Security

Testing is an important activity not only during development of a hardware circuit but also after its manufacturing. Typically, not all microchips that have been manufactured are working properly. This has various reasons, for example, varying process parameters during production or imperfections of material and masks. The yield, which is the ratio between the number of working chips and the overall number of manufactured chips, should be as high as possible to maximize the profit. In order to separate faulty chips from working chips, tests have to be applied.

Releasing a faulty chip causes tremendous costs. Imagine the following simple example: A company manufactures 100,000 chips, and sells them at the price of $1 per chip. We assume that one percent of the chips (i.e., 1,000 chips) are faulty. When the faulty chips are immediately detected after production through tests before they get sold, costs of $1,000 will arise. When the faulty chips get detected after they have been sold and soldered on a board, costs will already result in $50,000 if repairing a malfunctioning board costs $50. Even worse, when the failing parts get detected after integration into a whole system, costs will boost to $1,000,000 when repairing a non-working system costs $1,000. This simple example clearly emphasizes the need of detecting faulty parts as early as possible after production.

In order to get confidence about proper operation of a microchip after production, reliable tests are necessary. For realizing such reliable tests, the underlying test concepts that are used have to be planned and included at the design time of a hardware circuit. This so-called "design-for-test" approach integrates additional test structures to a circuit to allow fast and comprehensive analysis of a chip after production. The more internal details of a chip can be accessed, the more comprehensive tests can be conducted, lowering the chance that malfunctioning parts remain undetected.

A powerful and widely-used test concept uses scan chains that provide access to the values internally stored in the flip flops of a hardware circuit. For cryptographic devices, giving access to internal values can be problematic. As shown in the work of Yang et al. [29, 30], test structures based on scan chains can be easily used to mount attacks against cryptographic devices. In order to prevent such attacks, test structures of security-related devices are typically deactivated after successfully testing the chip (e.g. by blowing a fuse) or even totally removed by cutting them off [11].

An alternative to scan-chain approaches are built-in self tests (BISTs). The NIST suggests to use BISTs instead of scan-chains for cryptographic devices [14]. For a BIST, necessary test data and test cases are generated within the evaluated chip. The only information that is returned after conducting the tests is whether all tests have been passed or not. This is advantageous from a security point of view but comes at cost of a lower fault-detection rate, since comprehensive tests as with scan-chain approaches are not possible. Moreover, generating test data within the chip causes significant hardware overhead.

5.8 Hardware Trojans

Test structures implemented by a designer to be able to verify the correct functionality of an IC after production, may unintentionally weaken or bypass security measures implemented on that device. In contrast, hardware Trojans describe security-threatening hardware structures in ICs, intentionally implemented at some point in the fabrication chain. For cost reasons, more and more semiconductor companies outsource the chip fabrication process to cheaper facilities. Hence, hardware Trojans may be implemented without the designers' knowledge or notification (assuming that the designer is not an adversary). Moreover, they may remain undetected even if the designer receives the ICs from the wafer factory and performs conventional functionality tests, as hardware Trojans may or may not be activated as soon as the IC is powered up. Once activated, the malicious actions a hardware Trojan may perform are incredibly powerful:

- simply shut down the device the Trojan is running on or disable connected devices,
- disable security mechanisms in a system,
- transmit critical data via various interfaces (e.g. radio-frequency emission),
- open a backdoor for an adversary and provide access to a system,
- or bypass implemented hardware countermeasures against side-channel attacks.

Wang et al. [26] classified hardware Trojans into three classes according to their *physical, activation,* and *action* characteristics. The *physical characteristics* describe how the Trojan is introduced in a digital circuit (addition/deletion of cells or modification of existing cells), the size and the location of the Trojan (how many cells are involved), as well as if the insertion of the Trojan entails significant changes of the physical layout of the digital circuit. The *activation characteristics* describe whether the Trojan is "always-on" or has to be activated externally, e.g. via antenna, or internally. Wolff et al. [28] further divided internally activated Trojans into three categories based on their trigger behavior: rare value triggered, time triggered, and both value and time triggered. Wang et al.introduced three *action characteristics* that describe whether an activated Trojan modifies a function (addition/deletion of logic cells) or specification (e.g. modification of wires changes the timing specifications), or directly transmits critical information over various channels.

Preventing hardware Trojans is a very complex issue, as Trojans may be implanted in different phases, e.g. during the high level or hardware-level design phase of a system, during synthesis/place/route of hardware modules, or even during the fabrication process of an IC. One possible protection against hardware Trojans is a chip developer would have to somehow establish a chain of trust, starting at the designer, continuing with the hardware experts, up to the IC production facility and the package house.

Another possibility to detect hardware Trojans is based on SCA [2]. In this approach a few ICs from one IC family (produced with the same mask) are first subjected to sufficient I/O tests to verify all parts of the circuitry. During these tests, some side-channel signals are also collected to build a side-channel fingerprint. In a next step, the ICs are destructively reverse engineered in order to thoroughly ensure that these few samples are free of Trojans. All other ICs from the same family can then be checked by comparing the fingerprints. This example shows that security-threatening side-channel attacks can also be used to some degree to detect hardware Trojans in ICs.

5.9 Conclusion and Summary

Implementing efficient, secure, and reliable ICs is a highly sophisticated task that provides manifold possibilities, but requires to be performed with great care. In this chapter we have shown that a hardware designer has almost indefinite possibilities to basically add specific functionality to an IC as well as to optimize the implementation to reach various design goals. By means of several examples of cryptographic hardware implementations we have illustrated the degrees of freedom that are at a designer's disposal. We also pointed out the importance of performing tests and simulations throughout the whole design cycle in order to minimize the possibility of errors as well as to decrease the effort if an error occurs during the design phase.

With regard to security-threatening side-channel attacks we described the fundamental characteristics of modern CMOS circuits and pointed out the reason for the vulnerability of ICs against such attacks by means of a conventional CMOS inverter. We discussed basic approaches for implementing countermeasures against side-channel attacks and introduced some particular countermeasures in more detail. We also pointed out the possibility to verify the effectiveness of various types of countermeasures to some degree in early design phases. This minimizes the costs as well as the effort to perform changes in the implementation.

We combined the topics of efficient hardware implementations, cryptographic ISEs, and hardware countermeasures against side-channel attacks and presented a sophisticated concept with custom instructions and countermeasures on a modern SPARC V8-compliant 32-bit processor platform. It turned out that such a concept benefits from both the efficiency and compactness of ISEs as well as the security gain achieved by a secure logic style.

In a last part of this chapter, we contrasted the testability with the security of ICs. It has been shown that test structures implemented by designers to be able to comprehensively verify the correct functionality of a design after production may lead to significant security leaks. We also discussed the insertion and some possible effects of hardware Trojans, which represent a worrying yet interesting and currently evolving research topic.

References

1. Aeroflex Gaisler. The Aeroflex Gaisler Website. http://www.gaisler.com/.
2. D. Agrawal, S. Baktir, D. Karakoyunlu, P. Rohatgi, and B. Sunar. Trojan Detection using IC Fingerprinting. In *IEEE Symposium on Security and Privacy (SP '07), Berkeley, Californie, USA, May 20–23 2007*, pages 296–310, 2007.
3. D. Canright and L. Batina. A Very Compact "Perfectly Masked" S-Box for AES. In *Applied Cryptography and Network Security - ACNS 2008, New York, USA, June 3–6, 2008, Proceedings*, volume 5037 of *Lecture Notes in Computer Science*, pages 446–459. Springer, 2008.
4. M. Feldhofer, J. Wolkerstorfer, and V. Rijmen. AES Implementation on a Grain of Sand. *IEE Proceedings on Information Security*, 152(1):13–20, October 2005.
5. Gaisler Research. LEON2 Processor Users Manual. XST Edition. [Online] http://www.gaisler.com/doc/leon2-1.0.30-xst.pdf, July 2005. Version 1.0.30.
6. D. Gajski and R. H. Kuhn. New VLSI Tools - Guest Eidtors' Introduction. *IEEE Computer*, 16(12):11–14, 1983.
7. A. Hodjat and I. Verbauwhede. Interfacing a High Speed Crypto Accelerator to an Embedded CPU. In *Conference Record of the Thirty-Eighth Asilomar Conference on Signals, Systems, and Computers, 2004*, volume 1, pages 488–492. IEEE, November 2004.
8. H. Kaeslin. *Digital Integrated Circuit Design - From VLSI Architectures to CMOS Fabrication*. Cambridge University Press, 2008. ISBN 978-0-521-88267-5.
9. P. C. Kocher. Timing Attacks on Implementations of Diffie-Hellman, RSA, DSS, and Other Systems. In N. Koblitz, editor, *Advances in Cryptology - CRYPTO '96, 16th Annual International Cryptology Conference, Santa Barbara, California, USA, August 18–22, 1996, Proceedings*, number 1109 in Lecture Notes in Computer Science, pages 104–113. Springer, 1996.
10. P. C. Kocher, J. Jaffe, and B. Jun. Differential Power Analysis. In M. Wiener, editor, *Advances in Cryptology - CRYPTO '99, 19th Annual International Cryptology Conference, Santa Barbara, California, USA, August 15–19, 1999, Proceedings*, volume 1666 of *Lecture Notes in Computer Science*, pages 388–397. Springer, 1999.
11. O. Kömmerling and M. G. Kuhn. Design Principles for Tamper-Resistant Smartcard Processors. In *Proceedings of the 1st USENIX Workshop on Smartcard Technology (Smartcard '99), Chicago, Illinois, USA, May 10–11, 1999*, pages 9–20, McCormick Place South, May 1999. USENIX Association. ISBN 1-880446-34-0.
12. S. Mangard, M. Aigner, and S. Dominikus. A Highly Regular and Scalable AES Hardware Architecture. *IEEE Transactions on Computers*, 52(4):483–491, April 2003.
13. S. Mangard, T. Popp, and B. M. Gammel. Side-Channel Leakage of Masked CMOS Gates. In A. Menezes, editor, *Topics in Cryptology - CT-RSA 2005, The Cryptographers' Track at the RSA Conference 2005, San Francisco, CA, USA, February 14–18, 2005, Proceedings*, volume 3376 of *Lecture Notes in Computer Science*, pages 351–365. Springer, February 2005.
14. National Institute of Standards and Technology (NIST). FIPS PUB 140–1: Security Requirements for Cryptographic Modules, 1994. [Online] http://www.itl.nist.gov/fipspubs/.
15. National Institute of Standards and Technology (NIST). FIPS-197: Advanced Encryption Standard, November 2001. [Online] http://www.itl.nist.gov/fipspubs/.
16. T. Popp, M. Kirschbaum, T. Zefferer, and S. Mangard. Evaluation of the Masked Logic Style MDPL on a Prototype Chip. In P. Paillier and I. Verbauwhede, editors, *Cryptographic Hardware and Embedded Systems - CHES 2007, 9th International Workshop, Vienna, Austria, September 10–13, 2007, Proceedings*, volume 4727 of *Lecture Notes in Computer Science*, pages 81–94. Springer, September 2007. ISBN 978-3-540-74734-5.
17. T. Popp and S. Mangard. Masked Dual-Rail Pre-Charge Logic: DPA-Resistance without Routing Constraints. In J. R. Rao and B. Sunar, editors, *Cryptographic Hardware and Embedded Systems - CHES 2005, 7th International Workshop, Edinburgh, UK, August 29–September 1, 2005, Proceedings*, volume 3659 of *Lecture Notes in Computer Science*, pages 172–186. Springer, 2005.

18. J. M. Rabaey. The SPICE Home Page. http://bwrc.eecs.berkeley.edu/Classes/IcBook/SPICE/.
19. J. M. Rabaey. *Digital Integrated Circuits - A Design Perspective.* Electronics and VLSI Series. Prentice Hall, 1st edition, 1996. ISBN 0-13-178609-1.
20. P. Schaumont and K. Tiri. Masking and Dual-Rail Logic Dont Add Up. In P. Paillier and I. Verbauwhede, editors, *Cryptographic Hardware and Embedded Systems - CHES 2007, 9th International Workshop, Vienna, Austria, September 10–13, 2007, Proceedings,* volume 4727 of *Lecture Notes in Computer Science,* pages 95–106. Springer, September 2007.
21. D. Suzuki, M. Saeki, and T. Ichikawa. Random Switching Logic: A New Countermeasure against DPA and Second-Order DPA at the Logic Level. *IEICE Transactions on Fundamentals of Electronics, Communications and Computer Sciences,* E90-A(1):160–168, 2007. ISSN 0916-8508.
22. S. Tillich and J. Großschädl. Instruction Set Extensions for Efficient AES Implementation on 32-bit Processors. In L. Goubin and M. Matsui, editors, *Cryptographic Hardware and Embedded Systems - CHES 2006, 8th International Workshop, Yokohama, Japan, October 10–13, 2006, Proceedings, volume 4249 of Lecture Notes in Computer Science,* pages 270–284. Springer, 2006.
23. S. Tillich and J. Großschädl. Power-Analysis Resistant AES Implementation with Instruction Set Extensions. In P. Paillier and I. Verbauwhede, editors, *Cryptographic Hardware and Embedded Systems - CHES 2007, 9th International Workshop, Vienna, Austria, September 10–13, 2007, Proceedings,* volume 4727 of *Lecture Notes in Computer Science,* pages 303–319. Springer, September 2007.
24. S. Tillich, M. Kirschbaum, and A. Szekely. SCA-Resistant Embedded Processors - The Next Generation. In C. Gates, M. Franz, and J. P. McDermott, editors, *26th Annual Computer Security Applications Conference (ACSAC 2010), 6–10 December 2010, Austin, Texas, USA,* pages 211–220. ACM Press, 2010.
25. K. Tiri and P. Schaumont. Changing the Odds against Masked Logic. In E. Biham and A. M.Youssef, editors, *Selected Areas in Cryptography, 13th International Workshop, SAC 2006, Montreal, Quebec, Canada, August 17–18, 2006, Revised Selected Papers,* volume 4356 of *Lecture Notes in Computer Science,* pp. 134–146. Springer, 2007. [Online] http://rijndael. ece.vt.edu/schaum/papers/2006sac.pdf.
26. X. Wang, M. Tehranipoor, and J. Plusquellic. Detecting Malicious Inclusions in Secure Hardware: Challenges and Solutions. In M. Tehranipoor and J. Plusquellic, editors, *Hardware-Oriented Security and Trust (HOST 2008), Anaheim, CA, June 9 2008, Proceedings,* pages 15–19, 2008.
27. N. H. E. Weste and D. Harris. *CMOS VLSI Design—A Circuits and Systems Perspective.* Addison-Wesley, 3rd edition, May 2004. ISBN 0-321-14901-7.
28. F. G. Wolff, C. A. Papachristou, S. Bhunia, and R. S. Chakraborty. Towards Trojan-Free Trusted ICs: Problem Analysis and Detection Scheme. In *Design, Automation and Test in Europe (DATE), 10–14 March, 2008,* 2008.
29. B. Yang, K. Wu, and R. Karri. Scan Based Side Channel Attack on Dedicated Hardware Implementations of Data Encryption Standard. In *Proceedings of the International Test Conference on International Test Conference,* CCS '05, pages 139–146, New York, NY, USA, 2005. ACM.
30. B. Yang, K. Wu, and R. Karri. Secure Scan: A Design-for-Test Architecture for Crypto Chips. *IEEE Trans. on CAD of Integrated Circuits and Systems,* 25(10):2287–2293, 2006.
31. P. Yu and P. Schaumont. Secure FPGA circuits using controlled placement and routing. In *Proceedings of the 5th IEEE/ACM international conference on Hardware/software codesign and system synthesis, Salzburg, Austria, September 30 - October 5, 2007,* pages 45–50. ACM Press, September 2007. ISBN 978-1-59593-824-4.

Part II
Generic Security and Processing Platforms

Chapter 6
Information Security Best Practices

Keith Mayes and Konstantinos Markantonakis

Abstract We are increasingly reliant on the use of IT systems in our normal day-to-day business and personal activities. It is of paramount importance that these systems are sufficiently secure to protect sensitive, valuable and private data, and associated storage, communications and transactions. Therefore, the design and use of such systems should be in accordance with best practices for information security that have been developed by industry, government and the worldwide expert community. This chapter emphasises the need for system security and goes on to explain technical choices such as algorithms, key size and trust management, and concludes with a real-world case study.

6.1 Introduction

We live in an age in which the tools of information technology (IT) and the Internet have become ubiquitous and essential for all manner of good reasons and responsible uses. We are able to harness the creativity, skills, enthusiasm and cooperation of the people for greater and positive achievements. Unfortunately, these same tools have a darker side and can be used just as effectively for negative purposes. This is certainly true in the area of Information Security. On the one hand, we have designers and system owners trying to safeguard sensitive data and functionality using security protection, whilst on the other hand we have attackers (and worse!) trying to undermine the protection. This is no clear cut, good against evil struggle and whilst system owners will condemn the action of attackers, the attackers may rally to

K. Mayes (✉) · K. Markantonakis
Information Security Group, Smart Card Centre, Royal Holloway, University of London,
London, United Kingdom
e-mail: keith.mayes@rhul.ac.uk

K. Markantonakis
e-mail: k.markantonakis@rhul.ac.uk

K. Markantonakis and K. Mayes (eds.), *Secure Smart Embedded Devices,*
Platforms and Applications, DOI: 10.1007/978-1-4614-7915-4_6,
© Springer Science+Business Media New York 2014

a cause because of perceived failings of the system owners. What is clear is that if you plan to implement a system you should assess the likely risks and impacts of attacker activity and design, implement, test, operate and manage your system in-line with information security industry best practices. There is no real excuse to be ignorant of these practices as guidelines are published by several authoritative sources, although it is also important to realise that there are often practical and financial constraints, especially with legacy systems. A lot of effort in information security is working with imperfect solutions; however, if you initially design to best-practices, you at least know where the imperfections and risks lie and are better able to manage them. This chapter provides an overview of some aspects of best-practice that are applicable to embedded devices and where possible links the topic to relevant published exploits.

6.1.1 What is Information Security and Who are the Adversaries?

A system typically makes use of data, functionality and communication, and this supports a data centric view of security. The data may be sensitive or private, can have direct linkage to money and value or can control configuration, ownership and privilege. Preserving the integrity, confidentiality and access privileges of data is therefore essential for the system to work as planned. Functions act upon data and hence the integrity of the functionality must also be maintained, so that it cannot be modified or disrupted and that it does not leak the sensitive information. Data is communicated between valid entities and so it is equally important that integrity, confidentiality and entity authentication is maintained.

Another viewpoint is that security is a barrier to keep attackers at bay. There are in fact many different types of active attackers and examples are listed below.

- Academics/researchers
- Private "enthusiasts"
- Petty criminals
- Activists
- Organised - well equipped criminals

As there are a variety of attacker types it is not surprising that there is also a diversity of motivations.

- Fame
- Curiosity/Challenge/Fun
- Reduce own costs
- Political agenda/embarrassment
- Money

In the not too distant past, when conducting a risk review it was customary to always consider the business case of an attack activity. Basically for a given attack target, what would be a reasonable amount of time, money, effort to commit in order for the attacker to succeed? The implication being that if the target's value was lower than the attack investment it would probably be left alone. It is unwise to hold too close to this way of thinking, because the Internet, communications and social networking is used effectively by the attacker community allowing efforts to be co-ordinated, tasks to be shared, with software and hardware designs being freely published (see [1]). The idea of attacking systems that are poorly designed, implemented or that are suspected of abusing power in some way is accepted in such groups and indeed bringing these things to the public domain has become a fun "sport" [2]. Predicting what targets an attacker group would like to attack might even be impossible; there actually does not need to be much of reason other than the group wants to do it. It should be noted that a lot of the pioneering activity is driven by curiosity and has no real malicious objective; however, it is dangerous if it serves to educate groups and organisations with ruthless and criminal intent.

When designing a system, it is therefore prudent to assume that you are up against a host of experts and criminals who are as least as smart as you, that have almost unlimited human resources, and have specialist computing equipment many times more powerful than yours. Therefore, don't go looking for trouble! Translated to best practice, this means reduce the chances of becoming a target. If it is known that your system relies on secretive, proprietary and unreviewed techniques or algorithms, it is equivalent to painting a large target on the system; just the thing that some attacker groups like. Whereas, if it is publicly declared that your system adheres to best practices then there is less "fun" in attacking it; so you will be overlooked by a large proportion of your potential adversaries. You reduce the target still further by minimising the potential prize from a successful attack. For example, if a difficult attack reveals a key that can be used for one second, in one location, by one person, for one purpose then you might deter all but the most persistent attackers. Conversely, if an attack reveals a global secret (e.g. common key or secret proprietary algorithm) that undermines the security of the entire system, there will be a larger group of adversaries to resist.

6.2 Security Objectives

Paranoia about security attacks is a good starting point, but you then need to carry out a security risk review of your system or assets. This should be done independently of suppliers as not surprisingly they will rarely point out vulnerabilities in the products they are trying to sell to you! Similarly you would be wise to require independent evidence from certification labs or your own trusted experts, that a product or system component meets your security requirements. This of course means that you will have determined your requirements in sufficient detail. This is not so simple, but one way to approach it, is to identity the security critical data assets and functions that are relied on by the system.

6.2.1 Data Assets

With a sensitive data asset we may be concerned about integrity, access control and privacy. There are different types of data assets that may need protection.

- Personal data: This could be information on customers such as names and addresses, ages, perhaps even health records and criminal convictions etc.
- Monetary data: This could include the contents of an e-wallet, tickets, vouchers, credit/debit card data and bank details.
- Transaction data: This could be all manner of access to on-line services e.g. shopping history, tax returns, health-site logs as well as conventional POS transactions.
- Cryptographic data: This is the data used to provide security protection so includes account IDs, passwords, PINs, keys, signatures, certificates, etc.

As a general guide, we are concerned about critical data assets when they are in storage, communication or being acted upon by critical functions.

6.2.2 Critical Functions

Protecting data assets alone is not sufficient. The functions that act upon the data assets are also attack targets. Example functions are included below.

- Service and data access functions.
- Payment transaction handlers.
- Data handling functions.
- Cryptographic functions: Encryption/Decryption, Hash, MAC, random number generation, signing and verification.

6.2.3 The Range of Security Protection

Having identified the critical data assets and functions we can set about protecting them; however, effective protection requires appreciation of the full range of security measures. We can split this range into logical security which relates mainly to design aspects, implementation security and critical process security.

Logical Security: This is the traditional design of a system and where you will find a lot of the best practice advice. It covers aspects including.

- Algorithms and security functions.
- Cryptographic keys.
- Protocols.
- Random number generation.

Implementation is equally important, as the logical security design has to be part of some hardware and usually software combination in order to be used in practice and attackers target implementations just as well as designs. This area can be split as follows.

- Platform security.
- Physical security.
- Side-channel security.[1]

When thinking of critical processes, most people thing of the operational transactions, which are of course vital. However, security solutions do not magically start working and continue without any management attention. System components require initialisation so they have the common data and functions needed by the system and then personalisation to have them appropriately configured for individuals or as different system components. An important part of this is making sure that all the entities have the correct keys for use in the crypto operations. Keys have to be generated programmed and distributed, and sometimes updated and/or withdrawn from use, all of which is covered by the general term of Key Management. There will also be operational management that might be another target for attack. For example, a transaction website address may occasionally need changing so it is important that only authorised parties can do this. Managing the system and its components throughout their normal useful life, including the data, the keys and operations is often referred to as lifecycle management.

This chapter will focus mainly on an overview of logical security best practice; however, the reader should be quite clear that this alone does not make a secure system. Equal attention is needed to prevent attacks against the implementation as well as critical processes; see Chap. 9 of [3] for some overview information.

6.3 Cryptographic Algorithms

To safeguard our data assets, in storage, communication or processing we can use cryptographic algorithms that have useful security properties, and can be employed for a variety of purposes including the following.

- Encryption/decryption: This is effectively disguising the data assets so that only an authorised party can determine the source content.
- Integrity protection: This is used to ensure that a data asset has not been modified or corrupted.
- Signing/Verification: This can be used (rather like a written signature) to determine the authenticity of data and transactions.

---

[1] Side-channel security is the ability for an implementation to prevent information leakage (e.g. from timing and power consumption variations and radio emissions) when running critical functions. Attackers may analyse leakage to recover sensitive information and cryptographic keys [15].

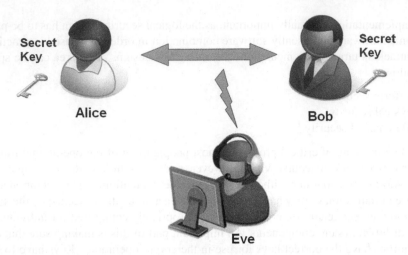

Fig. 6.1 Symmetric algorithm

The cryptographic algorithms can be used in combination as part of a security protocol. For example, they could be used in an authentication protocol to determine that the communicating parties are legitimate (e.g. a real bank and its real genuine customer), to establish session keys for communication privacy and verify the integrity of exchanged information.

There are some common cryptographic algorithms that can be used for a variety of purposes. They are split into two main classes.

- Symmetric (e.g. DES [4], AES [5])
- Asymmetric (e.g. RSA [6])

6.3.1 Symmetric Algorithms

In a symmetric algorithm, the transaction participants share the same key that is kept secret from all other parties.

If the algorithm is designed without structural weaknesses, then the most efficient way to attack it is via brute-force key recovery. This means that Eve must use trial and error to find the key that Alice and Bob have shared. In practice, a key may be considered as a pattern of N binary digits (bits). This means that there may be at most 2N unique keys. Eve could be very lucky and get the right key first try, or very unlucky and find the key on the last attempt. The normal design assumption is that on average Eve will guess the key after about 2N-1 attempts, i.e. after having tried half of the possible key values. Therefore to have a useful symmetric algorithm we need to have an N that is sufficiently large to make it impractical for Eve to guess the key, despite having lots of computing power at her disposal. In fact we have to assume that she has a key-cracker machine. A key-cracker is a specialist machine that can run an algorithm very many times in a short period, usually employing parallel processing, Historically one would purchase a physical machine [7] to do this although similar power may now be obtained from an Internet based virtual computer or a grid of computer resource shared between a hacker "enthusiast" group.

The obvious question of course is how large is enough when choosing the keysize N for a symmetric algorithm? The more correct question is what is the minimum recommended effective keysize (in bits) for a symmetric algorithm? The distinction between the number of bits used to represent the key N and the effective keysize will be explained later. The answer to the keysize question changes over time as the keysize will need to increase to allow for technology advances which will improve attacker resources. Fortunately there are respected sources of expert advice such as NIST [8, 9] and ECRYPT [10] that can suggest an appropriate keysize for a given target lifetime of your system. Although the advice from different sources is not identical, it is similar, and so roughly speaking a legacy system in say 2011 would be just about OK with an effective keysize of 80 bits, but if you were introducing a new system you should have at least 128 bits, and if it is particularly security sensitive then play safe with 256 bits. With these guidelines in mind, we can analyse a common symmetric algorithm, namely the Data Encryption Standard (DES) and its TripleDES variant.

6.3.1.1 The DES Example

DES [4] is referred to as a Feistal Cipher and has 16 sequential "rounds" of processing. A single round is shown in Fig. 6.2 and generally the output from the previous round is subject to expansion, sub-key addition, substitution and permutation.

The algorithm has an input data block size of 64 bits and a keysize of 56 bits. Clearly, the keysize falls way below current best practice and DES keys have been well within the range of brute force key-crackers for many years. It is therefore not surprising that DES is considered obsolete and should be phased out from legacy systems. There are variants known as Two Key Triple DES (2key-3DES) and Three Key Triple DES (3key-3DES) that have stronger logical security and have been used

Fig. 6.2 Round of DES [4]

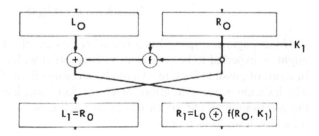

Fig. 6.3 Triple DES
processing   I $\rightarrow$ DES E_{K1} $\rightarrow$ DES D_{K2} $\rightarrow$ DES E_{K3} $\rightarrow$ O

Table 6.1 DES key-sizes and best practice [10]

DES variant	Actual bey bits	Effective key bits	Best-practice view
1-key DES	56	56	Not recommended for use - obsolete
2-key 3DES	112	80	Not recommended for use after 2012
3-key 3DES	168	112	Recommended for use until 2030

in relatively recent systems. Both variants simply run the algorithm multiple times
as shown in Fig. 6.3.

In 3key-3DES, the keys K1, K2 and K3 are all different. In 2key-3DES, K1 and K3
are the same. If you set all keys to be the same then you have backward compatibility
to normal DES, albeit in an inefficient implementation. There is no performance
advantage in using 2key-DES over 3key-DES so its use is often a throw-back to past
issues of compatibility, memory limitation and practical key management.

6.3.1.2 Effective Keysize

When judging an algorithm, it is the effective key-bits that are important as these
relate to the brute-force attack difficulty. If you are curious about the estimation of
effective keysize and can follow equations then read on, otherwise skip to the next
section. The reason that triple DES does not increase the effective key-size as much as
one might initially think is that the iterative (3-stage) processing may be exploited in
attack optimisation. For example, 3-key 3DES has a lower effective keysize due to a
"meet-in-the-middle" [11] attack strategy. This is classically described by successive
encryption (E) by two keys; K1 and K2 of an input plaintext P1 to generate and output
cipher text C_1.

$$C_1 = E_{K2}(E_{K1}(P_1))$$

Generally for the $m^t h$ plaintext and ciphertext.

$$C_m = E_{K2}(E_{K1}(P_m))$$

If the single encryption has a key of N bits length, then the double encryption
might be expected to be equivalent to encryption with a key of 2N bits in length.
In terms of possible keys, this would be a change from 2^N to 22^N key possibilities
which should increase the effort for a brute force attack enormously. Unfortunately,
the meet-in-the-middle attack brings us back close to the single key brute force
difficulty. It works as follows.

1. Compute $E_{K1}(P_1)$ under all possible values of K1 and store in a table.
2. Compute $D_{K2}(C_1)$ under possible values of K2 until you find a matching result in the table.
3. Check another $P_m \& C_m$ pair to see if the correct K1 and K2 found (if not go back to step 2).

If the table look-up time is considered negligible, then the brute force effort is the encryption under all possible K1 (N operations) and the decryptions under K2 (worst case N operations). The number of cryptographic processes (which related to brute-force attack difficulty) is therefore.

$$Processes = 2^N + 2^N = 2^{N+1}$$

Even under worst case conditions (K2 last of keys tried) this is equivalent to adding just one bit to the key length rather than doubling it and so the double encryption has made little improvement.

In the case of 3-key 3DES, the encryption is represented as.

$$Cm = E_{K3}(E_{K2}(E_{K1}(P_m)))$$

where each key has size N bits.

If we assume that the combined second and third encryptions $E_{K3}(E_{K2}(.))$ are equivalent to a single encryption of a key K23 which has 2N bits, we can rewrite the equation as.

$$C_m = E_{K23}(E_{K1}(P_m))$$

If we follow the meet-in-the-middle approach, the computation is now.

$$Processes = 2^{2N} + 2^N \approx 2^{2N}$$

So for 3key-3DES N = 56 and so the effective key size = 2N = 112 bits[2].

This "trick" does not come for free and you need a large memory to store the results from the original stage of encryption. The amount depends on N, but also on the blocksize of the algorithm. For DES variants the blocksize is 64 bits (= 8bytes). The memory requirement in bytes is therefore.

$$2^N \times 8 = 2^{N+3} = 259 \approx 5 \times 10^{17} bytes \approx 500,000\ Terabytes = half\ an\ Exabyte$$

This sounds like a lot of memory, but according to [32] the Internet is reckoned to be storing around 300 Exabytes at the moment and new satellite TV boxes are already shipping to thousands of homes with Terabyte hard drives. Note that attackers don't need to own all this memory, just share and use it for a while and so developments

---

[2] In this case, the example calculation results in the same effective keysize as appears in the recommendations. However, in the more general case, other considerations of positive or negative impact may be taken into account when defining effective keysizes.

like the Cloud are also interesting. The likelihood is that half an Exabyte will not sound "big" for very long.

The reduction in effective key size for the 2-key 3DES is not due to the meet-in-the-middle attack (as the same key is used in the outer stages), but rather another attack strategy that makes use of captured plaintext/ciphertext pairs and a memory table. The effective keysize is reduced the more pairs that it is feasible to collect. If the keys are used rarely, then the effective key size would be closer to the actual keysize. According to [12] the attack process difficulty can be estimated as follows.

If n is the number of collected plaintext/ciphertext pairs compute.

$$r = \log_2 n$$

Then the number of effective key bits b for estimating attack difficulty is given by.

$$b = 2^{120-r}$$

Just to illustrate the evolving dynamic nature of best-practice, the attack which was published back in 1990 [13] estimated the cost of the table memory at $10–20 million. The table size was 2^{40} bits = 2^{37} bytes $\approx 10^{11}$ bytes (100 GB). Today you would struggle to buy a hard disk drive this small and a 500 GB disk would cost around $100.

6.3.1.3 Choosing a Symmetric Algorithm Today

Although we have used DES as an example, the general issues are relevant to all similar algorithms. In fact despite the known attack optimisations, DES has proved itself to have a robust design, sufficient to justify its long life-time of practical use. However, if we wish to remove temptation for mounting brute-force attacks over the longer term we need an algorithm that has bigger effective key sizes and does not have to be run multiple times to raise its security. Fortunately, there is a readymade solution in the form of the Advanced Encryption Standard (AES) [5], the successor to DES. AES has a multi-round style of construction (DES is also multi-round) and is standardised in 128, 192 and 256 bit key options, and even the smallest key is expected to be secure beyond 2030. The block size is increased from 64 to 128 bits which makes it less vulnerable to some attack strategies. It is also fast like its DES predecessor.

A very important point to note about AES is that it was evaluated and selected by the world's expert community and is published openly. This is very important because of the following.

- An algorithm's security should not rely on the secrecy of the algorithm (Kerckhoffs' principle [14]).
- Violation of this principle is called "Security by Obscurity".

In principle, a secret and proprietary algorithm might be "secure," but it may be hard to convince anyone of this and history has taught us that most proprietary algorithms are shown to be flawed once their design information is disclosed.

The main message here is that the rational choice for a symmetric algorithm today would be AES, unless you have a resource limited implementation and/or demanding performance requirements. One important point to note is that we have only commented on the logical design choice. The construction of both DES and AES requires execution of multiple rounds of processing, each of which may be analysed using side-channel techniques. If you do not implement the algorithms in a way that prevents side channel leakage, the individual rounds may be attacked in sequence [15] and the "good" design of the algorithm can be completely undermined. The reader is again referred to Chap. 9 of [3].

6.3.1.4 Symmetric Algorithm Modes

Choosing the symmetric algorithm is not the end of the design process either. You need to decide which mode you will use it in. If we are interested in encryption, we know that both DES and AES are block ciphers which gives us several mode options.

The simplest mode is referred to as electronic codebook (ECB) in which each block of plaintext is encrypted using the same secret key(s). There is a weakness with this approach if your plaintext blocks are meant to form part of a contiguous sequence. An adversary could remove or insert cipher blocks as part of an attack strategy. To counter this threat the Cipher Block Chaining (CBC) mode is commonly used as illustrated in Fig. 6.4.

The basic principle is that the input plaintext is EXORed with the ciphertext result from the preceding block encryption, thereby chaining the encryptions together. The initial encryption is a special case as there is no preceding ciphertext result, so an Initialisation Vector (IV) is used instead. Determining the value of the IV is a design choice. In some applications it has been zero, a fixed value, the result of a fixed calculation, but for best practice it should be a random dynamic value. In the last

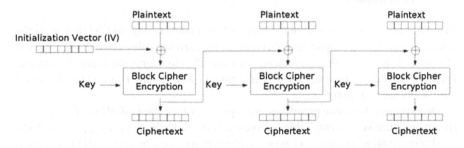

Cipher Block Chaining (CBC) mode encryption

Fig. 6.4 CBC

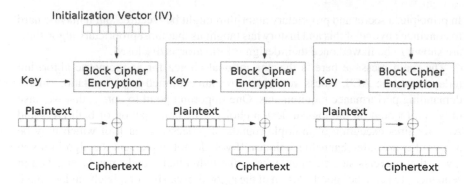

Output Feedback (OFB) mode encryption

Fig. 6.5 OFB mode to create a cipher stream [33]

case, an additional mechanism is needed to ensure that both the encryption and decryption processes are able to agree a common IV value. Although the IV does not have the same secrecy requirements as the symmetric key(s) it is important that its integrity is protected.

Just because you are using a block cipher does not mean you have to use a block mode for encryption. You can use a block cipher to generate a keystream that is then simply EXORed with the plaintext bit stream to create the ciphertext bitstream. Figure 6.5 gives an example of the Output feedback mode that generates keystream blocks for EXORing with the plaintext. Cipher Block Feedback is another variant that generates a keystream.

If you decide to take the stream cipher approach, then extra care is needed due to particular vulnerabilities and in particular bit-flipping attacks. You should also not reuse the IV; otherwise the output from the first block cipher encryption will be a constant value.

6.3.1.5 Bit-Flipping

The basic idea is that because of the way that the plaintext bits are EXORed with the keystream, changing a ciphertext bit, changes the corresponding bit in the deciphered plaintext. This means that an attacker can modify the decrypted plaintext without knowing the algorithm or key used to generate the keystream! This can be illustrated with reference to Fig. 6.6.

Decryption is the same process as encryption, just simple EXORing of the cipher-stream with the keystream. Table 6.2 provides an example of a bit-flipping exploit.

While the attacker may not know the algorithm or key, he may well have knowledge of the plaintext structure. For example, the plaintext could have been an instruction to load an e-wallet with $25, and so the attacker may have in fact loaded $89 by simply flipping a bit (underlined in Table 6.2). In an attempt to maintain the integrity

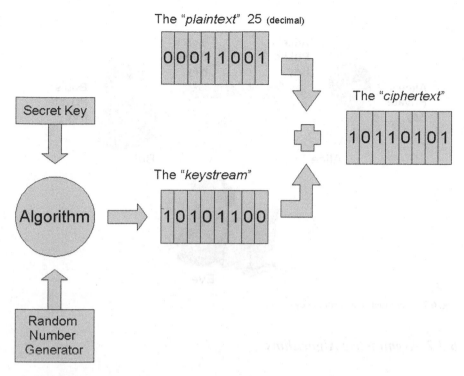

Fig. 6.6 Keystream example

Table 6.2 The effect of bit-flipping on stream ciphering		Normal case	Bit-flipping attack
	Ciphertext	10110101	11110101
	Keystream	10101100	10101100
	Plaintext	00011001	01011001
	Value (base 10)	25(Correct)	89 (Modified)

of the payload data, some protocols add a Cyclic Redundancy Check (CRC), however this does not help if the data structure is known, as the attacker simply flips the CRC bits to fit the modified payload. The use of Message Authentication Codes is far more effective as the attacker does not have the key to generate new MACs for his modified payload; although this creates another key to manage. Key management is discussed in a later section, however for now it is sufficient to note that the difficulties of managing secret keys leads to interest in the use of asymmetric algorithms.

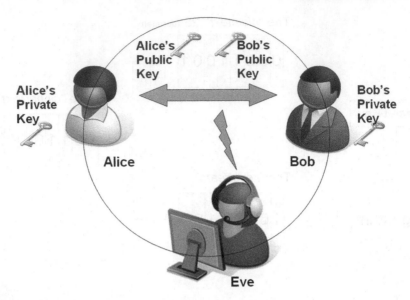

Fig. 6.7 Asymmetric algorithm keys

6.3.2 Asymmetric Algorithms

In an asymmetric algorithm, each party has a pair of keys. The "private" key is kept secret and only known to the owning party. The corresponding public key can be given to anyone. An example is shown in Fig. 6.7.

Eve is allowed to know the public keys of Alive and Bob, but cannot determine the corresponding private keys. Solutions of this kind need measures to prevent Eve from presenting her own public key and claiming to be someone else, e.g. Bob, as then Alice may send information that Eve can decrypt thinking it is destined for Bob.

All such algorithms rely on a mathematical problem (e.g. factoring or discrete logarithm), such that knowing the public key does not allow someone to work out the private key. RSA [6] is the most popular asymmetric algorithm and can be used both for encryption and signing, although you should use a different key-pair for each. It is the nearest thing to a "safe bet" for new systems designs. In simple terms, the mathematical problem faced by an attacker is to factor a very large integer (n) into its prime factors (p, q), which is very difficult. If this fact can be accepted, then the mechanics of setting up and using RSA are not hard to understand even for non mathematicians. If you have no interest in the equations (that use modulo[3] operations) then skip to Table 6.3, otherwise consider the following steps to create a key-pair and use it for encryption and decryption:

• In the key generation process, two large prime numbers (p, q) are selected.

---

[3] In modular division the result of a mod (b) would be the remainder after dividing the integer a by integer b, so for example 1 = 13 mod 6.

Table 6.3 RSA best practice key sizes [10]— practical/rough approximation

Keybits	Symmetric equivalent	Best-practice
1024	Similar to 2-key 3DES	No longer recommended for use—obsolete
2048	Similar to 3-key 3DES	Recommended for use until 2030
4096	Similar to 128 bit AES	Recommended for use until about 2040

- pq = n is called the modulus and we denote the product (p-1)(q-1) by $\varphi(n)$.
- Choose a positive integer e $(1 < e < pq)$ that has no common factors with p-1 and q-1. This is equivalent to having $gcd^4(e, \varphi(n)) = 1$.
- Your public key is then (e, n).
- Compute a positive integer d, such that $ed = 1$ mod $\varphi(n)$. The number d is the private key.
- Encryption of plaintext (m) is $c = m^e$ mod n.
- Decryption of ciphertext (c) is $m = c^d$ mod n.

RSA can alternatively be used for signing and signature verification as illustrated by the following steps;

- A digital signature (s) on plaintext (m) is $s = m^d$ mod n

  - Here m is usually a message digest

- Signature verification is $m = s^e$ mod n

  - This is compared with the recipients message digest calculation

Note that the effective keysize of the algorithm relates to the size of p and q.

Whilst it is possible to encrypt and decrypt bulk data using RSA, normally the algorithm is used in a protocol to ensure the trusted establishment of a session key that can be used in a faster symmetric algorithm. Table 6.4 provides some comparative performance figures. Note that RSA decryption and signature generation are much slower than encryption and than signature verification. RSA decryption tends to be the biggest overhead as this has to be computed for all blocks in the plaintext/message, whereas an RSA signature is normally a one-off function on the message digest.

It is also important to note that the plaintext should undergo specialized padding before RSA encryption. Failure to do this is sometimes referred to (in a non complimentary manner) as textbook RSA. There are several reasons why padding is necessary and one can be illustrated by a simple example. We know that encryption is $c = m^e$ mod n and that the reason it is relatively fast is that e can be quite small; values such as 3, 17 are often used. We also know that n must be large as p and q are of the order of 2048 bits. The data to be encrypted is considered as a binary number m also of size 2048 bits. Depending on the actual data represented, m could map to a very small binary number, e.g. with mainly zero/null content. So it is feasible for the encrypted data $m^e < n$. In this case it is possible to compute the inverse of me to

---
4 gcd means greatest common divisor.

Table 6.4 RSA v AES performance Crptopp lib for Pentium 4 - 2.93 GHz on 256 byte message [16]

Operation	Execution time (us)	Compared to AES
AES 128 Encrypt/Decrypt	3.03	1
RSA 2048 Encrypt	220	x73
RSA 2048 Decrypt	10530	x3475
RSA 2048 Sign	10640	x3512
RSA 2048 Signature Verify	220	x73

recover m. Best practice advice on RSA encryption is to use padding in accordance with Optimal Asymmetric Encryption Padding (OAEP) [17].

6.3.3 Other Algorithms/Modes

Cyclic Redundancy Check (CRC): This is a simple/small computed integrity check value from a message. It is primarily intended for the detection of accidental or transmission errors rather than those created by an attack. A CRC calculation uses a shift register and a divisor (which is usually called the polynomial of the CRC). There are many variants and the common CRC-16-CCITT is represented below as an example.

$$x^{16} + x^{12} + x^5 + 1$$

Because a CRC calculation only uses shift and EXOR functions is simple and fast, but it has limited security use as it easy to compute, modify and to find collisions. A collision is when two or more messages produce the same CRC and it may allow an attacker to modify the source message, yet still pass the CRC check.

Hash Function: This function has a special one-way property so that it is simple to compute a hash value H from a message m i.e. H = hash(m), yet it should be practically infeasible to find m from H. This is sometimes referred to as pre-image resistance and similar to the arguments on symmetric keys the size of the hash needed for this quality would at first appear to be 80–128 bits. However, there is an important additional requirement that the hash function should have good collision resistance so that the hash of one message is unlikely to be the same as the hash of another message. Because of the potential for finding collisions via a birthday attack [18] the hash needs to be twice the size required for pre-image resistance. Best practice recommendation is to use the SHA-256 (256 bits) [19] algorithm in new applications. SHA-1 (160 bits) [19] is still in use, but there are concerns (not fully substantiated) that it may be vulnerable to attack. MD5 should no longer be used as it is obsolete. Because a hash value can be regarded as a compact way to represent the integrity of an entire message it is often used as and referred to as a message digest.

Message Authentication Code (MAC): Because a hash value can be easily calculated it is could be possible for an attacker to modify a message and then calculate
and substitute the original hash function. To prevent this we use the hash value and a
suitable secret key (known to the message originator and recipient) to create a MAC
which is effectively an authentic hash. A commonly used MAC is the HMAC which
is computed as follows [20].

$$HMAC(K, m) = H((K \oplus opad) \bullet H((K \oplus ipad) \bullet m))$$

The parameters opad and ipad are padding constants designed to make the key K
different in both operations.

6.4 Key/Trust Management

In the discussion so far, we have mentioned cryptographic keys and their importance,
however we have said little about how they are generated and managed. In much the
same way that an ill considered implementation can undermine good security design
choices, poor key management can negate the intended cryptographic protection and
indeed determine whether a system is likely to be attacked. We of course need to
generate keys of the correct size and in a manner and location that does not leak secret
or private key values. This can be achieved in a highly secure manufacturing and/or
personalisation facility and sometimes within a user device. For the latter case, it is
extremely important that the device is tamper resistant and not prone to side channel
leakage of information under normal or fault conditions.

Critical to key generation (and indeed to many protocols) is a random number
source. For example, a random number generator could be used to produce an N bit
pattern used as a secret key. However, it is important that the generator has sufficient
entropy otherwise the number of possible bit patterns may be far less than 2N. In
this instance, entropy is the number of information bits per physical bit and so this
will be close to one for a good random generator. Best-practice guidance on random
number generation can be found in [21].

A key should be used for only one purpose. For example, an encryption key should
not be used for a digital signature. This limits the exposure of the key and restricts the
impact of any vulnerability or secret/private key discovery. Key exposure can also be
reduced by deriving working keys from long term secret or private keys, noting that
a working key cannot be more secure than the original key from which it is derived.

It is a very bad idea indeed to choose a common key for many user accounts. This
is called a "global secret" and if discovered it means that the security for all users
is compromised. It also increases the likelihood of the system being attacked in a
sophisticated and perhaps intrusive manner as discovery of the global secret is often
of significant value. To avoid this, all user account keys should be diversified, so that
an attack can only compromise a single account; making a sophisticated attack of

very limited value. Note that a global secret need not be a cryptographic key, but possibly a proprietary unpublished algorithm.

Once a device is outside of a physically secure and trusted environment, then further measures are required to keep secret or private keys safe. For example they need to be securely,

- transferred/communicated to the authorised parties,
- kept in attack/tamper resistant storage,
- protected during operation.

For transfer via electronic means, the keys should be encrypted using a transport key that is as least as strong as the keys that it protects. A notable exception to this is when you are transporting a public key, as by definition there is no need for secrecy.

6.4.1 Asymmetric Key Management

Although public keys are not secret, it does not mean that key management for asymmetric algorithm systems is straightforward, as they exchange the requirement for secrecy for one of authenticity. This is usually achieved via a Certification Authority and associated hierarchy as shown in Fig. 6.8. This is sometime referred to as a Public Key Infrastructure (PKI).

The basic system relies on the fact that everyone directly or indirectly trusts the Certification Authority (CA), so that when they are presented with a certificate that has been endorsed (signed) by the CA they assume it is authentic. For this assumption

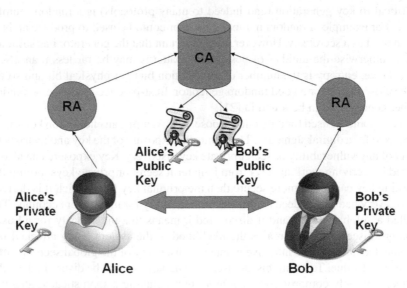

Fig. 6.8 Public key/certificate hierarchy

to be valid, the CA should only sign certificates from identifiable, legitimate entities, which is normally determined by a Registration Authority (RA). For example, if Alice would like to have a signed certificate she must apply to the RA. The RA should carry out appropriately rigorous checks to ensure that it is dealing with the legitimate Alice and her genuine data and keys. Assuming that the RA is satisfied, it requests the CA to create and digitally sign a certificate for Alice that is bound to her data and her public key. The CA creates the signature using its own secret key, so anyone verifying the signature on Alice's certificate must have some way of accessing the genuine CA public key certificate, which could be via a server or pre-installed with transactional equipment. Note there can be an interconnection of multiple CAs, so a verifier can follow a chain of certification to reach the necessary level of security assurance. EMV [22] credit cards are a good example of multi-level certification. The issuing bank signs data held on the card and then Visa or MasterCard sign the bank's certificate also stored on the card. The Visa/MasterCard public keys can be pre-stored in the Point of Sale (POS) card readers, allowing the full chain of trust to be checked.

Of course things are never quite this simple when we look at the detail. Recalling the performance examples given in Table 6.4 there may be a general desire to reduce the number of PKI operations that are carried out. Furthermore, the certificates would not be expected to have indefinite validity and can be revoked, e.g. as a result of a service provider fraud/security control or user wishes. A verifier is normally able to determine if a certificate is still valid by reference to the Certificate Revocation List (CRL). In principle, you should always check the CRL to determine if a certificate is still valid prior to use, however with mass market applications the CRL can become very large and so the temptation is to skip the check and rely on other monitoring or controlling facilities to disable an application. This is particularly true in time critical transactions on resource limited devices such as smart cards, where it may be simpler to disable the entire card rather than revoke a certificate.

6.4.2 Trust and Management

Regardless of whether we are considering symmetric or asymmetric solutions for major systems, we tend to find that an entity or organisation is at the top of the trust hierarchy. A good example is SIM card management in mobile communications. The network operators issue SIM cards to customers, yet retain all the necessary keys and parameters to manage them. Bank cards are similar although the banks have scheme operators (e.g. Visa/Mastercard) for the interoperability of their services and security. Card technology has the technical capability for delegated security management via security domains (see GlobalPlatform [23]) however this is rarely used due to business and competition issues.

As technology moves on, the trust management roles are far less clear-cut. For example, a Near Field Communication (NFC) [24] phone allows a mobile phone to emulate multiple smart cards. As the phone is not issued by the bank and not necessarily by any one network operator, the ownership of trust management is not well defined. If we also consider that handset manufacturers, party trust service providers and a wide range of influential commercial organisations have ambitions in this area, then it is difficult to predict who will dominate trust management.. What is clear is that whilst there are multiple security management solutions and multiple competing business parties there is a danger for incompatible and vulnerable solutions to be rushed into use. One might imagine that agreement for minimum security requirements would be in the common interest, however even this is in doubt when it comes to security evaluation.

6.5 Security Evaluation and Common Criteria

Consider that a banking card used for EMV debit or credit transactions must satisfy strict security evaluation requirements set by the banking industry. This means the card (and mainly the secure chip and its applications) has to have undergone common criteria evaluation [25]. This is a best-practice approach to establishing design and implementation security to levels that are internationally recognised. Evaluated smart card chips are usually EAL4+ or better, which should mean that not only is the logical design to a good standard, but that the implementation is strongly resistant to all known attacks. This is clearly a beneficial security step; however it comes at a price. This is not just financial, but also the time it takes for the evaluation. Table 6.5 gives a very rough indication of the relative time/cost of the evaluation levels.

For banking cards that have an expected lifetime of about 5 years and whose functionality changes relatively slowly, the EAL4+ evaluation time may be acceptable, but this has not been the case in other more fast moving industries. Mobile communi-

Table 6.5 Comparison of common criteria evaluation levels (Chap. 8 of [3])

CC Level	Description	Time (months)	Cost
EAL1	Functionally tested	2–3	Several 10s of 000s
EAL2	Structurally tested		
EAL3	Methodically tested and checked		
EAL4[a]	Methodically designed, tested and reviewed	6–9	Many 10s of 000s to low 100s of 000s
EAL5	Semi-formally designed and tested		
EAL6	Semi-formally verified, designed and tested		
EAL7	Formally verified design and tested	7–18+	Several 100s of 000s

[a] Note that the '+' such as in EAL4+ means that EAL4 is satisfied, plus some extra requirements from higher levels

cations is a good example where SIM lifetimes can be very short and the application evolution rapid. Historically this industry has not CC evaluated its SIMs, but rather put contractual pressure (penalty clauses) on their suppliers to make sure they carry out their own evaluations, perhaps using private test labs or in-house capability. This may have to change with the deployment of NFC phones in which SIM hosted Security Elements (SE) will have to satisfy the security requirements of multiple parties. Alternatively a SE chip can be used, that is embedded within the phone hardware, although phone innovation and replacement is very fast compared to bank cards and the CC evaluation process. There does not seem to be anything better than CC at the moment, but critics say it is too slow and cumbersome and that there should be more attention on vulnerabilities rather than compliance.

6.6 Handling Imperfection

Lack of best-practice security rarely stops the introduction of a new service or application, although it can herald its early demise. Practical information security has much to do with trying to make best use of imperfect solutions which can take the form of back-office monitoring and controls, plus old favourites at the client end such as passwords and PINs. Back office solutions are most useful when you have on-line transactions, but have also been helpful in periodic batch reporting of off-line transactions. This latter mode of usage is becoming stressed by modern client technology and back office systems may struggle to sufficiently constrain the window of opportunity to make attacks unattractive, especial if faced with widespread attack attempts. The perils of passwords and PINs are well known as are the techniques to make the best of them; however, it is worth noting that the increasing use of touch-screen devices could undermine security a little further. For example, a PIN challenge is a request for a four digit static data pattern. The thinking is that there are 10,000 combinations and so an attacker would on average have to try 5,000 to a 50:50 chance of getting the right PIN. So a best-practice countermeasure could be a block after a number of tries much less than 5000, although then you also need an unblocking mechanism as users often forget PINs anyway. It has been observed that on a cleaned touch screen device (e.g. smart phone) the PIN-entry leaves visible grease marks on the screen that can be used to reduce the PIN possibilities. For example, if the marks tell us four different values there are only 24 PIN possibilities, and so the attacker gets to the 50:50 point after 12 tries even if the electronic implementation is secure. Furthermore, some devices prompt the user to use their finger to draw a pattern on the screen as a convenient alternative to a PIN. A simple continuous pattern can probably only be drawn forwards or backwards and so just one or two attempts are required. The best-practice message is to consider the security of the entire solution and where possible avoid reliance on static data in authentication, but of course recognise that the system must still be easy to use.

6.7 Case Study the MIFARE Classic

Having suggested various aspects of a security system that might require attention to best practice, a good way to remember them is with a public case study in which several of these wise principles were seemingly ignored!

The case study is based on a contactless smart card called the MIFARE Classic that was released by Philips (now NXP) in 1994. This is now an old design and it is easy to criticise, however when it was introduced there was probably nothing more secure around that provided even comparable functionality and speed. It was never intended for high security or high value use and the fact that this was seemingly forgotten at a time when the product was quite aged is pivotal to its demise.

Recall from earlier in this chapter that a good objective of best-practice system design is to avoid becoming an attractive target and so choosing the obsolete MIFARE Classic was not a great idea. It was a target for attackers/enthusiast even when first introduced as a product, because it used 48 bit keys at a time when DES (with its 56 bit key) was coming into range of practical key-cracker equipment and when GSM authentication algorithms had been introduced with 128 bit keys. A key-cracker would need to try 256 times less keys for MIFARE Classic than for DES and 232 less keys than for today's best-practice minimum key size of 80 bits. Another reason to be a target was that the algorithm was proprietary and secret, and within a product that had not been evaluated to a recognisable standard such as common criteria. Furthermore, the product relied on that secrecy to prevent the algorithm being implemented in key-crackers. Recall that this violated Kerckhoffs principle; "An algorithm's security should not rely on the secrecy of the algorithm". This is sometime referred to (in a disparaging way) as security by obscurity which is an information security sin and a rallying cry for all kinds of attackers and researchers. There were also other trigger events including the planned use of the ageing MIFARE Classic product in new and high value national systems.

History has shown that technical secrets usually leak out. This can be due to reverse engineering or by less direct, but effective means including, bribery, leakage disgruntled employees, breaking of confidentiality, IT attack or robbery. In fact for many years it was considered that the product design had been reverse engineered as there were tales of unauthorised MIFARE Classic type products being sourced in Asia and one datasheet was dated 2004. The design was not openly published at that time as commercial gain seemed to be the motive.

The design of the proprietary algorithm (known as CRYPTO1) did not start to leak into the public domain until a meeting of the Computer Chaos Club 2007, when Nohl et al. [26] presented their work on reverse engineering. Although not all details were provided at this stage it opened the flood gates and other researchers (notably Dutch) rushed to fill in the gaps. There was some debate as to whether the initial reverse engineering was really research, bearing in mind it just reproduced information known to the NXP designers, at least 15 years earlier. Whatever label the work was given the effect was the same in that CRYPTO1 was in the public domain and so available for analysis. In common with most algorithms exposed in this way the proprietary design of CRYPTO1 started to fall apart under scrutiny.

One of the first problem areas was the random number generator. Recall that for best practice, a random number should be large to make it difficult to guess. Today a reasonable choice would be 128 bits from a quality generator. The MIFARE Classic used 32-bit values, but on closer inspection they were generated by a Linear Feedback Shift Register that had only a 16-bit state and so the maximum variation was only 216. However, worse was to come as the state was reset to a fixed value on start-up and then dependent on time of operation, which was found to be predictable.

It even proved unnecessary to get access to a key-cracker as researchers discovered structural weaknesses, which led to attack optimisations [27–29] that were far more efficient than exhaustive key search, bringing attacks into the range of amateur hackers. The first attacks extracted the keys for a victim card's identity from a legitimate reader and then subsequent attacked the actual card, reading out all the data for modification or use within clones. In the final form of exploit it was possible to attack a legitimate card directly using PC/reader equipment to extract all keys and data. Armed with this information an attacker could in principle reprogram the device much like the Issuer, or increase the value of the wallet or a recorded receipt or privilege; or create a non-standard implementation in a clone emulator, e.g. a wallet that never reduces after purchase.

With such effective and well publicised attacks, it is easy to overlook the fact that the MIFARE Classic also uses stream ciphering for encryption (with integrity only protected by CRC), which means that it is also vulnerable to the bit-flipping attacks described earlier. Furthermore as we know that the so called random number generator has predictable output, the route is open for conventional replay attacks.

There were also cases where common keys were used in certain applications, although this was really lack of best-practice from the application providers, rather than anything to do with the product itself.

6.7.1 Impact

The attacks on the MIFARE Classic were seized upon by the press and presented in a sensationalistic manner, especially in the Netherlands. The Dutch transport ministry was in the process of overseeing the roll-out of a Smartcard ticket solution (OV-Chipkaart [30]) for national travel and unfortunately the MIFARE Classic was chosen for this. There was an element of bad luck and fate about the focus on this particular country; a Dutch project, using a Dutch smartcard product that violated a critical Dutch (Kerckhoff) security principle that was enthusiastically attacked mainly by Dutch researchers. The immediate result was front page newspaper coverage over a prolonged period and threats to political/ministerial careers. Views were polarised on what the developments actually meant for the future of the OV-Chipkaart. The system owners initially claimed there was little to worry about in practice, whereas some hackers/researchers claimed that the end of the World was nigh! To try and get a balanced expert view the Dutch Government asked the ISG Smart Card Centre at Royal Holloway University of London to carry out a review. A report was eventually produced that confirmed the attacker claims of security vulnerabilities and urged the system owners to make migration plans for a move to an improved solution. The reason for this was that the system security was considered as "fragile" as it was almost totally reliant on the back-office systems that were not designed to deal with widespread technical attacks and cloning. The general recommendation was to get ready to migrate as soon as possible, so in case of rapidly growing and widespread fraud, the improved system could be introduced swiftly (but at high cost). If measured or predicted fraud was low, a slower (and more cost-efficient) pace of migration could be adopted. To date, some proof-of-concept attacks have been detected by the back office and a suspected criminal seems to have bought batches of legitimate cards to sell on with inflated wallet values [31], but generally fraud seems to have been low. It is important to remember that security is not all about technology and transport system owners have a lot of experience of managing imperfect solutions and fraud in general. Recall that before e-tickets one would only have paper or mag-stripe tickets that can be targeted with very low tech attacks.

6.8 Concluding Remarks

Anyone who is responsible for the security and fraud prevention of a new or existing system should have awareness of best-practices. There are very good sources of guidance that can, for example, advise on the choice of algorithms and key-sizes for a required system lifetime. However, it is important to remember that the implementation, processes, management and application configuration/personalisation are equally important for a secure solution. An independent expert risk assessment is always advised and should go into sufficient analysis detail. Products and components can be Common Criteria evaluated so that their security levels and capabilities are internationally recognised. If this is not deemed practical then independent lab-testing for vulnerabilities is probably the next best thing. Even if it is likely that a less than perfect solution will be implemented, a system owner should scope the entire problem as if a perfect solution was sought, so that the risk and potential impact of any compromises can be fully appreciated. This advice can of course be ignored, but when your company ends up in the newspapers, your brand is damaged, money is lost, damages are claimed and you perhaps face government enquiries, you would need a very good explanation as to why you ignored the expert best practice advice from the information security community.

Looking to the future, one should also be aware of certain disruptive technologies and trends that might challenge our current thoughts on best practices and the resources available to attackers and criminals. Firstly, the recommendations of algorithm best-practice have been traditionally linked to Moore's Law effects, i.e. the increase in computer processing with time (roughly doubling every 18 months in recent years). However, as this is usually achieved by reducing the size of the chip fabrication technology, the current trends are predicted to end around 2015–2020 due to limitations of physics and manufacturing. So it maybe that massive parallelism provides the next leap in processing power. We have already seen initiatives in shared and grid computing and are now being encouraged to use powerful "Cloud" computing resources (which is another major security challenge). There is also talk of quantum computers used against conventional security algorithms; for example solving a problem in time N instead of 2N. This is still in the very early days and if, or when, the technology for a quantum key-cracker becomes readily available then it is not unreasonable to suppose that so too would the technology for new quantum algorithms. There is much research in future chip technology, Clouds and quantum computing and so time will tell what all this means for the long-term future of information security best-practice. For the time being, following the published recommendations of NIST [8, 9] and ECRYPT [10] is a very prudent strategy.

References

1. Hacking at Random, [Online Available] http://www.wiki.har2009.org/page/Main_Page
2. Hack a Day site, [Online Available] http://hackaday.com/
3. K. Mayes, K. Markantonakis, "Smart Cards, Tokens, Security and Applications", Springer Verlag, 2007
4. Federal Information processing Standards, Data Encryption Standard (DES), FIPS publication 46–3 [Online Available] http://csrc.nist.gov/publications/fips/fips46-3/fips46-3.pdf
5. Federal Information processing Standards, Advanced Encryption Standard (AES), FIPS publication 197. [Online Available] http://csrc.nist.gov/publications/fips/fips197/fips-197.pdf
6. Rivest, R.; A. Shamir; L. Adleman (1978). "Method for Obtaining Digital Signatures and Public-Key Cryptosystems". Communications of the ACM 21 (2): 120–126.
7. Jan Petzl (2006), "Cryptanalysis with a low cost FPGA Cluster", IPAM Workshop Special Purpose Hardware for Cryptography [Online Available] http://www.copacobana.org/paper/IPAM2006_slides.pdf
8. SP 800-57 Recommendation for Key Management - Part 1: General, and Part 2: Best Practices for Key Management Organizations, NIST, March 2007
9. SP 800-131, Recommendations for the Transitioning of Cryptographic Algorithms and Key Lengths. NIST, drafted June 2010
10. ECRYPT II Yearly Report on Algorithms and key-sizes (2009–2010), Revision 1.0, ECRYPT II, 30th March 2010
11. Diffie, Whitfield; Hellman, Martin E. (June 1977). "Exhaustive Cryptanalysis of the NBS Data Encryption Standard". Computer 10 (6): 74–84
12. Ralph Merkle, Martin Hellman: On the Security of Multiple Encryption (PDF), Communications of the ACM, Vol 24, No 7, pp 465–467, July 1981
13. Paul van Oorschot, Michael J. Wiener, A known-plaintext attack on two-key triple encryption (PDF), EUROCRYPT'90, LNCS 473, 1990, pp 318–325
14. Auguste Kerckhoffs, "La cryptographie militaire", Journal des sciences militaires, vol. IX, pp. 5–83, Jan. 1883, pp. 161–191, Feb. 1883
15. P. Kocher, J. Jaffe, B. Jun, "Differential Power Analysis", technical report, 1998; later published in Advances in Cryptology - Crypto 99 Proceedings, Lecture Notes In Computer Science Vol. 1666, M. Wiener, ed., Springer-Verlag, 1999
16. Crypto++, Benchmarks, [Online Available] http://www.cryptopp.com/benchmarks-p4.html, April 2011

17. M. Bellare, P. Rogaway. Optimal Asymmetric Encryption - How to encrypt with RSA. Extended abstract in Advances in Cryptology - Eurocrypt'94 Proceedings, Lecture Notes in Computer Science Vol. 950, A. De Santis ed, Springer-Verlag, 1995
18. Birthday attack, [Online Available] http://en.wikipedia.org/wiki/Birthday_attack
19. FIPS 180–2: Secure Hash Standard (SHS) (PDF,) - Current version of the Secure Hash Standard (SHA-1, SHA-224, SHA-256, SHA-384, and SHA-512), 1 August 2002, amended 25 February 2004.
20. rfc2104, [Online Available] http://tools.ietf.org/html/rfc2104
21. NIST SP 800–90. Recommendation for Random Number Generation, March 2007.
22. EMV Books 1–4 Version 4.1 2004, [Online Available] http://www.emvco.com/specifications
23. GlobalPlatform, [Online Available] http://www.globalplatform.org.
24. NFC Forum http://www.nfc-forum.org
25. Common Criteria Portal [Online Available] http://www.commoncriteriaportal.org/
26. Nohl K, Starbug, Plotz H. MIFARE, little security, despite obscurity. Presentation on the 24th Congress of the Chaos Computer Club (CCC); December 2007
27. Courtois NT, Nohl K, O'Neil S. Algebraic attacks on the crypto-1 stream cipher in MIFARE Classic and oyster cards, vol. 166. Cryptology ePrint Archive, [Online Available] http://eprint.iacr.org/2008/166; 2008. Report.
28. Gans GK, Hoepman JH, Garcia FD. A practical attack on the MIFARE Classic. Proceedings of the 8th Smart Card Research and Advanced Application Workshop (CARDIS 2008). LNCS 5189, pp. 267–282. Heidelberg: Springer; 2008.
29. Garcia FD, Gans GK, Muijrers R, Rossum P, Verdult R, Schreur RW, et al. Dismantling MIFARE Classic. Proceedings of ESORICS 2008, LNCS 5283. Springer; 2008. pp. 97–114.
30. OV-Chipkaart System, [Online Available] http://www.ov-chipkaart.nl/
31. Dutch News Public transport smart card fraud under investigation 6th July 2011 [Online Available] http://www.dutchnews.nl/news/archives/2011/07/public_transport_smart_card_fr.php
32. M. Hilbert and P. Lopez, "The world's technological capacity to store, communicate and compute information", Science Express: Feb. 10, 2011. [Online Available] http://www.physorg.com/news/2011-02-world-scientists-total-technological-capacity.html
33. [Online Available] http://en.wikipedia.org/wiki/Block_cipher_modes_of_operation

Chapter 7
Smart Card Security

Michael Tunstall

Abstract In this chapter, a description of the various attacks and countermeasures that apply to secure smart card applications is described. This chapter focuses on the attacks that could affect cryptographic algorithms, since the security of many applications is dependent on the security of these algorithms. Nevertheless, how these attacks can be applied to other security mechanisms is also described. The aim of this chapter is to demonstrate that a careful evaluation of embedded software is required to produce a secure smart card application.

Keywords Embedded software · Fault analysis · Side channel analysis · Smart card security

7.1 Introduction

The implementation of secure applications on smart cards is different to the development on other platforms. Smart cards have limited computing power, comparatively small amounts of memory and are reliant on a smart card reader to provide power and a clock. There are security considerations that are specific to smart cards, that need to be taken into account when developing a secure smart card-based application.

In this chapter, attacks that are specific to smart cards and other devices based around a secure microprocessor will be described. There are other considerations that need to be taken into account when implementing a secure application, but these are generic and beyond the scope of this chapter.

M. Tunstall (✉)
Department of Computer Science, University of Bristol, Bristol, United Kingdom
e-mail: tunstall@cs.bris.ac.uk

K. Markantonakis and K. Mayes (eds.), *Secure Smart Embedded Devices,*
Platforms and Applications, DOI: 10.1007/978-1-4614-7915-4_7,
© Springer Science+Business Media New York 2014

There are three main types of attack that are considered in smart card security. These are:

1. **Invasive Attacks:** These are attacks that require the microprocessor in a smart card to be removed and directly attacked through a physical means. This class of attacks can, at least in theory, compromise the security of any secure microprocessor. However, these attacks typically require very expensive equipment and a large investment in time to produce results. Invasive attacks are therefore considered to be primarily in the realm of semiconductor manufacturers and students at well-funded universities. An example of such an attack would be to place probes on bus lines between blocks of a chip (a hole needs to be made in the chip's passivation layer to allow this). An attacker could then attempt to derive secret information by observing the information that is sent form one block to another.

   At its most extreme, this type of attack could make use of a focused ion beam to destroy or create tracks on the chips surface. In theory, this could, for example, be used to reconnect fuses. Traditionally, chip manufacturers typically used a test mode where it was possible to read and write to all memory addresses whilst a fuse was present. Once the fuse was blown inside the chip (before the chip left the manufacturer's factory), this mode was no longer available. In modern secure microprocessors, this test circuit is typically removed when the chip is cut from the die and this attack is no longer possible.

   Further information on invasive attacks is available in [3, 28].

2. **Semi-Invasive Attacks:** These attacks require the surface of the chip to be exposed. An attacker then seeks to compromise the security of the secure microprocessor without directly modifying the chip.

   Some examples of this type of attack include observing the electromagnetic emanations using a suitable probe [16, 40] and injecting faults using laser light [6] or white light [43]. More details on these attacks are given in later sections.

   A description of numerous semi-invasive attacks is available in [44].

3. **Non-Invasive Attacks:** These attacks seek to derive information without modifying a smart card, i.e. both the secure microprocessor and the plastic card remain unaffected. An attacker will attempt to derive information by observing information that leaks during the computation of a given command, or attempt to inject faults using mechanisms other than light.

   Some examples of this type of attack would be to observe the power consumption of a microprocessor [27], or to inject faults by putting a glitch into the power supply [3].

   Further descriptions of power analysis attacks can be found in [29], and fault attacks in [6].

This chapter will focus on semi-invasive and non-invasive attacks, as the equipment required to conduct these attacks is more readily available. Invasive attacks are of interest but are extremely expensive to conduct. This chapter will focus more on what can be achieved in a reasonably funded laboratory. However, some information is given on invasive attacks where relevant.

Organisation

Section 7.2 contains a description of the cryptographic algorithms that will be used in later sections to give examples of attacks. Section 7.3 describes certain hardware security features that are typically included in a smart card. Section 7.4 describes the different forms of side channel analysis and how they can be applied to smart card implementations of cryptographic algorithms. Section 7.5 describes how fault attacks can be applied to smart cards. Section 7.6 describes how the techniques given in Sects. 7.4 and 7.5 can be applied to other security mechanisms. Section 7.7 summarises the chapter.

Notation

The base of a value is determined by a trailing subscript, which is applied to the whole word preceding the subscript. For example, FE_{16} is 254 expressed in base 16 and $d = (d_{\ell-1}, d_{\ell-2}, \ldots, d_0)_2$ gives a binary expression for d.

In all the algorithms described in this chapter, ϕ represents Euler's totient function, where $\phi(N)$ equals the number of positive integers less than N which are coprime to N. In particular, if $N = p \cdot q$ is an RSA modulus then $\phi(N) = (p - 1)(q - 1)$.

7.2 Cryptographic Algorithms

Some of the attacks detailed in later sections will assume a detailed knowledge of some of the commonly used cryptographic algorithms. Specifically, the Data Encryption Standard (DES) and RSA are detailed in this section to provide a reference, and to describe the notation that will be used.

7.2.1 Data Encryption Standard

The DES was introduced by NIST in the mid 1970s [37], and was the first openly available cryptography standard. It has since become a worldwide *de facto* standard for numerous purposes. It is only in recent years that it has been practically demonstrated that an exhaustive search of the keyspace is possible, leading to the introduction of triple DES and the development of the Advanced Encryption Standard (AES) [38].

DES can be considered as a transformation of two 32-bit variables (L_0, R_0), i.e. the message block, through 16 iterations of a round function, as shown in Fig. 7.1, to produce a ciphertext block (L_{16}, R_{16}). The Expansion permutation selects eight overlapping six-bit substrings from R_n. The P-permutation is a bitwise permutation

Fig. 7.1 The DES round
function for *round n*

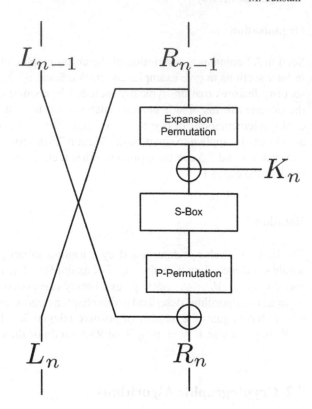

on the 32-bit output of the S-box function. For clarity of expression, these permutations will not always be considered and the round function will be written as:

$$R_n = S(R_{n-1} \oplus K_n) \oplus L_{n-1}$$
$$L_n = R_{n-1}$$

(7.1)

where S is the S-box function. The subkeys K_n, for $1 \leq n \leq 16$, are each 48 bits generated from the 56-bit secret key, by choosing 48 bits from the 56-bit key. This is done by initially conducting a bitwise permutation on the key, referred to as Permuted Choice 1 (PC1). Each round bit shifts are conducted on the key, and 48 bits are chosen from this shifted key to form each subkey using the Permuted Choice 2 (PC2) function.

Eight different S-boxes are applied in each round to sets of six bits, thereby reducing the 48-bit output of the XOR with K_n to 32-bits. Each S-box is a substitution table that is addressed using six bits of information, and each entry is a 4-bit number.

The algorithm also includes an initial and final permutation (these permutations are referred to as IP and IP^{-1}, respectively), where the final permutation is the inverse of the initial permutation. More precisely the permutation at the end of DES is conducted on (R_{16}, L_{16}) rather than (L_{16}, R_{16}). These permutations will be ignored in this chapter, as they do not contribute to the security of the algorithm. The permutations IP and IP^{-1} were included since this was the most convenient way of introducing the bits into the chip use to calculate the DES algorithm (at the time software implementations were not considered because of the complexity of the algorithm) [33].

7.2.1.1 Triple DES

In order to mitigate the key length problem, a modification to DES was proposed to make an exhaustive key search prohibitively complex. Triple DES is a construction that uses two different DES keys and is defined in [37]. In the algorithm below, these are labelled K_1 and K_2, and in order to generate a ciphertext block C from a plaintext block M the following calculation is performed:

$$C = \text{DES}\big(\text{DES}^{-1}\big(\text{DES}(M, K_1), K_2\big), K_1\big) \qquad (7.2)$$

where $\text{DES}(M, K)$ denotes the output of the DES encryption algorithm applied to message block M with key K. Deciphering the ciphertext block C uses the function,

$$M = \text{DES}^{-1}\big(\text{DES}\big(\text{DES}^{-1}(C, K_1), K_2\big), K_1\big) \qquad (7.3)$$

The structure of triple DES allows for backward compatibility as if K_1 and K_2 are equal the resulting ciphertext will the equivalent to that produced with a single DES. The triple DES requires that three instantiations of the DES algorithm are used, since it has been shown that two instantiations of DES only increase the security of the algorithm by one bit (see meet-in-the-middle attacks [31]).

Another version of triple DES is proposed in [37], in which three different keys are used rather than two.

7.2.2 RSA

RSA was first published in 1978 [41], and was the first published example of a public key encryption algorithm. The security of RSA depends on the difficulty of factorising large numbers. This means that RSA keys need to be quite large, because of advances in factorisation algorithms and the constantly increasing processing power available in modern computers.

To generate a key pair for use with RSA, two prime numbers, p and q, typically of equal bit length, are generated; they are then multiplied together to create a value N, the modulus, whose bit length is equal to that desired for the cryptosystem.

That is, in order to create a 1,024-bit modulus, $2^{511.5} < p, q < 2^{5121}$ (if values of p or q are chosen from $\{2^{511} + 1, \ldots, 2^{511.5}\}$ the product of p and q is not guaranteed to have a bit length of 1,024-bits). A public exponent, e, is chosen that is coprime to both $(p - 1)$ and $(q - 1)$.

A private exponent, d, is generated from the parameters previously calculated, using the formula:

$$
\begin{aligned}
e \cdot d &\equiv 1 \quad (\text{mod } (p - 1)(q - 1)), \text{ or equivalently} \\
e \cdot d &\equiv 1 \quad (\text{mod } \phi(N))
\end{aligned}
\tag{7.4}
$$

where ϕ is Euler's Totient function.

7.2.2.1 The RSA Cryptosystem

In the RSA cryptosystem, to encrypt a message, M, and create ciphertext, C, one calculates:

$$C = M^e \bmod N \tag{7.5}$$

The value of e is often chosen as 3 or $2^{16} + 1$, as these value are small, relative to N, and have a low Hamming weight, which means that the encryption process is fast (see below). To decrypt the ciphertext, the same calculation is carried out but using the private exponent, d, which generally has the same bit length as N:

$$M = C^d \bmod N \tag{7.6}$$

7.2.2.2 The RSA Signature Scheme

The RSA digital signature scheme involves the reverse the operations used in the RSA cryptosystem. The generation of a signature, S, uses the private exponent d. By convention, this is expressed as:

$$S = M^d \bmod N \tag{7.7}$$

The verification therefore uses the public exponent and is expressed as:

$$M = S^e \bmod N \tag{7.8}$$

---

[1] This is possibly overly strong, as it typically recommend that the bit lengths of p and q are approximately equal. However, it will provide the most security for a modulus of a given bit length assuming that $p - q$ is sufficiently large to prevent an attacker from guessing their values by calculating $\sqrt{N}$.

7.2.2.3 Padding Schemes

Applying the RSA primitive to a message, as described above, will not yield a secure signature or encryption scheme (for reasons beyond the scope of this chapter). To achieve a secure scheme, it is necessary to apply the RSA operation to a transformed version of the message, e.g., as can be achieved by hashing the message, adding padding, and/or masking the result. This process is termed padding, and the interested reader is referred to [31] for a treatment of padding schemes.

Some of the attacks presented in this chapter will not be realistic when a padding scheme is used, since padding schemes mean that an attacker cannot entirely control a message. However, it is important that an implementation of RSA is secure against all possible attacks. If a given implementation does not use padding or, more realistically, contains a bug that allows an attacker to remove the padding function, the implementation should still be able to resist all the attacks described in this chapter.

7.2.2.4 Computing a Modular Exponentiation

Many different algorithms can be used to calculate the modular exponentiation algorithm required for RSA. In practice, a large number of algorithms cannot be implemented on smart cards, as the amount of available memory does not usually allow numerous intermediate values to be stored in RAM. The manipulation of large numbers is usually performed using a coprocessor (see Sect. 7.3), as implementing a multiplication on an 8-bit platform would not give a desirable performance level.

The simplest exponentiation algorithm is the square and multiply algorithm [31], and is given below for an exponent d of bit length ℓ:

The Square and Multiply Algorithm
Input:
$M, d = (d_{\ell-1}, d_{\ell-2}, \ldots, d_0)_2, N$
Output:
$S = M^d \bmod N\{$
    $A := 1$
    for $i = \ell - 1$ **to** 0 {
        $A := A^2 \bmod N$
        if $(d_i = 1)$ {
            $A := A \cdot M \bmod N$
        }
    }
    return A
$}$

The square and multiply algorithm calculates $M^d \bmod N$ by loading the value one into the accumulator A and d is read bit-by-bit. For each bit a squaring operation modulo N takes place on A, and when a bit is equal to one A is subsequently multiplied by M. It is because of this multiplication that e is typically chosen to as 3 or $2^{16} + 1$, as both values only have two bits set to one; therefore minimising the number of multiplications required. It is not possible to only have one bit set to one as it is necessary for e to be an odd number in order for it to have an inverse modulo N. The most significant bit of a number will always be set to one, and the least significant bit will need to be set to one to produce an odd number.

7.2.2.5 Using the Chinese Remainder Theorem

The RSA calculation using the private exponent (i.e. where $S = M^d \bmod N$ and $N = p \cdot q$) can be performed using the Chinese remainder theorem (CRT) [25]. Initially, the following values are calculated,

$$
\begin{aligned}
S_p &= (M \bmod p)^{(d \bmod (p-1))} \bmod p \\
S_q &= (M \bmod q)^{(d \bmod (q-1))} \bmod q
\end{aligned}
\tag{7.9}
$$

which can be combined to form the RSA signature S using the formula $S = aS_p + bS_q \bmod N$, where:

$$
\begin{aligned}
a &\equiv 1 \quad (\bmod\ p) \\
a &\equiv 0 \quad (\bmod\ q)
\end{aligned}
\quad \text{and} \quad
\begin{aligned}
b &\equiv 0 \quad (\bmod\ p) \\
b &\equiv 1 \quad (\bmod\ q) \ .
\end{aligned}
$$

This can be implemented in the following manner:

$$
S = S_q + \left(\left(S_p - S_q \right) q^{-1} \bmod p \right) \cdot q
\tag{7.10}
$$

This provides a method of calculating an RSA signature that is approximately four times quicker than a generic modular exponentiation algorithm, i.e. two exponentiations, each of which can be computed eight times faster than an exponentiation using d (the bit length of $d \bmod (p-1)$ and $d \bmod (p-1)$ will be half that of d). This advantage is offset by an increase in the key information that needs to be stored. Rather than storing just the value of d and N, the values of $(p, q, d \bmod (p-1), d \bmod (q-1), q^{-1} \bmod p)$ need to be precalculated and stored.

7.3 Smart Card Security Features

This section will detail some of the features of smart cards that are pertinent when considering their security. Smart cards have traditionally been based on 8-bit complex instruction set computer (CISC) architectures [35]. Usually built around a Motorola 6805 or Intel 8051 core, often with extensions to the instruction set. More sophisticated smart cards are emerging based on 32-bit reduced instruction

set computer (RISC) architecture chips, containing dedicated peripherals (cryptographic coprocessors, memory managers, large m memories, ...) [34].

7.3.1 Communication

A smart card has five contacts that it uses to communicate with the outside world
defined in the ISO/IEC 7816-2 standard [22]. Two of these are taken by the power
supply (usually 3 or 5 V), referred to as Vcc, and the ground used to power the
chip. Another contact is used to supply a clock, which is allowed to vary between 1
and 5 MHz but is typically set to 3.57 MHz. The remaining two contacts are used to
communicate with the microprocessor. A sixth contact was originally used to provide
a higher voltage to program the EEPROM (referred to as Vpp), but is no longer in
use for reasons described in Sect. 7.6. The location of the different contacts is shown
in Fig. 7.2.

One of the contacts, called the I/O, is used for communication and to send commands to the chip in a smart card. The protocols used to communicate with a smart
card are referred to as $T = 0$ and $T = 1$ and are defined in the ISO/IEC 7816-3
standard [21]. This section will describe both protocols, as they are nearly identical.
The extra requirements of $T = 1$ are detailed where relevant.

The remaining contact is used to reset the smart card (there are a further two
contacts defined in the ISO/IEC 7816-3 standard but they are not currently used).
This is a physical event (i.e. moving the voltage applied to this contact from 0 to 1)
that will always provoke a response from the smart card. A user can apply the reset
at any time. The smart card will respond by sending an answer to reset (ATR) to the
I/O contact, which is a string of bytes that defines the protocols the smart card can
use, the speeds at which the smart card can communicate and the order in which bits
are going to be sent during the session (i.e. most or least significant bit first).

Fig. 7.2 The contacts used to
power and communicate with
a smart card

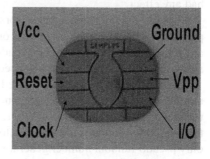

7.3.2 Cryptographic Coprocessors

Traditionally, smart cards have been based around 8-bit architectures. In order to manipulate large numbers, e.g., to calculate the RSA algorithm, dedicated coprocessors can be appended to the CPU. In more modern 32-bit chips [34] this is no longer necessary, as efficient software implementations can be achieved. DES is also often implemented in a coprocessor to help increase performance, and AES implementations should be available in the near future. These coprocessors can increase the smart card's performance, as hardware implementations of secret key algorithms can be expected to require one or two clock cycles per round of the block cipher. However, the inclusion of coprocessors also increases the size of the chip and the overall power consumption. This means that chips with coprocessors are usually more expensive and are not ideal in environments where the amount of available current is limited.

7.3.3 Random Number Generators

Random number generators are usually included in smart cards, as unpredictable numbers are an important element in many secure protocols. A true random number generator is typically based on a signal generated by an analogue device which is then treated to remove any bias that may exist, or has been induced, in the bits generated. The correct functioning of all aspects of a smart card chip under varied environmental conditions is important, but is critical for random number generation because the quality of the generated random values can have a profound effect on cryptographic schemes. Random number generators are therefore designed to function correctly in a large range of environmental conditions, including temperature, supply voltage, and so on. However, if an attacker succeeds in modifying the environmental conditions such that the physical source of randomness is affected, the subsequent treatment is included so that an attacker will not be able to determine if the change in conditions had any effect.

Pseudo-random number generators are also often included in a secure microprocessor. These are typically based on linear feedback shift registers (LFSRs) that are able to generate a new pseudo-random value every clock cycle, but are deterministic over time and are not usually used for critical security functions.

Where random values are required in cryptographic algorithms, a true random number generator is used when the quality of the random value is important, e.g. for use in a cryptographic protocol. Where the quality of the random value is less important, a pseudo-random number generator can be used. In some secure microprocessors only pseudo-random number generators are available. In this case, mechanisms that combine a random seed (that can be inserted into the chip during manufacture) with pseudo-random values can be used to provide random values.

An example of this latter type of random number generator is given in the ANSI X9.17 [2, 19] standard, that uses DES to provide random values based on a random seed generated during the initialisation of a given device and another source of pseudo-random information. This functions by taking a 64-bit pseudo-random input (X), a 64-bit random seed (S) and a DES key (K). X is usually generated by calculating $X = \text{DES}(D, K)$, where D is a the date and/or time, but this information is not available to a smart card and is therefore replaced with values provided by a pseudo-random number generator. To output a random value R, the following calculation takes place:

$$R = \text{DES}(X \oplus S, K), \qquad (7.11)$$

and the random seed is updated using:

$$S = \text{DES}(R \oplus X, K) . \qquad (7.12)$$

For increased security, the DES function can be replaced with triple DES, as the key length used by DES has proven to be too short to entirely resist an exhaustive key search.

7.3.4 Anomaly Sensors

There are usually a number of different types of anomaly detectors present in smart cards. These are used to detect unusual events in the voltage and clock supplied to the card, and the environmental conditions (e.g. the temperature). These enable a smart card to detect when it is exposed to conditions that are outside the parameters within which it is known to function correctly. When unusual conditions are detected, the chip will cease to function until the effect has been removed (i.e. initiate a reset or execute an infinite loop when the sensor is activated). However, it is considered prudent not to rely solely on these sensors and to implement further countermeasures (see Sect. 7.5).

7.3.5 Chip Features

The surface of the chip used in a smart card can be uncovered by removing the plastic body of the card and using fuming nitric acid to remove the resin used to protect the microprocessor. Once the chip has been revealed the easiest form of analysis is to simply look at it under a microscope. The various different blocks can often be identified, as shown in Fig. 7.3.

Reverse engineering can target the internal design to understand how a given chip or block functions. An attacker can use such information to improve their knowledge of chip design and find potential weaknesses in the chip, which may allow them to compromise the chip's integrity.

Fig. 7.3 A chip surface with
readily identifiable features

In modern smart cards, various features used to inhibit reverse engineering are
implemented using glue logic: important blocks are laid out in a randomised fashion
that makes reverse engineering difficult. This technique increases the size of the
block, and is therefore not used in the design of large blocks such as ROM and
EEPROM.

Another common technique to prevent this sort of identification and targeting is
to overlay the chip with another metal layer that prevents the chip's features being
identified. This can be removed using hydrofluoric acid that eats through the metal
layer; this reaction is then stopped using acetone before further damage is done and
the chip surface can be analysed. The chip becomes non-functional but the layout of
the chip can be determined, so that other chips of the same family can be attacked.
The result of such a process is shown in Fig. 7.4.

Discovering the layout and functioning of a chip is particularly important when
using a laser as a fault injection mechanism (see Sect. 7.5). Different areas of a chip
can be targeted through the metal layer once the layout of a chip is known.

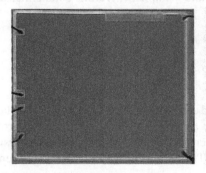

Fig. 7.4 A chip with a shield present and removed

7.4 Side Channel Analysis

Side channel attacks are a class of attacks where an attacker will attempt to deduce what is occurring inside a device by observing information that leaks during the normal functioning of the device. If this information can be related to any secret information being manipulated the security of the device can be compromised. It should be noted that side channel analysis is a passive form of attack, i.e. an attacker will simply observe what is occurring when a device is functioning. In the case of smart cards, the message being manipulated can be controlled, but this is not necessary to construct a valid side channel attack.

The first publication that mentions a side channel attack is described in [45]. In 1956, MI5 mounted an operation to decipher communications between Egyptian embassies. The communications were enciphered using Hagelin machines [24]. These machines did not function by using a key value as described in Sect. 7.2. Enciphering occurred by routing electronic signals from a keyboard through seven rotating wheels to generate a ciphertext. The "key" was the initial setting of these seven wheels. The machine was reset every morning by the clerk who would be sending messages. MI5 managed to plant a microphone in close proximity to one of these machines. This allowed the initial settings to be determined by listening to the initial settings being made every morning. This would have allowed them to decipher intercepted communications with another Hagelin machine set to the same key. In practice, MI5 was only able to determine a certain amount of wheel settings because of the difficulty of distinguishing the noise of the wheels being set from background noise. This made deciphering more complex, but not impossible, as the number of possible keys was significantly reduced by the partial information.

7.4.1 Timing Analysis

The first modern example of a side channel attack was proposed in [26]. This involved observing the differences in the amount of time required to calculate a RSA signature for different messages to derive the secret key. This attack was conducted against a PC implementation but a similar analysis could potentially be applied to smart card implementations. It would be expected to be more efficient against a smart card as more precise timings can be achieved with an oscilloscope or proprietary readers. An example of a trace acquired with an oscilloscope that would provide this sort of information is shown in Fig. 7.5. The I/O events on the left hand side of the figure represent the reader sending a command to the smart card. The I/O events on the right hand side of the figure show the response of the smart card. The time taken by a given command can be determined by observing the amount of time that passes between these two sets of events.

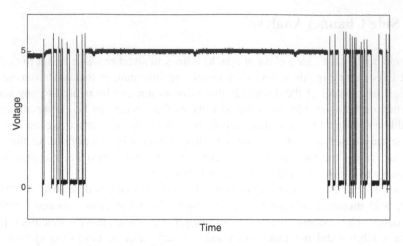

Fig. 7.5 The I/O of a smart card command

7.4.2 Power Analysis

The most common form of side channel attack, when considering smart cards, is the analysis of the instantaneous power consumption [27]. This is measured by placing a resistor in series with a smart card and the power supply (or ground), and measuring the potential difference across the resistor with an oscilloscope. The acquired information can be analysed *a posteriori* to attempt to determine information on what is occurring within a secure microprocessor. There are two main types of power attack; these are simple power analysis (SPA) and differential power analysis (DPA).

7.4.2.1 Simple Power Analysis

A powerful form of power analysis is to search for patterns within the acquired power consumption. An attacker can attempt to determine the location of individual functions within a command. For example, Fig. 7.6 shows the power consumption of a smart card during the execution of DES. A pattern can be seen that repeats 16 times, corresponding to the 16 rounds that are required during the computation of DES.

This analysis can be further extended by closely inspecting the power consumption during the computation of one round, to attempt to determine the individual functions within each round. This is shown in Fig. 7.7 where the functions in the second round of a DES implementation are evident from the power consumption trace. This may not be immediately apparent, as close inspection of the trace's features is necessary to identify the individual functions. For example, if an attacker is seeking to determine where in a command the compression permutation

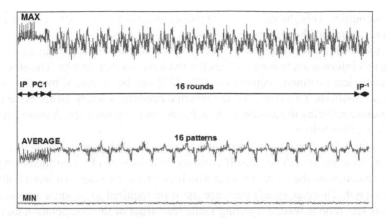

Fig. 7.6 The power consumption of a DES implementation showing the rounds of the algorithm

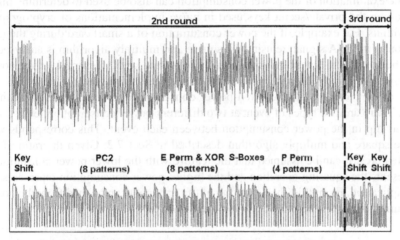

Key: Key Shift A bitwise shift applied to the key each round.
PC2 PC2 used to generate a 48-bit round key each round.
E Perm Expansion permutation applied to R_i,
for $1 \le i \le 16$, to produce a 48-bit output.
XOR The XOR with the round key.
S-boxes Eight substitution tables reducing 48 bits to 32 bits.
P Perm The P permutation, a bitwise transformation.

Fig. 7.7 The power consumption of a DES implementation showing the round functions

(PC2) is computed. An attacker will look for eight patterns of six events. This is because the compression permutation selects 48-bits from the 56-bit DES key, where the 48-bit result is divided into segments of 6-bits (for use in the S-box function). The natural method of implementing this permutation will, therefore, be to construct a loop that will repeat eight times. Each loop will move six bits from the DES key to the 48-bit output. This should therefore produce eight patterns of six events because of the individual bits being selected and written.

The use of this information is not necessarily immediately apparent; an attacker can use this information to improve the effectiveness of other attacks. The efficiency of the statistical treatment required for DPA [27] can be increased by taking more precise acquisitions. This is because the area that needs to be analysed can be defined, and therefore reducing the amount of data that needs to be acquired. A more detailed analysis is given below.

The same is true for fault injection techniques, detailed in Sect. 7.5, as it is often necessary to target specific events. If arbitrary functions can be identified using the power consumption, the point in time at which an attacker wishes to inject a fault can be discovered. This can greatly decrease the time required to conduct a successful attack, as less time is wasted injecting faults into areas of the computation that will not produce the desired result.

The examination of the power consumption can also be used to determine information on the private/secret keys used in naïve implementations of cryptographic algorithms. For example, if the power consumption of a smart card during the generation of an RSA signature using the square and multiply algorithm is analysed, it may be possible to determine some bits of the private key. An example of the power consumption during the generation of an RSA signature is shown in Fig. 7.8.

Looking closely at the acquired power consumption, a series of events can be seen. There are two types of event at two different power consumption levels, with a short dip in the power consumption between each event. This corresponds well to the square and multiply algorithm described in Sect. 7.2. Given the ratio of the two features, it can be assumed that the feature with the lower power consumption represents the squaring operation and the higher power consumption represents the multiplication. From this, the beginning of the exponent can be read from the power consumption, in this case the exponent used is $F00F000FF00_{16}$.

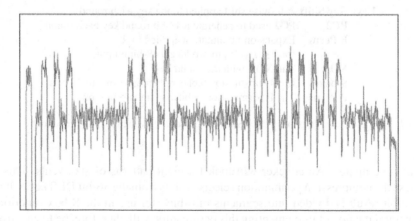

Fig. 7.8 The power consumption of an RSA implemented using the square and multiply algorithm

It should be noted that all the examples given in this section have been taken from chips that display the differences in an obvious manner. Modern secure microprocessors rarely display the functions being executed as clearly as in the examples given.

7.4.2.2 Differential Power Analysis

The idea of statistically treating power analysis traces was first presented to the cryptographic community in [27], and is referred to as DPA. DPA is based on the relationship between the power consumption and the Hamming weight of data being manipulated at a given point in time. The differences in power consumption are potentially extremely small, and cannot be interpreted individually, as the information will be lost in the noise incurred during the acquisition process. The small differences produced can be seen in Fig. 7.9, where traces were taken using a chip where the acquisition noise is exceptionally low.

DPA can be performed on any algorithm in which an intermediate operation of the form $\beta = S(\alpha \oplus K)$ is calculated, where α is known and K is the key (or some segment of the key). The function S is typically a non-linear function, usually a substitution table (referred to as an S-box), which produces an intermediate output value β.

The process of performing the attack initially involves running a microprocessor N times with N distinct message values M_i, where $1 \leq i \leq N$. The encryption of the message M_i under the key K to produce the corresponding ciphertext C_i will result in power consumption waveforms w_i, for $1 \leq i \leq N$. These waveforms are captured with an oscilloscope, and sent to a computer for analysis and processing.

To find K, one bit of β is chosen which we will refer to as b. For a given hypothesis for K this bit will classify whether each trace w_i is a member of one of two possible sets. The first set S_0 will contain all the traces where b is equal to zero, and

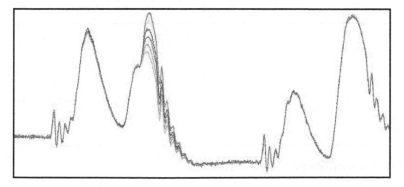

Fig. 7.9 Overlaid acquisitions of the power consumption produced by the same instruction but with varying data

the second set S_1 will contain all the remaining traces, i.e. where the output bit b is equal to one.

A differential trace Δ_n is calculated by finding the average of each set and then subtracting the resulting values from each other, where all operations on waveforms are conducted in a pointwise fashion, i.e. this calculation is conducted on the first point of each acquisition to produce the first point of the differential trace, the second point of each acquisition to produce the second point of the differential trace, etc.

$$\Delta_n = \frac{\sum_{w_i \in S_0} w_i}{|S_0|} - \frac{\sum_{w_i \in S_1} w_i}{|S_1|}$$

A differential trace is produced for each value that K can take. In DES, the first subkey will be treated in groups of six bits, so 64 (i.e. 2^6) differential traces will be generated to test all the combinations of six bits. The differential trace with the highest peak will validate a hypothesis for K, i.e. $K = n$ corresponds to the Δ_n featuring a maximum amplitude. An example of a differential trace produced by predicting one bit of the output a DES S-box, with a correct key guess, is shown in Fig. 7.10.

The differential trace in Fig. 7.10 shows a large difference in the power consumption at five different points, which are referred to as DPA peaks. The first peak corresponds to the output of the S-box, i.e. where the output of the S-box function is determined and written to memory. The four subsequent peaks correspond to the same bit being manipulated in the P-permutation. This occurs because the output of each S-box consists of four bits, and the memory containing those four bits will be accessed each time one of those bits is required in the output of the P-permutation.

A more complete version of this attack uses Pearson's correlation coefficient to demonstrate the correlation between the Hamming weight and the instantaneous power consumption. This can be used to validate key hypotheses in an identical manner to DPA. Details of this method are beyond the scope of this chapter, but the interested reader is referred to [11].

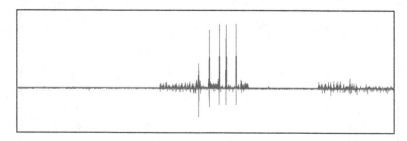

Fig. 7.10 A differential trace

Fig. 7.11 Electromagnetic
probing of a chip

7.4.3 Electromagnetic Analysis

An alternative side channel to measuring the power consumption of a smart card is to measure the instantaneous electromagnetic emanations as a cryptographic algorithm is being computed [16, 40]. This is typically implemented using a small probe, an example of which can be seen in Fig. 7.11. Such probes can measure the electromagnetic emanations for different blocks of a chip, as such probes are an equivalent size to the chip's features. This means that the probe can be placed just above a given feature to try and get a strong signal from that part of the chip, e.g. the bus between two areas of the chip, while excluding noise from other areas of the chip.

Unfortunately, it is a much more complicated attack to realise, as the chip needs to be open, as shown in Fig. 7.11. If the chip surface is not exposed the signal is not usually strong enough for any information to be deduced. The tools required to capture this information are also more complex, as the power consumption can be measured by simply reading the potential difference over a resistor in series with a smart card. To measure the electromagnetic field involves building a suitable probe (the probe in Fig. 7.11 is handmade) and the use of amplifiers, so that an oscilloscope can detect the signal.

The signals that are acquired using this method are also different to those acquired by reading the instantaneous power consumption. The signals acquired during two executions of a selected command by an 8-bit microprocessor is shown in Fig. 7.12. The black traces show the acquired power and electromagnetic signals when the chip manipulates FF_{16}, and the grey traces shows the same command where the microprocessor is manipulating 00_{16}. The difference in the black and grey traces representing the power consumption can be seen as a increase in the power consumption for short periods. The difference in the traces representing the electromagnetic emanations is caused by sudden changes in the electromagnetic field, shown by spikes in the signal at the same moment in time the difference in the power consumption can be observed.

The traces acquired from measuring the instantaneous electromagnetic emanations can be treated in exactly the same way as power consumption acquisitions [16, 40]. The acquisitions can be analysed individually, referred to as simple electromagnetic analysis (SEMA), or treated statistically, referred to as differential electromagnetic analysis (DEMA).

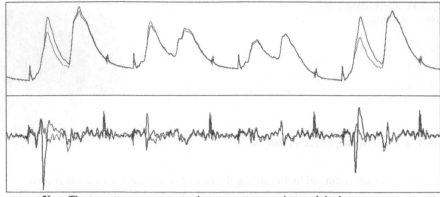

Key: The upper traces represents the power consumption, and the lower traces
represent the electromagnetic emanations during the same command. The
black traces were taken were FF_{16} is being manipulated, and the grey traces
where 00_{16} is being manipulated.

Fig. 7.12 Power and electromagnetic measurements

7.4.4 Countermeasures

There are several countermeasures for protecting cryptographic algorithms against
side channel attacks. Some countermeasures can either be implemented in hardware
or software; only software implementations are considered here for simplicity. These
countermeasures are listed below:

Constant Execution can be used to fix the time taken by an algorithm, so that no
deductions on secret information can be made though timing analysis or SPA. This
extends to individual processes being executed by a smart card. If a process takes
different lengths of time depending on some secret information and the difference in
time is made up by a dummy function, there is a good chance that this will be visible
in the power consumption or electromagnetic emanations. It is therefore important
that an algorithm is written so that the same code is executed for all the possible
input values.

Random Delays can be inserted at different points in the algorithm being executed,
i.e. a dummy function that takes a random amount of time to execute can be called.
The algorithm can no longer be said to comply with the constant execution criteria
given above, but any variation is completely independent of any secret information.
This does not prevent any attacks, but creates an extra step for an attacker. In order to
conduct any power analysis an attacker needs to synchronise the power consumption
acquisitions *a posteriori*. The problem of attempting to conduct side channel attacks
in the presence of random delays is described in [14].

Randomisation (or data whitening) is where the data is manipulated in such a way
that the value present in memory is always masked with a random value. This randomisation remains constant for one execution, but will vary from one acquisition to

another. This mask is then removed at the end of the algorithm to produce the cipher-text. Some ideas for implementing this countermeasure were proposed in [12], and an example of this sort of implementation applied to block ciphers can be found in [1].

The size of the random value used in block ciphers is generally limited as S-boxes need to be randomised before the execution of the block cipher. This is generally achieved by creating an alternative S-box in memory for each execution of the cryptographic algorithm using the algorithm given below:

Randomising S-box Values
Input:
$S = (s_0, s_1, s_2, \ldots, s_n)_x$ containing the S-box, $\mathbf{R}$ a random $\in \{0, 1, \ldots, n\}$,
and r a random $\in \{0, 1, \ldots, x - 1\}$
Output:
$RS = (rs_0, rs_1, rs_2, \ldots, rs_n)_x$
{
    for $i = 0$ **to** n {
        $rs_i := s_{(i \oplus \mathbf{R})} \oplus r$
    }
    return RS
}

The random value used for masking the input data can be no larger than n, and the random used for the output value can be no larger that x. In an implementation of DES $\mathbf{R} \in \{0, 1, \ldots, 63\}$ and $r \in \{0, 1, \ldots, 15\}$. The rest of the algorithm needs to be a carefully designed to produce values masked with R, and to be able to manipulate returned values masked with r.

This is not possible in the case of RSA, where the calculation methods do not facilitate the method described above. A method for randomising the calculation of an RSA signature is given in [23], where the signature generation can be calculated using the formula:

$$S = \left((M + r_1 \cdot N)^{d + r_2 \cdot \phi(N)} \bmod (r_3 \cdot N) \right) \bmod N \qquad (7.13)$$

where ϕ is Euler's totient function and, r_1, r_2 and r_3 are small random values. The effect of the each of the small random values does not change the outcome, but the order of the squaring operations and multiplications required to compute S is randomised. This does not provide a totally secure algorithm as the modular exponentiation itself also has to be secured against SPA attacks. A discussion of these algorithms is given in [13].

Randomised Execution is the manipulation of data in a random order so that an attacker does not know what is being manipulated at a given moment in time. If, for example, n bytes are being XORed with n key bytes then it is prudent to do it in a random order. If an attacker wishes to determine which byte has been XORed at any particular time this will be infeasible given that the order that bytes are being manipulated is unknown.

This also inhibits any statistical analysis of a side channel (i.e. using DPA), as this relies on the same unknown variable being treated at the same point in time. As an attacker cannot know the order in which the data has been treated, this provides an extremely efficient countermeasure when combined with randomisation. An discussion of this technique applied to DES is described in [32].

7.4.4.1 Remarks

The above list gives the countermeasures that would need to be applied to a cryptographic algorithm to render it secure against side channel analysis. An attacker would, therefore, have to overcome the combination of all these countermeasures. For an extensive treatment of side channel analysis, the interested reader is referred to [29].

7.5 Fault Analysis

The problem of faults occurring in microprocessors has existed for a relatively long time. One of the initial observations of faults being provoked in microprocessors was accidental. It was observed that radioactive particles produced by elements naturally present in packaging material caused faults in chips [30]. Specifically, these faults were caused by Uranium-235, Uranium-238 and Thorium-230 residues present in the packaging decaying to Lead-206 and releasing α particles. These particles were energetic enough to cause bits stored in RAM to change.

Further research involved the analysis of the effect of cosmic rays on semiconductors [46]. While cosmic rays are very weak at ground level, their effect in the upper atmosphere and outer space is important for the aero-spacial industry. This provoked research into integrity measures that need to be included in semiconductors utilised in the upper atmosphere and space.

In 1997, it was pointed out that a fault present in the generation of an RSA signature, computed using the CRT, could reveal information on the private key [10] (this attack is detailed below). This led to further research into the effect of faults on the security of implementations of cryptographic algorithms in secure microprocessors, and the possible mechanisms that could be used to inject faults in a microprocessor.

7.5.1 Fault Injection Mechanisms

There are a variety of different mechanisms that can be used to inject faults in microprocessors. These are listed here:

Variations in Supply Voltage [3, 9] during execution may cause a processor to misinterpret or skip instructions.

Variations in the External Clock [3, 4, 28] may cause data to be misread (the circuit tries to read a value from the data bus before the memory has time to latch out the correct value) or an instruction miss (the circuit starts executing instruction $n + 1$ before the microprocessor has finished executing instruction n).

Extremes of Temperature [10, 18] may cause unpredictable effects in microprocessors. When conducting temperature attacks on smart cards, two effects can be obtained [6]: the random modification of RAM cells due to heating, and the exploitation of the fact that read and write temperature thresholds do not coincide in most non-volatile memories (NVMs). By tuning the chip's temperature to a value where write operations work but read operations do not, or the other way around, a number of attacks can be mounted.

Laser Light [15, 20, 39] can be used to simulate the effect of cosmic rays in microprocessors. Laser light is used to test semiconductors that are destined to be used in the upper atmosphere or space. The effect produced in semiconductors is based on the photoelectric effect, where light arriving on a metal surface will induce a current. If the light is intense, as in laser light, this may be enough to induce a fault in a circuit.

White Light [3] has been proposed as an alternative to laser light to induce faults in microprocessors. This can be used as a relatively inexpensive means of fault induction [43]. However, white light is not directional and cannot easily be used to illuminate small portions of a microprocessor.

Electromagnetic flux [42] has also been shown to be able to change values in RAM, as eddy currents can be made strong enough to affect microprocessors. However, this effect has only been observed in insecure microprocessors.

7.5.2 Modelling the Effect of a Fault

The fault injection methods described above may have many different effects on silicon. They can be modelled in ways that depend on the type of fault injection that has been used. The following list indicates the possible effects that can be created by these methods:

Resetting Data: an attacker could force the data to the blank state, i.e. reset a given byte, or bytes, of data back to 00 or FF_{16}, depending on the logical representation.

Data Randomisation: an attacker could change the data to a random value. However, the adversary does not control the random value, and the new value of the data is unknown to the adversary.

Modifying Opcodes: an attacker could change the instructions executed by the chip's CPU, as described in [3]. This will often have the same effect as the previous two types of attack. Additional effects could include removal of functions or the breaking of loops. The previous two models are algorithm dependent, whereas the changing of opcodes is implementation dependent.

These three types of attack cover everything that an attacker could hope to do to an implementation of an algorithm. It is not usually possible for an attacker to create all of these possible faults in any particular implementation. Nevertheless, it is important that algorithms are able to tolerate all types of fault, as the fault injection methods that may be realisable on a given platform are unpredictable. While an attacker might only ever have a subset of the above effects available, if that effect is not taken into account then it may have catastrophic consequences for the security of a given implementation.

In the literature, one-bit faults are often considered. This is a useful model for developing theoretical attacks, but has proven to be extremely difficult to produce on a secure microprocessor. The model given above is based on published descriptions of implementations of fault attacks.

7.5.3 Faults in Cryptographic Algorithms

The faults mechanisms and fault model described above can be used to attack numerous cryptographic algorithms. Two examples of fault attacks on cryptographic algorithms are described below.

7.5.3.1 Faults in RSA Signature Generation

The first published fault attack [10], proposed an attack focused on an implementation of RSA using the Chinese Remainder Theorem (CRT). The attack allows for a wide range of fault injection methods, as it only requires one fault to be inserted in order to factorise the RSA modulus.

The technique requires an attacker to obtain two signatures for the same message, where one signature is correct and the other is the result of the injection of a fault during the computation of S_p or S_q (see above). That is, the attack requires that one of S_p and S_q is computed correctly, and the other is computed incorrectly.

Without loss of generality, suppose that $S' = aS_p + bS'_q \bmod N$ is the faulty signature, where S_q is changed to $S'_q \neq S_q$. We then have:

$$
\begin{aligned}
\Delta &\equiv S - S' \pmod{N} \\
&\equiv (aS_p + bS_q) - (aS_p + bS'_q) \pmod{N} \\
&\equiv b(S_q - S'_q) \pmod{N} .
\end{aligned}
\tag{7.14}
$$

As $b \equiv 0 \pmod{p}$ and $b \equiv 1 \pmod{q}$, it follows that $\Delta \equiv 0 \pmod{p}$ (but $\Delta \not\equiv 0 \pmod{q}$) meaning that Δ is a multiple of p (but not of q). Hence, we can derive the factors of N by observing that $p = \gcd(\Delta \bmod N, N)$ and $q = N/p$.

In summary, all that is required to break RSA is one correct signature and one faulty one. This attack will be successful regardless of the type or number of faults injected during the process, provided that all faults affect the computation of either S_p or S_q.

Although initially theoretical, this attack stimulated the development of a variety of fault attacks against a wide range of cryptographic algorithms. A description of an implementation of this attack is given in [5].

7.5.3.2 Faults in DES

A type of cryptanalysis of ciphertext blocks produced by injecting faults into DES was proposed in [8], based on using techniques used in differential cryptanalysis [31]. One-bit faults were assumed to occur in random places throughout an execution of DES. The ciphertext blocks corresponding to faults occurring in the 14th and 15th round were taken, enabling the derivation of the key. This was possible as the effect of a one-bit fault in the last three rounds of DES is visible in the ciphertext block when it is compared with a correct ciphertext block. This allowed the key to be recovered using between 50 and 200 different ciphertext blocks. It is claimed in [8] that, if an attacker can be sure of injecting faults towards the end of the algorithm, the same results could be achieved with only 10 faulty ciphertext blocks, and that, if a precise fault could be induced, only three faulty ciphertext blocks would be required.

This algorithm was improved upon in [17]. When searching for a key, the number of times a given hypothesis is found is counted. This means that faults from earlier rounds can be taken into account. It is claimed in [17] that faults from the 11th round onwards can be used to derive information on the key, and that in the ideal situation only two faulty ciphertext blocks are required.

The simplest case of a fault attack on DES involves injecting a fault in the 15th round, and such an attack is well-known within the smart card industry. The last round of DES can be expressed in the following manner:

$$\begin{aligned} R_{16} &= S(R_{15} \oplus K_{16}) \oplus L_{15} \\ &= S(L_{16} \oplus K_{16}) \oplus L_{15} \end{aligned}$$

If a fault occurs during the execution of the 15th round, i.e. R_{15} is randomised by a fault to become R'_{15}, then:

$$\begin{aligned} R'_{16} &= S(R'_{15} \oplus K_{16}) \oplus L_{15} \\ &= S(L'_{16} \oplus K_{16}) \oplus L_{15} \end{aligned}$$

and

$$\begin{aligned} R_{16} \oplus R'_{16} &= S(L_{16} \oplus K_{16}) \oplus L_{15} \oplus S(L'_{16} \oplus K_{16}) \oplus L_{15} \\ &= S(L_{16} \oplus K_{16}) \oplus S(L'_{16} \oplus K_{16}) \ . \end{aligned}$$

Table 7.1 The expected
number of hypotheses per
S-box for one faulty
ciphertext block

S-box	E_k
1	7.54
2	7.67
3	7.58
4	8.36
5	7.73
6	7.41
7	7.91
8	7.66

This provides an equation in which only the last subkey, K_{16}, is unknown. All of the other variables are available from the ciphertext block. This equation holds for each S-box in the last round, which means that it is possible to search for key hypotheses in sets of six bits, i.e. the 48-bit output after the XOR is divided into eight groups of six bits before being substituted with values from the S-boxes.

All 64 possible key values corresponding to the XOR just before each individual S-box can be used to generate a list of possible key values for these key bits. After this, all the possible combinations of the hypotheses can be searched though, with the extra eight key bits that are not included in the last subkey, to find the entire key.

If R'_{15} is randomised by a fault, then the expected number of hypotheses that are generated can be predicted using the methods given in [7]. The Table 7.1 shows the statistically expected number of key hypotheses E_k that would be returned by a fault producing a difference across each S-box in the last round. This is an average of the non-zero elements in the expected number of hypotheses that are generated using the tables defined in [7].

The expected number of hypotheses for the last subkey will be the product of all eight expected values E_k; this gives an expected number of around 2^{24}. This is just for the last subkey, an actual exhaustive search will need to take into account the eight bits that are not included in the last subkey, giving an overall expected keyspace size of 2^{32}.

This substantially reduces the number of possible keys that would need to be tested to try and determine the secret key used. The size of the keyspace can be further reduced if the fault attack is repeated and the intersection of the two resulting keyspaces is determined.

The same attack can also be applied if small faults occur in the last five rounds of DES, but the treatment is statistical in nature and requires many more faults to determine information on the key. Further details of this attack, and a brief description of an implementation, are given in [17].

7.5.4 Countermeasures

The countermeasures that can be used to protect microprocessors from fault attacks are based on methods previously employed for integrity purposes. However, countermeasures only need to be applied in processes where an attacker could benefit from injecting a fault, although a careful analysis of a given application is required to determine where countermeasures are required. This has proven to be true even where algorithms are based on one-time random numbers, as it has been shown that the manipulation of the random number can compromise the security of an cryptographic algorithm [36]. The list of countermeasures is given below:

Checksums can be implemented in software or hardware. This prevents data (such as key values) being modified by a fault, as the fault can be detected followed by appropriate action (see below).

Execution Randomisation can be used to change the order in which operations in an algorithm are executed from one execution to another, making it difficult to predict what the machine is doing at any given cycle. For most fault attacks this countermeasure will only slow down a determined attacker, as eventually a fault will hit the desired instruction. However, this will thwart attacks that require faults in specific places or in a specific order.

Random Delays can be used to increase the time required to attack. As with execution randomisation a determined attacker can attempt to inject a fault until the moment the fault is injected coincides with the target. However, this can take significantly more time than would otherwise be required, especially if an attacker is able to identify a target through a side channel (e.g. using SPA).

Execution redundancy is the repeating of algorithms and comparing the results to verify that the correct result is generated. This is most effective when the second calculation is different to the first, e.g. the inverse function, to prevent an attacker form trying to inject an identical fault in each execution.

Variable redundancy is the reproduction of a variable in memory. When a variable is tested or modified the redundant copy is also tested or modified. This is most effective when the copy is stored in a different form to the original, e.g. the bitwise complement, to avoid a fault being applied to each variable in the same way.

Ratification counters and baits can be included to prevent an attacker from successfully completing a fault attack by rendering a microprocessor inoperative once a fault attack is detected. Baits are small (<10 byte) code fragments that perform an operation and test it's result. A typical bait writes, reads and compares data, performs XORs, additions, multiplications and other operations whose results can be easily checked. When a bait detects an error it increments a counter in non-volatile memory (NVM), and when this counter exceeds a tolerance limit (typically three) the microprocessor ceases to function.

7.5.4.1 Remarks

Many of the countermeasures in this list can be implemented in either hardware or software. A more complete list of the countermeasures (in hardware and software), along with a description of certain fault attacks, is given in [6].

7.6 Embedded Software Design

The attacks described in the previous sections of this chapter have focused on attacking cryptographic algorithms to determine a secret or private key. In this section, some example of how the attack methods presented in Sects. 7.4 and 7.5 can be applied to other security mechanisms are described. This is to demonstrate that implementing a secure application on a smart card is not trivial, and requires the careful evaluation of every implemented function. It should also be noted that the attacks described below are only possible where no specific countermeasures are implemented.

7.6.1 PIN Verification

As described in Sect. 7.3, the first smart cards included a contact that was called Vpp used to supply power to the microprocessor so that it could program the EEPROM present in the chip. At the time, the voltage supply (Vcc) did not supply enough power to allow a microprocessor to modify EEPROM and a higher voltage needed to be applied to the Vpp contact. The Vpp contact is no longer used as it led to security problems, as described below.

If the Vpp contact was masked (e.g. covered with nail varnish), then no power would be available for the microprocessor to program the EEPROM, but power would be available through the Vcc to run every other function of the smart card. This meant that an attacker could try every single PIN number without decrementing the PIN counter (typically a PIN counter is set to three and decremented with every false PIN presentation, once the PIN number is zero a smart card will render itself non-functional). This process could be automated using a standard PC and a smart card reader to determine a PIN number in matter of minutes.

After the Vpp contact was removed, further problems were encountered. The most natural way to implement a PIN verification would be as follows:

where the PIN entered by a user is returned by the RequestPIN function and compared with the PIN in NVM. If the entered PIN is not equal to the stored PIN, the PINcounter will be decremented.

It was observed that the power consumption increased when a smart card modified the value of the PIN counter, i.e. it was visible using the SPA techniques described in Sect. 7.4 as an increase in the power consumption. Attackers then developed tools

Insecure PIN Verification Algorithm
```
{
    if (PINcounter > 0) {
        PIN := RequestPIN();
    }
    else {
    return false;
    }
    if (PIN ≠ UserPIN) {
        PINcounter := PINcounter − 1;
        return false;
    }
    else {
        return true;
    }
}
```

to cut the power being supplied to a microprocessor once this increase in power consumption was detected. This allowed automated tools to attempt every PIN number until the correct PIN was found. The correct PIN would be the only value where the command would finish as the microprocessor would not attempt to modify the NVM.

This made it necessary to change the algorithm used to verify a PIN number. Typically, a secure PIN verification will be implemented in the following manner:

Secure PIN Verification Algorithm
```
{
    if (PINcounter > 0) {
        PIN := RequestPIN();
    }
    else {
    return false;
    }
    PINcounter := PINcounter − 1;
    if (PIN = UserPIN) {
        PINcounter := PINcounter + 1;
        return true;
    }
    else {
        return false;
    }
}
```

where the PINcounter is decremented before it is tested, and only incremented if the PIN is entered correctly. The power supply can be removed at any point during the command without producing a security problem. However, further modifications would need to be made to render it resistant to fault attacks. An example of a fault attack against an operating system is described below.

7.6.2 File Access

Another possible target within a smart card operating system is the file structure. All personalisation information, e.g. PIN numbers etc., is stored in a file structure situated in NVM. Each file will have a set of access conditions that determine who can read and write to each file or directory. For example, a user's PIN number on a SIM card, unless intentionally disabled, will grant access to the files that contain SMS messages once verified. If the PIN is not verified, access to these files will be denied. There are often administrator identification codes (essentially eight-digit PIN numbers) that grant access to more files and allow the modification of files that the end user is not able to directly modify, e.g. the user's PIN number.

If an attacker wishes to attempt to access information stored in files without any of the codes mentioned above, a fault attack could be attempted. An attacker can attempt to inject a fault at the moment a smart card is evaluating whether the right to read a file, for example, should be granted. If successful, the evaluation will be erroneous and the right to access the file will be temporarily granted.

In order to determine the point in which a fault would need to be injected, an attacker can use SPA (see Sect. 7.4). An attacker could compare a trace of the power consumption where file access is granted with a trace where file access has been denied. An example of this is shown in Fig. 7.13. The black trace represents the power consumption where access has been denied. The grey trace represents the power consumption where access has been granted. It can be see at the point indicated in the figure that the two traces diverge. This should represent the moment at which the access conditions are evaluated and will, therefore, be the targeted area for a fault attack.

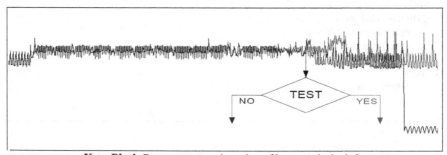

Key: **Black** Power consumption where file access is denied.
Grey Power consumption where file access is granted.

Fig. 7.13 Determining the moment file access is granted using the power consumption

On a smart card, there are typically files that contain serial numbers and such information, which can be read by anyone. Finding two files to attempt to read in order to generate traces, as shown in Fig. 7.13, should be straightforward.

This type of fault attack means that the access conditions to files, and other security mechanisms such as PIN verification, need to include redundancy in their tests to ensure that a fault attack is not possible. The various countermeasures that can be implemented are described in Sect. 7.5.

7.7 In Conclusion

This chapter presents the particular security considerations that need to be taken into account when implementing a secure smart card-based application. Implementations of all the commands on a smart card need to be subjected to careful analysis to prevent power analysis and fault injection techniques from compromising the security of the smart card.

Research in the domain of smart card security is typically a cyclic process. New attacks are developed against algorithm implementations, standards, etc. and countermeasures are proposed. The modifications are then reviewed for potential vulnerabilities and further countermeasures proposed if required. The aim of this process is to remain sufficiently ahead of what can be achieved by an individual attacker that smart cards remain secure throughout the period that they are active.

References

1. Akkar, M.-L. and Giraud, C. (2001). An implementation of DES and AES secure against some attacks. In Koç, C. K., Naccache, D., and Paar, C., editors, *Cryptogaphic Hardware and Embedded Systems – CHES 2001*, volume 2162 of *Lecture Notes in Computer Science*, pages 309–318. Springer-Verlag.
2. American National Standards Institute (1985). *Financial Institution Key Management (Wholesale)*. American National Standards Institute.
3. Anderson, R. and Kuhn, M. (1996). Tamper resistance – a cautionary note. In *Proceedings of the Second USENIX Workshop of Electronic Commerce*, pages 1–11.
4. Anderson, R. and Kuhn, M. (1997). Low cost attacks on tamper resistant devices. In Christianson, B., Crispo, B., Lomas, T. M. A., and Roe, M., editors, *Security Protocols*, volume 1361 of *Lecture Notes in Computer Science*, pages 125–136. Springer-Verlag.
5. Aumüller, C., Bier, P., Hofreiter, P., Fischer, W., and Seifert, J.-P. (2002). Fault attacks on RSA with CRT: Concrete results and practical countermeasures. In Kaliski, B. S., Koç, C. K., and Paar, C., editors, *Cryptographic Hardware and Embedded Systems – CHES 2002*, volume 2523 of *Lecture Notes in Computer Science*, pages 260–275. Springer-Verlag.
6. Bar-El, H., Choukri, H., Naccache, D., Tunstall, M., and Whelan, C. (2006). The sorcerer's apprentice guide to fault attacks. *Proceedings of the IEEE*, 94(2):370–382.
7. Biham, E. and Shamir, A. (1991). Differential cryptanalysis of DES-like cryptosystems. In Menezes, A. and Vanstone, S., editors, *Advances in Cryptology – CRYPTO '90*, volume 537 of *Lecture Notes in Computer Science*, pages 2?-21. Springer-Verlag.

8. Biham, E. and Shamir, A. (1997). Differential fault analysis of secret key cryptosystems. In Kaliski, B. S., editor, *Advances in Cryptology – CRYPTO '97*, volume 1294 of *Lecture Notes in Computer Science*, pages 513–525. Springer-Verlag.

9. Blömer, J. and Seifert, J.-P. (2003). Fault based cryptanalysis of the advanced encryption standard (AES). In Wright, R. N., editor, *Financial Cryptography – FC 2003*, volume 2742 of *Lecture Notes in Computer Science*, pages 162–181. Springer-Verlag.

10. Boneh, D., DeMillo, R. A., and Lipton, R. J. (1997). On the importance of checking computations. In Fumy, W., editor, *Advances in Cryptology – EUROCRYPT '97*, volume 1233 of *Lecture Notes in Computer Science*, pages 37–51. Springer-Verlag.

11. Brier, E., Clavier, C., and Olivier, F. (2004). Correlation power analysis with a leakage model. In Joye, M. and Quisquater, J.-J., editors, *Cryptographic Hardware and Embedded Systems – CHES 2004*, volume 3156 of *Lecture Notes in Computer Science*, pages 16–29. Springer-Verlag.

12. Chari, S., Jutla, C. S., Rao, J. R., and Rohatgi, P. (1999). Towards approaches to counteract power-analysis attacks. In Wiener, M., editor, Advances in *Cryptology – CRYPTO '99*, volume 1666 of *Lecture Notes in Computer Science*, pages 398–412. Springer-Verlag.

13. Chevallier-Mames, B., Ciet, M., and Joye, M. (2004). Low-cost solutions for preventing simple side-channel analysis: Side-channel atomicity. *IEEE Transactions on Computers*, 53(6):760–768.

14. Clavier, C., Coron, J.-S., and Dabbous, N. (2000). Differential power analysis in the presence of hardware countermeasures. In Koç, C. K. and Paar, C., editors, *Cryptographic Hardware and Embedded Systems – CHES 2000*, volume 1965 of *Lecture Notes in Computer Science*, pages 252–263. Springer-Verlag.

15. Fouillat, P. (1990). *Contribution a l'etude de l'interaction entre un faisceau laser et un milieu semiconducteur, Applications a l'etude du Latchup et al l'analyse d'etats logiques dans les circuits integres en technologie CMOS*. PhD thesis, University of Bordeaux.

16. Gandolfi, K., Mourtel, C., and Olivier, F. (2001). Electromagnetic analysis: Concrete results. In Koç, C. K., Naccache, D., and Paar, C., editors, *Cryptographic Hardware and Embedded Systems – CHES 2001*, volume 2162 of *Lecture Notes in Computer Science*, pages 251–261. Springer-Verlag.

17. Giraud, C. and Thiebeauld, H. (2004). A survey on fault attacks. In Deswarte, Y. and Kalam, A. A. El, editors, *Smart Card Research and Advanced Applications VI – 18th IFIP World Computer Congress*, pages 159–176. Kluwer Academic.

18. Govindavajhala, S. and Appel, A. W. (2003). Using memory errors to attack a virtual machine. In *IEEE Symposium on Security and Privacy 2003*, pages 154–165.

19. Gutmann, P. (2004). *Security Architecture*. Springer-Verlag.

20. Habing, D. H. (1992). The use of lasers to simulate radiation-induced transients in semiconductor devices and circuits. *IEEE Transactions On Nuclear Science*, 39:1647–1653.

21. International Organization for Standardization (1997). *ISO/IEC 7816-3 Information technology - Identification cards - Integrated circuit(s) cards with contacts - Part 3: Electronic signals and transmission protocols*. International Organization for Standardization.

22. International Organization for Standardization (1999). *ISO/IEC 7816-2 Identification cards - Integrated circuit cards - Part 2: Cards with contacts - Dimensions and location of the contacts*. International Organization for Standardization.

23. Joye, M. and Olivier, F. (2005). Side-channel attacks. In van Tilborg, H., editor, *Encyclopedia of Cryptography and Security*, pages 571–576. Kluwer Academic Publishers.

24. Kahn, D. (1997). *The Codebreakers: The Comprehensive History of Secret Communication from Ancient Times to the Internet*. Simon & Schuster Inc., second edition.

25. Knuth, D. (2001). *The Art of Computer Programming*, volume 2, Seminumerical Algorithms. Addison-Wesley, third edition.

26. Kocher, P. (1996). Timing attacks on implementations of Diffie-Hellman, RSA, DSS, and other systems. In Koblitz, N., editor, *Advances in Cryptology – CRYPTO '96*, volume 1109 of *Lecture Notes in Computer Science*, pages 104–113. Springer-Verlag.

27. Kocher, P., Jaffe, J., and Jun, B. (1999). Differential power analysis. In Wiener, M. J., editor, *Advances in Cryptology – CRYPTO '99*, volume 1666 of *Lecture Notes in Computer Science*, pages 388–397. Springer-Verlag.
28. Kommerling, O. and Kuhn, M. (1999). Design principles for tamper resistant smartcard processors. In *USENIX Workshop on Smartcard Technology*, pages 9–20.
29. Mangard, S., Oswald, E., and Popp, T. (2007). *Power Analysis Attacks – Revealing the Secrets of Smart Cards*. Springer-Verlag.
30. May, T. and Woods, M. (1978). A new physical mechanism for soft erros in dynamic memories. In *16th International Reliability Physics Symposium*.
31. Menezes, A., van Oorschot, P., and Vanstone, S. (1997). *Handbook of Applied Cryptography*. CRC Press.
32. Messerges, T. S. (2000). *Power Analysis Attacks and Countermeasures for Cryptographic Algorithms*. PhD thesis, University of Illinois, Chicago.
33. Meyer, C. (2000). Private communication. Carl Meyer was one of the designers of the DES algorithm.
34. MIPS-Technologies (2001). MIPS™architecture for programmers volume I: Introduction to the MIPS32™architecture. Technical Report MD00082, Revision 0.95.
35. Murdocca, M. and Heuring, V. P. (2000). *Principles of Computer Architecture*. Addison-Wesley.
36. Naccache, D., Nguyen, P. Q., Tunstall, M., and Whelan, C. (2005). Experimenting with faults, lattices and the DSA. In Vaudenay, S., editor, *Public Key Cryptography – PKC 2005*, volume 3386 of *Lecture Notes in Computer Science*, pages 16–28. Springer-Verlag.
37. NIST (1999). *Data Encryption Standard (DES) (FIPS-46-3)*. National Institute of Standards and Technology.
38. NIST (2001). *Advanced Encryption Standard (AES) (FIPS-197)*. National Institute of Standards and Technology.
39. Pouget, V. (2000). *Simulation experimentale par impulsions laser ultra-courtes des effets des radiations ionisantes sur les circuits integres*. PhD thesis, University of Bordeaux.
40. Quisquater, J.-J. and Samyde, D. (2001). Electromagnetic analysis (ema): Measures and counter-measures for smart cards. In Attali, I. and Jensen, T. P., editors, *Smart Card Programming and Security, International Conference on Research in Smart Cards – E-smart 2001*, volume 2140 of *Lecture Notes in Computer Science*, pages 200–210. Springer-Verlag.
41. Rivest, R., Shamir, A., and Adleman, L. M. (1978). Method for obtaining digital signatures and public-key cryptosystems. *Communications of the ACM*, 21(2):120–126.
42. Samyde, D., Skorobogatov, S. P., Anderson, R. J., and Quisquater, J.-J. (2002). On a new way to read data from memory. In *Proceedings of the First International IEEE Security in Storage, Workshop*, pp. 65–69.
43. Skorobogatov, S. and Anderson, R. (2002). Optical fault induction attacks. In Kaliski, B. S., Ç. K. Koç, and Paar, C., editors, *Cryptographic Hardware and Embedded Systems – CHES 2002*, volume 2523 of *Lecture Notes in Computer Science*, pages 2–12. Springer-Verlag.
44. Skorobogatov, S. P. (2005). *Semi-Invasive Attacks – A New Approach to Hardware Security Analysis*. PhD thesis, University of Cambridge. available at http://www.cl.cam.ac.uk/TechReports/
45. Wright, P. (1987). *Spycatcher*. Heineman.
46. Ziegler, J. (1979). Effect of cosmic rays on computer memories. *Science*, 206:776–788.

Chapter 8
Graphics Processing Units

Peter Schwabe

Abstract This chapter introduces graphics processing units (GPUs) for general-purpose computations. It describes the highly parallel architecture of modern GPUs, software-development toolchains to program them, and typical pitfalls and performance bottlenecks. Then it considers several applications of GPUs in information security, in particular in cryptography and cryptanalysis.

Graphics Processing Units (GPUs) are coprocessors that traditionally perform the rendering of 2-dimensional and 3-dimensional graphics information for display on a screen. In particular, computer games request more and more realistic real-time rendering of graphics data and so GPUs became more and more powerful highly parallel specialist computing units. It did not take long until programmers realized that this computational power can also be used for tasks other than computer graphics. For example already in 1990 Lengyel, Reichert, Donald, and Greenberg used GPUs for real-time robot motion planning [43]. In 2003, Harris introduced the term general-purpose computations on GPUs (GPGPU) [28] for such nongraphics applications running on GPUs. At that time, programming GPGPUs meant expressing all algorithms in terms of operations on graphics data, pixels, and vectors. This was feasible for speed-critical small programs and for algorithms that operate on vectors of floating-point values in a similar way as graphics data are typically processed in the rendering pipeline. The programming paradigm shifted when the two main GPU manufacturers, NVIDIA and AMD, changed the hardware architecture from a dedicated graphics-rendering pipeline to a multi-core computing platform, implemented shader algorithms of the rendering pipeline in software running on these cores, and explicitly supported GPGPUs by offering programming languages and software-development toolchains. This chapter first gives an introduction to the architectures of these modern GPUs and the tools and languages to program them. Then it highlights several applications of GPUs related to information security with a focus on applications in cryptography and cryptanalysis.

P. Schwabe (✉)
Digital Security Group, Radboud University Nijmegen, Nijmegen, The Netherlands
e-mail: peter@cryptojedi.org

K. Markantonakis and K. Mayes (eds.), *Secure Smart Embedded Devices,*
Platforms and Applications, DOI: 10.1007/978-1-4614-7915-4_8,
© Springer Science+Business Media New York 2014

8.1 An Introduction to Modern GPUs

GPUs have evolved to coprocessors of a size larger than typical CPUs. While CPUs use large portions of the chip area for caches, GPUs use most of the area for arithmetic logic units (ALUs). The main concept that both NVIDIA and AMD GPUs use to exploit the computational power of these ALUs is executing a single instruction stream on multiple independent data streams (SIMD) [23]. This concept is known from CPUs with vector registers and instructions operating on these registers. For example, a 128-bit vector register can hold four single-precision floating-point values; an additional instruction operating on two such registers performs four independent additions in parallel. Instead of using vector registers, GPUs use hardware threads that all execute the same instruction stream on different sets of data. NVIDIA calls this approach to SIMD computing "single instruction stream, multiple threads (SIMT)". The number of threads required to keep the ALUs busy is much larger than the number of elements inside vector registers on CPUs. GPU performance, therefore, relies on a high degree of data-level parallelism in the application.

To alleviate these requirements on data-level parallelism, GPUs can also exploit task-level parallelism by running different independent tasks of a computation in parallel. This is possible on all modern GPUs through the use of conditional statements. Some recent GPUs support the exploitation of task-level parallelism also through concurrent execution of independent GPU programs. Each of the independent tasks again needs to involve a relatively high degree of data-level parallelism to make full use of the computational power of the GPU, but exploitation of task-level parallelism gives the programmer more flexibility and extends the set of applications that can make use of GPUs to accelerate computations.

The remainder of this section gives an overview of the hardware architectures of modern GPUs, introduces the relevant programming languages, and discusses typical performance bottlenecks and GPU benchmarking issues. The section focuses on NVIDIA GPUs, because most of the implementations of subsequent sections target these GPUs.

8.1.1 NVIDIA GPUs

In 2006, NVIDIA introduced the Compute Unified Device Architecture (CUDA). Today all of NVIDIA's GPUs are CUDA GPUs. CUDA is *not* a computer architecture in the sense of a definition of an instruction set and a set of architectural registers; binaries compiled for one CUDA GPU do not necessarily run on all CUDA GPUs. More specifically, NVIDIA defines different *CUDA compute capabilities* to describe the features supported by CUDA hardware. The first CUDA GPUs had compute capability 1.0. In 2011, NVIDIA released GPUs with compute capability 2.1, which is known as "Fermi" architecture. Details about the different compute capabilities are described in [50, Appendix F].

A CUDA GPU consists of multiple so-called *streaming multiprocessors* (SMs). The threads executing a GPU program, a so-called *kernel*, are grouped in *blocks*. Threads belonging to one block all run on the same multiprocessor but one multiprocessor can run multiple blocks concurrently. Blocks are further divided into groups of 32 threads called *warps*; the threads belonging to one warp are executed in lock step, i.e., they are synchronized. As a consequence, if threads inside one warp diverge via a conditional branch instruction, execution of the different branches is serialized. On GPUs with compute capability 1.x all SMs must execute the same kernel. Compute capability 2.x supports concurrent execution of different kernels on different SMs.

Each SM contains several so-called CUDA cores, 8 per SM in compute capability 1.x , 32 per SM in compute capability 2.0, and 48 per SM in compute capability 2.1. One could think that for example a reasonable number of threads per SM is 8 for compute capability 1.x GPUs or 48 for compute capability-2.1 GPUs. In fact, it needs many more threads to fully utilize the ALUs; the reason is that concurrent execution of many threads on one SM is used to hide arithmetic latencies and up to some extent also memory access latencies. For compute capability 1.x, NVIDIA recommends to run at least 192 or 256 threads per SM. To fully utilize the power of compute capability 2.x, GPUs even more threads need to run concurrently on one SM. For applications that involve a very high degree of data-level parallelism, it might now sound like a good idea to just run as many concurrent threads as possible. The problem is that the register banks are shared among threads; the more threads are executed, the fewer registers are available per thread. Finding the optimal number of threads running concurrently on one SM is a crucial step to achieve good performance.

Aside from registers, each thread also has access to various memory domains. Each SM has several KB of fast shared memory accessible by all threads on this multiprocessor. This memory is intended to exchange data between the threads of a thread block, latencies are as low as for register access but throughput depends on access patterns. The shared memory is organized in 16 banks. If two threads within the same half-warp (16 threads) load from or store to different addresses on the same memory bank in the same instruction, these requests are serialized. Such requests to different addresses on the same memory bank are called *bank conflicts*, for details see [50, Sect. 5.3.2.3]. Graphics cards also contain several hundred MB up to a few GB of device memory. Each thread has a part of this device memory dedicated as so-called local memory. Another part of the device memory is global memory accessible by all threads. Access to device memory has a much higher latency than access to shared memory or registers. For details on latencies and throughput see [50, Sects. 5.3.2.1, 5.3.2.2]. Additionally, each thread has cached read-only access to constant memory and texture and surface memory. Loads from constant cache are efficient if all threads belonging to a half-warp load from the same address; if two threads within the same half-warp loaded from different addresses in the same instruction, throughput decreases by a factor equal to the number of different load addresses. Another decision (aside from the number of threads per SM) that can have huge impact on performance is what data are kept in which memory domain. Access from threads to different memories is depicted in Fig. 8.1.

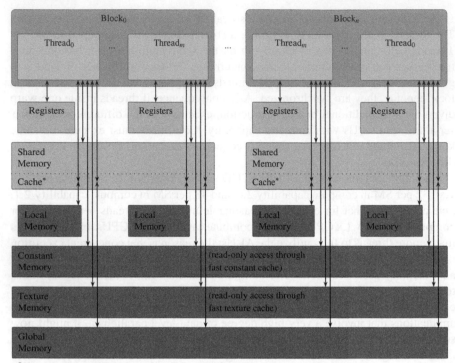

*Since compute capability2.0 : configurable amount of shared memory
serves as transparent cache to local and global memory

Fig. 8.1 Access to different memories from threads on NVIDIA CUDA devices. *Light-red* memories are fast; *dark-red* memories are parts of device memory with high-latency access

Communication between CPU and GPU is done by transferring data between the host memory and the GPU device memory or by mapping page-locked host memory into the GPUs address space. Asynchronous data transfers between *page-locked* host memory and device memory can overlap with computations on the CPU. For some CUDA devices since compute capability 1.1 they can also overlap with computations on the GPU. For details on data transfers to and from NVIDIA GPUs see [50, Sects. 3.4, 3.5]. Since CUDA 4.0 NVIDIA simplifies data exchange between host memory and device memory of Fermi GPUs by supporting a unified virtual address space. For details see [50, Sects. 3.2.7, 3.3.9]. The unified virtual address space is particularly interesting in conjunction with peer-to-peer memory access between multiple GPUs. This technique makes it possible to access the memory of one GPU directly from another GPU without data transfers through host memory. For details see [50, Sects. 3.2.6.4, 3.2.6.5].

8.1.2 AMD GPUs

The hardware and software technologies that allow programmers to use AMD GPUs for general-purpose computations are called AMD accelerated parallel processing (APP), formerly known as ATI Stream. For a detailed description of the architecture and the programming environment see [3].

Each APP device consists of multiple so-called *compute units*, each compute unit contains multiple *stream cores*, which, in turn, contain multiple *processing elements*. Multiple instances of a GPU program (kernel) are executed concurrently on different data, one such instance of a kernel is called a *work-item*. Multiple work-items are executed by all stream cores of one compute unit in lock step, one such group of work items executed together is called *wavefront*. The number of work items in a wavefront is hardware dependent. The programmer decides how many work items are scheduled to one compute unit in a so-called *workgroup*. Best performance is obtained, if this number is a multiple of the size of a wavefront.

In principle, different compute units can execute different kernels concurrently. However, the number of different kernels running on one APP device may be limited. All stream cores of one compute element execute the same instruction sequence consisting of very-large-instruction-word (VLIW) arithmetic instructions, control-flow instructions, and memory load and store instructions. Then up to four or five (depending on the device) instructions inside a VLIW instruction word are co-issued to the processing elements.

Similar to NVIDIA GPUs, AMD GPUs have various memories with different visibility to work items and different latencies and throughputs. The private memory is specific to each work item and is kept in a register file with very fast access. Work items inside one workgroup, i.e., running on the same compute unit, can communicate through local memory. This "local memory" is not a part of the device memory as on NVIDIA GPUs. In fact, it is very similar to what NVIDIA calls shared memory, a relatively small memory with fast access for efficient exchange of data between work items. Access to local memory is about an order of magnitude faster than access to device memory. Furthermore, all work items executing in one context have access to the global device memory and cached read-only access to a part of the device memory called constant memory. Access from work items to different memories is depicted in Fig. 8.2.

Communication with the host is done through DMA transfers between host and device memory. Computation on both the CPU and the GPU can overlap with DMA transfers.

8.1.3 Programming GPUs in High-Level Languages

With the CUDA architecture NVIDIA introduced language extensions to the C programming language that allowed to write programs that are partially executed on the GPU. The resulting programming language is called "C for CUDA". Note that

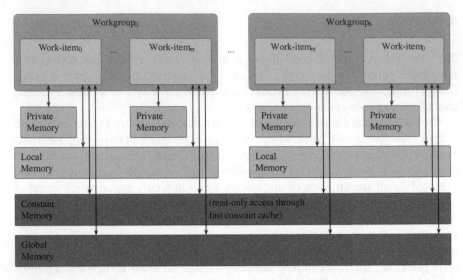

Fig. 8.2 Access to different memories from work items on AMD APP devices. *Light-red* memories are fast; *dark-red* memories are parts of device memory with high-latency access

depending on the compute capability some restrictions apply for the part of the program that is executed by the GPU, for example compute capability 1.x does not support recursive function calls. For details on C for CUDA see [50].

The first software-development tool that AMD offered for GPGPUs was called Close-to-Metal (CTM) which gave low-level access to the native instruction set of the GPU. High-level-language support was first offered in the ATI Stream SDK v1 with the ATI Brook+ language, which is based on BrookGPU developed at Stanford University [14].

Both solutions, C for CUDA and Brook+ could only be used to implement software for the respective manufacturer's GPUs. As a more portable approach, both NVIDIA and AMD now also support the OpenCL programming language and API developed by the Khronos group. This programming language is designed for the development of software for parallel computations on arbitrary heterogeneous systems. Two versions of the language have been released, OpenCL 1.0 in November 2008 [26] and OpenCL 1.1 in June 2004 [27].

Today, the recommended way to program NVIDIA GPUs is using either C for CUDA [50] or OpenCL for CUDA [51]. AMD recommends OpenCL as high-level programming language for their GPUs in their latest APP SDK [3].

The compilation process is very similar for all of the high-level languages. In a first step, the compiler separates the parts of the program that run on the CPU from the parts that run on the GPU. The CPU part is further compiled using native C or C++ compilers for the respective host architecture. The GPU part is first translated to an intermediate low-level language. For NVIDIA this language is called PTX, for AMD it is called IL. The advantage of this intermediate language is that it is somewhat device independent. More specifically, PTX code is compatible across minor revisions of the compute capability; IL code is forward compatible. The GPU driver contains a just-in-time compiler for this intermediate language. Code that needs to run on GPUs with different hardware capabilities can thus be translated only to intermediate language, final compilation to binary code is performed by the respective driver. This last compilation step can also be done off-line to produce binaries for a specific GPU architecture.

8.1.4 Programming GPUs in Assembly

Most software today is written in high-level languages, but some areas of computing still employ hand-optimized assembly routines to achieve best performance. One of these areas is high-performance computing—in computations that run for weeks or months even small performance gains are typically worth the effort of implementing parts of the software in assembly. Now that GPUs explicitly support applications in high-performance computing one would expect that the manufacturers also provide assemblers. However, this is not the case. Until the CUDA 4.0 toolkit was released in May 2011, NVIDIA offered neither an assembler nor a disassembler for their GPUs, an assembler is still not provided by NVIDIA. To fill this gap, van der Laan reverse-engineered the binary format and developed the cubin utilities [57] consisting of the disassembler decuda and the assembler cudasm.

For the Fermi GPUs (compute capability 2.0 and 2.1) NVIDIA includes the cuobjdump disassembler in the CUDA 4.0 toolkit. An assembler for Fermi GPUs is being developed by the asfermi project [33].

AMD documents the instruction-set architecture of their recent GPUs, for example in [2] for the Radeon R600 series, in [5] for the Radeon R700 series, and in [4] for the Evergreen series. AMD does not document the complete ELF format of the binaries and does not provide an assembler for their GPUs. Similar to NVIDIA, community projects work on assemblers that support different families of AMD GPUs [48, 53].

8.1.5 GPU Performance Bottlenecks

What makes GPUs a very interesting computing platform for many algorithms is their pure computing power. For example, an NVIDIA GTX 295 graphics card containing two GT200b GPUs can dispatch a total of 745 billion single-precision floating-point operations per second. For comparison, all four cores of a 2.4 GHz Intel Core 2 Quad CPU can dispatch a total of 57.6 billion single-precision floating-point operations, more than one order of magnitude less. One might thus expect that GPUs speed up computations by a factor of 10 or more, but as the examples in the following sections show this is not the case for many applications. The reason is that in order to make use of the computational power of GPUs, applications need to fulfill two conditions:

- The degree of data-level parallelism required to keep hundreds of threads busy is much larger than the degree of data-level parallelism that is required for the SIMD implementations of current CPUs. For example, keeping 192 threads on each of the 30 multiprocessors of 2 GPUs on an NVIDIA GTX 295 graphics card busy needs 11,520 independent data streams. Keeping the four cores of a CPU busy working on 128-bit vector registers needs just 16 such independent streams. Less data-level parallelism typically requires multiple threads to work on the same data which involves communication and thread synchronization overhead.
- GPU performance depends on memory-access patterns much more than CPU performance does. The reason is that GPUs spend most of their chip area on ALUs while CPUs spend a large part of the chip area on fast caches that reduce load and store latencies. Computations that can keep the active set of data in the available registers benefit from the large computational power of the ALUs, but the high latencies of device-memory loads and stores typically incur huge performance penalties in applications that cannot. Some applications can use the shared memory on NVIDIA GPUs or the local memory on AMD GPUs as cache, Fermi GPUs make this easy by using a configurable amount of shared memory as transparent cache. If the same data are required by all threads this is indeed a very good solution. However, if each thread requires different data in cache (for example, register content temporarily spilled to memory), the amount of shared memory per thread is typically too small. Compilers therefore use device memory for register spills. Another way to deal with high memory latencies is to run more threads and thus hide the latencies. Note that this comes at the price of a smaller number of registers per thread and even higher requirements on data-level parallelism.

Another potential bottleneck is data transfer between host memory and device memory. All modern graphics cards are connected through PCI Express. Throughput rates highly depend on the version of PCI Express, and the number of lanes. For example, the theoretical throughput of PCI Express 2.0 with 16 lanes (commonly denoted x16) is 8 GB/s in both directions. The throughput obtained in practice is considerably lower and depends on the size of data packets transmitted over the bus. For details see, for example, [20]. More serious than throughput limitations can be the latency incurred by data transfers over PCI Express, at least for applications that require frequent communication and cannot interleave communication with computations.

With these limitations in mind, it is interesting to see that GPU advertisements and also various scientific papers claim speedups by a factor of 100 and more of software running on a GPU compared to software running on a CPU. In most of the cases, a careful look at how these speedups were achieved reveals that the CPU implementation is far from state of the art, for example, it does not use the SIMD computing capabilities of modern CPUs, and the CPU implementation is not set up to run on multiple cores.

Despite these misleading comparisons found in many places, GPUs are very powerful computing devices and with careful optimization GPUs can speed up many computations considerably compared to the same computations running on a CPU. The following sections give examples of applications of GPU computing in information security and try to put the performance numbers in a meaningful perspective in comparison to state-of-the-art CPU implementations.

8.2 GPUs as Cryptographic Coprocessors

Cryptographic computations such as encryption and decryption, hashing, signature generation, and signature verification rely on high performance in software for many applications. Furthermore, most of the algorithms involved can be implemented in relatively small code size and it is feasible to hand-optimize code on the assembly level.

This is why for example advanced encryption standard (AES) and RSA encryption were among the algorithms that were implemented using shader instructions of the graphics rendering pipeline of traditional GPUs [29, 47, 62].

In 2006, before CUDA was introduced by NVIDIA, Cook and Keromytis published a book on cryptography on graphics cards [21]. This book claims that using GPUs for cryptography has two additional advantages aside from speeding up computations:

- The authors suggest that GPU implementations may be more resistant to (at least existing) side-channel attacks. They do not claim that GPU implementations are inherently protected against any side-channel attacks that work against CPU implementations. In fact, there is no immediate reason to believe that GPUs generally offer better protection against any side-channel attacks than CPUs. Certainly one of the most relevant class of attacks, namely cache-timing attacks (see, e.g., [56]) will not work on GPUs that have uncached access to memory, but at least the most recent NVIDIA GPUs use part of their shared memory as transparent cache for access to the GPUs main memory [49].
- Chapter 3 of [21] describes a video-streaming service that uses GPUs to decrypt video data that shall only be displayed but never be stored or modified. The system uses the GPU as "the only trusted component in a spyware-safe system".
  This idea starts from the assumption that GPUs and graphics drivers are more trustworthy than the operating system for computations involving sensitive data such as cryptographic keys. This is a dangerous assumption to make, attackers controlling the operating system can also exchange the graphics driver, there is not even a guarantee that any code really runs on the GPU.

When using GPUs for cryptographic computations, one should keep in mind that GPUs and graphics drivers are not designed for computations on sensitive data and should be used for such computations only with precaution. For instance, on various graphics cards it is possible for a computing kernel to read out parts of the memory content left behind by a previously executed kernel. Keeping cryptographic keys in these parts of the memory can be used to speed up computations—for example, a key can be expanded once and be left in constant memory for all subsequent kernel launches as suggested in [52]. On the other hand, this can also be a serious security threat in multi-user environments if one user manages to launch a GPU kernel that reads out the key of another user.

In environments where data in GPU memory can be protected, for example on a single-user server, or with careful protections to avoid memory readout, modern GPUs can be used as powerful cryptographic coprocessors for throughput-oriented applications.

8.2.1 AES on GPUs

In particular, the possibility to implement the AES, the most widely used symmetric encryption algorithm, on GPUs has attracted a lot of attention. AES is a block cipher with supported key sizes of 128, 192, and 256 bits and a block size of 128 bits. Most implementations focus on AES with 128-bit keys. In this setting, the key is first expanded into 11 round keys $K_0, \ldots, K_{11}$. Each 128-bit input block (*state*) is then transformed into 10 rounds, each round involving one of the 11 round keys. The first round key K_0 is xored to the block before the first round. The most common implementation technique for AES, described in [22, Sect. 5.2.1], operates on 32-bit words and uses 4 lookup tables T0, T1, T2, and T3 of size 1 KB (256 32-bit words) each. The 128-bit state is represented as 4 such 32-bit words. The operations of one round of AES in C notation is given in Listing 1.

Listing 1 One round of AES encryption in C, the 128-bit input state is in 32-bit unsigned integers y0, y1, y2, y3, the output state is in 32-bit unsigned integers z0, z1, z2, z3; the 128-bit round key is in 32-bit unsigned integers k0, k1, k2, k3.

```
z0 = T0[ y0 >> 24          ] ^ T1[(y1 >> 16) & 0xff] \
      ^ T2[(y2 >>  8) & 0xff] ^ T3[ y3          & 0xff] ^ k0;
z1 = T0[ y1 >> 24          ] ^ T1[(y2 >> 16) & 0xff] \
      ^ T2[(y3 >>  8) & 0xff] ^ T3[ y0          & 0xff] ^ k1;
z2 = T0[ y2 >> 24          ] ^ T1[(y3 >> 16) & 0xff] \
      ^ T2[(y0 >>  8) & 0xff] ^ T3[ y1          & 0xff] ^ k2;
z3 = T0[ y3 >> 24          ] ^ T1[(y0 >> 16) & 0xff] \
      ^ T2[(y1 >>  8) & 0xff] ^ T3[ y2          & 0xff] ^ k3;
```

To achieve the required degree of parallelism, GPU implementations of AES typically either consider many independent streams that are encrypted in parallel or they use a parallel mode of operation such as ECB or CTR that allows to encrypt blocks of a single stream independently. The most important decision to make for high-performance AES encryption on GPUs is how to use the available memory domains. CPU implementations store lookup tables and expanded keys in RAM, after some rounds of AES the tables will be in level-1 cache and lookups are fast. On most GPUs a straightforward adaptation of this approach, placing tables and expanded keys in device memory, will incur high latency penalties, because access to device memory is uncached (except for NVIDIA Fermi GPUs where part of the shared memory is used as transparent cache). A better approach is to place the lookup tables in the fast shared memory of NVIDIA GPUs or the local memory of AMD GPUs. Recall that loads from shared memory on NVIDIA GPUs can be as fast as register access, but that throughput and latency depend on the access pattern. AES table lookups have an unpredictable access pattern, so one must expect penalties due to memory bank conflicts. One solution to avoid these penalties is to store multiple copies of the lookup tables in the fast memory, such that each entry is available on each memory bank. If shared memory is not large enough to hold these copies of the tables, it may still be possible to store copies of only one of the tables and obtain entries of the other tables through rotations (see, e.g., [22, Sect. 5.2.1]). The best combination of optimization techniques depends on the target GPU.

Not only the decision about location and layout of the lookup tables is important, also handling of the round keys influences performance. This is relatively easy if one big stream is encrypted in a parallel mode of operation. In this case, all threads use the same key and it can be stored in constant memory. Unlike lookups from the tables, the round keys are accessed in a completely predictable pattern; they are broadcasted to all threads which is exactly what the constant memory is made for. The situation is different for the encryption of many independent streams under different keys. If each thread needs different round keys, there is not enough fast memory on most GPUs to store all these round keys. Instead of loading round keys from slow device memory, it may be a better choice to expand the key on the fly. Again, the best solution highly depends on the specific target GPU.

A completely different approach to implement AES is bitslicing. This technique was first introduced for the Data Encryption Standard (DES) by Biham in [12] and has also been used for various AES implementations [37, 41, 45]. The idea of this technique is transposition of data: Instead of storing a 128-bit state in, e.g., 4 32-bit registers, it uses 128 registers, 1 register per bit. This representation of data allows to simulate a hardware implementation, logical gates become bit-logical instructions. For just one computation this is not efficient, but if all n bits of registers are used to perform computations on n independent streams, this can be very efficient. Note that on top of the high degree of parallelism required for GPU computations, bitslicing requires another factor of n of parallelism, n being the register width.

Various GPU implementations of AES are described in the literature. In [63], Yang and Goodman describe different implementations of AES for AMD GPUs. Their bitsliced implementation aims at key search, so keys need to be expanded into round keys on the fly. On an AMD HD 2900 XT GPU this implementation performs encryption of one block under 145 million keys per second, this corresponds to a throughput of 18.5 Gbit/s. For the lookup-table-based implementation, they report an AES encryption throughput of 3.5 Gbit/s on an AMD HD 2900 XT GPU.

The implementation by Manavski described in [44] uses a lookup-table-based approach to achieve a peak throughput of 8.28 Gbit/s on an NVIDIA 8800 GTX graphics card (G80 GPU); to achieve this peak throughput at least 8 MB of data need to be encrypted under the same key. This implementation exploits parallelism inside AES, four threads perform the transformation of one 128-bit block. Harrison and Waldron report a throughput of 15.423 Gbit/s in [30] for their lookup-table-based implementation of AES on an NVIDIA G80 GPU. This peak performance is achieved for input messages of $\geq$ 65 MB, overhead from data transfers to and from the GPU are not included in the benchmarks. Both the implementation in [44] and the implementation in [30] achieve a significantly lower throughput when data transfers are included in the benchmarks: 2.5 Gbit/s for [44] and 6.9 Gbit/s for [30].

Two more recent papers report speeds beyond 30 Gbit/s on NVIDIA GPUs. Osvik, Bos, Stefan, and Canright in [52] describe an implementation of AES with 128-bit keys that achieve 30.9 Gbit/s throughput on one GPU of an NVIDIA GTX 295 graphics card (containing 2 GT200b GPUs). The implementation interleaves data transfers with computations by using page-locked host memory. Interleaving data transfers with kernel execution were not possible for the GPUs used for benchmarking in [30, 44]. This throughput is achieved for encryption under one key in constant memory, but the paper also describes an implementation with on-the-fly key schedule suitable for key-search applications, that achieve a throughput of 23.8 Gbit/s. Jang, Han, Han, Moon, and Park present a GPU-accelerated SSL proxy in [36]. For the AES implementation included in this proxy, they report 32.8 Gbit/s on an NVIDIA GTX 285 graphics card (GT200b GPU), not including data transfers. They also report detailed performance numbers of AES encryption in the nonparallel CBC mode for different numbers of independent streams on an NVIDIA GTX 580 graphics card (GF110 GPU).

Note that these high throughputs of AES on GPUs can only be achieved by performing AES encryption on thousands of blocks in parallel. This amount of data-level parallelism can certainly be found for some database applications or when writing large amounts of data to an encrypted hard disk. The encryption of typically small Internet packages in applications that do not just need high throughput but also low latency will still do better with a CPU-based approach, not only when using CPUs that support AES in hardware. For example, the bitsliced implementation for Intel processors presented in [37] encrypts 1,500-byte packets in 7.27 cycles per byte on a 2,668 MHz Intel Core i7 920 CPU. This corresponds to a throughput of more than 11.7 Gbit/s on 4 cores.

8.2.2 Asymmetric Cryptography on GPUs

Asymmetric cryptographic primitives can be accelerated by laying off the computations from the CPU to the GPU. As for symmetric primitives like AES one way to obtain the necessary degree of parallelism is to consider operations on many independent messages. However, there is another source for parallelism inherent in the algorithms. Most state-of-the-art asymmetric algorithms involve operations on large integers, for example RSA signature generation is the computation of m^d mod n, where m, d and n are integers of 1,024 bits or larger. Arithmetic on such integers, in particular multiplication, squaring and modular reduction, needs to be decomposed in many operations on machine words. Elliptic-curve cryptography involves modular arithmetic on integers of smaller size, typically between 160 and 256 bits, but arithmetic on those integers still decomposes into many operations on machine words. For example, when using a multiplier with 32-bit output, schoolbook multiplication of two 256-bit integers requires 256 multiplications of 16-bit limbs and 240 additions of the 32-bit multiplication outputs. Most of these operations are independent and can be done in parallel by multiple threads. Exploiting such parallelism inside one computation has some obvious advantages. If multiple threads process one input stream together, fewer independent input streams are required to make use of the computational power of the GPU. This makes GPU computations attractive also for applications that require low latency rather than high throughput. Furthermore, when multiple threads carry out one computation together the overall amount of data involved in the computations is smaller; this can be used to fit all data into memory domains that offer low-latency access. However, exploiting data-level parallelism inside computations like big-integer multiplication comes with the disadvantage that it involves overhead from thread synchronization and exchange of data between treads.

Several papers describe implementations of RSA on modern graphics cards. In [55], Szerwinski and Güneysu describe a CUDA implementation that performs 813 modular exponentiations (RSA encryption) of 1,024-bit integers on a NVIDIA 8800 GTS graphics card. This paper furthermore reports a throughput of 104.3 modular exponentiations for 2,048-bit RSA encryption. Harrison and Waldron in [31] focus on RSA decryption and report 5536.75 RSA-1024 decryptions per second on an NVIDIA 8800 GTX graphics card. This computation can make use of the Chinese Remainder Theorem to perform arithmetic on half-size integers. The RSA implementation included in the SSL proxy described in [36] can perform for example 74732 RSA-1024 encryptions or 12044 RSA-2048 encryptions per second on an NVIDIA GTX 580 graphics card. What is particularly interesting about this implementation is that it does not purely focus on throughput but also needs to keep the latency low enough for the application in the SSL proxy. For RSA-1024 the latency is at 3.8 ms, for RSA-2048 it is 13.83 ms.

To put this into perspective to what is currently possible on CPUs, the eBACS benchmarking project [11] reports, for example, more than 11,000 1,024-bit integer exponentiations per second on all six cores of an AMD Phenom II X6 1090T. Again, this speed does not require the large number of independent parallel computations that GPU implementations need and although it is much slower from a pure throughput perspective, it may be the better choice for applications that do not process multiple messages in parallel.

Elliptic-curve cryptography has been implemented on GPUs. Szerwinski and Güneysu report 1,412 scalar multiplications on the NIST P-224 elliptic curve on an NVIDIA 8800 GTS graphics card in [55]. On the same curve but the more recent

NVIDIA GTX 285 graphics card Antão, Bajard, and Sousa report 9990 scalar multiplications per second. More than an order of magnitude slower at significantly lower
security is the implementation of scalar multiplication on an elliptic-curve over a
binary field described in [19]. Cohen and Parhi report only 96.5 scalar multiplications per second.

Elliptic-curve scalar multiplication has received more attention on CPUs, for
example [10] reports 226872 cycles for a scalar multiplication on a 255-bit elliptic
curve on an Intel Xeon E5620 CPU running at 2.4 GHz. This corresponds to more
than 40,000 scalar multiplication per second on all four cores. Even faster speeds for
CPU implementations are reported in [34] for scalar multiplication on elliptic curves
with efficiently computable endomorphisms. These comparative numbers may suggest that GPU implementations of elliptic-curve cryptography cannot compete with
state-of-the-art CPU implementations, not even in throughput-oriented applications.
However, the next section describes implementations of elliptic-curve operations on
GPUs for cryptanalysis that outperform CPU implementations. The reason that there
are no faster GPU implementations targeting constructive applications may be that
there are simply not many applications that require only throughput and can ignore
latency.

An asymmetric cryptosystem that appears to be much better suited for implementation on GPUs than elliptic-curve cryptography or RSA is NTRU. The central
operation for encryption and decryption is convolution which can be carried out
by many threads without significant communication or synchronization due to its
parallel structure. In [32], Hermans, Vercauteren, and Preneel describe an implementation of NTRU with a set of parameters that aims at the 256-bit security level.
This implementation is able to perform 218,000 encryption operations per second
on an NVIDIA GTX 280 graphics card (GT200 GPU).

8.3 GPUs in Cryptanalysis

Cryptanalytical computations are in many ways similar to cryptographic computations. In many cases, breaking a cryptographic system means executing the same or
very similar computations that are used in the constructive use of the cryptosystem.
One example is brute-force key recovery of symmetric ciphers that simply performs
encryption with many different keys. Another example is hash-function collision
search with the computationally most expensive part being computing hashes. An
example in the cryptanalysis of asymmetric systems is Pollard's rho algorithm to
solve the discrete logarithm problem (DLP). Again the computationally most expensive part are the same or very similar operations in the same mathematical structures
that are involved in the legitimate use of the DLP-based system.

In three very important points, cryptanalytical computations are different from
cryptographic computations and all three make them even better suited for GPUs.
First, they typically involve an arbitrary amount of data-level parallelism, the same
computations are carried out on huge amounts of independent data; this is exactly

the sort of computations that GPUs are best at. Second, many of these computations do not care about latency, they are purely throughput oriented. Third, there is no confidential data involved that needs to be protected, one could say that the opposite is true, revealing the confidential data is the target of the computation.

The most obvious applications of GPUs for cryptanalysis are attacks against symmetric encryption and hash functions. Various commercial solutions for password recovery already include GPU implementations to speed up the computations. These tools typically try out many different passwords from a given word list and either compare with given hash values or derive symmetric keys from a list of known passphrases to recover the content of encrypted files.

The power of GPUs was also used by the winner of Engineyard's SHA-1 programming contest: The task was to find an input to SHA-1 that has minimal Hamming distance to a given hash value. Lange in [42] reports that code by Bernstein is able to compute more than 328 million hashes per second on an NVIDIA GTX 295 graphics card. Each of these hashes required computation of only one 64-byte block of input, so this corresponds to a throughput of more than 167 Gbit/s. As a comparison, all four cores of a 2.4 GHz Intel Core 2 Quad Q6600 CPU involved in the same computation computed 47 million hashes per second. Also the SHA-3 candidates have been implemented on GPUs, password recovery being the most obvious application. In [13], Bos and Stefan describe implementations of all of the SHA-3 round-2 candidates on NVIDIA GT200 GPUs. The reported throughputs reach from 0.9 Gbit/s for Cubehash 16/1 up to 36.8 Gbit/s for Blake-32 and BMW-256 on one GPU of an NVIDIA GTX 295 graphics card. Again to put this into perspective, on a recent CPU, the Intel Core i7-2600K, hashing with Blake-32 takes 6.68 cycles/byte [11]; this corresponds to a throughput of 16.29 Gbit/s.

These applications in password recovery are quite straightforward, but GPUs have also been used for cryptanalysis of asymmetric systems. One of the most famous problems closely related to the RSA cryptosystem is the factorization of large numbers. A critical step inside the factorization of large RSA numbers with the number-field sieve is the factorization of many smaller numbers using the elliptic-curve factorization method (ECM). In [9], Bernstein, Chen, Cheng, Lange, and Yang describe an implementation of ECM for 280-bit numbers. This implementation running on both GT200b GPUs of an NVIDIA GTX 295 graphics card outperforms a state-of-the-art CPU implementation running on all 4 cores of an Intel Core 2 Quad Q9550 by a factor of more than 2.8. The GPU implementation tries 400.7 curves per second, the CPU implementation 142.17 curves per second. A much higher ECM throughput for slightly smaller numbers is reported in [8]. For example, for 210-bit numbers a GTX 295 graphics card is reported to try 4,928 curves per second. Although these numbers are not as impressive as the speedups achieved by using GPUs in symmetric cryptanalysis, the results show that GPUs can also be used to speed up elliptic-curve arithmetic.

This is confirmed for elliptic curves over binary fields in [7]. As part of a large effort to solve Certicom's elliptic-curve discrete-logarithm-problem (ECDLP) challenge ECC2K-130 [15, 16], this paper presents an implementation of Pollard's rho algorithm for GT200b GPUs. On the two GPUs inside the GTX 295 graphics card, this implementation is able to perform 63 million Pollard rho iterations per second. As a comparison, the CPU implementation computing the same iteration function described in [6] performs 22.45 million iterations per second on all 4 cores of an Intel Core 2 Extreme Q6850 CPU.

GPUs have also been considered for solving the DLP on elliptic curves over large prime fields. The implementation described in [17] targets an ECDLP on a 109-bit prime curve and is reported to "have generated about 320.000 points/second" on an NVIDIA 8800 GTS graphics card with a G92 GPU. This probably means 320,000 iterations per second, but it is unclear what the exact performance of the implementation is.

8.4 Malware Detection on GPUs

Similar to cryptographic applications, malware-detection software is expected to operate in the background with as little influence on the system's performance as possible. A large computational task of virus detection is pattern matching of byte sequences found in files with known signatures of malware. This task is highly parallel, so it is an application that can run at high speed on GPUs.

Seamans and Alexander describe an implementation of parallel virus signature matching for NVIDIA GPUs in [54]. The authors integrated this implementation into the ClamAV virus scanner [18] and compare the performance of this implementation running on an NVIDIA GTX 7800 graphics card to the original CPU implementation running on an unspecified 3-GHz Intel Pentium 4 CPU; the authors do not specify the number of CPU cores used for this comparison. The speedup obtained by running the pattern matching on the GPU depends on the number of matches, because matches need to be communicated back to the CPU. If no matches are found, the GPU implementation is 27 times faster than the CPU implementation; this factor drops to 17 at a match rate of 1 % and further to 11 at a match rate of 50 %.

In [59], Vasiliadis and Ioannidis describe an implementation of virus-signature pattern matching targeting more recent NVIDIA GPUs. Their implementation filters out clean, unsuspicious regions, it is included as a preprocessing step into the ClamAV [18] virus scanner. The authors achieve a 100-times higher throughput with this approach running on an NVIDIA GTX295 graphics card compared to the CPU-only virus scanner running on 1 core of an Intel Xeon E5520 CPU. Compared to the CPU implementation running on 8 cores of 2 CPUs, the speedup is still 10-fold.

The approach of using the GPU as a coprocessor for malware detection is not purely academic. In December 2009, Kaspersky announced that they incorporated an implementation of the "similarity service" for NVIDIA Tesla cards into their infrastructure. The press release [38] does not give much detail but claims a 360-times speedup of the GPU implementation running on an NVIDIA Tesla S1070 compared to the a CPU implementation running on a 2.6 GHz Intel Core 2 Duo processor. This comparison does not give details about the number of CPU cores used, it also does not say whether the speedup is obtained from running the GPU code on one or all four GPUs included in the Tesla S1070.

Signature matching is also one of the main performance bottlenecks of network-intrusion-detection systems. Consequently, GPUs can also be used to speed up such systems. This was first described by Jacob and Brodley who use a traditional GPGPU approach targeting the NVIDIA 6800 GT graphics card in [35]. They conclude that with their GPU pattern-matching extension to the open-source intrusion detection system Snort "there was no appreciable speedup in packet processing under normal-load conditions". A more efficient approach targeting the NVIDIA 8600 GT graphics card is described in [58]. Vasiliadis, Antonatos, Polychronakis, Markatos, and Ioannidis present a GPU pattern-matching extension of Snort that increases the overall Snort throughput capacity by a factor of two compared to CPU-only Snort running on a 3.4 GHz Intel Pentium 4 processor. The most comprehensive solution for intrusion detection involving GPUs to date is presented in [61]. Vasiliadis, Polychronakis, and Ioannidis describe a Snort-based intrusion detection solution that exploits parallelism on multiple levels. The system makes use of multiple GPUs and multiple CPU cores and copes with a network throughput of 5.2 GBit per second. This performance number was achieved on a system with two NVIDIA GTX 480 graphics cards and two Intel Xeon E5520 CPUs. The pure pattern-matching step reaches a peak performance of more than 70 GBit per second on the two graphics cards.

8.5 Malware Targeting GPUs

GPUs can not only be used to accelerate malware detection, malware itself can also use GPUs to hide from virus scanners. In [60], Vasiliadis, Polychronakis, and Ioannidis describe an implementation of a malware unpacker running on an NVIDIA GPU. The complete malware package consists of two parts, the unpacker running on the GPU and the actual malware that runs on the CPU. These two parts communicate through host memory mapped into the GPUs address space.

Unpackers are one of the most common techniques to hide malware from scanners: The malware code is packed or encrypted in some way and gets unpacked (decrypted) only when it is actually executed. The advantage from the malware author's perspective of using GPU code for the unpacker is twofold as it offers better protection against detection by both static and dynamic malware-detection systems. Static systems try different known unpacking techniques to recover the original malware. This becomes harder if the computational power of the GPU is used for computationally more expensive unpacking algorithms. Dynamic unpacking tools use the unpacker that is included in the malware, for example inside a sandbox or virtual machine. At least existing dynamic tools do not support GPU binaries and would thus fail.

As a second step [60], Yang and Goodman also describe GPU-assisted run-time polymorphism on the function level. The malware binary is never fully decrypted, only the currently executed function resides in memory, when returning from a function call the function is encrypted again and the next function context is decrypted.

The implementations are still just a proof of concept and there have been no reports of real-world malware using the GPU to hide from scanners. Some of the claimed advantages of using the GPU to hide malware can obviously be addressed by malware-detection tools also using the GPU. Others will require better tools for static and dynamic analysis of GPU code. It will be interesting to see whether or how much GPUs become a new battlefield in the everlasting fight between malware and malware detection.

8.6 Accessing GPUs from Web Applications

Software becomes more and more Web centric; programs such as office suites, image-processing software, and games, which traditionally run directly on a computer, are now implemented as applications running inside a Web browser. The most consistent implementation of this approach is Google's Chromium OS, an operating system that is designed to run a Web browser as only application—all other software is Web applications running inside this browser.

As a consequence of higher demands for advanced graphics in Web applications, various technologies have been developed to let those applications access the GPU. The most prominent three approaches are WebGL developed initially by Mozilla and now by the Khronos group [39], Silverlight 5 developed by Microsoft [46], and Flash 11 developed by Adobe.

All of these approaches have in common that they expose the graphics driver and hardware to software originating from the Internet and thus from typically untrusted sources. The implications for security of this approach have so far been discussed primarily for WebGL. In March 2011, version 1.0 of the specification of WebGL was released by the Khronos group. Browsers supporting this specification include Mozilla's Firefox and Google's Chrome. Only about 2 months later, Forshaw publicized an article [24] that describes several security issues in these implementations and claims that these are actually caused by design flaws in WebGL. One of these issues is the possibility to remotely exploit vulnerabilities in the graphics driver to crash or freeze the system. Another one is a cross-site timing attack that extracts image data processed on the GPU. A follow-up article by Forshaw, Stone, and Jordon [25] describes an attack targeting the WebGL implementation of Firefox. In this attack, a malicious website can take screenshots of arbitrary applications running on the client computer.

Khronos has reacted to these articles in a WebGL security whitepaper [40] that describes approaches to address the security issues. These approaches can not all be implemented only on the browser side but need support on the graphics-driver side.

Even without any vulnerabilities in the framework, the computational power of GPUs enables attacks that would otherwise be infeasible. For example, the JavaScript bitcoin miner of bitp., it has been discontinued because "Javascript is just too slow to mine bitcoins" [1]. This would certainly be different with the computing power of GPUs open to Web applications. Mining bitcoins on the GPU in the background while a user is visiting a website could on the one hand be a legitimate new way of funding websites (if the user is asked for permission), on the other hand it would most likely also be done silently and thus become sort of a Web Trojan.

The discussion about WebGL security and more general security issues related to exposing the GPU and the graphics driver to untrusted code from the Internet is still ongoing. On the one hand WebGL, Silverlight 5, and Flash 11 are still very young technologies and maybe some of the vulnerabilities are just teething troubles. On the other hand, the concept of letting Web applications access the driver layer of a client's operating system flies in the face of conventional wisdom that tells us that untrusted code should be kept as far away from any critical parts of a system as possible. The future will have to show what changes are required to browsers, operating systems, and drivers to deal with current and future security vulnerabilities and whether it is actually possible to establish these technologies without exposing their users to severe risks.

References

1. "1bitc0inplz". Edited forum post on bitcointalk.org, 2011. https://bitcointalk.org/index.php?topic=9042.0.
2. Advanced Microdevices *Inc. R600-Family Instruction Set Architecture*, 2008. http://developer.amd.com/gpu_assets/r600isa.pdf.
3. Advanced Microdevices Inc. *AMD Accelerated Parallel Processing OpenCL Programming Guide, rev. 1.3f*, 2011. http://developer.amd.com/sdks/AMDAPPSDK/assets/AMD_Accelerated_Parallel_Processing_OpenCL_Programming_Guide.pdf.
4. Advanced Microdevices Inc. *Evergreen Family Instruction Set Architecture Instructions and Microcode*, 2011. http://www.amddevcentral.com/sdks/AMDAPPSDK/assets/AMD_Evergreen-Family_Instruction_Set_Architecture.pdf.
5. Advanced Microdevices Inc. *R700-Family Instruction Set Architecture*, 2011. http://developer.amd.com/gpu_assets/R700-Family_Instruction_Set_Architecture.pdf.
6. Daniel V. Bailey, Lejla Batina, Daniel J. Bernstein, Peter Birkner, Joppe W. Bos, Hsieh-Chung Chen, Chen-Mou Cheng, Gauthier Van Damme, Giacomo de Meulenaer, Luis Julian Dominguez Perez, Junfeng Fan, Tim Güneysu, Frank Gürkaynak, Thorsten Kleinjung, Tanja Lange, Nele Mentens, Ruben Niederhagen, Christof Paar, Francesco Regazzoni, Peter Schwabe, Leif Uhsadel, Anthony Van Herrewege, and Bo-Yin Yang. Breaking ECC2K-130. Cryptology ePrint Archive, Report 2009/541, 2009. http://eprint.iacr.org/2009/541/.
7. Daniel J. Bernstein, Hsieh-Chung Chen, Chen-Mou Cheng, Tanja Lange, Ruben Niederhagen, Peter Schwabe, and Bo-Yin Yang. ECC2K-130 on NVIDIA GPUs. In Guang Gong and Kishan Chand Gupta, editors, *Progress in Cryptology - INDOCRYPT 2010*, volume 6498 of LNCS, pp. 328–346. Springer, 2010. http://cryptojedi.org/papers/#gpuev11.
8. Daniel J. Bernstein, Hsueh-Chung Chen, Ming-Shing Chen, Chen-Mou Cheng, Chun-Hung Hsiao, Tanja Lange, Zong-Cing Lin, and Bo-Yin Yang. The billion-mulmod-per-second pc. In *Workshop Record of SHARCS'09: Special-purpose Hardware for Attacking Cryptographic Systems*, pp. 131–144, 2009. http://www.hyperelliptic.org/tanja/SHARCS/record2.pdf.
9. Daniel J. Bernstein, Tien-Ren Chen, Chen-Mou Cheng, Tanja Lange, and Bo-Yin Yang. ECM on graphics cards. In Antoine Joux, editor, *Advances in Cryptology - EUROCRYPT 2009*, volume 5479 of LNCS, pp. 483–501. Springer, 2009. http://cr.yp.to/papers.html#gpuecm.
10. Daniel J. Bernstein, Niels Duif, Tanja Lange, Peter Schwabe, and Bo-Yin Yang. High-speed high-security signatures, 2011. http://cryptojedi.org/papers/#ed25519.

11. Daniel J. Bernstein and Tanja Lange. eBACS: ECRYPT benchmarking of cryptographic systems. http://bench.cr.yp.to (Accessed Nov. 3, 2011).
12. Eli Biham. A fast new DES implementation in software. In Eli Biham, editor, *Fast Software Encryption*, volume 1267 of *LNCS*, pp. 260–272. Springer, 1997. http://www.cs.technion.ac.il/users/wwwb/cgi-bin/tr-get.cgi/1997/CS/CS0891.pdf.
13. Joppe W. Bos and Deian Stefan. Performance analysis of the SHA-3 candidates on exotic multicore architectures. In Stefan Mangard and François-Xavier Standaert, editors, *Cryptographic Hardware and Embedded Systems - CHES 2010*, volume 6225 of *LNCS*, pp. 279–293. Springer, 2010. http://www.ee.cooper.edu/stcfan/pubs/conference/ches2010.pdf.
14. BrookGPU. http://graphics.stanford.edu/projects/brookgpu/, Accessed Nov. 5, 2011.
15. Certicom ECC Challenge, 1997. http://www.certicom.com/images/pdfs/cert_ecc_challenge.pdf, Accessed Nov. 6, 2011.
16. ECC Curves List, 1997. http://www.certicom.com/index.php/curves-list, Accessed Nov. 6, 2011.
17. Marta Chinnici, Salvatore Cuomo, Maurizio Laporta, Alberto Pizzirani, and Silvio Migliori. CUDA based implementation of parallelized Pollard's rho algorithm for ECDLP. In *Final Workshop of Grid Projects, "Pon Ricerca 2000–2006, Avviso 1575"*, 2009. http:///www.cresco.enea.it/Documenti/web/presentazioni/ProceedingsCatan%ia2009/7chinnici.pdf.
18. Clam AntiVirus. http://clamav.net, Accessed Nov 1, 2011.
19. Aaron E. Cohen and Keshab K. Parhi. GPU accelerated elliptic curve cryptography in $GF(2^m)$. In *53rd IEEE International Midwest Symposium on Circuits and Systems (MWSCAS)*, pages 57–60. IEEE, 2010.
20. James Coleman and Perry Taylor. Hardware level IO benchmarking of PCI Express. White Paper, Intel Corporation, 2008. ftp://download.intel.com/design/intarch/PAPERS/321071.pdf.
21. Debra L. Cook and Angelos D. Keromytis. *CryptoGraphics: Exploiting Graphics Cards For Security*, volume 20 of *Advances in Information Security*. Springer, 2006.
22. Joan Daemen and Vincent Rijmen. AES proposal: Rijndael, version 2, 1999. http://csrc.nist.gov/archive/aes/rijndael/Rijndael-ammended.pdf.
23. Michael J. Flynn. Very high-speed computing systems. *Proceedings of the IEEE*, 54(12):1901–1909, 1966. http://ieeexplore.ieee.org/iel5/5/31091/01447203.pdf.
24. James Forshaw. WebGL - a new dimension for browser exploitation. Blog entry on the Context Information Security Ltd. blog, 2011. http://www.contextis.com/resources/blog/webgl/.
25. James Forshaw, Paul Stone, and Michael Jordon. WebGL - more WebGL security flaws. Blog entry on the Context Information Security Ltd. blog, 2011. http://www.contextis.com/resources/blog/webgl2/.
26. Khronos OpenCL Working Group. *The OpenCL Specification, Version* 1.0, 2008. http://www.khronos.org/registry/cl/specs/opencl-1.0.pdf.
27. Khronos OpenCL Working Group. *The OpenCL Specification, Version* 1.1, 2010. http://www.khronos.org/registry/cl/specs/opencl-1.1.pdf.
28. Mark Harris. *Real-Time Cloud Simulation and Rendering*. Ph.D. thesis, University of North Carolina at Chapel Hill, 2003. http://www.markmark.net/dissertation/index.html.
29. Owen Harrison and John Waldron. AES encryption implementation and analysis on commodity graphics processing units. In Pascal Paillier and Ingrid Verbauwhede, editors, *Cryptographic Hardware and Embedded Systems - CHES 2007*, volume 4727 of *LNCS*, pages 209–226. Springer, 2007.
30. Owen Harrison and John Waldron. Practical symmetric key cryptography on modern graphics hardware. In *USENIX Security Symposium*, pages 195–209. Usenix Association, 2008.
31. Owen Harrison and John Waldron. Efficient acceleration of asymmetric cryptography on graphics hardware. In Bart Preneel, editor, *Progress in Cryptology - AFRICACRYPT 2009*, volume 5580 of *LNCS*, pages 350–367. Springer, 2009.
32. Jens Hermans, Frederik Vercauteren, and Bart Preneel. Speed records for NTRU. In Josef Pieprzyk, editor, *Topics in Cryptology - CT-RSA 2010*, volume 5985 of LNCS, pages 73–88. Springer, 2010.

33. Yunqing Hou. asfermi: An assembler for the NVIDIA Fermi instruction set, 2011. http://code. google.com/p/asfermi/, Accessed Nov. 1, 2011.
34. Zhi Hu, Patrick Longa, and Maozhi Xu. Implementing 4-dimensional GLV method on GLS elliptic curves with j-invariant 0. Cryptology ePrint Archive, Report 2011/315, 2011. http:// eprint.iacr.org/2011/315.
35. Nigel Jacob and Carla Brodley. Offloading IDS computation to the GPU. In *Proceedings of the 22nd Annual Computer Security Applications Conference*, pp. 371–380. IEEE Computer Society, 2006. http://www.acsac.org/2006/papers/74.pdf.
36. Keon Jang, Sangjin Han, Seungyeop Han, Sue Moon, and KyoungSoo Park. SSLShader: cheap SSL acceleration with commodity processors. In David G. Andersen and Sylvia Ratnasamy, editors, *Proceedings of the 8th USENIX Symposium on Networked Systems Design and Implementation (NSDI '11)*. ACM Press, 2011. http://www.usenix.org/events/nsdi11/tech/full_papers/ Jang.pdf.
37. Emilia Käsper and Peter Schwabe. Faster and timing-attack resistant AES-GCM. In Christophe Clavier and Kris Gaj, editors, *Cryptographic Hardware and Embedded Systems - CHES 2009*, volume 5747 of LNCS, pp. 1–17. Springer, 2009. http://cryptojedi.org/papers/#aesbs.
38. Kaspersky Lab. Kaspersky Lab utilizes NVIDIA technologies to enhance protection, 2009. http://www.kaspersky.com/about/news/business/2009/Kaspersky_Lab_utilizes_ NVIDIA_technologies_to_enhance_protection.
39. WebGL - OpenGL ES 2.0 for the web, 2011. http://www.khronos.org/webgl/.
40. WebGL security, 2011. http://www.khronos.org/webgl/security/, Accessed Nov. 4. 2011.
41. Robert Könighofer. A fast and cache-timing resistant implementation of the AES. In Tal Malkin, editor, *Topics in Cryptology* - CT-RSA 2008, volume 4964 of LNCS, pages 187–202. Springer, 2008.
42. Tanja Lange. CodingCrypto's page on Engineyard's programming contest, 2009. http://www. win.tue.nl/cccc/sha-1-challenge.html, Accessed Nov. 5, 2011.
43. Jed Lengyel, Mark Reichert, Bruce R. Donald, and Donald P. Greenberg. Real-time robot motion planning using rasterizing computer graphics hardware. *SIGGRAPH Computer Graphics*, 24(4):327–335, 1990. http://dl.acm.org/citation.cfm?id=97915&CFID=67002344& CFTOKEN=69396923.
44. Svetlin A. Manavski. CUDA compatible GPU as an efficient hardware accelerator for AES cryptography. In 2007 *IEEE International Conference on Signal Processing and Communications (ICSPC 2007)*, pages 65–68. IEEE, 2007. http://www.manavski.com/downloads/PID505889. pdf.
45. Mitsuru Matsui. How far can we go on the x64 processors? In Matthew Robshaw, editor, Fast *Software Encryption*, volume 4047 of LNCS, pp. 341–358. Springer, 2006. http://www.iacr. org/archive/fse2006/40470344/40470344.pdf.
46. What's new in Silverlight 5, 2011. http://www.silverlight.net/learn/overview/what%27s-new- in-silverlight-5.
47. Andrew Moss, Daniel Page, and Nigel P. Smart. Toward acceleration of RSA using 3d graphics hardware. In Steven D. Galbraith, editor, *Cryptography and Coding*, volume 4887 of LNCS, pp. 364–383. Springer, 2007. http://www.cs.bris.ac.uk/Publications/Papers/2000772.pdf.
48. Ruben Niederhagen. Calasm, 2011. http://www.polycephaly.org/projects/calasm/.
49. NVIDIA Corporation. *Tuning CUDA Applications for Fermi, Version 1.0*, 2010. http://developer.download.nvidia.com/compute/cuda/3_0/toolkit/docs/NVIDIA_ FermiTuningGuide.pdf.
50. NVIDIA Corporation. *NVIDIA CUDA - NVIDIA CUDA C Programming Guide, Version 4.0*, 2011. http://developer.download.nvidia.com/compute/cuda/4_0/toolkit/docs/CUDA_C_ Programming_Guide.pdf.
51. NVIDIA Corporation. *OpenCL Programming Guide for the CUDA Architecture*, 2011. http://developer.download.nvidia.com/compute/DevZone/docs/html/OpenCL/doc/OpenCL_ Programming_Guide.pdf.
52. Dag Arne Osvik, Joppe W. Bos, Deian Stefan, and David Canright. Fast software AES encryption. In Seokhie Hong and Tetsu Iwata, editors,*Fast Software Encryption*, volume 6147 of LNCS, pages 75–93. Springer, 2010.

53. Ádám Rák. AMD-GPU-Asm-Disasm, 2011. https://github.com/rakadam/AMD-GPU-Asm-Disasm/, Accessed Nov. 1, 2011.
54. Elizabeth Seamans and Thomas Alexander. Fast virus signature matching on the GPU. In Hubert Nguyen, editor,*GPU Gems* 3, pp. 771–784. Addison-Wesley, 2007. http://http.developer.nvidia.com/GPUGems3/gpugems3_ch35.html, Accessed Nov. 1, 2011.
55. Robert Szerwinski and Tim Güneysu. Exploiting the power of GPUs for asymmetric cryptography. In Elisabeth Oswald and Pankaj Rohatgi, editors, *Cryptographic Hardware and Embedded Systems -CHES 2008*, volume 5154 of *LNCS*, pages 79–99. Springer, 2008.
56. Eran Tromer, Dag Arne Osvik, and Adi Shamir. Efficient cache attacks on AES, and countermeasures. *Journal of Cryptology*, 23(1):37–71, 2010. http://people.csail.mit.edu/tromer/papers/cache-joc-official.pdf.
57. Wladimir J. van der Laan. Cubin utilities, 2007. https://github.com/laanwj/decuda/wiki, Accessed Nov. 1, 2011.
58. Giorgos Vasiliadis, Spiros Antonatos, Michalis Polychronakis, Evangelos P. Markatos, and Sotiris Ioannidis. Gnort: High performance network intrusion detection using graphics processors. In Richard Lippmann, Engin Kirda, and Ari Trachtenberg, editors, *Recent Advances in Intrusion Detection*, volume 5230 of *LNCS*, pp. 116–134. Springer, 2008. http://www.ics.forth.gr/_pdf/brochures/gnort.raid08.pdf.
59. Giorgos Vasiliadis and Sotiris Ioannidis. GrAVity: A massively parallel antivirus engine. In Somesh Jha, Robin Sommer, and Christian Kreibich, editors, Recent Advances *In Intrusion Detection, LNCS*, pp. 79–96. Springer, 2010. http://dcs.ics.forth.gr/Activities/papers/gravity-raid10.pdf.
60. Giorgos Vasiliadis, Michalis Polychronakis, and Sotiris Ioannidis. GPU-assisted malware. In Jean-Yves Marion, Noam Rathaus, and Cliff Zou, editors, *Proceedings of the 5th International Conference on Malicious and Unwanted Software (MALWARE)*. IEEE, 2010. dcs.ics.forth.gr/Activities/papers/gpumalware.malware10.pdf.
61. Giorgos Vasiliadis, Michalis Polychronakis, and Sotiris Ioannidis. MIDeA: A multi-parallel intrusion detection architecture. In George Danezis and Vitaly Shmatikov, editors, *Proceedings of the 18th ACM/SIGSAC Conference on Computer and Communications Security*, pp. 297–308. ACM Press, 2011. http://dcs.ics.forth.gr/Activities/papers/midea.css11.pdf.
62. Takeshi Yamanouchi. AES encryption and decryption on the GPU. In Hubert Nguyen, editor, *GPU Gems 3*, pp. 785–804. Addison-Wesley, 2007. http://http.developer.nvidia.com/GPUGems3/gpugems3_ch35.html, Accessed Nov. 1, 2011.
63. Jason Yang and James Goodman. Symmetric key cryptography on modern graphics hardware. In Kaoru Kurosawa, editor, *Advances in Cryptology - ASIACRYPT 2007*, volume 4833 of *LNCS*, pages 249–264. Springer, 2007.

Chapter 9
A Survey of Recent Results in FPGA Security and Intellectual Property Protection

François Durvaux, Stéphanie Kerckhof, Francesco Regazzoni and
François-Xavier Standaert

Abstract Field programmable gate arrays (FPGAs) are reconfigurable devices which have emerged as an interesting trade-off between the efficiency of application-specific integrated circuits (ASICs) and the versatility of standard microprocessors [81]. Progresses over the last 10 years have improved their capabilities to the point where they can hold a complete system on a chip (SoC) and thus become an attractive platform for an increasing number of applications (e.g., signal processing, image processing, aerospace, etc.). In view of the important data manipulated by these devices, but also of the high amount of intellectual property (IP) they may contain, security-related questions have arisen. First, can we use FPGAs as security devices for example, securely and efficiently encrypting sensitive data (in particular when compared to software solutions)? Second, how can we guarantee that the IP corresponding to FPGA designs is protected (i.e., cannot be easily counterfeited)? Such questions have been the target of a large number of papers in the literature,

François Durvaux: PhD student funded by the Walloon region MIPSs project.
Stéphanie Kerckhof: PhD student funded by a FRIA grant, Belgium.
François-Xavier Standaert: Associate Researcher of the Belgian Fund for Scientific Research
(FNRS-F.R.S.).
François-Xavier Standaert: Work funded in part by the ERC project 280141 (acronym CRASH).

F. Durvaux (✉)· S. Kerckhof · F. Regazzoni · F. X. Standaert
UCL Crypto Group, Université catholique de Louvain, Louvain-la-Neuve, Belgium
e-mail: francois.durvaux@uclouvain.be

S. Kerckhof
e-mail: stephanie.kerckhof@uclouvain.be

F. Regazzoni
e-mail: regazzoni@alari.ch

F. X. Standaert
e-mail: fstandae@uclouvain.be

F. Regazzoni
ALaRI Institute, University of Lugano, Lugano, Switzerland

K. Markantonakis and K. Mayes (eds.), *Secure Smart Embedded Devices,*
Platforms and Applications, DOI: 10.1007/978-1-4614-7915-4_9,
© Springer Science+Business Media New York 2014

including several surveys, example [13, 71, 83]. In this chapter, we take another look at them and review a number of important recent results related to security IPs and IP security in modern reconfigurable devices. The chapter is structured in three main sections. First, we briefly describe the structure of recent FPGAs. Next, we discuss security IPs in FPGAs, taking the example of symmetric encryption with the AES Rijndael, and including their performance evaluations and resistance against physical attacks. Finally, we emphasize recent trends for improving IP security in FPGAs, including bitstream security, the use of code watermarking techniques and the exploitation of physically unclonable functions (PUFs).

9.1 FPGAs: An Overview

In this section, we introduce the major features of field programmable gate arrays (FPGAs) for the non-familiar reader. First, we present the overall structure of the devices. Next, we briefly describe the different steps of an FPGA design flow. Finally, we discuss the different technologies of reconfigurable devices that are publicly available.

9.1.1 Structure

For convenience, we will focus on the two main FPGA manufacturers: Altera [3] and Xilinx [84].[1] The configurable logic block and the logic array block (CLB and LAB) are the basic logic cells for the most recent Xilinx (e.g., Virtex-7) and Altera (e.g., Stratix-V) FPGAs. Such devices typically contain an array of these cells connected together through a configurable routing matrix. The CLB is composed of multiple slices (Fig. 9.1) and the LAB is composed of multiple adaptive logic modules (ALMs, Fig. 9.2). The main components of slices and ALMs are the look-up tables (LUTs) and the registers. In the latest FPGAs (Xilinx Virtex-7 or the Altera Stratix-V), the LUTs are 6-bit input and 2-bit output functions generators. They can also be configured as small embedded memories or shift registers. In older technologies, the LUT input was generally limited to 4 bits. Slices and ALMs combine the LUTs with a chaining logic in order to allow efficient arithmetic operations. Next, this combinatorial part of the logic cells is followed by registers used to generate synchronous logic. Efficient FPGA designs essentially try to take advantage of these resources in the best manner in order to perform some algorithmic task. Hence, slices, LABs, ALMs, LUTs, and registers are the typical figures of merit used to evaluate the performances of FPGA implementations (see the next section).

---

[1] These two manufacturers produce mainly "volatile" FPGAs in which the configuration is stored in nonvolatile memory devices like EEPROM or Flash. Nonvolatile FPGAs also exist but are out of the scope of this chapter.

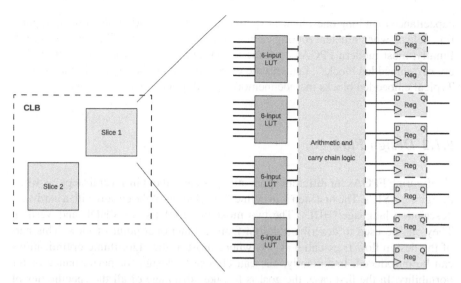

Fig. 9.1 Xilinx configurable logic block and detailed slice (Dashed registers are optional, they can be bypassed)

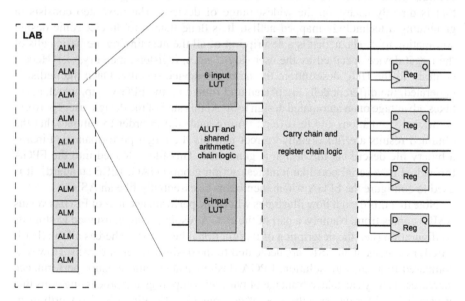

Fig. 9.2 Altera logic array block and detailed adaptive logic module

In addition, it is worth mentioning that the routing matrix of the FPGAs also has a major impact on final performances. These routes, allowing the connection between the different computational blocks and memories, are generally structured according to the connection lengths. Quite naturally, small wires have much lower effective capacitance than long ones, hence explaining significant variations of the computation delays, when efficient routing cannot be ensured with "local" connections [69]. Finally, most modern FPGAs combine reconfigurable logic elements with a variety of "embedded blocks", i.e., specific elements that are hardwired in the devices. Typical embedded blocks include memories, multipliers, and processors.

9.1.2 Design Flow

Configuring FPGAs for different applications is carried out in several steps, as when designing ASICs. The first step is to define the behavior of the circuit with a hardware description language (HDL). The two most usual HDLs are VHDL and Verilog. They can be used to describe a design from a functional point of view. This part of the design flow is essential as it is where most of the algorithmic optimizations can be introduced. Importantly, one can choose to design for performance or for portability. In the first case, the goal is to take advantage of all the specificities of the target platform, in order to increase performances (e.g., particular logic element configurations, embedded blocks, ...). In the second case, the goal is to have a code that is directly usable on the widest range of devices. The next step consists in generating a technology-mapped netlist. It is done thanks to an electronic design automation tool. The netlist is a description of all the nets linking the basic cells of the target device specified by the user (logic gates, registers, memory, ...). Hence, contrary to the HDL description, the netlist is device-specific. Once the netlist is generated, the different cells are placed and routed on the FPGA map, which again takes advantage of an automated design tool. At this step of the design, users perform additional simulations and tests (e.g., timing analysis) in order to validate that the obtained results are functionally correct and fits the target performances. Finally, a binary file describing the design is generated. This file is loaded into the FPGA through, e.g., a serial port like joint test action group (JTAG, IEEE standard). It is used to configure the FPGA which then behaves essentially like an ASIC.

Note that this design flow illustrates where the performance loss of FPGAs versus ASICs comes from. Namely, a part of the FPGA resources are consumed to store its configuration (i.e., the description of its functional behavior), whereas all the (hardwired) resources of an ASIC are dedicated to the design processing tasks. However, compared to a software solution, FPGA designs usually allow major performance increases, as they can take advantage of parallel computing and processing units that are specialized for one specific type of computation. Finally, it is also worth mentioning that designing with recent FPGAs offer more and more facilities in terms of predesigned blocks. That is, a large number of primitives (e.g., embedded memories, or embedded processors like the Microblaze for Xilinx and Nios for Altera) are now made available by manufacturers, which can essentially be used in a HDL design as black boxes.

9.1.3 Technologies

Different categories of FPGAs are available from different vendors. In the first place, high-end (more expensive) FPGAs are generally distinguished from lower-end ones. The first category features all the latest developments of the manufacturers (e.g., the Xilinx Virtex and Altera Stratix devices), while the second one is essentially optimized for cost (e.g., the Xilinx Spartan or Artix and Altera Cyclone devices). These FPGAs also differ in their fabrication technology ranging from 130 to 28 nm in 2011. As an illustration, we next describe a few examples of recent reconfigurable devices.

The most recent high-end FPGA from Xilinx is the Virtex-7. It is part of a new generation of devices built from a 28 nm technology and designed for maximum power efficiency. Virtex-7 FPGAs contain up to 1,954,560 logic cells, which correspond to 305,400 slices. Each slice contains four LUTs and eight flip-flops, some of them being usable as distributed RAM, for up to 21.55 Mb. The largest device in the family also contains 2,160 DSP slices, which include a pre-adder, an adder, an accumulator, and a 25×18 multiplier. It additionally contains 46.51 Mb of RAM blocks that can be instantiated as 18 or 36 Kb blocks. Finally, these FPGAs contain 24 clock management tiles, 4 interface blocks for PCI express, 36 low-power 12.5-Gbps transceivers, 1 analog-to-digital converter, 24 I/O banks, and 1,200 user I/O.

The latest family of high-end FPGAs developed by Altera is denoted as the Stratix-V. They are based on a similar 28-nm high-performance process optimized for low power and contain up to 358,000 ALMs. Each ALM is based on two combinational adaptive LUTs (ALUTs) and four registers. Some ALMs can be configured as distributed SRAM, for up to 12.12 Mb. This largest version of the Stratix-V includes 352 27×27 DSP blocks, 704 18×18 multipliers, 52 Mb of RAM blocks, which can be instantiated as 20 Kb blocks, 4 PCI express hard IP blocks, and 48 14.1 Gbps transceivers.

9.2 Security IPs

Modern reconfigurable devices are complex and complete platforms which provide an appealing and cost-effective solution to implement high performance, low to medium volume custom integrated circuits. FPGA designs are characterized by reduced nonrecurring engineering costs and reduced time to market. The cost for a typical mask set to fabricate an ASIC using modern CMOS technology runs in the range of \$500–\$700K. By contrast, a system designer can purchase an off-the-shelf FPGA and program it for only a fraction of the cost. Quite naturally, the resulting circuit will be slower, consume more power and utilize significantly more silicon resources than its ASIC equivalent. Still, FPGAs are an attractive platform for several applications nowadays.

In general, and as discussed in the previous section, designing with FPGAs shares a number of similarities with ASIC development. In the first place, the clear definition of the architectural choices and performance goals is a prerequisite in both cases. The requirements of the target application also determine the main figures of merit which will be optimized by the designer. But once these decisions are specified, the task of efficiently designing for FPGAs is different from its ASIC counterpart. In the latter, the designer has full control over the components to implement. By contrast, in the FPGA case, he is forced to use the components that the FPGA vendor selected as the most suitable for a majority of applications. As a result, the strategy to maximize the exploitation of the available resources may significantly depend on the selected FPGAs.

In the remainder of this section, we illustrate this discussion in the context of security IPs. In particular, and as a case study, we first review different implementations of the Advanced Encryption Standard. Next, we take advantage of these examples in order to underline the problem of fairness in the comparison among different architectures, and the meaningfulness of the metrics currently used for this purpose. Finally, we discuss the specificities of security IPs in FPGAs regarding so-called physical attacks in which an adversary either observes physical emanations of the target devices (side-channel attacks), or tries to induce faults during the cryptographic computations.

9.2.1 The AES Case

The Rijndael algorithm was adopted as the advanced encryption standard (AES) in 2001 [54]. The standard supports a block size of 128-bit and key sizes of 128, 192, and 256 bits. The encryption process, which is illustrated in Fig. 9.3, starts with the first key addition, followed by a number of round functions which depends on the key size. The round function is composed of four transformations applied to a state of 16 bytes. *ShiftRows* cyclically shifts to the left, the bytes in the last three rows of the state, using different offsets; *SubBytes* is a nonlinear byte substitution and operates independently on each byte of the state; *MixColumns* multiplies modulo $x^4 + 1$ the columns of the state by the polynomial $\{03\}x^3 + \{01\}x^2 + \{01\}x + \{02\}$; finally, *AddRoundKey* adds a round key to the state. All the round keys are generated by a *key schedule* routine, which takes the secret key and expands it as specified in the standard. The decryption algorithm is similar to the encryption one and uses the inverted versions of the basic transformations used during the encryption. The key schedule for decryption is identical to the one used for encryption, but it starts using the last round key and generates the round keys in reverse order. In this context, the typical design decisions that have to be taken include:

- Which key size should be supported (128, 192, or 256)?
- Does the implementation have to compute encryption only, decryption only, or both of them?

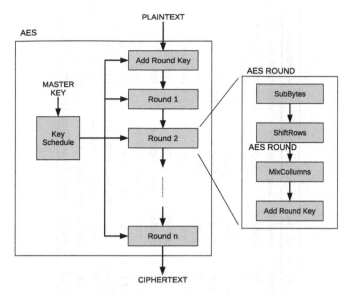

Fig. 9.3 AES encryption block diagram

- How is the key scheduling computed ("on-the-fly", precomputed on chip before each encryption, or precomputed off chip)?
- Is the algorithm supposed to run in a specific encryption mode (with feedback, without feedback, …), which would prevent parallelization?

Additionally, depending on the application requirements, the designer also has to select a number of architectural parameters, including the datapath size, the type of architecture (loop or unrolled), the target throughput (high performances or not), the portability of the design among different platforms, the usable area (low-cost or not), and, more specific to the reconfigurable world, the type of resources which can be used (BRAMs, DSPs, only LUTs, …), and the target FPGA.

In the remainder of the section, we describe a representative subset of designs targeting several of these goals. Reported architectures range from high-speed to low-cost, also including designs which maximize the exploitation of the inner structure of the target FPGA, summarized in Table 9.1.

Area efficient designs of AES were implemented using reduced datapath. The 8-bit architecture proposed by Good and Benaissa [21] features a datapath consisting of two processing units, one to perform the SubBytes transformation and the second to compute the multiply and accumulate operations needed by MixColumn. The processor has an instruction set composed of 15 instructions and relies on a pipeline to execute a new instruction every cycle. Low-cost implementations were also targeting 32-bit datapaths; possible examples are those of Chodowiec and Gaj [10] and Rouvroy et al. [63].

Multiple AES implementations were proposed by Helion Technology [78], targeting different purposes: Standard encryption, fast encryption, and fast encryption

Table 9.1 FPGA implementations of the AES algorithm

Device	Datapath	Logical element	Memory blocks	Frequency (MHz)	Thr. (Gbps)	enc/dec	Architecture type	Key scheduling
Xilinx Spartan-II [21]	8	124 (slices)	2	–	0.0022	enc/dec	loop + pipeline	precomputed on chip
Xilinx Spartan-III [63]	32	163 (slices)	3	71.5	0.208	enc/dec	loop	precomputed on chip
Altera Cyclone-III [78]	32	314 (LEs)	3	170	0.45	enc	loop	precomputed off chip
Altera Cyclone-III [78]	32	603 (LEs)	3	170	0.45	enc	loop	on-the-fly
Xilinx Spartan-II [10]	32	222 (slices)	3	60	0.166	enc/dec	loop	precomputed on chip
Altera Cyclone-III [78]	128	906 (LEs)	10	174	2.02	enc	loop	on-the-fly
Altera Stratix-IV [78]	128	651 (ALUTs)	10	300	3.49	enc	loop	on-the-fly
Altera Stratix-IV [78]	128	1652 (ALUTs)	18	285	3.32	enc/dec	loop	on-the-fly
Altera Cyclone-II [35]	128	3039 (LEs)	18	198.9	2.5	enc/dec	loop + pipeline	on-the-fly
Xilinx Virtex-E [75]	128	1767 (slices)	0	167	2.085	enc	loop + pipeline	on-the-fly
Xilinx Virtex-5 [7]	128	400 (slices)	0	350	4.1	enc	loop + pipeline	on-the-fly
Xilinx Virtex-II Pro [29]	128	5.177 (slices)	84	168.3	21.54	enc	unrolled + pipeline	precomputed off chip
Xilinx Virtex-II 2000 [30]	128	10.750 (slices)	0	139.1	17.8	enc	unrolled + pipeline	on-the-fly
Xilinx Virtex-II Pro [9]	128	3.513 (slices)	80	271	34	enc/dec	unrolled + pipeline	precomputed off chip
Xilinx Virtex-5 [15]	128	321 (slices)	80	413	52.8	enc/dec	unrolled + pipeline	precomputed off chip

and decryption. The user also has the possibility to choose whether the expansion of the key is performed on-the-fly or precomputed off chip. The fast encryptor (and decryptor) are based on a high-throughput 128-bit datapath version, and are evaluated both on Altera Cyclone-III and Stratix-IV FPGAs, i.e., a low-cost and a high-end FPGA. Several implementations of AES were also proposed by Standaert et al. [75], the efficiency of their architectures was evaluated at different stages of the design process and the structure of the pipeline which better considered the place and route constraints was discussed.

Kenney [35] proposed an energy efficient 128-bit implementation in his Ph.D. thesis. It is based on a complete unroll of the rounds and deeply exploits the pipeline. Examples of this kind of implementation are also reported by Järvinen et al. [30], Hodjat and Verbauwhede [29], and Chaves et al. [9].

Designs which maximize the specific resources of the target FPGA have also been proposed. Bulens et al. [7] proposed a design which deeper exploits the 8-bit look-up table structure of the Xilinx Virtex-E. Drimer et al. [15] proposed an AES design which is largely implemented on the additional components of the FPGA, such as DSPs and BRAM, attempting to leave the majority of the programmable logic available for other applications.

To conclude, let us mention that very similar considerations and choices apply to the implementation of other cryptographic algorithms. Typical examples include the implementation of Elliptic Curve Cryptography and hash functions. We refer the interested reader to [11, 17, 26–28, 87].

9.2.2 Performance Evaluation

Evaluating the performances of a design are a very natural goal. As introduced in the previous sections, each of the resources of an FPGA (slices, LABs, ALMs, LUTs, registers, ...) can be used as figure of merit to carry on the comparison. Unfortunately, producing fair comparisons for FPGA implementations are limited by a number of difficulties that we discuss in this section.

Eventually, and as previously mentioned, the very goal of optimizing an implementation is highly dependent on the need for portability. Namely, a designer always has the possibility to deeply exploit the inner structure of his target FPGA, in order to achieve higher speed or smaller occupation. However, this comes at the price of a reduced portability, since the inner structure is specific to the device, model, and vendor. Hence, considering such a decision is important when performing comparative analyzes. In this respect, it is worth underlining that academic publications generally tend to focus on device-specific optimizations more than found in industrial IP cores, where portability is usually appreciated for the cost-reduction it allows.

Of course, the different limitations discussed here do not mean that it is impossible to compare different designs. They simply underline that any comparison should be carried out with care and the results of the comparison should be well understood. In this context, the typical comparison metrics include working frequency (measured in GHz or MHz), throughput (measured in Mbit/second), hardware occupation (measured in LUTs, registers, ...), and the previously mentioned throughput over area ratio. In addition, improved comparisons can take the reproducibility of the synthesis results into account. For this reason, Drimer encouraged the academic and scientific community to favor the publication of implementations which are presented together with the source code [14].

This issue of fair comparisons for FPGA designs has recently received attention, as FPGA implementations are one of the criteria for the selection of the next hash standard [53]. In this context, Gaj et al. proposed a series of guidelines including the definition of suitable performance metrics and the development of uniform interfaces [18], also used in [36]. Furthermore, their evaluation of the candidates was completed on several representative FPGA platforms, from the two major vendors discussed in the first section of this chapter.

9.2.3 Side-Channel Attacks and Countermeasures

The previous section discussed different implementations of the AES algorithm. However, when the target application involves security IPs, it is important to consider also the security of the designs against various types of physical attacks. Physical attacks exploit the characteristics of the hardware platform on which the algorithm is implemented in order to acquire sensitive information. They are usually classified between two axes: Invasive or non-invasive and active or passive. Side-channel attacks are a particular class of passive and non-invasive physical attacks which use the information leaked, while data are being processed in order to break some security guarantee, e.g., by deriving the secret key of a cryptographic algorithm [46]. Common examples of this leaking information are the time employed for the computation [38], the corresponding power consumed [39], or the electromagnetic emissions [2, 58].

Among the different types of power-based attacks available in the literature, the most common ones are simple power analysis (SPA) and differential power analysis (DPA). In a SPA, an attacker measures the power consumed by a device while performing cryptographic operations and, by observing the traces, attempts to deduce the secret information, e.g., by distinguishing different operations. SPA attacks are typically powerful in context where distinguishing operations directly allows recovering secret information, e.g., in public key cryptographic computations. DPA attacks extend this principle toward data dependencies and try to recover secret information with some statistical processing. A typical DPA attack consists of four steps. At first, an intermediate key dependent result is selected as the target. Then, the attacker encrypts (decrypts) a certain number of known plaintexts (ciphertexts) and measures the corresponding power consumption traces. Subsequently, hypothetical intermediate values are calculated, based on a key guess. Finally, the hypothetical intermediate values (and thus the corresponding secret keys) are verified against the measured power traces. If the attack is successful, the right key hypothesis will be clearly visible in correspondence of the time frame where the information is leaked [46].

First investigations of power analysis attacks on FPGA devices were carried out in [56, 76] in 2003. These initial results have been followed by several papers, e.g., [72–74] which improved the efficiency of the attacks and analyzed them in deeper details, considering also the specificity of reconfigurable devices. Examples of tackled problems are the dependency between the attack and used resources or the effects that a pipelined architecture might have on side-channels.

Counteracting side-channel attacks are difficult since the attack is caused by an intrinsic feature of the transistors. Hence, physical security is generally guaranteed by a combination of countermeasures, acting at different abstraction levels (e.g., hardware, algorithm, protocol). Among these solutions, so-called hiding and masking techniques have attracted significant attention for FPGAs. Hiding aims at obtaining independence between the power consumed by a cryptographic device and the secret data being processed, e.g., by ensuring constant power consumption for all inputs. Masking is a technique inspired by cryptographic secret-sharing schemes; the original message to be encrypted is divided into parts called shares that are then encrypted separately. This aims to ensure that only attacks combining the leakage of different shares can be successful. The ciphertext is eventually reconstructed by combining the output shares.

As an illustration, to implement logic function resistant against power analysis attacks, Tiri and Verbauwhede proposed WDDL [79], a differential logic style with precharge, which can be designed from FPGA elements. WDDL essentially translates every gate on a design into "protected gates" having the goal to produce constant power consumption. For this purpose, each gate uses four inputs that are pairwise complementary, and always computes the two complementary outputs. On the Virtex-II FPGAs, WDDL requires two LUTs to generate a gate with two differential outputs.

Masked implementations of different algorithms were also proposed for FPGAs, incurring different performances and area overheads. Concerning the AES algorithm, an implementation which combines Boolean and multiplicative masking was proposed by Mentens et al. [49]. The area overhead of their secured core compared with the reference unsecured version is approximately 20 %, while the speed is degraded by 30 %. Kamoun et al. [32] implemented a masked AES S-box on Virtex-4 FPGA which incurs an area overhead of 44 % and a frequency decrease of 31 %. Specific features of state-of-the-art FPGAs were also exploited for implementing masking; the larger input size of the basic block of Xilinx Virtex-5 FPGA was combined with optimization techniques for S-boxes to obtain an efficient FPGA implementation of the AES algorithm, masked against side-channel attacks [62].

Note that none of the countermeasures investigated so far can guarantee perfect security. In particular, a certain number of physical effects, like glitches in integrated circuits for masking, or early propagation effects for hiding, can compromise the security of these countermeasures [47, 48, 77]. Overall, providing security against physical attacks remains an active scope of research. For example, one recent trend is to investigate FPGA-dedicated countermeasures [25], taking advantage of the specificities of reconfigurable devices.

Still concerning the power analysis attacks, it is worth mentioning the effort done by the Research Center for Information Security (RCIS) of AIST and Tohoku

University in the direction of developing a common platform for standardizing the evaluation of attacks and the comparison of countermeasures. The outcome was the Side-channel Attack Standard Evaluation BOard (SASEBO) [65], which was distributed together with design information to research institutes as common experimental platforms. To date, there are five types of SASEBO boards, based on both Xilinx and ALTERA FPGAs. The FPGA boards have microprocessor features, and thus side-channel attack experiments against cryptographic software can also be performed. Additionally, SASEBO-R is extended with a custom cryptographic LSI that supports all of the block ciphers adopted by ISO/IEC 18033-3, as well as the public-key cipher RSA. The advantage of SASEBO board, besides eliminating the high engineering costs, is to provide a common platform which allows results to be reproduced.

9.2.4 Fault Attacks and Countermeasures

More recently, fault attacks have also been applied to FPGAs. Since it is a relatively new area of research, they are not as common as power analysis attacks and there are only few works discussing them, all focusing on attacks against the AES algorithm. In a nutshell, a fault attack consists of a deliberate injection of a fault into a target device, in our case the FPGA where the cryptographic routine is executed. The fault is injected by actively tampering with the device itself. Once the error required by the attack model is produced, an adversary can analyze the differences between the correct and the faulty outputs of the device and extract sensitive information, e.g., about a secret key. Depending on the specific attack used, it can be required that the fault has to appear at a specific point of the computation [16]. Several methods for inducing a fault were proposed and verified to be effective in FPGA. Some of them, such as clocking the circuit at a different speed, or reducing the supply voltage, can be exploited using very inexpensive devices.

Examples of successful attacks mounted on FPGAs were presented by Saha et al. [64], Khelil et al. [37], and Selmane et al. [68]. In Saha et al. [64], the attack induces faults into a diagonal of the AES state matrix that is input to the $8th$ round. Since, to successfully mount a fault attack, it is necessary to introduce exactly the error required by the fault model, the adversary has to perform a preliminary exploration in order to find the sweet spot for the attack. A sweet spot is the area where the degradation induced in the device is sufficient to generate only the needed fault without resulting in a complete corruption of the circuit behavior. In this case, the fault is injected by switching the clock to a faster frequency when the $8th$ round of the encryption function begins.

Other works such as [37, 68] also evaluate the feasibility on FPGAs of the attack proposed by Piret and Quisquater [57]. The attack requires the injection of a fault on one byte of the state before the computation of MixColumn in the $9th$ round of the AES algorithm. In both works, the fault is injected by reducing the supply voltage to induce a setup time violation. This is possible since the propagation delay is inversely proportional to the power supply; when the voltage is reduced, the signals which are propagated into the circuit require more time to stabilize. As a consequence, the values stored in the register might not be computed in time and they produce a faulty result.

Considering the relatively recent appearance of fault attacks in FPGAs, no countermeasures have been developed so far, specifically targeting reconfigurable devices. However, initial work has been performed to evaluate the resistance of power analysis countermeasures against fault attacks. In particular, Selmane et al. [67] used the lowering voltage technique to evaluate the intrinsic resistance of WDDL against fault attacks on FPGA implementations. Furthermore, techniques for fault tolerance were already successfully adapted to the needs of fault attack prevention in ASICs [6, 33]. Hence, it is possible to envision the use of similar schemes in reconfigurable devices. As a final note, it is important that the designer also considers the interaction between countermeasures against one attack and the vulnerability to another attack, since it has been shown that several error correcting and detection codes increase the vulnerability of a cryptographic circuit to power analysis attacks [60, 61].

9.3 IP Security

Since FPGAs are volatile and generic platforms, a large number of designs can be implemented on them, ranging from essentially hardware to on chip combinations of hardware and software. Most of the time, these implementations are developed by different designers and the inherent value of their IP can be high. As a result, various design houses base their business on the selling of IPs, and their protection against various types of counterfeiting has become an important issue. In the first place comes the problem of bitstream security. That is, given an FPGA board running an application, how to guarantee that no adversary can recover the bitstream and, from it, clone or reverse engineer a design? Next, and more critical, comes the problem of design security. That is, given an IP that is sold by a design house, how to guarantee that this IP is not used beyond the terms of a license? And how to combine this requirement with the flexibility constraints of the client (e.g., the need to integrate an IP in a large design, and to simulate it)? In this section, we briefly tackle these two problems and describe some recent trends that are considered to solve them.

9.3.1 Bitstream Security

In this case, the most frequent solution is bitstream encryption. That is, the bitstream is encrypted by the CAD tool with user-defined symmetric (secret) keys. The same keys are stored on the FPGA, e.g., in a volatile memory with an external battery. During configuration, an on-chip decryption circuit is used to recover the original configuration file. Readback is not allowed when encrypted bitstreams are used. The main drawback of bitstream encryption is the need for an external battery to maintain the keys, and the difficult key management. For example, if a single key is used for all the boards, then a system designer has no opportunity to update the configuration files for only a part of them. Ideally, it should be possible to update the symmetric keys remotely. This could be achieved either by the use of a symmetric master key (but the system security would then depend on this single key) or a public-key mechanism in which each FPGA would come with a private/public key pair stored in a nonvolatile memory. Note also that the importance of authenticated encryption for FPGA bitstreams has been discussed in [12, 80]. Finally, let us mention that the on-chip circuitry that is used to perform decryption also has to be secure against physical attacks. As recently discussed in [50], security against side-channel attacks was not considered in certain Xilinx designs.

9.3.2 Design Security

9.3.2.1 Encrypted Netlists

A first solution to protect IP providers' work is exchanging encrypted netlists and simulation models with the system designer. This encryption is done through the Xilinx development tools which embed the key in its software. So, the system designer can easily instantiate the IP core as a black box, i.e., without having access to the implementation details. The problem is that this solution only protects the IP integrity and does not prevent from reusing the IP many times. Moreover, the IP is only protected by a key stored in the CAD software. Hence, it may be potentially recovered by reverse engineering this tool.

9.3.2.2 Security Chips

Another solution is proposed in the Xilinx documentation [43]. It works as follows: The IP provider sends a preprogrammed security chip with the encrypted netlist to the system manager. To use the IP, the system manager instantiates it and connects the security chip to the FPGA. The encrypted netlist contains the IP core itself and a security module embedding the same identification number as the security chip. When the system starts, the security module checks whether the right security chip is present or not. This solution allows managing the quantity of IP cores sold. However, it faces similar limitations for the key management as bitstream encryption (as the key is contained in an encrypted netlist). Furthermore, this solution still relies on the key of the encrypted netlist. Hence, its main advantage is to allow per device licensing.

9.3.2.3 Physically Unclonable Functions

In view of the limitations of encrypted netlists and the use of security chips, the development of device-specific identification tools has become an intensive research topic in the recent years. Physically unclonable functions are among the frequently considered solutions for this purpose. They are challenge-response systems relying on uncontrollable random features inherent to the manufacturing processes. Their particularities lie in their output, which is easy to measure but assumably hard to predict if it has not been previously queried. Many kinds of PUF have been proposed in the literature. The following lines describe some of them, that are suitable for FPGAs. Implementation details are given in the related references.

1. *SRAM-based PUF* initially proposed by Guajardo et al. [23], the SRAM-based PUF uses the initialization values of dedicated SRAM blocks. They consider a range of memory locations as challenges and start-up values at these locations as responses. These values depend on the small asymmetry between two cross-coupled inverters, ensuring that the start-up values will always be the same with high probability. Guajardo et al. define this kind of PUF as *intrinsic*, because the PUF generating circuit is directly present in the design to protect. The main drawback of SRAM-based PUF with FPGAs is that most FPGA manufacturers initialize the embedded memory blocks to zero before loading the bitstream to avoid shortcuts in the reconfigurable circuitry.
2. *Flip-flop PUF* proposed by Maes et al. [44], the flip-flop PUF uses flip-flops start-up values as responses similarly to the SRAM-based PUF. Maes et al. imagined this PUF because it is possible to prevent flip-flops from being reset. Hence, this allows having an efficient PUF suitable for every FPGA.
3. *Butterfly PUF* proposed by Kumar et al. [40], the butterfly PUF is another solution to overcome the SRAM PUF reset drawback. It consists in two cross-coupled latches initialized with two different values to have an unstable operating point. The latches are initialized on an external signal. When this one is released, the stable state depends on the slight differences between the connecting wires which are designed using symmetrical paths on the FPGA matrix. The butterfly PUF needs manual routing to have symmetric paths and its performance highly depends on the targeted FPGA [51].
4. *LUT-based PUF* proposed by Anderson [4], the LUT-based PUF harnesses the FPGA's LUT structure. It uses LUTs from the same basic logic block (slice or ALM), configured in shift-register, and the carry-chain logic. This PUF relies on delays introduced by the LUTs and the multiplexers. It uses the presence or absence of glitches along the carry chain to determine the output bit. This PUF has the advantage to be completely described in HDL.
5. *Ring oscillator PUF* introduced by Gassend [19], Gassend et al. [20], it mainly relies on a self-oscillating circuit and a counter. The ring oscillator produces an oscillating signal with a delay-dependent frequency. Besides, the counter measures the number of positive edges over a period of time. The obtained value is a good representation of the ring oscillator intrinsic delay. The main drawbacks of

this kind of PUF are the limited number of possible challenges and the significant dynamic power consumption.

6. *Time-bounded PUF* introduced by Majzoobi et al. [45], the time-bounded PUF relies on three flip-flops placed around the *circuit under test* (CUT) : The *Launch* FF, the *Sample* FF, and the *Capture* FF. Initially, the flip-flops are set to zero. Then, the *Launch* FF is set to one on the rising edge of the clock. This signal propagates through the CUT and is sampled by the *Sample* FF on the falling edge of the clock. The CUT adds a challenge-dependent delay which may be greater than the half of the clock period. Hence, the sampled value depends on it and is xor-ed with the true launched value to be captured by the *Capture* FF.

As the PUFs lead to unique and device-dependent challenge-reponse pairs, they can be used in many identification-related applications including IP protection. Simpson and Schaumont [70] are among the few to have proposed FPGA-specific IP protection protocols using PUFs (Fig. 9.4). Their work relies on the assumption that the FPGA manufacturer has embedded a standard security module which contains two different hardware blocks: The PUF itself, used for hardware authentication and key generation, and a block cipher used for symmetric encryption and software authentication.

During a preliminary step, the FPGA manufacturer and the IP provider have sent authentication information to the trusted third party (TTP) including a challenge-response pair vector from the PUF ($\overline{\text{CRP}}$). Then, the protocol is a 3-player game among the system designer (SYS), the IP provider (IPP), and a TTP. First, the SYS sends a request to the TTP to obtain a particular IP. Then, the TTP forwards this

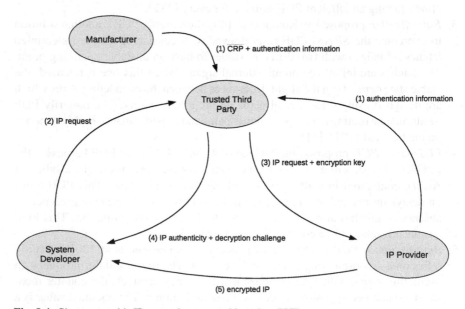

Fig. 9.4 Simspon et al.'s IP protection protocol based on PUFs

request to the IPP attached with a PUF response that is used as encryption key. In the same time, the TTP sends the IP authenticity information attached with a PUF challenge that allows the SYS to recover the PUF response (i.e., the encryption key). Finally, IPP sends his encrypted IP to the SYS which is able to decrypt it into the device without having access to the IP itself. Hence, the SYS cannot use the data for another platform without asking the IP provider. The strength of using PUFs in this kind of IP protection protocol comes from the hidden, secret, and nonvolatile aspects of PUFs.

Starting from the Simpson et al.'s work, Guajardo et al. [23] proposed a protocol where the TTP cannot access the IP block exchanged between the IPP and the SYS. Indeed, as the TTP knows all the challenge-response pairs and since the SYS–IPP channel is public, he is able to access to the IP core. To avoid this, they introduced a public-key based operation. In a following work [24], the same team studied the advantages that asymmetric cryptography provides in this context. It allows that secret information from PUF never has to leave the FGPA unlike in previous works. This results in increased security guarantees.

In general, PUFs are interesting objects for dealing with security questions in nonvolatile FPGAs, in particular when these devices do not contain any nonvolatile memory that could be used to store a key. As far as IP protection is concerned, it can also allow per device licensing, without the need of any security chip. However, it also faces limitations, as the protected designs have to include the processing of the PUFs, which could possibly be removed by an adversary, e.g., if the security of the netlists is compromised.

9.3.2.4 IP Watermarking

Digital watermarking is the process of embedding an evidence of ownership into all types of digital content. The embedded information must be very difficult to remove. Hence, each copy of the content will also include the watermark information. This process is largely used in multimedia content by slightly modifying the data in an unperceivable way from a user point of view (e.g., hiding a signature in the highest frequencies of a picture). Similar techniques have been proposed for IP protection. However, watermarking for IP protection is more difficult as most of the times, when a chip needs to be tested, it is not available before returning completely assembled and packaged from the manufacturing process. Hence, e.g., the constraint-based watermarking proposed by Kahng et al. [31] and Narayan et al. [52] cannot capture such postproduction features. Also, the main difference with other IP protection techniques lies in the fact that the watermarking is an *a posteriori* solution. That means that, unlike other solutions, the watermarking is searched and checked only if the IP owner has doubts on the IP authenticity (i.e., it allows detecting counterfeiting, but does not directly prevent it).

When watermarking IPs, care must first be taken that the functional correctness of the core is preserved. Also, in order to be efficient, an IP watermarking strategy should only use the usual design tools, not affect the performances of the core, be robust to removals or modifications, and be sufficient as proof-of-ownership. In this context, the two main goals of IP watermarking are the detectability and the proof-of-ownership [5]. The detectability means that, given an integrated circuit (IC), the IP core designer is able to determine whether his IP core is used. The proof-of-ownership means that, given an IC, the IP core designer is able to prove to a third party that he owns the IP core used. A general survey and analysis of watermarking techniques are provided by Abdel-Hamid et al. in [1], and complete states-of-the-art for IP watermarking can be found in Drimer's and Ziener's Ph.D. thesis [14, 85].

A watermark can be embedded at the three different levels of the design flow, with pros and cons: The behavioral HDL description, the netlist generation, and the bitfile generation. Including the watermark in the HDL or the netlist allows protection of individual IP cores as they can be handled independently. However, if the adversary is able to break the netlist encryption, he can then (as for bitstream encryption and PUFs) easily remove the watermark. Including it in the bitfile may prevent the netlist encryption issue, but can only protect the whole system as the bitfile is generated by the system designer and not by the IP provider.

Next, different ways to recover a watermark signature exist, among which we find:

1. *Bitfile* the bitfile can be read either using the readback command if it is activated, either by wire tapping the bus between the PROM (Programmable Read-Only Memory) and the FPGA if the bitfile is not encrypted.
2. *Ports* some ports of the FPGA can be dedicated to the watermark reading. However, in the case of IPs integrated in larger systems, the system designer could deliberately choose to disconnect those ports from the IP.
3. *Power* introduced by Ziener and Teich [86], as in cryptographic side-channel attacks, the clock frequency and the toggling logic can be extracted from the measured power traces.
4. *Electromagnetic radiations* the EM radiations are easy to measure and offer a similar information as the power traces if the chip is not protected.
5. *Temperature* proposed by Kean et al. [34], the temperature just needs a thermocouple to measure it through the chip package without having access to the power pins. Although it is a rather slow process, it is currently the only commercially available approach.

As this chapter focuses specifically on FPGAs, we now report a few examples of watermarking techniques that are or could be applicable in this case.

Lach et al. [41, 42] propose to include a watermark signature in the unused LUTs at the bitstream level. Furthermore, Schmid et al. [66] improve it by tightly integrating the watermark with the LUTs of the design, so that simply removing the mark carrying components would damage the IP core. The watermark extraction is done knowing their positions and performing a readback of the bitfile.

Oliveira [55] proposes a Finite-State-Machine-based watermarking, while Castillo et al. [8] propose a technique using unused parts of the distributed RAM memory. These two techniques offer the ease to be implemented in HDL but the main drawback is that they need FPGA ports to extract the watermark.

A side-channel based technique is introduced by Becker et al. [5] and consists in hiding a watermark signature below the noise floor in the power traces, by including leakage generating circuit depending on an identification number. The signature can be easily recovered with DPA. This solution is well suited to tag netlist cores but it can hardly be applied to HDL IP cores, since the identification and the suppression of the watermarking circuit would then be easy.

Finally, as previously mentioned, Kean et al. [34] propose a solution which consists in hiding the watermark signature in the chip's temperature. The underlying working is quite the same as the Becker et al.'s work but, instead of measuring power consumption leakages, they measure the heat generated by their circuit. This one is composed of multiple ring oscillators that are switch on/off according to the identification number. The tag signature can be easily found with cross-correlation.

9.4 Conclusions

FPGAs are useful for the implementation of security algorithms, because of the significant performance gains that they provide compared with software solutions. As they allow customizing and specializing the processing blocks, they also offer interesting opportunities for the design of countermeasures against noninvasive (passive) physical attacks. In particular, their inherent parallelism generally allows noisy measurements in side-channel attacks and less precise fault insertions in active attacks. Overall, cryptographic implementations in FPGAs can be seen as the result of a flexibility versus efficiency and security trade-off. That is, for security and efficiency, it is best to fully take advantage of each given device architecture. But for flexibility, it is more attractive to have designs able to run on a large amount of devices. A similar statement can be made for the important problem of IP protection in reconfigurable devices. For flexibility reasons, it is desirable that the security relates to the netlists, so that IPs can be easily simulated and integrated in larger designs. But for security reasons, the best solution would be to deal directly with bitstreams. IP and bitstream securities are also limited by the difficult key management problem. Ideally, the integration of nonvolatile keys and public key cryptography facilities in each device would be the best solution to allow the "per device" licensing of the IPs based on bitstreams. But present devices do not offer such facilities. Alternative solutions exist, based on the detection of a security chip, or physically unclonable functions (PUFs), but are then limited by some other assumptions (e.g., the difficulty to remove the "detection mechanisms from the design"). Watermarking-based techniques are yet another way to detect IP theft a posteriori. These questions illustrate the rapidly evolving nature of security issues in reconfigurable computing, for which several important research problems remain open.

References

1. Amr T. Abdel-Hamid, Sofiène Tahar, and El Mostapha Aboulhamid. Ip watermarking techniques: Survey and comparison. In IWSOC, pages 60–65. IEEE Computer Society, 2003
2. Dakshi Agrawal, Bruce Archambeault, Josyula R. Rao, and Pankaj Rohatgi. The EM side-channel(s). In Burton S. Kaliski Jr., Çetin Kaya Koç, and Christof Paar, editors, CHES, volume 2523 of Lecture Notes in Computer Science, pages 29–45. Springer, 2002.
3. Altera. http://www.altera.com/
4. Jason H. Anderson. A PUF design for secure FPGA-based embedded systems. In Design Automation Conference (ASP-DAC), 2010 15th Asia and South Pacific, pages 1–6, jan. 2010.
5. Georg T. Becker, Markus Kasper, Amir Moradi, and Christof Paar. Side-channel based watermarks for integrated circuits. In Hardware-Oriented Security and Trust (HOST), 2010 IEEE International Symposium on, pages 30–35, june 2010.
6. Guido Bertoni, Luca Breveglieri, Israel Koren, Paolo Maistri, and Vincenzo Piuri. Error analysis and detection procedures for a hardware implementation of the advanced encryption standard. IEEE Trans. Computers, 52(4):492–505, 2003.
7. Philippe Bulens, François-Xavier Standaert, Jean-Jacques Quisquater, Pascal Pellegrin, and Gaël Rouvroy. Implementation of the AES-128 on Virtex-5 FPGAs. In Serge Vaudenay, editor, AFRICACRYPT, volume 5023 of Lecture Notes in Computer Science, pages 16–26. Springer, 2008.
8. Encarnación Castillo, Luis Parrilla, Antonio García, Antonio Lloris-Ruíz, and Uwe Meyer-Bäse. IPP watermarking technique for IP core protection on FPL devices. In FPL, pages 1–6, 2006.
9. Ricardo Chaves, Georgi Kuzmanov, Stamatis Vassiliadis, and Leonel Sousa. Reconfigurable memory based AES co-processor. In IPDPS. IEEE, 2006.
10. Pawel Chodowiec and Kris Gaj. Very compact FPGA implementation of the AES algorithm. In Walter et al. [82], pages 319–333.
11. Guerric Meurice de Dormale, Philippe Bulens, and Jean-Jacques Quisquater. Collision search for Elliptic Curve Discrete logarithm over GF(2^m) with FPGA. In Pascal Paillier and Ingrid Verbauwhede, editors, CHES, volume 4727 of Lecture Notes in Computer Science, pages 378–393. Springer, 2007.
12. Saar Drimer. Authentication of fpga bitstreams: Why and how. In Pedro C. Diniz, Eduardo Marques, Koen Bertels, Marcio Merino Fernandes, and João M. P. Cardoso, editors, ARC, volume 4419 of Lecture Notes in Computer Science, pages 73–84. Springer, 2007.
13. Saar Drimer. Security for volatile FPGAs. PhD dissertation, University of Cambridge Technical, Report UCAM-CL-TR-763, 2009.
14. Saar Drimer. Security for volatile FPGAs. Technical Report UCAM-CL-TR-763, University of Cambridge, Computer Laboratory, November 2009.
15. Saar Drimer, Tim Güneysu, and Christof Paar. DSPs, BRAMs, and a pinch of logic: Extended recipes for AES on FPGAs. TRETS, 3(1), 2010.
16. Pierre Dusart, Gilles Letourneux, and Olivier Vivolo. Differential fault analysis on AES. CoRR, cs.CR/0301020, 2003.
17. Junfeng Fan, Daniel V. Bailey, Lejla Batina, Tim Güneysu, Christof Paar, and Ingrid Verbauwhede. Breaking Elliptic Curve Cryptosystems using reconfigurable hardware. In FPL, pages 133–138. IEEE, 2010.
18. Kris Gaj, Ekawat Homsirikamol, and Marcin Rogawski. Fair and comprehensive methodology for comparing hardware performance of fourteen round two SHA-3 candidates using FPGAs. In Stefan Mangard and François-Xavier Standaert, editors, CHES, volume 6225 of Lecture Notes in Computer Science, pages 264–278. Springer, 2010.
19. Blaise Gassend. Physical Random Functions. Master's thesis, MIT, USA, 2003.
20. Blaise Gassend, Dwaine Clarke, Marten van Dijk, and Srinivas Devadas. Silicon physical random functions. In ACM Conference on Computer and Communications Security, pages 148–160, New York, NY, USA, 2002. ACM Press.

21. Tim Good and Mohammed Benaissa. AES on FPGA from the fastest to the smallest. In Rao and Sunar [59], pages 427–440.
22. Louis Goubin and Mitsuru Matsui, editors. Cryptographic Hardware and Embedded Systems - CHES 2006, 8th International Workshop, Yokohama, Japan, October 10–13, 2006, Proceedings, volume 4249 of Lecture Notes in Computer Science. Springer, 2006.
23. Jorge Guajardo, Sandeep S. Kumar, Geert Jan Schrijen, and Pim Tuyls. FPGA intrinsic PUFs and their use for IP protection. In Cryptographic Hardware and Embedded Systems Workshop, volume 4727 of LNCS, pages 63–80, September 2007.
24. Jorge Guajardo, Sandeep S. Kumar, Geert Jan Schrijen, and Pim Tuyls. Physical unclonable functions and public-key crypto for FPGA IP protection. In Field Programmable Logic and Applications, 2007. FPL 2007. International Conference on, pages 189–195, Aug. 2007.
25. Tim Güneysu and Amir Moradi. Generic side-channel countermeasures for reconfigurable devices. In Bart Preneel and Tsuyoshi Takagi, editors, CHES, volume 6917 of Lecture Notes in Computer Science, pages 33–48. Springer, 2011.
26. Tim Güneysu and Christof Paar. Ultra high performance ECC over NIST primes on commercial FPGAs. In Elisabeth Oswald and Pankaj Rohatgi, editors, CHES, volume 5154 of Lecture Notes in Computer Science, pages 62–78. Springer, 2008.
27. Mohamed N. Hassan and Mohammed Benaissa. Efficient time-area scalable ECC processor using μ-coding technique. In M. Hasan and Tor Helleseth, editors, Arithmetic of Finite Fields, volume 6087 of Lecture Notes in Computer Science, pages 250–268. Springer Berlin / Heidelberg, 2010.
28. Mohamed N. Hassan and Mohammed Benaissa. Small footprint implementations of scalable ECC point multiplication on FPGA. In Communications (ICC), 2010 IEEE International Conference on, pages 1–4, May 2010.
29. Alireza Hodjat and Ingrid Verbauwhede. A 21.54 Gbits/s fully pipelined AES processor on FPGA. In FCCM, pages 308–309. IEEE Computer Society, 2004.
30. Kimmo U. Järvinen, Matti Tommiska, and Jorma Skyttä. A fully pipelined memoryless 17.8 Gbps AES-128 encryptor. In FPGA, pages 207–215, 2003.
31. Andrew B. Kahng, Darko Kirovski, Stefanus Mantik, Miodrag Potkonjak, and Jennifer L. Wong. Copy detection for intellectual property protection of VLSI designs. In Computer Aided Design, 1999. Digest of Technical Papers. 1999 IEEE/ACM International Conference on, pages 600–604, 1999.
32. Najeh Kamoun, Lilian Bossuet, and Adel Ghazel. SRAM-FPGA implementation of masked S-Box based DPA countermeasure for AES. In Design and Test Workshop, 2008. IDT 2008. 3rd International, pages 74–77. IEEE, 2009.
33. Ramesh Karri, Kaijie Wu, Piyush Mishra, and Yongkook Kim. Concurrent error detection schemes for fault-based side-channel cryptanalysis of symmetric block ciphers. IEEE Trans. on CAD of Integrated Circuits and Systems, 21(12):1509–1517, 2002.
34. Tom Kean, David McLaren, and Carol Marsh. Verifying the authenticity of chip designs with the DesignTag system. In Hardware-Oriented Security and Trust, 2008. HOST 2008. IEEE International Workshop on, pages 59–64, June 2008.
35. David Kenney. Energy efficiency analysis and implementation of AES on an FPGA. Master's thesis, University of Waterloo, Canada, 2008.
36. Stéphanie Kerckhof, François Durvaux, Nicolas Veyrat-Charvillon, Francesco Regazzoni, Guerric Meurice de Dormaele, and François-Xavier Standaert. Compact fpga implementations of the five sha-3 finalists. ECRYPT II Hash Workshop, Talinn, Estonia, May 2011.
37. Farouk Khelil, Mohamed Hamdi, Sylvain Guilley, Jean-Luc Danger, and Nidhal Selmane. Fault analysis attack on an FPGA AES implementation. In NTMS'08, pages 1–5, 2008.
38. Paul Kocher. Timing attacks on implementations of Diffie-Hellman, RSA, DSS, and other systems. In Neal I. Koblitz, editor, Advances in Cryptology-CRYPTO '96, volume 1109 of LNCS, pages 104–13. Springer, Berlin, September 1996.
39. Paul Kocher, Joshua Jaffe, and Benjamin Jun. Differential Power Analysis. In Michael Wiener, editor, Advances in Cryptology-CRYPTO '99, volume 1666 of LNCS, pages 398–412. Springer, Berlin, August 1999.

40. Sandeep S. Kumar, Jorge Guajardo, Roel Maes, Geert Jan Schrijen, and Pim Tuyls. Extended abstract: The butterfly PUF protecting IP on every FPGA. In Hardware-Oriented Security and Trust, 2008. HOST 2008. IEEE International Workshop on, pages 67–70, June 2008.
41. John Lach, William H. Mangione-Smith, and Miodrag Potkonjak. Signature hiding techniques for FPGA intellectual property protection. In ICCAD, pages 186–189, 1998.
42. John Lach, William H. Mangione-Smith, and Miodrag Potkonjak. Robust FPGA intellectual property protection through multiple small watermarks. In DAC, pages 831–836, 1999.
43. Bernhard Linke. Xilinx FPGA IFF copy protection with 1-wire SHA-1 secure memories. http:// www.maxim-ic.com/app-notes/index.mvp/id/3826, June 2006
44. Roel Maes, Pim Tuyls, and Ingrid Verbauwhede. Intrinsic PUFs from flip-flops on reconfigurable devices. In 3rd Benelux Workshop on Information and System Security (WISSec 2008), page 17, Eindhoven, NL, 2008.
45. Mehrdad Majzoobi, Ahmed Elnably, and Farinaz Koushanfar. Information Hiding, volume 6387 of Lecture Notes in Computer Science, pages 1–16. Springer Berlin / Heidelberg, 2010.
46. Stefan Mangard, Elisabeth Oswald, and Thomas Popp. Power Analysis Attacks: Revealing the Secrets of Smart Cards. Advances in Information Security. Springer, New York, 2007.
47. Stefan Mangard, Norbert Pramstaller, and Elisabeth Oswald. Successfully attacking masked AES hardware implementations. In Rao and Sunar [59], pages 157–171.
48. Stefan Mangard and Kai Schramm. Pinpointing the side-channel leakage of masked AES hardware implementations. In Goubin and Matsui [22], pages 76–90.
49. Nele Mentens, Lejla Batina, Bart Preneel, and Ingrid Verbauwhede. An FPGA implementation of Rijndael: Trade-offs for side-channel security. In IFAC Workshop-PDS, pages 493–498. Citeseer, 2004.
50. Amir Moradi, Alessandro Barenghi, Timo Kasper, and Christof Paar.
51. Sergey Morozov, Abhranil Maiti, and Patrick Schaumont. An analysis of delay based PUF implementations on FPGA. In Phaophak Sirisuk, Fearghal Morgan, Tarek El-Ghazawi, and Hideharu Amano, editors, Reconfigurable Computing: Architectures, Tools and Applications, volume 5992 of Lecture Notes in Computer Science, pages 382–387. Springer Berlin / Heidelberg, 2010.
52. Naveen Narayan, Rexford D. Newbould, Jo Dale Carothers, Jeffrey J. Rodriguez, and W. Timothy Holman. IP protection for VLSI designs via watermarking of routes. In ASIC/SOC Conference, 2001. Proceedings. 14th Annual IEEE, International, pp. 406–410, 2001.
53. NIST. http://csrc.nist.gov/groups/st/hash/sha-3/index.html
54. NIST. Announcing the Advanced Encryption Standard (AES). Federal Information Processing Standards Publication 197, November 2001.
55. Arlindo L. Oliveira. Techniques for the creation of digital watermarks in sequential circuit designs. IEEE Trans. on CAD of Integrated Circuits and Systems, 20(9):1101–1117, 2001.
56. Siddika Berna Örs, Elisabeth Oswald, and Bart Preneel. Power-analysis attacks on an FPGA - first experimental results. In Walter et al. [82], pages 35–50.
57. Gilles Piret and Jean-Jacques Quisquater. A differential fault attack technique against SPN structures, with application to the AES and KHAZAD. In CHES'03, pages 77–88, 2003.
58. Jean-Jacques Quisquater and David Samyde. Electromagnetic analysis (ema): Measures and counter-measures for smart cards. In Isabelle Attali and Thomas P. Jensen, editors, E-smart, volume 2140 of Lecture Notes in Computer Science, pages 200–210. Springer, 2001.
59. Josyula R. Rao and Berk Sunar, editors. Cryptographic Hardware and Embedded Systems - CHES 2005, 7th International Workshop, Edinburgh, UK, August 29–September 1, 2005, Proceedings, volume 3659 of Lecture Notes in Computer Science. Springer, 2005.
60. Francesco Regazzoni, Thomas Eisenbarth, Luca Breveglieri, Paolo Ienne, and Israel Koren. Can knowledge regarding the presence of countermeasures against fault attacks simplify power attacks on cryptographic devices? In Cristiana Bolchini, Yong-Bin Kim, Dimitris Gizopoulos, and Mohammad Tehranipoor, editors, 23rd IEEE International Symposium on Defect and Fault-Tolerance in VLSI Systems (DFT 2008), pages 202–210. IEEE Computer Society, 2008.
61. Francesco Regazzoni, Thomas Eisenbarth, Johann Großschädl, Luca Breveglieri, Paolo Ienne, Israel Koren, and Christof Paar. Power attacks resistance of cryptographic S-boxes with added

error detection procedures. In Cristiana Bolchini, Yong-Bin Kim, Adelio Salsano, and Nur A. Touba, editors, 22nd IEEE International Symposium on Defect and Fault-Tolerance in VLSI Systems (DFT 2007), pages 508–516. IEEE Computer Society, 2007.

62. Francesco Regazzoni, Yi Wang, and François-Xavier Standaert. FPGA implementations of the AES masked against power analysis attacks. In COSADE 2011, 2011.

63. G. Rouvroy, F.-X. Standaert, J.-J. Quisquater, and J.-D. Legat. Compact and efficient encryption/decryption module for fpga implementation of the aes rijndael very well suited for small embedded applications. In Information Technology: Coding and Computing, 2004. Proceedings. ITCC 2004. International Conference on, volume 2, pages 583–587 Vol. 2, April 2004.

64. Dhiman Saha, Debdeep Mukhopadhyay, and Dipanwita RoyChowdhury. A diagonal fault attack on the Advanced Encryption Standard. Cryptology ePrint Archive, Report 2009/581, 2009. http://eprint.iacr.org/

65. Sasebo. http://staff.aist.go.jp/akashi.satoh/SASEBO/en/

66. Moritz Schmid, Daniel Ziener, and Jürgen Teich. Netlist-level IP protection by watermarking for LUT-based FPGAs. In Proceedings of IEEE International Conference on Field-Programmable Technology (FPT 2008), pages 209–216, Taipei, Taiwan, December 2008.

67. Nidhal Selmane, Shivam Bhasin, Sylvain Guilley, Tarik Graba, and Jean-Luc Danger. WDDL is protected against setup time violation attacks. In Fault Diagnosis and Tolerance in Cryptography (FDTC), 2009 Workshop on, pages 73–83, Sept. 2009.

68. Nidhal Selmane, Sylvain Guilley, and Jean-Luc Danger. Practical setup time violation attacks on AES. In Proceedings of the 2008 Seventh European Dependable Computing Conference, pages 91–96, Washington, DC, USA, 2008. IEEE Computer Society.

69. Li Shang, Alireza S. Kaviani, and Kusuma Bathala. Dynamic power consumption in virtex-II FPGA family. In Proceedings of the 2002 ACM/SIGDA tenth international symposium on Field-programmable gate arrays, FPGA '02, pages 157–164, New York, NY, USA, 2002. ACM.

70. Eric Simpson and Patrick Schaumont. Offline hardware/software authentication for reconfigurable platforms. In Louis Goubin and Mitsuru Matsui, editors, Cryptographic Hardware and Embedded Systems - CHES 2006, volume 4249 of Lecture Notes in Computer Science, pages 311–323. Springer Berlin/Heidelberg, 2006.

71. François-Xavier Standaert. Secure and efficient symmetric encryption using FPGAs. Cryptographic Engineering. Chapter 11, pp 295–320, Springer, 2009.

72. François-Xavier Standaert, François Macé, Eric Peeters, and Jean-Jacques Quisquater. Updates on the security of FPGAs against power analysis attacks. In Koen Bertels, João M. P. Cardoso, and Stamatis Vassiliadis, editors, ARC, volume 3985 of Lecture Notes in Computer Science, pages 335–346. Springer, 2006.

73. François-Xavier Standaert, Siddika Berna Örs, and Bart Preneel. Power analysis of an FPGA: Implementation of Rijndael: Is pipelining a DPA countermeasure? In Marc Joye and Jean-Jacques Quisquater, editors, CHES, volume 3156 of Lecture Notes in Computer Science, pages 30–44. Springer, 2004.

74. François-Xavier Standaert, Eric Peeters, Gaël Rouvroy, and Jean-Jacques Quisquater. An overview of power analysis attacks against field programmable gate arrays. Proceedings of the IEEE, 94(2):383–394, 2006.

75. François-Xavier Standaert, Gaël Rouvroy, Jean-Jacques Quisquater, and Jean-Didier Legat. Efficient implementation of rijndael encryption in reconfigurable hardware: Improvements and design tradeoffs. In Walter et al. [82], pages 334–350.

76. François-Xavier Standaert, Loïc van Oldeneel tot Oldenzeel, David Samyde, and Jean-Jacques Quisquater. Power analysis of fpgas: How practical is the attack? In Peter Y. K. Cheung, George A. Constantinides, and José T. de Sousa, editors, FPL, volume 2778 of Lecture Notes in Computer Science, pages 701–711. Springer, 2003.

77. Daisuke Suzuki and Minoru Saeki. Security evaluation of dpa countermeasures using dual-rail pre-charge logic style. In Goubin and Matsui [22], pages 255–269.

78. Helion Technology. http://www.heliontech.com/

79. Kris Tiri and Ingrid Verbauwhede. A logic level design methodology for a secure DPA resistant ASIC or FPGA implementation. In DATE, pages 246–251. IEEE Computer Society, 2004.

80. Stephen Trimberger, Jason Moore, and Weiguang Lu. Authenticated encryption for fpga bit-streams. In Proceedings of the 19th ACM/SIGDA international symposium on Field programmable gate arrays, FPGA '11, pages 83–86, New York, NY, USA, 2011. ACM.
81. Frank Vahid. The softening of hardware. Computer, 36:27–34, April 2003.
82. Colin D. Walter, Çetin Kaya Koç, and Christof Paar, editors. Cryptographic Hardware and Embedded Systems - CHES 2003, 5th International Workshop, Cologne, Germany, September 8–10, 2003, Proceedings, volume 2779 of Lecture Notes in Computer Science. Springer, 2003.
83. Thomas Wollinger, Jorge Guajardo, and Christof Paar. Security on FPGAs: State-of-the-art implementations and attacks. ACM Trans. Embed. Comput. Syst., 3:534–574, August 2004.
84. Xilinx. http://www.xilinx.com/
85. Daniel Ziener. Techniques for Increasing Security and Reliability of IP Cores Embedded in FPGA and ASIC Designs. Dissertation, University of Erlangen-Nuremberg, Germany, July 2010. Verlag Dr. Hut, Munich, Germany.
86. Daniel Ziener and Jürgen Teich. Power signature watermarking of IP cores for FPGAs. Signal Processing Systems, 51(1):123–136, 2008.
87. The SHA-3 Zoo. http://ehash.iaik.tugraz.at/wiki/the_sha-3_zoo

Part III
Applications and Platform Embedded Security Requirements

Part III

Chapter 10
Mobile Communication Security Controllers

Keith Mayes and Konstantinos Markantonakis

Abstract Cellular communication via a traditional mobile handset is a ubiquitous part of modern life and as device technology and network performance continues to advance, it becomes possible for laptop computers, Personal Digital Assistants (PDAs) and even electrical meters to better exploit mobile networks for wireless communication. As the diverse demands for network access and value added services increase, so does the importance of maintaining secure and consistent access controls. A critical and well-proven component of the Global System for Mobile Communications (GSM) and Universal Mobile Telecommunications System (UMTS) security solution is the smart card in the form of the Subscriber Identity Module (SIM) or USIM, respectively. However, with the enlarged range of communications devices, some manufacturers claim that the hardware selection, chip design, operating system implementation and security concepts are different from traditional mobile phones. This has led to a suggestion that types of "Software SIM" should be used as an alternative to the smart card-based solution. This paper investigates the suggestion.

10.1 Introduction

Mobile communication and computing technology have evolved at a remarkable pace and it is now possible to propose new functionality and services that even a few years ago would have been dismissed as impossible. Compared to today, the mobile telephony pioneers of the 1980s worked in a much more restricted technical environment and with a different set of design priorities and associated assumptions.

K. Mayes (✉) · K. Markantonakis
Information Security Group, Smart Card Centre, Royal Holloway,
University of London, London, UK
e-mail: keith.mayes@rhul.ac.uk

K. Markantonakis
e-mail: k.markantonakis@rhul.ac.uk

K. Markantonakis and K. Mayes (eds.), *Secure Smart Embedded Devices,*
Platforms and Applications, DOI: 10.1007/978-1-4614-7915-4_10,
© Springer Science+Business Media New York 2014

However, it is not only the underlying technology that has changed, but also the market's attitude and expectation toward communication and computing, largely fuelled by ubiquitous Internet connectivity, the boom in e-mail, on-line purchasing and increasingly the exchange of digital content and sensitive information. Whilst there are many initiatives aimed to offer new services to individual customers, there is growing interest in machine-to-machine communications and telemetry systems. Even boundaries between business segments are breaking down as the distinction between wireless and fixed communications is increasingly blurred and if we consider modern Near Field Communication (NFC)[1] we will see mobile phones acting as credit cards, train tickets or smart poster readers.

Whilst many of the changes are positive and exciting, the long standing problem of securing systems and services against criminal mis-use remains and indeed has become more acute as electronic applications and transactions have become complex and remote. What is also evident is that the great technological advances that are enabling modern services are also being exploited by attackers. In the case of the GSM mobile communications standard (GSM), the attackers have been successfully repelled (with a few avoidable exceptions) by a device known as the Subscriber Identity Module (SIM). The SIM is essentially an application hosted by a specialised microcontroller (optimised for security) that is normally housed within a plug-in format smart card. There were very good reasons for introducing the SIM for GSM and its design has now been elegantly upgraded to an advanced USIM application for UMTS[2] third generation networks. There is also a smart card called a RUIM defined in some competing cellular standards.[3] However, despite its proven track record there will always be parties that question the necessity of a separate and removable hardware component to support security and/or whether the SIM is really the best solution for the twenty-first century. This paper will address such questions in an attempt to provide guidance to the industry.

The discussion will begin in Sect. 10.2 by recapping on the reasons why the SIM was adopted into standards and the main role that it takes in a mobile communications network. Section 10.3 first considers the trust relationships of the various parties involved in SIM/handset supply, usage and maintenance with respect to traditional and new categories of cellular usage. Section 10.4 goes on to extract the security fundamentals and critical capabilities that are necessary to carry out the range of trusted/secure operations in the various communication scenarios. As a fundamental reason for using hardware security modules such as the SIM is the ability to resist anticipated security attacks; Sect. 10.5 introduces the broad range of attacks that a smart card would be expected to defend against. Section 10.6 introduces some candidate software SIM solutions and compares them to the conventional SIM. Section 10.7 considers the Trusted Platform security element as a potential

---

[1] Near Field Communication (NFC) is similar to a contactless smart card/Radio Frequency Identification (RFID) interface for mobile phones.

[2] UMTS is the successor to GSM, initially standardised by ETSI and now by the Third Generation Partnership Project (3GPP).

[3] Standards defined by 3GPP2—relating to IS95/CDMA2000.

complementary or alternative technology to the SIM smart card. The overall findings are further discussed and summarised in Sect. 10.8, leading to the final concluding remarks.

Note that unless there is a need to show a distinction, the term SIM will imply a SIM or USIM application implemented on a conventional smart card platform. Similarly, the term SIM Application will refer to the SIM or USIM functionality, independent of the hardware platform. The term Mobile Equipment (ME) will be used to refer to all types of cellular communication device such as a mobile phone or data modem, unless there is a need to differentiate between devices.

10.2 An Overview of the SIM

Before considering the detailed trust issues surrounding the role of the SIM, it is essential to understand why the SIM exists in the first place. In fact, one of the main reasons was to overcome problems caused by weak security mechanisms implemented in early mobile phones. For example, before the GSM digital phone networks appeared in the UK, there was an analogue system called Total Access Communications (TACS). For a mobile to be allowed access to the network, it needed to transmit two identifiers; its telephone number (MSISDN) and a unique electronic serial number (ESN). If the MSISDN-ESN pairing matched the network's own records then the mobile was judged legitimate and was allowed access. Unfortunately, there was no algorithm available to provide confidentiality and so the phone transmissions were not encrypted, making it relatively easy for an attacker to eavesdrop the signalling exchanges and discover the MSISDN-ESN pairs. Taking a modern-day analogy for the network access procedure, the phone was acting as a very simple RFID [1]. Basic RFIDs can be observed and copied/cloned onto alternative platforms; however, a legitimate RFID should at least resist unauthorised changes to its critical data/ID; so it cannot be used to assume another identity. The TACS phones were meant to share this property, however in practice, it proved simple to re-program them as "clones". Whilst lack of encryption was a system design issue that resulted in eaves-dropping of signalling and phone calls, the weak handset security was a major factor in facilitating cloning.

Fig. 10.1 SIM authentication and session key generation

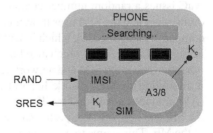

Therefore, when the European Technical Standards Institute (ETSI) [2] standardised its digital phone system (GSM) [3], it decided to remove the reliance on the handset as a security component and introduce a more secure solution based on a SIM smart card. There was still some reliance on phone-based security controls in the form of network locks, so that a subsidised mobile could not be moved from one network to another without appropriate authorisation. The fact that even today, so many "unauthorised" parties offer to unlock handsets does little to suggest that mobile-based security control is the correct approach. However, technology and application requirements have advanced dramatically, and so it is reasonable to suggest that the whole issue of security and trust be reconsidered from a complete system perspective.

10.2.1 The SIM in Operation

As the role and implementation of the SIM is core to the issues discussed within this paper, it is first necessary to obtain a basic understanding of its contribution to the system security. The early SIMs were regarded as a combination of smart card and application i.e. there was no logical decoupling between the SIM Application and the smart card hardware/platform. More modern cards are based on the Universal Integrated Circuit Card (UICC) smart card platform, and so the SIM is regarded more as an application rather than part of the underlying hardware. However, in real SIM products, the SIM Application is not necessarily a complete abstraction from the hardware, but rather a "special-case" application, exploiting lower level hardware and software functionality within the platform for both efficiency and resistance to attacks on its security. The SIM stores quite a lot of information and supports a range of functionality, but for clarity we will just focus on three critical components that are stored within the SIM and also in the mobile network's Authentication Centre (AuC).

- An Identity (International Mobile Subscriber Identity—IMSI).
- An Algorithm (A3/A8)
- A Secret Key (Ki)

When a ME is switched on and attempts to access a mobile network, there is a signalling exchange that transmits the SIM's identity (IMSI) to the AuC. The AuC issues a random number challenge (RAND) to the SIM (using the phone as a dumb pipe) and the SIM uses its authentication algorithm (A3) and secret key (Ki) to compute a result (SRES), which is sent back to the network. The process is illustrated in Fig. 10.1. The AuC (which also knows the Ki and algorithm) carries out a similar operation and if the SIM and network results match, the SIM is legitimate and the ME is allowed to use the network. In parallel with this, the SIM encryption key algorithm (A8) produces a session key (Kc) to cipher/scramble the communications between the ME and network. For performance reasons, the encryption is actually performed in the ME. This seems to go against the philosophy of not trusting the ME; however, the key has no long-term significance and is refreshed regularly. It is very important to note that whilst the SIM functionality for authentication and key generation is well standardised, the algorithms (A3/A8) are not and so every network could have its own proprietary algorithms. Furthermore, because the AuC can identify a SIM from its IMSI, it can not only obtain its secret Ki, but also determines the associated algorithm. There are numerous proprietary algorithms in use and most designs are kept strictly secret.

Regardless of the particular algorithm used in the above procedure there are some security weaknesses. The procedure only checks that the SIM is legitimate and not the network, nor does it check that the network challenge is new/fresh. Therefore, the USIM used in 3G networks [4] has an improved exchange that allows the USIM to also test the legitimacy of the messages/challenges received from the network. An example set of algorithms, the *milenage* set [5], which may be used for authentication and key generation has also been published. The milenage algorithm set was designed by the ETSI Sage group [6] and is based on the Advanced Encryption Standard (AES) [7]. Whilst it is expected that milenage will be quite widely used, proprietary algorithms are also likely. A complete description of 3G/UMTS authentication is beyond the scope of this report; however, the important changes with respect to 2G/GSM authentication can be appreciated with respect to Fig. 10.2.

There are some changes in nomenclature[4] and field sizes, but the main design difference is that accompanying the RAND challenge is an authentication token (AUTN). The token includes a Message Authentication Code (MAC) that can be checked by the USIM to determine that the challenge came from the authentic network. It also includes a sequence number (SQN) that can be used to thwart replay attacks and an anonymity key (AK) that can disguise the sequence number within

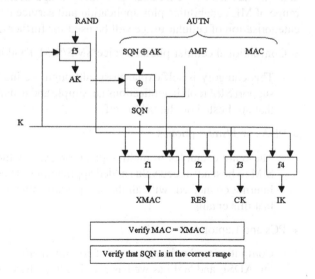

Fig. 10.2 USIM authentication overview

---

[4] K, CK and RES are similar to Ki, Kc and SRES in GSM authentication.

the legitimate challenge. Other enhancements that are beyond the scope of detailed discussion in this paper include the generation of an Integrity Key (IK) and an authentication management field (AMF). For any new system, 3G/UMTS authentication would be strongly recommended instead of the 2G/GSM method.

10.3 Security Analysis

The SIM has been at the heart of the technical measures that underpin the fundamental client-side mobile communication security. However, technology alone does not provide the required system security assurance and there are critical relationships between entities plus associated operational procedures that are necessary to establish and maintain trust. As technology has advanced and the role of mobile communications has expanded, the security requirements go beyond the need to simply safeguard the access to communication bearers. This added complexity not only impacts on technical solutions, but affects the relationship between operational entities. We can begin to examine the roles of these entities with respect to client side trust by considering a range of categories for cellular usage.

10.3.1 Categories of Cellular Usage

There are many ways that a modern mobile communications system can be exploited and this has lead to a number of cellular usage scenarios, supported by a wide range of ME capabilities plus application and service environments. The following categorisation of cellular usage will be used for further discussions.

- Conventional cellular phones (voice/SMS/SIM Toolkit)

  - This category is self explanatory, although we include within it the ability to support SIM toolkit applications i.e. simple and usually menu-based applications that are hosted on the SIM itself.

- PDAs and Smart phones

  - These are examples of multi-application MEs, which are more likely than the SIM to host advanced value added applications. These MEs are also regarded as Internet connected, with all the service advantages and potential security risks that this entails.

- PCs and Laptops

  - Computers tend to have a range of communication/connectivity options including ADSL and WiFi as well as cellular. It is, therefore, most likely that the ME is used as an added modem rather than the computer's application usage being strongly dependent on the MNO.

- Telematics and Metering

  - There is a wide range of possible telematics and metering applications, but considering vehicle telematics or electricity metering, one could imagine a fixed function application using a custom ME reporting to a central application server.

- General Machine to Machine cellular

  - This category is very diverse, but there is an implication that a custom and managed application exists at each "machine" and that the cellular network is used as a logical communications pipe between them. The applications could be hosted in the SIM, ME or a connected device/computer.

Due to their similarities, telemetry and machine to machine MEs will be referred to collectively as "T/M2M" devices. Similarly, PDAs and Smart phones will be referred to as "PDA/Smart" devices.

Having introduced an expanded set of cellular communications categories, it is now appropriate to introduce some generic definitions for the roles that one would typically find involved in and around a mobile communications system and/or a cellular-enabled computer.

10.3.2 The Roles in Communication Solutions

The following roles would be commonly recognised as present in mobile communications solutions, business and service provision activities. However, it is important to realise that for the SIM and network security viewpoint, we are most concerned about the hacker/attacker who would seek to undermine the system. There is no shortage of potential attackers ranging from academics developing proof-of-concept exploits to organised criminals seeking clones, counterfeit products and sensitive information for financial gain. Similarly, with over 2 billion MEs and SIM cards in circulation and active use, there is no difficulty in finding devices to attack or transactions to eavesdrop. Attack methods are described in detail within Sect. 10.5 and so for now we will focus on the more conventional roles in the mobile communication usage categories.

- Customer/User

  - A typical individual customer is likely to make use of communications services plus value added services that are accessible via his equipment. He is quite likely to change ME whilst keeping the same customer account; normally by transferring his SIM card. He may also sell his old phone and possibly include his pre-pay SIM. He may also change to a new MNO and perhaps keep his old ME. He may register for value added services either with the MNO or a third-party service provider. A MNO-enabled service might result in a SIM toolkit application being downloaded/enabled on the SIM. If he has a smarter ME he may download applications from the MNO, ME manufacturer or third-party service provider.

- The customer in a metering telemetry environment is likely to be a company such as an electricity supplier. This may require the use of a client-side application in the SIM, ME or computer and a network-based application server. Given the likely cost of custom deployment, upgrading the ME is unlikely, but changing MNO en masse is possible for commercial and service performance reasons.

- Mobile Network Operator

  - The owner of the mobile network, the servers and all its deployed SIM cards. Its primary business is to charge customers for use of its network, however, virtually all MNOs operate in the value added service domain. They may provide application servers and also corresponding client-side applications for SIMs and MEs. They are able to operate remote management servers for both SIMs and MEs. The large influential MNOs issue customised handsets to their customers.

- SIM Card Manufacturer/Supplier

  - The SIM card manufacturer provides the physical SIM cards to the MNO specification and usually the associated trust services such as initialisation and personalisation. SIM card manufacturers may also see themselves in an application development role and in the provision of operational trust services.

- Mobile Equipment (ME) Manufacturer/Supplier

  - Aside from manufacturing and supplying MEs, the ME manufacturer may see itself in the role of a value-added application/service provider. As most suppliers have some means to manage and control the data and applications on their products they may also see themselves in a security or trust services role.

- SIM Client Application Provider

  - A developer/provider of an application hosted on the SIM.

- ME Client Application Provider

  - A developer/provider of an application hosted on the ME.

- ME/SIM Service Provider

  - A provider of a network-hosted services matched to applications hosted on the ME or SIM.

- PC Service Provider

  - A provider of a network-hosted services matched to applications hosted on a conventional computer.

- PC/OS Manufacturer/Supplier

  - Aside from supplying the computer platform, these suppliers may also see themselves in the role of application/service provider. As most suppliers have

Fig. 10.3 Interaction of roles
in mobile service provision

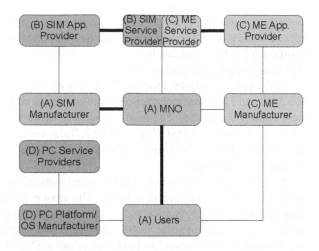

proprietary means to manage and control the data/applications on their MEs
they may also act in security or trust services roles.

The interaction between the various entities and roles is complex and as there are
many MNOs, SIM card manufacturers and ME manufacturers, one cannot cover all
possible combinations in this report. However, Fig. 10.3 is an attempt to show the
general interaction of the various entities/roles with respect to client-side trust for
various cellular usage categories.

An initial observation is that there has always been a very strong trust and security
relationship (A) between the MNO, the SIM manufacturer and the users (or at least
their SIM cards). This has underpinned every category of GSM cellular usage to
date.

In early phone systems, there was relatively little complexity as handsets were only
capable of basic functionality. As SIM capabilities improved, it was possible to have
value-added services that were implemented as SIM Toolkit (STK) [8] applications.
This implied a relationship between the SIM service provider and the MNO (B) who
had management control of the SIM card contents and functionality.

There was also a relationship between the STK application developer (SIM application provider) and the SIM card manufacturers (sometimes the same entity) in order
to implement the application on delivered SIMs. The reliance on the SIM manufacturer to load the applications has weakened over time due to the introduction of Java
Card technology [9] that allows remote loading (or more likely post-manufacture
loading) of additional applications.

With the introduction of PDA/Smart devices (case C), ME service providers and
ME client application developers/providers work closely together (or are the same
entity) to offer services that are less influenced by the MNO. In this scenario, the
need to establish some level of client-side trust and management may be leveraged
from the ME manufacturer e.g. Nokia signing of applications for download to Sym-
bian platforms. This may also serve to strengthen the ME manufacturer relationship

with the end user in the area of value-added applications and services. The ME manufacturer may not be totally free to control ME trust services as MNOs are major customers for MEs and can in some cases insist on customisations and controls on ME functionality and configuration.

In the case of cellular enabling PCs/laptops, there is much less influence from the MNO and ME manufacturer on client-side trust unless it is established by business/contractual means. The reason is simply that a laptop/PC user may normally work connecting to ADSL/WiFi and so cellular communications is often regarded as another connectivity channel rather than something that would dominate/control service trust and usage.

Telemetry services could in principle fit into any of the options shown in Fig. 10.3, although option (C) is the most likely. The reason is that a fixed function service and custom client application will probably require a customised ME, especially if the accuracy and integrity of the reported measurements are critical. Machine-to-machine applications would probably suit options (C) and (D) with more emphasis on the deployed applications/equipment and less influence and control from the MNO.

10.4 Security Fundamentals

So far we have considered trust and security in fairly general terms, but at this stage it is necessary to define some trust items that we will examine further in our cellular usage scenarios. First, we will introduce some information security fundamentals;

- Authentication
  - To ensure that entities involved in our trusted solution are legitimate/authentic.
- Confidentiality
  - Information, signals, commands or functionality that are restricted to certain authorised entities must be protected from disclosure/discovery by unauthorised entities.
- Integrity
  - Critical data and applications code should be protected from modification when in storage, operation or during communications/transactions.

  These fundamentals can in turn be underpinned by some practical capabilities;

- Cryptographic algorithm(s) plus supporting data for authentication/encryption/integrity
- Secure storage and verification of critical data, with strict access controls
- Secure verification and execution of algorithm(s) and other critical functions
- Secure communication protocols
- Controlled operating environment and isolated "security domains"

The word "Secure" has been used rather freely in the above bullet points and so we should be clear what it means in this context;

The functionality that embodies our security fundamentals has been correctly designed, implemented and tested to strongly resist the anticipated attacks[5] that may be made against it.

Having established some security fundamentals and supporting practical capabilities, the next step is to consider how they are used in critical operational processes.

10.4.1 Trust Operations

The concepts described thus far need to be applied to SIM usage and evolution, and so it is necessary to consider some relevant processes and operations that rely on security and trust. These entries will be split into two sections; Core SIM Operations, that are fundamental to all forms of mobile communication and Extended Operations that relate to value-added services. Although this paper focuses mainly on the Core SIM Operations, it is important to be aware of Extended Operations, because if the SIM's capabilities are not sufficient or easily accessible then alternative strategies and technology/management solutions will likely appear.

10.4.1.1 Core SIM Operations

- Initialisation, Personalisation and Key Management

  - Customised configuration of the solution prior to issue to a customer

- Authentication/ Encryption
- Management of SIM Data and Applications
- Migration

  - Change of MNO
  - Change of ME
  - Change of Computer/Peripheral
  - Change of Algorithm

10.4.1.2 Extended Operations/Value-Added Services Management

- MNO, ME Manufacturer and third-party Value-Added Services provision/ management
- Near Field Communication Management

---

[5] The attacks which are described more fully in the Sect. 10.5 include all known logical, physical, side-channel and fault classes.

10.4.2 Initialisation, Personalisation and Key Management

The actions under this heading are strongly protected under the trust relationship between the MNO and a few selected SIM card manufacturers; these parties are considered mutually authenticated. Initialisation prepares the SIM for issue to any of the MNO's customers; however, it is a significant stage as the MNO will have shared confidential details of its algorithm, services and data supported by the SIM. It is still the case that some MNOs keep their algorithms secret and whilst they may believe them to have been well designed, they are unlikely to disclose them; as it is not uncommon for proprietary algorithms to fail dramatically once exposed to widespread expert scrutiny. Personalisation is the next critical phase as the keys are generated/loaded to permit authentication/encryption. Normally, the keys are also loaded to permit future Over The Air (OTA) [10] management of the SIMs by the MNO and in the case of Java Cards the GlobalPlatform [11] key(s) are loaded to permit application loading/deletion. Various PIN codes are also set at this point to strictly control access to the SIM functionality. Clearly, initialisation, personalisation and the associated key generation and management processes are extremely security sensitive operations and reliant on the integrity of the information loaded onto the smart card. These process are intended only to be carried out in a secure environment by a party adhering to the highest physical, operational and IT security standards.

10.4.3 Authentication/Encryption

The basic GSM authentication/encryption support via the SIM was described in Sect. 10.3. For it to be secure we must be sure of the algorithm integrity; that it has not been modified and cannot be disclosed by anyone with access to the SIM, nor should it be revealed when in operation. The secret key must not be rewritable, or externally readable and may only be accessed by the algorithm with the condition that the key value cannot be inferred from the operation of the algorithm. To achieve this under anticipated attack conditions requires attention to design and implementation aspects in both software and hardware. Normally, these safeguards are provided by the SIM Card manufacturer (assisted by the specialist chip provider) who would normally also be the algorithm implementor. The session key generated by the SIM and passed to the phone to support confidentiality of radio transmissions is not a strong or long-term secret, but the SIM and phone should ensure that it cannot easily be extracted by an adversary.

10.4.4 Management of SIM Data and Application

The SIM contains numerous data files [12]. The integrity of the data critically affects the operation of the ME and the services/facilities offered to the customer. The data fall into a number of categories;

- System configuration (identity, SMS switching server, language preference, service table)
- User data (telephone numbers, SMS messages)
- Operational data (network lists, fixed dial numbers, customer-care telephone numbers, branding)
- Application data (menus)

The files have authentication access controls that were set during the initialisation/personalisation processes. Access permissions are controlled locally by a range of PIN codes. The most significant PINs are only known to the MNO (and SIM manufacturer), whereas PIN1 is trusted to the customer for locking the SIM. It is also possible to remotely access the files using an OTA server. This relies on a standardised protocol and another set of keys that have been pre-stored in the SIM during personalisation. The OTA server and keys are normally only available to the MNO; however in principle, some keys may be trusted to other parties for the management of application data on the SIM. SIM Toolkit applications may also be managed in this manner and the keys permit a secure/confidential channel to be established between the SIM application and the OTA server, independent and additional to any security offered by the basic GSM encryption. In the case of a Java Card supporting GlobalPlatform, it is possible to download/delete SIM applications in the form of Java applets. For this purpose, there is also a Card Manager key that is pre-stored on the SIM during personalisation. Clearly, the management capability for a modern SIM relies on a significant number of secret keys and PINs.

10.4.5 Migration

Migration is an important consideration in mobile networks and there are a number of scenarios that deserve consideration.

- Migration to a new MNO is usually achieved simply by SIM replacement. This ensures that the stored data, algorithms, keys, PINS and added functions are exactly as required for the new MNO. It also ensures that the SIM has the correct capabilities in terms of memory, speed, crypto-coprocessor and communications capabilities. There is usually quite a wide range, with some MNOs going for very low-cost limited devices, whereas others opt for the more sophisticated and expensive products. Variation may also be evident within SIMs for different customer segments used by a single MNO. There are also significant geographic variations driven by local market conditions and competition. Replacing the complete SIM card also provides assurance that security critical functionality and storage plus the associated attack resistant countermeasures in the underlying hardware and software have been implemented and tested to the MNOs standards.

- Migration to a new ME was one of the reasons that the SIM was introduced as a removable module. Providing that the new ME is not locked to a competing network, plugging in the customer SIM is all that is required. Networks are often obliged by regulation to unlock handsets (for a fee) when requested to do so by a customer, although most are unlocked by unauthorised means.
- Migration to a new PC/Laptop usually means swapping the whole ME + SIM, in the form of a PCMCIA or USB communications adaptor. A new driver may need to be loaded onto the computer. If the laptop has a built in ME then migration would normally be via a SIM and driver swap.
- Although less common, migration of algorithm is possible within mobile networks and indeed, multiple authentication algorithms can be simultaneously supported within the network, so that not all SIMs need to be replaced at the same time. At the client level, the migration could be achieved by a SIM swap, however, some networks implement back-up algorithms. These can be switched in, if the normal algorithm is compromised by some unexpected weakness or advance in attack techniques. Because of its emergency-use nature, the back-up algorithm is likely to be a confidential/proprietary design as is the authorisation method to enable it.

10.4.6 Extended Operations/Value-Added Service Management

For value-added service management, there are a number of scenarios to consider. First, an application may be hosted on the SIM or in the ME. SIM applications are normally controlled by the MNO and tend to be restricted by SIM technology and varying levels of standards support in MEs. On the positive side they benefit from mature SIM standards for authentication, confidentiality, integrity and management. ME applications can benefit from the much greater available processing and user interface resources and potentially may be managed by MNOs, ME Manufacturers and third-party application/service providers. The deployment/business strategies of these parties may be divergent and the security assurance, control and management requirements can vary for each application.

10.4.7 NFC Management

The prospect of MEs having Near Field Communication (NFC) capability and being able to emulate contact-less smart card readers and contactless smart cards is exciting and opens up new opportunities for mobile services. NFC capable MEs introduce the concept of a Security Element (SE) that might be used to host applications in a secure manner. The ownership and control of the SE and indeed whether it is an additional chip or added SIM functionality is still the subject of some debate. It is also feasible to drive the NFC functionality direct from the ME platform, thereby bypassing an additional or SIM hosted SE.

10.5 Generic Attacks on Smart Cards

Before considering whether the SIM card is any more at risk today compared to when it was launched, it is first necessary to review the range of attacks that have been successfully attempted against smart card solutions. It is a common characteristic of smart card-based systems that a critical part of the security solution is in the hand of the user and indeed often in the hands of millions of users. The opportunity for misuse/tampering is enormous and not only from criminals/hackers, but also from the legitimate holders and employees involved with card distribution, sales and operations. It is therefore comforting to note that the most advanced SIM cards are extremely capable of defending themselves, using a sophisticated arsenal of countermeasures against known attacks. It might be imagined that systems would only deploy these state-of-the-art smart cards; however, the need to deploy large numbers of cards that often have a limited lifetime, creates pressure on costs. What is actually deployed is the best business compromise, balancing price, risk, functionality and attack resistance. Advances in technology, new attacks and/or increasing the value of protected assets can all upset this balance, and so it is prudent to periodically re-evaluate a card's vulnerability to attack. Smart card attacks may be grouped within some general categories that typify the techniques and also the resources and expertise needed by the attackers. Commonly described categories are:

- Logical
- Physical
- Side Channel
- Fault[6]

10.5.1 Logical Attacks

Logical attacks are not unique to smart cards and are similar to attempts to hack into IT systems. The attack uses the normal interface to the device, but attempts to extract information via repeated guesses, invalid requests and stimulating error conditions etc. Logical attacks exploit things that have been done badly e.g.

- Weak design

---

[6] Fault attacks could be considered as combinations of other categories, but their importance merits separate mention.

- Bad implementation
- Poor testing
- Lack of monitoring/detection

Applied to a smart card, the logical attack only needs a card reader and a PC so there is no cost and/or equipment barrier to such an approach. The basic message formats and protocols used to communicate with smart cards are well standardised [13], and so the attacker can easily construct test messages of correct or intentionally incorrect formats and lengths to try and generate some revealing response.

Weak design is the biggest worry, as it affects not only the information that might be revealed to an attacker, but also its subsequent value and desirability. A great prize would be a global secret held on a legitimate smart card, which if revealed could be used for mass production of counterfeit cards or simply exposure to blackmail. A flawed security algorithm could also be exploited to extract card-specific secret keys, as in a well-reported logical attack [14] against an example GSM authentication algorithm called COMP128-1. Essentially, the attack involved repeated calls (many thousands) to the card authentication command, which eventually revealed sufficient information to allow the secret key (Ki) to be extracted. This should have been practically impossible from a brute-force/trial-and-error standpoint, and so the fact that the attack could succeed in a practical time frame is related to a weakness in the algorithm itself. There were also system design weaknesses in that the authentication command could be called from a PC device that had no provable authority and originally there was no monitoring or detection mechanism for the many authentication attempts. Stronger GSM algorithms exist which are less vulnerable and 3G systems overcome the mutual authentication problem by using a MAC to validate authentication requests. Where COMP128-1 cards are still deployed today they tend to incorporate an authentication counter that terminates the card after too many attempts—albeit with a corresponding reduction in the normal life of the card.

The best countermeasure against logical attack is quite simple, just design, implement and test things in a rigorous and best-practice manner. For good measure, it is also advisable to include monitoring methods to detect and react to attacks in progress. The good news is that modern smart cards bought from reputable manufacturers are generally well designed and tested, so that an attacker would be very fortunate to succeed with a logical attack alone. In response to this, more sophisticated attack methods have evolved including physical, side-channel and fault attacks.

10.5.2 Physical Attacks

It is a false assumption to consider a system secure because it uses a good algorithm. The algorithm alone may provide good logical security, but could be completely undermined by physical tampering. In the case of the smart card there is reliance on the chip to defend itself against intrusive physical attacks that seek direct access to, or modification of memories, buses, CPUs etc., thereby bypassing "logical" security defences.

Whereas logical attacks can be developed or copied by just about anyone, physical attacks normally require a high level of technical expertise and access to sophisticated and expensive laboratory equipment. Almost invariably the attacks require the decapsulation of the smart card chip i.e., its removal from the card in preparation for physical examination and/or modification.

Rendering the source card (or indeed many of them) unusable is not an important issue as physical attacks are most often used for reverse engineering the card implementation. The goal is to discover design information that could be used in some other type of attack or lead directly to the creation of card clones. Although physical probing techniques are tricky because the chip area is so tiny (typically less than 10 mm^2), the principle is quite simple i.e. attach conductive probes to interesting parts of the circuit. One of the favoured tools for investigating smart card chips was the probe station consisting of a microscope, a high precision mechanical platform and needle-like probes designed to make electrical contact with parts of the chip circuitry. A second-hand probe station can be bought for about €12,000; an example is shown in Fig. 10.4.

As chip technology has advanced, the smart card circuitry has got smaller and smaller, to the point where conventional probe stations are becoming impractical as the target contacts are very small compared to the size of the probe needles. A more modern and expensive tool (approx. €350,000) is the Focussed Ion Beam (FIB). The FIB can be used to examine and/or modify a circuit, and can connect additional circuitry or simply larger contact pads for access via a conventional probe station.

Fig. 10.4 Probe station (Wentworth labs)

A physical attack against a chip that has not been specially designed to resist it, will reveal its secrets quite rapidly, however smart cards can incorporate many countermeasure techniques which whilst not guaranteeing protection against physical attack, make it far more difficult for the attacker. A common countermeasure is to introduce a physical barrier that may simply be a slab of tough material or an active current carrying mesh. Removing either type of barrier without destroying the chip requires the type of expensive equipment and high-level expertise normally found in commercial test labs. This is not the end of the defences available to smart cards and indeed the circuit layout and layers may be scrambled to make it difficult to locate the desired attack points. Even if the barriers and layout confusion are eventually mastered there can still be low-level encryption methods that prevent direct reading of buses and memory contents. Whilst the attack/investigation is in progress there are also the environmental detectors to worry about. If the chip is exposed to light, extremes of temperature and/or voltage it will trigger a detector and the chip will cease to operate.

For a tiny piece of silicon, the smart card can be astonishingly good at resisting physical attack. However, one must always expect that any security device is only tamper resistant and not tamper proof as driven by enough motivation, expertise, money and time, a physical attack will succeed. The important question is why would anyone take the trouble? A company-commercial reason is that reverse engineering may permit the creation of counterfeit smart cards. A security reason is the belief that the card contains some secret information or technique that may directly or in combination with other attacks be used to exploit and or clone other such cards. It should also be appreciated that some researchers would embark on a sophisticated physical attack simply because of the academic challenge. Whilst proof-of-concept attacks are not usually motivated by financial gain, they unfortunately serve to educate other parties that may have malicious intent.

It is worth re-iterating that not all smart cards include all the physical attack countermeasures mentioned above and manufacturers are understandably coy about revealing what their chips will or will not do. It is also worth noting that there have been successful physical attacks that have been very simple and low cost [15] such as interrupting the power supply during critical processes. Generally, physical attacks are beyond the resources and capabilities of most attackers, whereas the same is not true of side channel attacks.

10.5.3 Side Channel Attacks

It is very difficult to keep a secret. Usually this statement might apply to some document, secret algorithm or key that is locked away; however it can be applied to the operation of electronic circuits. If a circuit such as a smart card chip is believed to be running an algorithm and using a secret key it may become an attack target. The logical attacker will actively try and trick the chip into revealing its secrets (but should fail), whereas the physical attacker will try and break in and perhaps destroy a few other chips as part of a learning process. What makes the side-channel attacker interesting is that he basically just waits and listens, extracts the secret with a modest amount of equipment and often leaves the original card undamaged. As an analogy, consider the attackers trying to get access to an interesting security lecture. The logical attacker tries to trick his way in, but cannot get past the security guard on the door, the physical attacker drives a bulldozer through the wall, whereas the side-channel attacker just listens at the door where the sound of the lecturer's voice "leaks" through the door (the side channel).

The basic principle of side-channel analysis is that secret information is always leaking despite the presence of logical and physical attack countermeasures, so the trick is to find the leakage and extract something useful from it. The smart card, in common with most other electronic devices, consists of many logic gates and transistors. An example is shown in Fig. 10.5. As the logic state of the gate changes e.g. from 1 to 0 or 0 to 1 there is a minute surge of electrical current accompanied by a spike of electromagnetic radiation. Therefore, if you can detect the current or radiation changes you can get an idea about the state of the logic gates. Now, if for example those gates are part of a register used by the security algorithm then at some stage, the transitions will be caused by the value of the key. To extract the key from the leakage may sound difficult, but the reality is that until countermeasures were put in place, side-channel attacks were effective against even high-end smart cards and they are still valid against other types of non-protected circuitry.

One of the major concerns is that side-channel attacks do not need a lot of expensive equipment and can be recreated with far less expertise than would be needed by a physical attacker. A typical set up for Simple Power Analysis [16] or Differential Power Analysis [17] requires little more than a digital oscilloscope and PC, whereas the electromagnetic emission version simply requires an added antenna/amplifier.

DES was an early target algorithm [18] as it uses multiple rounds of processing in which only a few key bits are used. The side-channel analysis is used to reveal key information at each round, so the attacker deals with a sequence of small key problems rather than the full DES key. It is of course known that with the right equipment, normal DES can be brute forced [19], so a common strategy is to recommend double or triple key DES. Whilst this is a logical defence, it does little to safeguard against a side-channel attack that can simply work its way through the small number of key variations of each round. The critical point to remember is that logically strong

Fig. 10.5 CMOS switching circuit

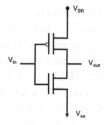

algorithms and keys may still be vulnerable to side-channel attack unless appropriate tamper-resistant measured are implemented.

Modern high-end smart cards are strongly tamper resistant and this extends to defences against side-channel attack. As attack analysis usually relies on statistical averaging, one approach is to add low-level timing jitter and high-level software variations to prevent alignment of the leakage waveforms. Another defence is to generate artificial noise on the leakage signals to disguise the target information. If the chip hardware is sophisticated enough to support differential switching (rare) then whenever a logic 1 changes to 0 there is a corresponding bit that changes a 0 to a 1, which snuffs out the leakage signal at source.

10.5.4 Fault Attacks

The basic principle behind a fault attack is that if you can induce a temporary fault into an algorithm the error can be used to reveal secret information. A necessary pre-requisite is therefore the means to introduce a fault in a non-destructive and controlled manner. This has been practically achieved using spikes on the card power supply and momentary exposure to bright light. The faults can be induced in theprocessing circuitry or possibly change the state of memory cells. A famous example of this attack class was used against a RSA implementation [20]. Because RSA is computationally intensive it is often implemented in two stages with the results being combined by the Chinese Remainder Theorem (CRT). If you can cause any error in the processing of a bit within one of the stages there is an elegant method that directly extracts the secret key by a simple mathematical calculation.

If a smart card device is likely to be subject to fault attacks then there are a number of countermeasures that can be used effectively. At the chip level, it is possible to add detectors that should trap the fault insertion attempt e.g. light or voltage glitch detectors. At the application level, the algorithm could be run multiple times and the results compared although this could have a significant effect on speed and usability.

10.5.5 Summary and Main Points

It is important to realise that the attacks mentioned in this chapter are not made against smart cards because their security is weak, but rather than they are identifiable and easily accessible security components. Except for very low-end devices, the SIM smart card's resistance to attack is quite admirable. When it was first introduced, the SIM's attack resistance provided far more assurance than the available mobile phones; however, we should not automatically assume that this is still the case. It is therefore worthwhile to consider various alternative options for SIM implementations with respect to our modern categories of cellular usage.

10.6 SIM Implementation Options

In this section, we will consider a range of implementation options for the SIM with respect to our core security requirements and various categories of cellular usage described in Sect. 10.4.

10.6.1 Pure Software SIM

Our definition of a Pure Software SIM (PSSIM) is a SIM application written in software running on a shared computer processing platform that cannot benefit from any hardware security component. Even before considering the detailed issues, it should be clear to most readers that this can only reduce rather than increase existing levels of security protection, and so it is important to try and understand why the PSSIM is proposed for some market segments.

10.6.1.1 Motivation

The root of the interest in the PSSIM lies in cost reduction and logistics. Although typical SIM cards are no longer very expensive (few Euros), the mobile industry deals with very large numbers of them and so even small reductions can result in saving of millions of Euros. With a PSSIM, it is not anticipated that there would be a significant saving to the ME manufacturer (apart from the SIM socket cost), although it may be a little easier to physically construct the product without access for SIM card insertion/removal. The beneficiary is therefore most likely to be the MNO (who typically buys/owns the SIM cards), but of course the MNO has most to lose from compromising the system security. The financial motivation is not only cost savings from smart card chips, but also the removal (or great simplification) of the procurement and distribution channels required to get SIMs from the factories and into mobile devices. The problem is worse for specialist MEs that are effectively embedded modems used in Telemetry/Machine-to-Machine (T/M2M) applications. For example, you may ship cars around Europe that have embedded communications to support local services and telematics, and so ideally you would insert the SIM for the country of destination. The destination may not be known during manufacture, so adding the SIM cards is a manual post-fit operation that may be costly and difficult in terms of physical access. There are some potential technical and tariffing solutions to this problem; however, they put more emphasis on strong security and it would be difficult to sell T/M2M equipment without full co-operation from an international MNO. In the case of general T/M2M, the operational environmental conditions may be more extreme than for conventional mobile telephony, and so a "special" SIM card may be needed in any case.

10.6.1.2 Analysis

It is relatively straightforward to write a software program that provides all the core functionality offered by a SIM. If the algorithm, key management and security policies are made available by the MNO, then the implementation will also include the main security functions. If our program has been designed, written and tested according to IT best practice then it should resist logical attacks, but alarmingly, none of the other categories described in Sect. 10.5. Furthermore, because the PSSIM runs on a non-secured shared platform we have additional security concerns. Other applications running on the platform may gain access to the memory used by the PSSIM and use this to monitor/modify critical data and functionality. The other platform applications may also cause (intentionally or accidentally) run-time performance and resource problems for the PSSIM that may compromise the SIM's real-time duties. Furthermore, if the platform has no boot protection there is no guarantee of integrity for any of the platform's operating system and application software. As the platform is also a communications device there are fears of fast-spreading remote attacks that typify PCs, such as viruses, worms and trojans. A successful attack on a PSSIM would seem almost inevitable and that would normally lead to clone devices.

Some MNOs, ill-advisedly, appear prepared to tolerate a level of cloning in their networks—as evidence from continued use of weak authentication algorithms. However, this is dangerous, as attacks can rapidly spiral out of control as particularly easy/lucrative techniques are discovered. PSSIMs would not only simplify attacks, but also provide convenient clone platforms for use with extracted keys. An attack on a PSSIM using a secret algorithm would be most severe as it would amount to the discovery of a "global secret" which may reveal weaknesses that could undermine all categories of cellular communication.

Aside from discarding almost all of the security measures that have protected the SIM application, the PSSIM also breaks the very important trust linkage between the MNO and the SIM manufacturer. For the PSSIM the MNO may have to share its algorithms, data, policies, secret keys and PINs with ME manufacturers. Whilst SIM manufacturers are set up as high security companies (and often generate much of the sensitive data for MNOs), this is not normally the role/capability of the ME manufacturer.

It is difficult to have any confidence in the personalisation and lifecycle management (including migration) of the PSSIM and its associated data. For example, personalisation of a conventional SIM is normally carried out in a very secure environment (usually by the SIM manufacturer) and relying on the integrity of the smart card platform. If we are using other parties in a different environment, to configure a platform that cannot guarantee its integrity, or resist simple attacks, we could completely compromise the core security solution.

10.6.1.3 PSSIM Summary

Despite the temptation to reduce costs and simplify logistics, it is strongly recommended that a PSSIM solution should not be used. It is vulnerable to a wide
range of security attacks and undermines proven trust relationships and management processes. PSSIMs would almost certainly result in clones and problems may
become fast-spreading once attacks exploit the communications capability of MEs.
The use of PSSIMs in some restricted categories of cellular usage is dangerous
because revealing secret information such as proprietary algorithms and security
policies may become a risk for all categories of usage. This risk is not just from
technical attack, but disclosure of this sensitive information to third parties that are
not specialist security/trust companies.

10.6.2 Hardware Shared Security Software SIM Solution (HS-SSIM)

The SIM is not the only application in an ME that may require security assurance
and following-on from the discussions on the PSSIM, it is clear that some specialist
hardware support is essential. The hardware must protect the particular application,
its operational processes and sensitive data, but to do this it must also protect the
integrity of the processing platform itself. The conventional SIM card and TPM (see
Sect. 10.7) approaches are examples of additional and specialisthardware security
processor modules that may help protect applications from attack. However, this
section focuses on the use of the existing/main ME processor to implement the SIM,
which we will refer to as a hardware shared software SIM (HS-SSIM). This differs
from the PSSIM because the chosen ME processor has some specialist security
enhancements. By way of example, this section will consider the "TrustZone" [21]
from Advanced RISC Machines Limited (ARM).

The ARM is a 32-bit RISC processor architecture that is used in a wide range of
embedded systems such as mobile phones and PDAs. The ARM architecture is complemented by a number of security extensions under the overall name of "TrustZone".
Key components of the TrustZone architecture are presented and summarised/quoted
below:

- a TrustZone CPU that is used to run trusted applications isolated from normal
  applications, and to access the memory space reserved for trusted applications,
- secure on-chip boot ROM to configure the system,
- on-chip non-volatile or one-time programmable memory for storing device or
  master keys,
- secure on-chip RAM used to store and run trusted code such as Data Rights Management (DRM) engines and payment agents, or to store sensitive data, such as
  encryption keys,

- other resources, peripherals, that can be configured to allow access by trusted applications only.

It offers two separate and parallel execution environments that run on the same processor. This is achieved through virtualisation which is responsible for deciding whether the currently executing application should run as normal or secure code. The decision controls the restrictions (e.g. which peripherals it can access) and privileges (e.g. access to certain memory locations) that apply to the application. As there is only one physical processor, a very close cooperation is required between the hardware and software components, in order to guarantee that the overall architecture behaves as a single well-defined system. The measures used to control the execution environment should make TrustZone more secure than the PSSIM solution, but only if the TrustZone software maintains its operational integrity and the underlying hardware functions correctly even when subject to attack.

A core element of the TrustZone functionality is the "Secure Monitor" entity which is responsible for performing the necessary checks and switches between secure and non-secure states. It is up to the processor to enforce the correct data and peripheral access policies depending on the actual execution state, e.g. secure or non-secure. The TrustZone platform also offers a secure boot process using cryptographically signed boot-strap code in ROM. The TrustZone software (TZSW) is at the centre of the platform's execution environment. It offers the ability to execute native code or interpreted applications. This interpreter is based on the Small Terminal Interoperability Platform (STIP) which is developed by GlobalPlatform [11] and in theory permits applications to run in a protected and sandboxed environment.

10.6.2.1 Motivation

The motivation to take this approach is similar to that of the PSSIM and for logistics reasons, it is most desirable for the T/M2M categories of cellular usage. From a cost minimisation viewpoint, it suggests improved security compared with the PSSIM whilst avoiding the cost of an additional security module. There may also be interest in using the method of software upgrades (re-flashing) to also correct/upgrade the SIM functionality.

10.6.2.2 Analysis

The HS-SSIM is a compromise solution that should offer improved security with respect to a PSSIM; however it may be difficult to achieve tangible assurance that this is actually the case. We are reliant on the HS-SSIM software to control secure/non-secure processing and reliant on the underlying hardware to implement the necessary execution and data storage controls. Unless these components have been evaluated to a known standard (e.g. Common Criteria) or lab-tested by a MNO, one must assume that these components may be vulnerable. The difficulty in trying to achieve a level of assurance is further compounded if software elements may be upgraded (or re-flashed) post-issue, as is often the case in a ME. This upgrade mechanism potentially provides an added security risk and the changes to the software could invalidate previous security evaluations. The hardware could be evaluated against known attacks, although resistance is usually a combination of hardware, operating system and application measures. However, the design of the processor is likely to be a compromise between the performance needed for normal use and the measures needed for secure execution. It might be expected that this compromise would lead to less protection and slower execution compared with a dedicated hardware security module such as a SIM card. The literature review reveals that the core of the Trust-Zone's functionality is not currently promoted (due to its increased size) for smart card products. In fact, the core functionality of TrustZone provides secure access to smart cards, suggesting that the technologies are expected to co-exist. Although all software may run in the secure core of the platform, it is advised that only security sensitive code is shielded for trusted execution as the extra checks increase the overall size and complexity of the platform software.

The last point is significant as one of the main motivations of the HS-SSIM was to provide added security without adding an extra chip. If as the TrustZone suggests, there may be added cost for the main processor and its memories and increased power consumption, the benefits of the HS-SSIM compared to adding a specialist hardware security module are eroded.

10.6.2.3 HS-SSIM Summary

The HS-SSIM should in practice be more secure than the PSSIM, because it has some logical features that are aimed to support secure execution. However, from a security assurance viewpoint there may be little difference unless the hardware, platform operating/system environment are evaluated to a known standard and then do not change post-issue. Without this assurance one has to assume that whilst the solution may resist some logical attacks, it will be vulnerable to other techniques. It is also questionable whether adding complexity to the main processor is the best solution and indeed whether it is possible to achieve effective assurance on a device that is shared by non-secured applications. The justification to use the HS-SSIM was to avoid an extra chip and the associated cost, however, adding secure execution capability to the main processor appears not without its own costs.

10.6.3 Standalone HW Security SIM Solution

Our definition of a Standalone Hardware Security SIM (SH-SIM) is the conventional SIM smart card used in GSM and UMTS communications. It relies on a specialised security microcontroller chip with a security optimised operating system that has an implementation of at least the SIM application and associated (MNO specific) algorithms and data. It is assumed that the whole platform and SIM application have been tested as resistant to all known attacks, to a level of assurance that satisfies the standards/requirements of individual MNOs. Furthermore, the SH-SIM is personalised in a secure environment prior to issue to customers and/or insertion in a ME. In key management terms, security critical keys (and PINs) have been pre-loaded onto the SIM during personalisation and these fields will not be changed thereafter. Post-issue re-personalisation is not possible and migration to a new network requires physical replacement of the SH-SIM. Changing to a new ME simply requires moving the SH-SIM to the new ME. A SH-SIM from a particular MNO might support algorithm migration, but the new algorithm and switching mechanism would have been pre-loaded onto the SH-SIM.

10.6.3.1 Motivation

The motivation to keep the existing SH-SIM is that it has done a remarkably good job of securing communication in mobile networks. Changing anything to do with the SH-SIM is therefore a risk that could have serious impact on a MNO's business and reputation. Moving from a well-proven security solution to an even better-proven security solution of course has merit. For example, upgrading from a 2G authentication method to the 3G milenage method is well advised as the introduction of mutual authentication and an improved open algorithm design would strengthen the system security.

10.6.3.2 Analysis

Although there are some arguments to avoid the SH-SIM to reduce costs there are also arguments to keep it. For example, the size of SH-SIM devices varies enormously; whilst a typical device might have a 32 kbyte Electrically Erasable Programmable Read-Only Memory (EEPROM), some MNOs are using Mbyte devices and highly advanced Gbyte SIM[7] devices are available [22]. If MNOs no longer supplied their own physical SIMs, but made use of a Hardware Security Module (HSM) built into a ME, the HSM would either be dimensioned very large (and expensive) to accommodate all MNOs, or would be too small for some. Advanced MNOs might want a HSM with USB interfaces [23] and Single-Wire-Protocol [24] capability to enable Near Field Communication [25] services; however other MNOs may not wish to pay for this. There are also cost factors around crypto-coprocessors that are necessary to support extra services that use public key cryptography. It is therefore important to realise that by providing its own SH-SIM the MNO can always ensure

---

[7] It is interesting to note that the traditional limitations of the SIM smart card (small memory, slow interface and restricted CPU) have been overcome by technology advances, albeit at added cost compared to a traditional SIM.

that the cost and capability of the SH-SIM matches its current and planned business requirements.

In the SH-SIM, the MNO is providing the complete package for its security. It can ensure that the combination of the chip, operating system, applications are not only functionally correct, but have been adequately tested against known attacks. There is sometimes criticism of SH-SIMs in that they claim to be tamper resistant, but are not evaluated to common criteria standards, and so the level of security offered to support cellular usage is unclear. The reason for this is partly due to cost, but also the rapid deployment rates in mobile communication that can mean a SH-SIM lifecycle is shorter than the time needed for a common criteria evaluation. However, while there remains no agreed and practical assurance level for a SH-SIM, it will be far more difficult (than it should be) to argue its superior security credentials compared to say a PSSIM. It would therefore be very useful to have MNOs adhere to a set of industry-wide best-practice criteria for SH-SIM devices (perhaps from the GSM Association [26]).

Any post-issue changes to the SH-SIM are completely controlled by the MNO and tend to affect data and added applications rather than the core SIM security and the OS/platform. The MNO can also ensure that all critical functionality, keys and data are personalised in a secure environment pre-issue, often by means of a highly trusted third party (e.g. SIM manufacturer).

It should be noted that all the good security properties of the SH-SIM arise from the chip and the associated operational and management processes. The smart card body has almost no useful role in normal operation and most of the card plastic is discarded before insertion into the ME. The body may help in production as standard smart card production machines may be used, which helps to keep costs down. The body also helps by providing portability which allows a SH-SIM to be swapped between MEs.

In principle, if the SH-SIM chip was prepared and personalised as if it were to go into a card body, but was actually soldered onto a ME circuit board and connected via its normal interface then the operational security would be identical to the current SH-SIM arrangement, except that the SH-SIM would not be removable. This could be a problem for conventional telephony as there can be legal requirements that allow a customer to migrate his ME to an alternative MNO; however, this might not be the case for T/M2M communications. Embedding the SH-SIM would result in a fixed network ME that would need to be replaced if there was any problem with the ME or the SH-SIM chip/account. Migration might still be possible if it were feasible to place the SH-SIM chip in a small socket. In this case, the MNO migration would be a job for a technician. This requirement may not however be unreasonable for T/M2M systems.

10.6.3.3 SH-SIM Summary

The conventional SH-SIM has proven itself as an effective security module for mobile communications and any change is a potential risk that must be properly understood and justified. One positive example of this is the SH-SIM migration from 2G to 3G security, as this has been rigorously investigated and improves system security with the addition of mutual authentication and an improved open algorithm. The large variation in the types and costs of smart cards used by different MNOs for various market systems suggests that any SH-SIM equivalent provided by the ME risks being over or under specified with corresponding impact on costs and services. An advantage of the SH-SIM is that operators supply the whole package including hardware/OS, applications and data, and so can control the level of security assurance and ensure resistance to all known attacks. It would perhaps be better for the industry if more effort was directed toward best-practice criteria for SH-SIMs, so comparisons with say PSSIM solutions could be more easily made in future. The conventional preparation and use of the SH-SIM ensures that personalisation is carried out by a trusted party in a secure environment and that the SH-SIM may be subsequently managed by the MNO in a secure manner. The importance of these last points is often overlooked when proposing alternative solutions to the SH-SIM. If a SH-SIM chip is prepared in the normal manner then in principle it could be included on a ME PCB for T/M2M communications and provide the same security as the removable SH-SIM. The disadvantage is that MNO migration may not be possible or require technician services.

10.7 Trusted Platform

It should be clear from the fore-going discussions that the security of mobile communications has relied to a significant extent on a specialised tamper-resistant microcontroller embedded within the SIM smart card module. Because this small computer platform has been specified, implemented and tested by the MNO and/or its suppliers, the MNO can trust it. Furthermore, as the SIM is owned and managed by the MNO this trust is not eroded during the SIM lifecycle. However, the need for manageable trust in a processing platform is not restricted to mobile communications. The PC world, via the Trusted Computing Group[8] [27] , has driven forward the Trusted Platform Module (TPM) concept. The main motivation behind the development of the TPM specifications was the difficulty in a modern/complex PC to verify whether it is running "uncorrupted" software. If for example the underlying operating system (e.g. Windows or Linux) cannot provide such assurance, there can be little confidence that any applications will perform correctly. Fundamentally, the TPM goal is to prove that the PC is in a state that can be trusted to run applications and process data. Note that in contrast with a SIM, the TPM is not intended to be portable or removable, but rather inextricably bound to one and only one platform (and usually in the form of a chip). It securely stores asymmetric keys which can be used in order to protect sensitive data like other keys, certificates and passwords. The TCG states

---

[8] Note the Trusted Computing Platform Alliance (TCPA) was the predecessor to the Trusted Computing Group (TCG).

that "Trust is the expectation that a device will behave in a particular manner for a specific purpose" [28]. The TCG specifications state that a TPM device" should provide at least three basic features: protected capabilities, integrity measurement and integrity reporting."

- Protected capabilities include specific commands and functionality that hold exclusive permission to specific "shielded" platform resources. These resources could for example be memory locations or registers (for holding sensitive data, keys and integrity measurements) and also management of cryptographic primitives and other keys. The protected capabilities play a crucial role in verifying the correct operation of the platform.
- The TPM should be in a position to reliably collect information that reflects the TPM host's software state (i.e. integrity measurements) and place it in the relevant integrity storage locations. This information will be used for subsequent verification.
- Platform attestation enables the integrity measurements (which reflect the TPM host's software state) to be reliably reported. This then enables a verifier to decide how much trust should be placed in the status of the platform.

10.7.1 Roots of Trust

To a great extent the required levels of assurance are achieved through the TPM's roots of trust.

- Root of Trust for Measurement (RTM)
- Root of Trust for Storage (RTS)
- Root of Trust for Reporting (RTR)

The RTM is a trusted entity that can generate a reliable integrity measurement for at least one process running on the underlying platform. The TPM specifications define the RTM as "the root in the chain of transitive trust" [28]. The RTM is typically implemented as the normal platform engine controlled by a particular instruction set (the so-called 'Core Root of Trust for Measurement' (CRTM)). On a PC, the CRTM may be contained within the BIOS or the BIOS Boot Block (BBB), and is executed by the platform when it is acting as the RTM [42]. The CRTM is the first component to be executed during an authenticated boot process (see Sect. 10.7.2).

The RTS is a trusted component that is responsible for providing confidentiality and integrity for stored TPM data, e.g. cryptographic keys and the Platform Configuration Registers (PCRs) used for storing integrity measurements.

The RTR is a trusted entity responsible for providing various integrity measurements (integrity digests) on information stored in the RTS. Secure storage involves both encryption and sealing. Sealed data is bound to a set of platform measurements (i.e. that define the state of the platform) that must be present in order for data to be decrypted. This will provide the necessary reassurance that the platform will obtain access to "sensitive information", i.e. the sealed message, only if it exists in a known configuration.

10.7.2 Authenticated Boot and Secure Storage

During the platform boot process the CRTM and TPM enable each of the components in the system (including hardware and software) to be reliably measured and the resultant measurements to be stored in a set of Platform Configuration Registers (PCR) located in the TPM. The integrity measurements simply involve a SHA-1 digest of the code to be loaded. The TPM is agnostic to the underlying operating systems (OS) or applications and provides no assurance regarding their inherent security. Verification simply reports pre-runtime configuration information and it is up to the OS, applications and external verifiers to judge the trustworthiness of the PC platform and permitted actions.

Providing secure storage functionality requires that the TPM should be tamper resistance and preferably be able to detect and report any tampering attempts. At the same time, secure storage requires the provision of cryptographic functionality (for the integrity and confidentiality) of data. For example, the concept can be further expanded to include the association of keys and data with passwords and/or metrics that depict the platform/software state). The TPM specifications make specific references to the required cryptographic functionality that should be offered, e.g. RSA [29], SHA-1 [30,31], random number generation [32]. A detailed review of TPM [33] is beyond the scope of this paper, but critical components can be seen in Fig. 10.6.

10.7.3 Ownership

The concept of TPM ownership is very important. Originally, some concerns were voiced as TPMs may have permitted a third party such as an equipment provider or

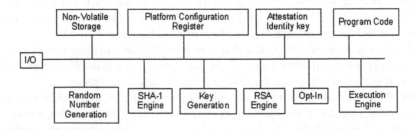

Fig. 10.6 TPM component architecture [28]

OS supplier taking control of a PC that was bought/owned by a user. The response was to put the user in control, by requiring the user to "opt-in" and take ownership of the device. The TPM may come with some pre-installed keys and certificates (which can be changed) that will provide the foundation for various platform operations. Opting-in means that the user is guided through a number of steps, e.g. generating further cryptographic keys, and formalising and defining the behaviour of the TPM.

Having, very briefly, examined the core functionality of TPM, it is becoming evident that in principle it could also be utilised by a range of applications requiring secure storage, enhanced authentication and additional platform/application security. Some of the proposed TPM usage scenarios are summarised in [34–36]. For a general purpose security component to be considered trustworthy by application providers it must clearly prove its security credentials and capabilities. The TCG proposes that the ISO-15408 [37] Common Criteria security evaluation standard should be utilised, in order to evaluate and certify TCG products and platforms. The TCG is working towards the development of common criteria protection profiles (PP) and guidelines that will assist the security evaluation of TPM devices. Although the TPM device is embedded into the PC motherboard, it could also be used in a variety of other devices[9] such as PDAs and mobile phones. In the next section, we examine the TCG's efforts to apply the TPM to mobile communications through the Mobile Trusted Platform (MTM) specification.

10.7.4 Mobile Trusted Platform (MTP)

Among the main missions of the TCG is the provision of trusted platform specifications for a number of devices (including MEs) that require information security assurance. The Mobile Trusted Module (MTM) [38] specification and the corresponding MTM reference architecture specifications were published in September 2006 and 2007 respectively. The TCG's definition of trusted computing (in terms of making sure that hardware and software behave as "expected") remains the driving force for the development of the MTM specifications. One of the quoted aims is to complement the existing functionality (of the operating system, hardware/software, SIM/USIM) with additional assurance that will enhance the overall security of the ME.

In common with the TPM, the MTM is responsible for protecting keys and other confidential information in secure storage. It might not be implemented on chip [39] and could even be a virtualisation layer making use of a TPM. The MTM defines two main types of MTM ownership, i.e. the Mobile Local-Owner Trusted

---

[9] https://www.trustedcomputinggroup.org/specs/

Module (MLTM), which is similar to the TPM model and the Mobile Remote-Owner Trusted Module (MRTM). The latter (MRTM) is a significant departure from the user controlled "opt-in" approach and allows remote entities (such as service providers, communication carriers, device manufacturers) to access and control their own space within the platform. In order to maintain clear boundaries between the entities and the operations that are allowed, the concept of "engines" is introduced. Each engine corresponds to an aforementioned owner and it has exclusive control over its data protection mechanisms. Some engines are considered mandatory (critical services) and other discretionary. The specifications state that mandatory engines are owned and controlled by a MRTM whereas the discretionary engines may be owned by a MLTM. There is also the role of Device Owner, which determines the remaining engines in the mandatory domain and determines all engines in the discretionary domain.

The significance of the change in ownership should not be overlooked. The user acceptance for TPM in the PC world is established through the "opt-in" process, however MRTM could see more management control in the hands of third parties. In the case of SIM card control, this has been justified by the MNOs remaining the legal owners, however the ME is normally bought/owned by the user.

Similar to the TPM, the MTM should be able to clearly demonstrate adequate levels of resistance to a wide range of attacks (see Sect. 10.5). However, the exact details of the fundamental threats, countermeasures and the planned Common criteria Protection Profile (PP) have not yet been fully finalised. Achieving an acceptable level of trust may prove more difficult for a MTM that uses a software abstraction layer.

10.7.4.1 MTP Use Cases

A relatively complete list of MTM use case scenarios is defined in [40]. In principle, the MTM can be implemented in many devices with relatively restricted processing capabilities including mobile phones and PDAs. Any application (e.g. Digital Rights Management, ticketing, payment, network access) requiring a level of platform integrity could in principle benefit from the underlying MTM specifications.

The MTM use cases will be influenced by the following three MTM characteristics. First, the specification introduces the secure boot of an MTM engine. This implies that on top of the standard functionality of data protection and platform attestation, the engine must boot into a pre-defined software state or not at all. Second, the MTM engine should have the ability to perform runtime checks on the integrity of software components. Finally, the local and remote TPMs essentially implement subsets of the TPM functionality. The TCG proposes these characteristics for MEs, but not for any other type of platform.

10.7.4.2 Comparison of MTP and UICC

The TCG developed the MTM specifications, in order to address specific needs of the mobile security. The fact that both the TPM and MTM standards are open and widely available encourages to some extent the acceptance of the technology and enables expert peer reviews. However, acceptance will not be sustainable in the long term if the technology does not deliver against its promises or any serious issues or flaws are identified. Fundamentally, the underlying hardware (which is the cornerstone of the security assurance) and software specifications must not only be properly defined, but correctly implemented by all suppliers of compatible MEs. Therefore, it remains to be seen whether all real ME products will be able to demonstrate, (through evaluation and operation), long-term sustainability against security threats and vulnerabilities. The latter is of particular importance, as many problem issues are discovered only after products are deployed.

In almost all the above issues, smart card technology is delivering its aims. This is partly because it is a complete customised and tested package (e.g. chip/OS/application), but also because its functionality is relatively restricted and simply, because it has proved itself over many years of practical use. Therefore, smart card design, operational and security requirements are well understood by chip developers, smart card manufacturers and MNOs.

The only reference known to the authors at of the time of writing this article, attempting to categorise the appropriate uses of smart cards and TPM (in terms of machine or user authentication scenarios) is presented in [41] and summarised in Table 10.1.

However, although the authors have not verified this analysis, it does appear that the two technologies can be considered as complementary. Although the table is not directly aimed at mobile communications there is a split between authentication of the user and the computer/machine. This split is not just related to the security

Table 10.1 Suitable uses of TPM and smart cards [41]

User/Machine authentication scenarios	Smart card	Trusted Platform Module (TPM)
User ID for virtual private network (VPN) access	Yes	No
User ID for domain logon	Yes	No
User ID for building access	Yes	No
User ID for secured e-mail	Yes	No
Host computer ID for VPN access	No	Yes
Host computer ID for domain access	No	Yes
Host computer ID for attestation (that is, authentication of software applications)	No	Yes

properties of the smart card or TPM, but also the fact that the user account, security and data tends to be personalised separately in a secure environment, independent of the PC/ME that is eventually used.

10.8 Summary

The role of the traditional SIM smart card has been described as a tamper-resistant module for supporting authentication and encryption in mobile networks. The capability to strongly resist known security attacks is fundamental to its existence and this is underpinned by both the design and implementation of the hardware, operating system and application software. The SIM memory size and crypto functionality for different MNOs (and market sectors) can vary significantly and cost sensitivity means there is no such thing as "one-size-fits-all". The SIMs are normally pre-personalised in a secure environment before issue to customers and the detailed requirements and related security processes tend to be MNO specific. It is common for there to be a strong trust relationship between the MNO and the SIM manufacturer as the supplier is trusted with and often generates, much of the most sensitive security data associated with SIMs. The personalisation of the SIM is not only vital to normal operation, but also the lifecycle management of the SIM. The authority to make changes to the SIM after issue is underpinned by an ownership model whereby the MNO always retains ownership of the SIM card and facilitated by management keys pre-stored on the SIM during personalisation. The removable nature of the SIM smart card means that a number of migration scenarios, which are sometimes a legal requirement, are supported. A user may migrate to a new ME, but keep an old SIM card, or migrate to a new MNO by swapping that network's SIM card. Occasionally, it is possible to migrate to a new account type or even to a back-up security algorithm without changing the SIM card.

The SIM card has proven itself, over a long period of time, to be capable of maintaining the security of mobile communications, and so it would take some very compelling reasons to take the considerable risk of changing to an alternative solution. One of the reasons might be to reduce the cost of SIM card supply, meaning not just the card itself, but the whole logistics, storage and distribution process. Another reason is that some newer cellular usage categories may have physical, distribution and environmental constraints that make it awkward to accommodate the normally personalised SIM card.

The two most interesting new cellular categories to consider are PDA/Smart and T/M2M. The former is an example of a high-end/expensive device, becoming virtually a connected and portable PC. Whilst the MNO is in control of communications security and may influence ME features and servers, the ME itself is often under some form of control from the ME manufacturer and perhaps more open to application developers independent of the MNO's influence. The T/M2M is a good contrast as it is not issued to "real users", may have fixed/custom functionality requiring a specialised ME and is likely to be very cost sensitive.

One might suggest for either of these categories that a Pure Software SIM (PSSIM) could be used, as the logical implementation of a SIM application on most processors would be possible. However, this would be extremely foolhardy as the implementation would be trivial to attack, risking exposure of not only secret keys, but details of confidential information and possibly secret algorithms. Clones would be simple to arrange even on the attacked MEs, dragging security levels down to those of old analogue systems. It is therefore strongly recommended that a PSSIM is not considered under any circumstances, but rather solutions that can leverage from some kind of tamper-resistant hardware.

The T/M2M MEs might be considered as candidates for the Hardware Shared Software SIM (HS-SSIM), in order to save the cost of a new hardware security module that may be difficult to insert/remove in practice. Intuitively, the HS-SSIM should offer more security than PSSIM, however, it may be difficult to provide sufficient assurance to MNOs on security, and also on vendor independence. Normally, the MNOs provide the whole SIM "package" i.e. an appropriately sized chip, OS, application and personalised data all tested against known attacks and competitively sourced from multiple suppliers. In the case of the HS-SSIM, the MNO has to fit onto a proprietary third-party platform that is a compromise design for supporting both normal and secure processing environments. There is likely to be reliance on critical software to handle the split between operation modes and it is difficult to see how the same levels of attack resistance could be achieved compared to a normal SIM card and indeed some product literature seems to deny support for smart card emulation. A formalised security evaluation would go along way to provide assurance; however, it is not clear if it is at all feasible to even consider this on a general purpose platform, especially if the processor code may need upgrading to support normal operation. The effort and time required for evaluations would probably not be merited and incompatible with the short lifecycle of processor chips used in ME production.

As an alternative to the HS-SSIM, the T/M2M could make use of an existing TPM chip. However, a T/M2M type of ME is likely to be of limited functionality and so would not be expected to already have a TPM. If one considers an important function of a TPM to be the support for an authenticated/secure boot, a T/M2M device may be installed and permanently powered and so may only boot once in its lifetime. Control over distribution and installation may obviate the need for any special control over the boot function.

It is difficult to identify a cost/logistics argument for investing in an embedded TPM for a T/M2M, rather than a conventional SIM smart card; however, it is conceivable that one might exist. The TPM is normally designed as a hardware security module and so in principle is capable of proving its security capabilities in a formal (or widely accepted) evaluation. However, this may not be the case if the MTM version is not fully implemented in the chip, but uses ME proprietary software to interface with a TPM. To progress the general argument further we will assume that sufficient assurance might one day prove possible and that the underlying TPM provides an appropriately dimensioned secure storage and execution environment that could be used for a SIM equivalent. The first observation to make is that the TPM would need to be dimensioned to accommodate the demands of the most demanding MNO and indirectly this may have a cost impact on the less demanding MNOs. A second and far more fundamental observation is that the normal processes and methods for secure SIM card personalisation could not be used. If personalisation is to be carried out in a secure environment then the MNO might need to either use ME manufacturers/suppliers (who are not usually security/trust specialists) or ship the MEs to a trusted party (such as a SIM manufacturer). The risk of the first option and cost of the second would seem to undermine the case for adding a TPM for cost saving and logistics simplification. Personalisation in a non-secure environment would be likely to contradict many MNOs' security policies and would not be recommended due to the risk of disclosing secret keys, sensitive data and secret algorithms/functionality. There is also the question of access to the initial management keys to personalise the platform and the fundamental question of ownership and management rights of a device that is deemed to be owned by the customer. Most migration scenarios would not be supported unless re-personalisation is possible, which MNOs may regard as a major security risk and there is an open question about the transfer of keys and management rights between competing parties.

The PDA/Smart could well have a TPM/MTM installed by default and so the chip cost justification would not be an issue; however, the significant problems around personalisation, ownership, management and migration would remain. The TPM could be seen as a complementary technology to the SIM card that could ensure the integrity of the PDA/Smart platform. This would be particularly useful as PDA/Smart devices would likely have Internet connectivity and require protection from common perils such viruses and trojans etc. The TPM/MTM would be a useful resource for application providers whose security requirements are less demanding as MNOs and might otherwise offer solutions on completed untrusted platforms. This may move some control from the MNOs towards the owner of the TPM. Who the owner should be is a matter for business debate rather than security analysis.

It would appear that the arguments for providing an alternative SIM implementation on the grounds of reducing costs and simplifying distribution logistics are not clearly compelling for any type of cellular usage. However, there is still the unsolved problem of coping with physical and environmental constraints (particularly in the T/M2M case) that would better suit an embedded chip. One solution is for the SIM suppliers to provide personalised chips, but not the smart card body. These would maintain SIM card functionality and security, but could be embedded directly onto a ME's PCB by a ME manufacturer. This could provide better temperature and physical characteristics and avoid the need for the SIM socket. Aside from the extra controls, management and cost at the ME manufacturer, the MEs would be permanently customised to a particular MNO, which might fragment a manufacturer's stock and production capability. Once deployed, the MEs would be bound to a single MNO/SIM and this lack of migration may not be acceptable to customers or MNOs in the long run. A variant would be to put the SIM chip in a socket (much smaller than a SIM card socket), so that a common batch of MEs could be subsequently configured for various MNOs. Migration would then be possible, but would probably require technician support which is not necessarily out of the question for

T/M2M deployments. In theory, one could use the personalised SIM chip approach in any phone, although the added migration difficulties suggest that the removable SIM smart card is still generally the more convenient option.

10.8.1 Value Added Service Management

The situation for value-added service management is less clear than the core SIM functionality and there are a number of different scenarios that need to be addressed in a standardised and consistent manner.

- For the MNO, the attraction of a SIM-hosted value-added service is that the SIM's capabilities and the functionality that it offers is completely specified and controlled by the MNO. A disadvantage is that the types of services have to date been fairly limited compared to the possibilities offered by modern MEs. Attempts to improve the sophistication of SIM applications have often been frustrated by limited support of the relevant standards by MEs. The MNO can decide to offer SIM security capabilities to ME hosted applications by means of APIs; however, considering the range of ME platform types and the many models in use (of varying levels of standards compliance), it is difficult to achieve a ubiquitous solution.
- The ME manufacturer sees the application management problem from another angle. The manufacturer will know its ME models intimately; however, there may be little interest in having an "open" application capable of running on MEs from competing manufacturers. If attempting to secure these applications through the SIM, the manufacturer might require a dialogue with all MNOs that use its MEs, to confirm that essential functionality is available. In fact, it is very unlikely that all MNOs could guarantee this support as SIM product capabilities are quite variable, and so the manufacturer may conclude that the most reliable and predictable strategy is to rely on the capabilities of the ME alone. For example, a developer can currently create an application and send it to Nokia for signature (based on a developer registration process), that is then checked when the application is loaded onto the ME. An obvious disadvantage of this approach is that the personalised and tamper-resistant storage and functional properties of the SIM card are not exploited to secure the applications.
- A third-party service provider may have security requirements; however, they may not share the ME manufacturer's confidence in the ME platform (and ideally would want the application to run on all MEs). The SIM is promoted by MNOs as a secure platform, however, without proof of a best-practice/independent security evaluation, it may be difficult to demonstrate adequate assurance levels to a third party. This problem could in principle be overcome, but there would also need to be more consistency/standardisation regarding the provision of SIM resources and third-party access to them. For example, a Java Card-based SIM could host an added application, but not if the MNO policy was to prevent third-party access.

Alternatively, the MNO could welcome third-party access, but then deploys low cost SIMs that have no spare memory to fit extra applications.

- The prospect of MEs having NFC capability and being able to emulate contact-less smart card readers and contactless smart cards has great potential for innovative new services. Some of these services will relate to financial transactions or the exchange of sensitive data and so security support is critical. In current NFC trials, the MEs have a separate Security Element (SE) that can be regarded as a Java capable smart card chip embedded in the ME. This again raises difficulties around personalisation and assurance. If a bank provided the SE then it may be satisfied with the security level and control/management of the chip; however, what if many banks or other types of application provider all insist on their own security levels and management methods? The MNOs are not keen on the separate SE chip and propose controlling the NFC functionality via the SIM card using Single Wire Protocol. This may satisfy MNOs, but may not be very helpful to other parties that wish to secure their NFC applications and services.

None of the value-added service security approaches described above do an ideal job. A possible way forward would be to use the SIM in a limited way that exploits its personalisation, key storage, remote management and security algorithms, without making great demands for memory or processing resources. An application developer would want to be assured that this support would be available from most SIMs and there should be no non-technical barriers (strategy/policy) that would prevent him using it. The applications themselves could sit in the ME providing there was an execution environment that was reasonably secured. If this environment was underpinned by some ME hardware then from an assurance viewpoint, the application provider would want a standardised and evaluated chip(s) that is used in most MEs, rather than a range of proprietary solutions.

10.8.2 Concluding Remarks

From a security and practical viewpoint, there seems little wrong with the long standing practice of using SIM applications implemented as easily replaceable smart card devices, to underpin core security requirements. The alternative approaches that were considered do not clearly or convincingly reduce costs or simplify logistics and come with great risk of compromising current and proven levels of security and assurance. The most promising alternative device is the TPM (and the MTM variant) as it may become a standard feature on many MEs and can in principle, provide some tangible (evaluated) level of security. However, the TPM seems most suited as a complementary solution, rather than a replacement for the SIM. The TPM is well suited to protecting the general integrity of the ME platform and adding some security support to value-added applications, particularly in high-end PDA/Smart MEs. An interesting further study would be to determine how the SIM, TPM and other ME security resources could best be used in combination.

There is some justification for doing away with the removable smart card in T/M2M applications due to physical/environmental restrictions. The suggested solution that does not compromise security and still provides means for migration (albeit with technician support) is to provide SIMs in the form of personalised chips for insertion into small sockets on the ME PCBs. It is suggested that SIM and ME manufacturers investigate the practicalities and costs associated with this approach.

The secure management of valued added services is not considered to be well handled by any of the individual technologies addressed in this paper. The existing SIM card approach should do more to help in this respect otherwise there will be little choice, but to seek other solutions even if they are less secure, more costly and difficult to manage. An important aspect for application/service providers is to have SIM security capabilities that they can rely on being present, regardless of the particular MNO and ME model. Therefore, the requirement is for an industry-wide/standards response rather than ad hoc solutions from particular MNOs. Specifically, the MNOs could be encouraged to agree an industry-wide best practice guide for the security evaluation of SIM cards or equivalents. For cost and time constraint reasons, it may not be practical to carry out independent security evaluations and so it might be sufficient for SIM Vendors to self certify their products against the industry guidelines. Similarly, the MNOs could consider establishing an industry wide, minimum set of functionality/APIs available to support applications in the ME. To be clear, this would not just be a standardisation exercise, but an initiative to ensure that the SIMs procured by MNOs should always support the minimal APIs. Furthermore, exploitation of the SIM capabilities should not be hindered by complex negotiations with MNOs, but perhaps by a simple registration and signing process, similar to those used for ME application developers.

Acknowledgments Originally published in Elsevier Information Security Report 13 (2008); reproduced with kind permission of Elsevier.

References

1. Anderson R (2008). Security engineering: a guide to building dependable distributed systems. John Wiley, New York.
2. German Federal Office for Information Security (2011). Protection Profiles. [Online Available] https://www.bsi.bund.de/DE/Themen/ZertifizierungundAnerkennung/ZertifizierungnachCC-undITSEC/SchutzprofileProtectionProfiles/schutzprofileprotectionprofiles_node.html .
3. EVITA project (20082011). E-Safety vehicle intrusion protected applications. http://www.evita-project.org.
4. Hersteller Initiative Software (HIS), Working Group Security (2010). SHE Secure hardware extension version 1.1.
5. ISO 11898 (20032007). Road vehicles Controller area network (CAN).
6. National Institute of Standards and Technology (2001). FIPS-140-2: Security requirements for cryptographic modules.
7. Trusted Computing Group (2011). TPM Main Specification Version 1.2. [Online Available] http://www.trustedcomputinggroup.org/resources/tpm_main_specification.

8. Russell R (2008). Virtio: Towards a de-facto standard for virtual I/O devices. ACM SIGOPS Operating Systems, Review (42).
9. Debian GNU/Linux FAQ (2011). Basics of the Debian package management system. [Online Available] http://www.debian.org/doc/FAQ/ch-pkg_basics.en.html.
10. RSA Laboratories (2004). Cryptographic Token Interface Standard 2.2.
11. Universitat Politecnica de Valencia (2012). XtratuM A hypervisor specially designed for realtime embedded systems. [Online Available] www.xtratum.org.
12. Standaert FX, Malkin T, Yung M (2009). A unified framework for the analysis of side-channel key recovery attacks. Springer-Verlag, Berlin.
13. IEEE 1609. Draft standards for wireless access in vehicular environments.
14. ISO 15408 (2007). Information technology Security techniques Evaluation criteria for IT security.
15. Scheibel M, Wolf M (2009). Security risk analysis for vehicular IT systems A business model for IT security measures. Embedded Security in Cars Workshop (escar 2009), Dsseldorf, Germany.
16. European Commission Information Society (2012). Emergency call (eCall). [Online Available] http://ec.europa.eu/information_society/activities/esafety/ecall/index_en.htm.
17. Poulsen K (2010). Hacker Disables More Than 100 Cars Remotely. The WIRED Magazine.
18. Eisenbarth T, Kasper T, Moradi A, Paar C et al. (2010). On the power of power analysis in the real world: A complete break of the KeeLoq code hopping scheme. Springer-Verlag, Berlin.
19. Koscher K et al. (2010). Experimental security analysis of a modern automobile. IEEE Symposium on Security and Privacy (SP).
20. Checkoway S et al. (2011). Comprehensive experimental analyses of automotive attack surfaces. USENIX association.
21. Rouf I et al. (2010). Security and privacy vulnerabilities of in-car wireless networks: A tire pressure monitoring system case study. USENIX association.
22. OVERSEE project (2009–2012). Open Vehicular Secure Platform. http://www.oversee-project.com.

Chapter 11
Security of Embedded Location Systems

G. P. Hancke

Abstract Determining the location and movement of objects or people is a core requirement in a number of embedded systems. To ensure that the location information gathered by embedded devices is accurate, the underlying method of location systems must be secure and reliable. This chapter provides an overview of the basic approaches for determining location information in embedded systems. The resilience of these methods against advanced attacks are discussed, along with proposals for securely verifying physical location estimates. Finally, the security aspects of global navigation space systems (GNSS) used for location information in embedded applications are briefly discussed.

11.1 Introduction

Technology is intended to make our lives easier. We are surrounded by a collection of devices with embedded computational intelligence that assists us in our daily tasks. Much has been written and spoken of over the years about the reliability and security of embedded systems, and with good reason. Embedded systems are often used to provide services that are crucial to our safety and security, such as an aircraft's autopilot or programmable logic controllers regulating industrial processes. System developers spend a significant amount of resources testing their software and hardware to ensure that systems such as these work reliably. It is, however, sometimes the case that these systems must interact with, and rely on, systems that are less secure.

The use of wireless communication and the introduction of mobile and pervasive computing have facilitated the growth of location-aware applications, which have the capability to recognise and react to the physical context of individual devices [1].

G. P. Hancke
Information Security Group, Smart Card Centre, Royal Holloway, University of London, London, UK
e-mail: gerhard.hancke@rhul.ac.uk

K. Markantonakis and K. Mayes (eds.), *Secure Smart Embedded Devices, Platforms and Applications*, DOI: 10.1007/978-1-4614-7915-4_11,
© Springer Science+Business Media New York 2014

There are an increasing number of embedded systems, used in safety and security sensitive applications, which incorporate location information into their core functionality. Embedded systems that rely on location information, either from internal or external sources, are responsible for aviation and maritime navigation, emergency response and rescue operations, high-value asset or vehicle tracking, and even rail signalling and train control.

As location information becomes more significant the interest in the topic of security of localisation schemes in embedded systems is increasing and numerous secure localisation schemes have been proposed [2, 3]. In these systems, location systems verify the position of individual devices, and by implication the events observed by these devices. Location metrics are also used as a basis not only for overarching system functionality, but also for other underlying technical procedures, such as determining routing paths and performing key management [4, 5]. A failure to obtain accurate and reliable information about a device's location could therefore significantly effect the performance of individual devices and the validity of the system as a whole. It is therefore important that embedded systems are designed to generate location information that is secure and reliable. The underlying components of an embedded location system should therefore be resistant to location fraud, whereby the system is made to generate incorrect location information by a malicious entity, and yet still function in a safe and reliable manner.

Embedded location information can be obtained in a number of ways. This chapter provides an overview of existing approaches for implementing location systems in embedded systems in Sect. 11.2, and evaluates whether the underlying location estimation methods are resilient against intentional attacks and operational errors. Current proposals for securely verifying physical position estimates are explained and the practical challenges that need to be addressed in implementing these methods are discussed in Sect. 11.4. As global navigation space systems (GNSS) are a particularly prominent location system, which often provides location information in embedded systems, we discuss the security issues in these systems in Sect. 11.5, with a particular emphasis on the global positioning system (GPS).

11.2 Embedded Location Systems

Location information systems are used in a variety of environments and applications, and as a result there are different opinions as to what should be defined as a location system in the context of embedded systems. For example, the ISO 24730 [6] standard states that a Real-Rime Locating System (RTLS) is a wireless system with the ability to determine the locations of an item anywhere in a defined space at a point in time by measuring the physical properties of the radio link. However, it also suggests that certain methods for locating an object that are based on radio frequency identification (RFID) should not be classified as RTLS. In practice, however, RFID is often used to track objects in a number of applications. For the purpose of this chapter a location system is defined as any scheme that could be used to infer the location or movement of entities at any given time.

Embedded location systems consist of two basic types of devices: the target nodes that the system is designed to track and reference nodes that have a known location. The location system must be able to determine the location of the nodes based on their interaction with the reference nodes. To accomplish this the location system needs to estimate the physical relation between the target node and the reference nodes using a position estimation method. The location of the node is then determined from the position estimates and the locations of the reference nodes. Boukerche et al. [2] define an additional component for location systems called the localisation algorithm, which could use the calculated node locations to determine the location of additional nodes. This third component is often found in systems with limited reference nodes where nodes rely on each other to build a network, such as wireless sensor networks (WSN), but is probably of less importance in embedded systems where reference nodes effectively covers all the areas of interest, e.g. GPS. This chapter concentrates on the latter approach, where all embedded devices have access to the required number of references.

A location system can be infrastructure based, i.e. the reference nodes collect physical estimates that are used by the system to determine a target node's location, or a location system can also be terminal, or target node, based. In the latter scenario the node determines its own location based on interaction with the reference nodes. A location system can also be built on a combination of these two approaches. To determine the relation between the tracked node and reference nodes the system can use range-dependent or range-independent methods, as shown in Fig. 11.1. In each case, the verifiers V are trying to determine whether the prover P is within distance d. *Range-dependent* methods measure properties of the physical communication channel that is influenced by the distance between the nodes. Examples of these properties are the amplitude of the received signal, and the direction or the time-of-flight of the communication. *Range-independent* methods do not measure physical properties of

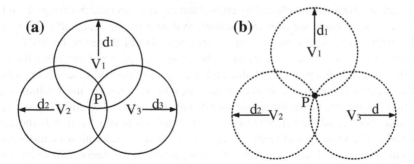

Fig. 11.1 Relationship between tracked node and reference nodes. The verifiers V use the estimated distance d to the prover P to determine its location. **a** Range-independent system with location area. **b** Range-dependent system with location point

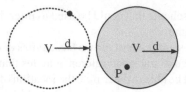

Fig. 11.2 Range-dependent and range-independent relative location systems with a single reference node. The verifier V is trying to determine whether the prover P is within distance d

the communication channel, instead relying on the concept of a 'location-limited' channel that only allows successful communication if the nodes have line-of-sight or are a short distance apart. RFID is an example of a range-independent technology because the token and the reader are assumed to be in close proximity if a transaction took place. This assumption is based on the fact that the operational range of the system is only up to a few meters.

If three or more reference nodes have determined their relation with regards to the node then the system can determine the *absolute location* of that node using multilateration, trilateration (with distance measurements) or triangulation (with communication angle measurement). If range-independent methods are use the exact location cannot be determined, but the system can identify an area in which the node is likely to be located. It is also common in some tracking systems to only have a single reference node and only determine the *relative location* of the embedded device. For example, the system only determines whether the device passed a specific reference node or is currently located close to a specific node. Examples of range-dependent and range-independent relative location systems are shown in Fig. 11.2.

11.3 Security and Resilience of Location Information

As location information gathered by embedded systems become more significant to sensitive and valuable services the location system must be designed to be secure and reliable. A system's underlying components should therefore be resistant to location fraud, i.e. an attacker causing the system to generate incorrect location information, and also provide reliable and accurate location information regardless of the security mechanisms implemented. Location systems are most reliant upon position estimation, the method used to estimate the physical relationship between the target nodes and the reference nodes. The analysis in this Section is therefore targeted at the security and practical aspects of position estimation methods most often used in current location systems. For the purpose of this section the operational environment of a system is assumed to be as follows:

- The embedded location system consists of a target node and reference nodes. The system tries to determine the locations of the target node based only on its interaction with the references nodes.
- The location of the target node can be calculated by the back-end or alternatively by the target node itself.
- The system can either calculate absolute location, based on the combination of measurements originating from multiple reference nodes, or relative location based on the observation resulting from a single reference. In the case of a single reference, the underlying physical measurement should be secure and reliable enough to provide useful location information without the need to collect collaborating measurements from other multiple reference nodes.

This Section assumes that the location system has been designed with security in mind. For example, we assume that nodes can be identified and authenticated. This prevents unauthorised parties from masquerading as either valid reference or target nodes. If an attacker could simply clone a target node or 'spoof' signals from reference nodes, without having to circumvent any security mechanisms it would clearly compromise the location calculations by both terminal and infrastructure-based systems regardless of which positioning methods are used. For example, an attacker could just introduce additional reference nodes linked with false locations into the system, as was practically illustrated by Tippenhauer et al. [7] on a wireless local area network (WLAN) location system. In reality, some systems generate location information without basic security services, as will be discussed in Sect. 11.5. However, this Section aims to introduce advanced attack strategies and illustrate that basic security is not enough to safeguard location information.

Having introduced the operational and security aspects of the location service it is time to describe the two basic attack threats that could undermine the integrity of location information, even if basic, application layer security mechanisms are in place. We introduce these attacks in the context of an infrastructure-based system, but these are equally relevant to terminal-based systems. If the location system is secured the reference nodes can be seen as *verifiers* (V), and the target node as a *prover* (P), i.e. the target node is providing proof of its location and the reference nodes should verify the validity of this information.

A *relay attack* involves an attacker attempting to misrepresent a prover's true location. The attacker uses a proxy-prover, i.e. fake target node, and a proxy-verifier, i.e. fake reference node, to relay the transactions between the real verifier and prover via another communication channel. It does not matter what application layer protocols or security algorithms are used as the attacker just relays all the application layer data, thereby ensuring that both the verifier and the prover always receive the data they expect [8]. As a result, the real verifier cannot distinguish between the prover and proxy prover and the verifier therefore concludes that the prover is located at the position occupied by the proxy. The effect of a relay attack on the distance measurement between a verifier and prover is illustrated in Fig. 11.3. Both the verifier and the prover are honest and unaware of the attack. This attack was first described

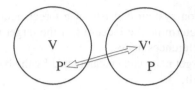

Fig. 11.3 A relay attack where the response of a legitimate target node P is forwarded by a proxy reference node V' to a proxy target node P'

Fig. 11.4 A compromised node P modifies the physical characteristics of its communication to convince a reference node that it is at a different positions P'

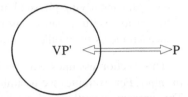

as 'mafia fraud' in [9] and is also known as a 'wormhole' attack in sensor-network literature [10].

Distance fraud occurs when the prover is fraudulent, i.e. the attacker is the device expected to prove its location, or the prover has been compromised, and tries to convince the verifiers that it is at a different location than is actually the case. In this case 'compromised' could indicate that the attacker has recovered the prover's secret key material, but it could also mean that the attacker has managed to tamper with the prover to such a degree that it is providing false position estimates. For example, a device's clock speed could be increased so that it replies faster or the communication channel can be manipulated without having to circumvent application layer security mechanisms [11, 12]. The effect of distance fraud on the distance measurement between a verifier and prover is illustrated in Fig. 11.4. Distance fraud is discussed in [13] and has mainly been described in the context of a infrastructure-based system where a compromised target attempts to misrepresent its location. It is, however, also applicable to terminal-based systems where it is possible that the reference nodes can be compromised. It should be noted that if the target node has access to at least one legitimate reference node then the possible fraudulent locations are reduced as the resultant position must correspond with the physical estimates provided by the legitimate node. For example, if two of the three nodes in the trilateriation process is compromised, as illustrated in Fig. 11.5, then the target node must still be located at a fixed distance from the remaining legitimate node. In some radio systems this distance may be large, and because of natural variations in the radio environment or the propagation time a system might disregard a node if it diverges notably from a majority of nodes.

Fig. 11.5 Two compromised reference nodes V_2' and V_3' generate incorrect distance measurements, which result in node P appearing at position P'

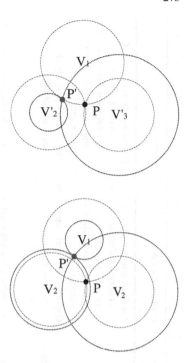

Fig. 11.6 A compromised node P appears to be at position P' by shortening the distance to reference node V_1 and enlarging the distances to reference nodes V_2 and V_3

11.3.1 Security and Resilience of Position Estimation Methods

A location system relies on the fact that the physical relation between reference nodes and the target node can be estimated. If these estimates are somehow modified by a malicious entity, or if the estimate cannot be accurately made due to operational errors, then the overall localisation process will be adversely affected. Čapkun and Hubaux [14] have shown that in the case of trilateration, and the principle extends to multilateration, a node located outside the triangle cannot prove that it is inside without shortening the distance measured to at least one of the reference nodes. This principle is illustrated in Fig. 11.6. Similarly, a node located inside the triangle cannot prove to be at a different position without shortening the distance measured to at least one of the nodes. An additional important consideration is that embedded location systems are required to be reliable in various environments while also taking into account limited resources as a result of practical constraints, such as cost, power and node size. Estimates should therefore be resilient against errors and practical to implement. In this Section we analyse three popular estimation methods against these requirements. A summary of the analysis is given in Table 11.1.

Table 11.1 Comparison of position estimate methods for RTLS

Method	Location type	Technology requirements	Security vulnerabilities	Resilience/accuracy	Comments
Angle of arrival (AoA)	Absolute	Directional antenna arrays in reference nodes. Accurate node orientation		Multipath errors in terminal-based systems, or if reference node is mobile, it would have to be orientated each time before measurement	Resistant to distance fraud if reference nodes are synchronised. Resistant to relay with unforgeable channel techniques
Received signal strength (RSS)	Absolute relative	Path-loss model or position calibration. Signal strength measurement	Relay attack distance fraud	Dependent on regular updates of path-loss/calibration	Easy to attack (amplify or attenuate signal)
Received signal ('In range')	Absolute relative	Compatible with any communication	Distance fraud	Dependent on communication range model/estimate. Unforgeable channel may introduce possible errors (false negatives)	Resistant to relay with unforgeable channel techniques
Time of arrival (ToA)	Absolute relative	Reference and target nodes need high frequency, synchronised clock. RF channel must allow measurement of propagation time	Distance fraud	Depends on clock synchronisation and frequency	Detects relay attack if transmission is authenticated, method relies on target node's time measurement so always vulnerable to distance fraud. Ultrasonic ranging not secure
Difference time of arrival (DToA)	Absolute	Reference nodes need high frequency, synchronised clocks. RF channel must allow measurement of propagation time	Distance fraud	Depends on clock synchronisation and frequency	Resistant to relay if transmission authenticated. Ultrasonic ranging not secure
Round trip time (RTT)	Absolute service	Reference node need high frequency clock. RF channel must allow secure measurement of round trip time		Depends on clock frequency distance bounding channels could introduce possible errors (false negatives)	Resistant to distance fraud and relay with distance bounding protocol. Ultrasonic ranging not secure

11.3.1.1 Angle of Arrival

Angle of Arrival (AoA) determines the direction from which a target node's transmission has been received. The reference node has a number of direction sensitive antennas measuring the angle between the direction from which the transmission was received and a reference direction. The position of a target node can then be calculated using triangulation using the measured angles from at least two reference nodes located at known locations. Position estimation using AoA requires the reference nodes to contain complex antenna arrays, which could increase costs if these are not already present in the system, such as in the case of cellular base stations. This method alone is not suited to relative location systems with only one reference node as a single angle only provides a vector of possible location points. In terms of security, simple AoA is vulnerable to both relay attacks and distance fraud if an attacker can alter the angle by reflecting or transmitting transmissions from different directions to individual reference nodes [12]. Distance fraud can be mitigated if the reference nodes share a synchronised clock. If this is the case, the reference nodes only accept measurements from a target node if the transmission was received at the same time, or within minimal time window, to ensure that the node could not change position between transmissions. However, this is difficult if the system normally has to deal with multi-path effects. Relay attacks can be mitigated by implementing unforgeable channels, which would allow the reference nodes to detect a proxy target node. AoA does not really allow for a single reference node to determine the relative location of another node. A single node can therefore only determine the direction, and not the distance, to another node. As such, it cannot be secured against relay it is therefore not suited to the requirements defined earlier for an RTLS.

11.3.1.2 Received Signal Strength

Received signal strength (RSS) position estimation is based on the principle that signal weakens as it is transmitted over a distance. In theory, a reference node could therefore estimate the distance to a target node if it knew the signal output power of that node. The target node's location can be determined using trilateration if at least three reference nodes, with known locations, obtain a distance estimate. RSS can also be used in relative location schemes, with one reference node determining if a target node is in close proximity, and in terminal-based schemes. Measuring the received signal is a relatively straight forward task for the reference nodes, but RSS distance estimation also requires an accurate path-loss model to estimate distance, especially if communication is not line-of-sight. Formulating such a model, or alternatively 'mapping' the signals strengths within an area of interest based on calibration measurements might be time consuming or not at all practical. For example, this is not practical if the system needs to deal with slow or fast fading effects, or if the environment changes often, e.g. shipping containers that are constantly moved around a warehouse obstruct line-of-sight or attenuate signals. RSS is vulnerable to simple distance fraud as it is straight forward for an attacker to amplify or attenuate the signal to commit distance fraud. RSS could also be used in a simplified range-independent systems, i.e. the target node is within communication range. This variant is less reliant on a path-loss model, an exact signal strength measurement is not required and can be made resistant to relay attacks when combined with device characterisation and unforgeable channel techniques, as will be discussed in Sect. 11.4.

11.3.1.3 Time of Flight

Time of flight (ToF) encompasses three distance estimation methods: Time of arrival (ToA), Time difference of arrival (TDoA) and Round trip time (RTT). All three these methods estimates distance between the target and reference node based on the propagation speed of the transmission medium and the propagation time between the target node and the reference node. All three methods can be used in terminal-based systems, although only ToA and RTT are suitable for relative location systems. One reference node using the TDoA method would not be able to estimate the distance to a target node as it needs multiple readings to determine a location. The propagation speed of radio waves in air approaches the speed of light. Apart from radio frequency (RF) ultrasound channels have been used in ToF systems, e.g. [15]. The propagation speed of sound is much slower than that of light, so it is easier to obtain high spatial resolution using simple hardware. This property, however, makes ultrasound vulnerable to a relay attack where messages are forwarded over a faster RF channel.

ToA requires that the target node records the time it transmits t_0 and that the reference node records the time it received the transmission t_1. The propagation time between the nodes $t_p = t_1 - t_0$ can then be used to estimate the distance between them. ToA is a costly method since each node must contain a reliable, high-frequency clock, which is synchronised with all the other node's clocks to ensure accurate distance estimates. The resolution of the distance estimate is dependent on the frequency of the clocks and how closely the are synchronised. ToA is also vulnerable to distance fraud as the distance estimate is dependent on the transmit time provided by the target node. The node can decrease the distance be increasing t_0 or increase the distance by decreasing t_0.

In TDoA the reference nodes record the time when the transmission was received. The reference node does not receive a transmit time from the target node, like is the case for ToA, but it does know that the same transmission was sent at the same time to the other reference nodes. If the target node has to transmit to each reference node in turn the delay between transmissions will be recorded. The reference nodes then estimate the position based on the difference in the time that the transmission takes to reach each of them. The DToA method also requires that each node contain a reliable, high-frequency clock, which is synchronised with all the other nodes' clocks to ensure accurate distance estimates. The resolution of the distance estimate is dependent on the frequency of the clocks and how closely the are synchronised. DToA is also vulnerable to distance fraud as the target cannot manipulate the time differences by controlling when it transmits to each reference node, taking into account that the system assumes that all transmissions were sent at the same time. Both TDoA and ToA methods could potentially detect simple relay attacks if the extra distance over which the communication is relayed will increase the propagation time beyond the accepted operational limits.

Reference nodes in RTT systems will measure the time from when they sent a transmission until it received a response from the target node. The distance estimate can then be calculated as $d = c \cdot \frac{t_m - t_d}{2}$ where c is the propagation speed, t_p is the one-way propagation time, t_m is the measured total round-trip time and t_d is the target node's processing delay between receiving a transmission and sending a response. Methods using RTT have the advantage that these do not require the nodes to have synchronised clocks and only the reference node needs a reliable, high-frequency clock. Combined with distance-bounding protocols the RTT method have the potential to be resistant to distance fraud and relay attacks, as will be discussed in Sect. 11.4.

11.4 Securing Position Estimation Methods

Secure location system proposals based on *statistical aggregation* methods often assume that there are multiple reference nodes [16, 17], or that a single reference node can obtain multiple types of position estimates [18], which can collaborate with each other to detect anomalies in target nodes' behaviour. For example, a node attempting distance fraud in an RSS system, by amplifying its signal, could have its transmission received by a number of reference nodes outside its normal range of communication. It is therefore possible that the node's transmission was received by two nodes whose observed areas should under legitimate circumstance not overlap. This approach is cannot be relied upon in all cases, especially if the attacker used a directional antenna to amplify the signal to a specific node. In a relay attack one reference node might receive the transmission from the proxy while another reference node, in another area, inadvertently receives the same transmission from the actual target node. Once the system starts to calculate the location of the node this situation will be detected and the fraudulent action identified. This method of making the underlying position estimates robust is a good solution if there is a high concentration of reference nodes, or if cost constraints allow for target nodes that can provide multiple types of positioning information, e.g. angle of arrival and time-of-flight. However, in industrial systems, where networks are more structured, the number of reference nodes might be the exact amount required to locate the object within a set area, with possibly only a few additional nodes to provide redundancy. This method is not suitable for relative location systems where there is only one reference node, and also cannot detect relay attacks that originate from outside the area covered, e.g. the attacker has stolen a device and has left the area covered by the location system, but left a proxy node in its place and none of the reference nodes will pick up transmissions from the real target node.

A common approach for preventing relay attacks is to construct *unforgeable channels*. If the reference node could physically verify whether the source of the transmission is the target node then relay attacks can be detected. The basic principle in this case is that the proxy target node will not be able to impersonate the real target node if it cannot exactly replicate the communication channel. For example, Alkassar et al.suggested that channel-hopping radio is difficult to track and thus difficult to relay [19]. The reference node can also try to uniquely identify the target node by using the physical characteristics of the channel. Rasmussen and Čapkun [20] proposed that a reference node can construct a unique 'fingerprint' for each target node by using the attributes of the received RF signal and there has been several practical examples of sensor network nodes and RFID tokens being identified by measuring unique characteristics at the physical communication layer [21–23]. These methods do not allow for accurate distance estimates, so are only suitable for verification within range-independent location services, and also do not protect against a fraudulent node. Further proposals hide additional information within the transmitted data. In a scheme by Hu et al. [10], geographical information, referred to as packet leashes, are added to transmitted data. Kuhn [24] proposed that the target node transmits a hidden 'watermark' along with the data, which is subsequently revealed, so that the reference node can retroactively check whether the positioning transmission it received was transmitted by the target node. An attacker would not be able to extract this information and relay the response without introducing a significant delay, which is then detected by the reference node. This method was suggested for GPS where the sender is trusted, and therefore it does not protect against a fraudulent target node committing distance fraud.

Distance-bounding protocols determine an upper bound for the physical distance between two communicating parties based on the Round-trip-time (RTT) of cryptographic challenge-response pairs. The format of the challenge-response pairs are specifically designed to allow for an accurate time measurement, e.g. choosing a response that takes a predictable or constant time to calculate. These protocols, if designed correctly, and implemented on a suitable communication channel can detect both relay and distance fraud attacks. Distance-bounding was first proposed by Brands and Chaum [13] in 1994 and since then distance-bounding protocols have become popular for doing secure neighbour detection [25] and proving proximity in relative location systems, such as in RFID and contactless smart card applications [8]. Time-of-flight distance-bounding protocols are dependent on time measurements made at the physical layer of the communication channel to accurately calculate the distance between the prover and verifier. This means that the security of the distance bound depends not only on the cryptographic protocol itself but also on the practical implementation and the physical attributes of the communication channel. Clulow et al. [12] show how an attacker can gain a timing advantage by exploiting the time allowed by the verifier for the transmission of redundant data, such as framing and error correction, at the packet level of the communication layer, and also define four principles that govern the practical implementation of secure distance-bounding protocols. Hancke et al. [11] also demonstrated how the attacker can achieve similar timing benefits at the physical level, i.e. by exploiting time delays in the coding and modulation stages of RF transceivers. Both these papers illustrate that systems planning to use distance-bounding protocols must implement special low-latency channels as conventional communication channels are designed for reliable data transfer and therefore feature redundancy and timing tolerances, which introduce timing uncertainty for an attacker to exploit. The practical implementation of a suitable distance-bounding channel still remains a technical challenge. Currently, there are two basic ideas proposed in literature. The first approach is to work with the current communication channel principles and try to add distance-bounding functionality, as Capkun et al. [26] has done with an Ultra-WideBand (UWB) system, while taking into account the possible security risks. The second approach is to construct a new channel with the correct security properties, e.g. [27, 28].

The aforementioned methods could provide the means to help secure positioning measurements. If the location system is based on methods using statistical aggregation the multiple position estimates also act as a measure for error correction. In other words, an erroneous estimate could be detected and rejected if it differs significantly from the other estimates. However, both unforgeable channels and distance bounding rely on physical characteristics of the communication channel or nodes. In certain environments the reliability of the location information is important and any security mechanism should be resilient to possible errors. If the system dealt with nodes of low complexity, then care must be taken that the proposed techniques do not cause operational failures due to node tolerances. Unforgeable channels require that the physical characteristics of nodes are characterised before being used, i.e. there has to be some prior calibration of the node's measurements and the system must store the resultant 'fingerprint'. This 'fingerprint' should take into account any outside influence on the channel but if the node is mobile, or functions in a environment that changes, the node's uniquely identifiable physical characteristics could change, in which case the reference node could fail to authenticate the target node. For example, the node's radio environment could dynamically change or the node could be moved a physical location with additional radio noise or objects that might affect communication, such as metal containers, could be placed in the vicinity. As a result, no position estimates for this target node is reported and the system loses track of its location.

Distance-bounding protocols provide some redundancy in terms of the distance estimate because the round trip exchange is done multiple times during a protocol run. However, if a communication error occurs during the timed exchange stage of the distance-bounding protocol the resultant cryptographic verification of challenge-response pairs will fail in most proposals and the system will not locate the target node. Some distance bounding proposals can be made resistant to communication errors by using either a bit-error threshold [27], where the verifier accepts a defined number of errors, or by applying an error-correction code (ECC) [29] to the challenge and response bit-streams, and then correcting any bit errors during the verification stage. Distance-bounding exchanges increase verification time and accurately timing exchanges might not be possible on limited devices, so this approach might not be usable in some operational environments.

11.5 Global Navigation Satellite Systems

A major source of location information in embedded systems is GNSS, such as GPS (US), GLONASS (Russia) and the forthcoming Galileo (EU) and Co mpass (China). The large-scale use of GNSS for position, navigation and time (PNT) data is the result of such systems' ubiquitous availability, i.e. it is available everywhere on earth, accuracy and the relatively low cost of use. There are numerous proposals for location-based services incorporating non-GNSS technology, which include the use of mobile network base stations or terrestrial radio (LORAN), but the issue of availability inhibits these systems reaching the scale of GNSS. Even though mobile infrastructure offers the advantage that it could work indoors it is generally limited to developed areas, for example it is of limited use to a shipping vessel out at sea, and its continued operation is dependent on private enterprise. LORAN has been largely decommissioned and it is not yet certain how widely the enhanced version (eLORAN) will be deployed.

11.5.1 GPS Security

The US GPS is currently the most widely used GNSS. A number of GPS satellites continuously broadcasts their location, together with the time the message was sent. GPS receivers then use TOA to estimate the distance to each observed satellite.

Figure 11.7 illustrates the GPS based location sensing and triangulation method. In the figure, d_i represents the distance of i_{th} satellite from Earth. c is the speed of light (299,792,458 m/s). ΔT is the time difference of signal sent from the satellite and received on the Earth.

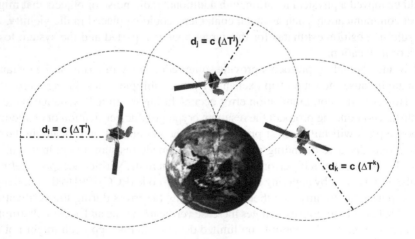

Fig. 11.7 GPS based location sensing and triangulation

$$d_i{}^2 = (x^i - x)^2 + (y^i - y)^2 + (z^i - z)^2 \qquad (11.1)$$

$$d_j{}^2 = (x^j - x)^2 + (y^j - y)^2 + (z^j - z)^2 \qquad (11.2)$$

$$d_k{}^2 = (x^k - x)^2 + (y^k - y)^2 + (z^k - z)^2 \qquad (11.3)$$

By solving (11.3), and after error corrections we get, $[X, Y, Z]$ where X = longitude, Y = latitude, and Z = altitude.

If a receiver can observe at least four different satellites it can calculated its own location, global co-ordinates and elevation. GPS satellites transmit their information on two basic signals:

- The precision P(Y) signal used by military receivers features an encryption mechanism and therefore only a transmitter with the correct shared key can generate it and only a receiver with the correct key can track it.
- The crude access (C/A) signal used by civilian receivers provides no security mechanisms.

Both signals are transmitted using direct-sequence spread-spectrum (DSSS) modulation. In DSSS the data is multiplied with a pseudo-random sequence, the spreading code, of much higher frequency than the original signal. This 'spreads' the signal power across the frequency spectrum, making it difficult to clearly distinguish the transmitted signal without knowledge of the spreading code. However, a receiver that knows the spreading code can recover the original signal by correlating the incoming signal with the code. In GPS systems the C/A signal is transmitted using a short, publicly known spreading code. This allows the signal to be demodulated by anyone. The P(Y) signal is, however, transmitted using a long, secret spreading code. Only receivers that knows the secret code will be able to reliably reconstruct the P(Y) signal. Similarly, an attacker without the code will not be able to generate and transmit a valid signal.

The security of GPS became a very public matter because of its use in the navigation of both military and civilian unmanned aerial vehicles (UAV) [30]. However, the security vulnerabilities of the GPS system and the increasing reliance of sensitive applications on global positioning were issues already pointed out more than a decade before. In 2001 the US Department of Transportation highlighted vulnerabilities in transport infrastructure relying on GPS in the 'Volpe' report [31]. This report also make recommendations to mitigate these security issues, including cryptographic authentication and verifying signals' angle-of-arrival. A decade later security vulnerabilities remain while the use of GPS in critical applications have increased dramatically, a situation affirmed in a 2011 report by the Royal Academy of Engineering [32] emphasising the reliance on, and the vulnerabilities of, GNSS in general.

Three main methods were identified of what was termed 'intentional interference', i.e. malicious entities trying to destabilise the system.

- Jamming: Wholesale disruption of P(Y) and C/A signals by transmitting radio interference.

- Meaconing: Rebroadcasting of legitimate P(Y) and C/A signals that results in an incorrect position being reported, essentially a relay attack whereby legitimate signals are selectively delayed. By delaying the signals the attacker makes the distance between device and the satellite appear larger than it really is.
- Spoofing: Creation of C/A signals with the purpose of causing a incorrect location to be reported.

At the time of the Volpe report the equipment needed to perform meaconing or spoofing was expensive and bulky. Technology has moved on from then and with the advent of software defined radios the transmission of GPS signals has become significantly easier and cheaper. These days GPS signals could feasibly be received, relayed or created with open source software, such as GNU Radio [33], and a small, generic software radio platform such as the Ettus Universal Software Radio Platform (USRP) [34]. There are several publications in the public domain detailing practical strategies for spoofing GPS signals [35, 36]. In practice, spoofing a location to a receiver that is locked onto legitimate sources is not trivial. Simply transmitting a stronger signal will not necessarily convince a receiver to lock onto this new signal. The most efficient method for successfully spoofing a location was first presented by Humprhreys et al. [36]. Figure 11.8 illustrates this attack with A being a signal generated by the attacker, G being the genuine signal and the triangle indicating which signal the receiver is currently locked onto. When the attacker initially transmits his spoofed signal to receiver remains locked on the genuine signal. The attacker then aligns his signal with the legitimate signal, with the receiver locking onto the larger combined signal. The attacker then slowly increases the amplitude of his signal and shifts his signal, causing the receiver to remain locked on his signal.

Although it is not as elegant an exploit, jamming is just as big a threat as meaconing and spoofing. The loss of GPS services potentially disrupt a wide range of services usually taken for granted. In October 2011, the Royal Navy jammed GPS signals off the coast of Scotland as part of pre-planned military exercises, but had to stop doing so because of safety considerations when the navigation systems of fishermen, who had not received the advanced warning, stopped functioning [37]. Small scale GPS jammers are widely available for on-line purchase and marketed as personal

Fig. 11.8 Sequence for spoofing a GPS channel. A attack signal, G genuine signal and Δ showing the signal locked onto by the receiver

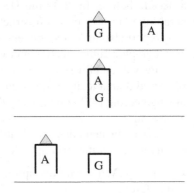

privacy devices (PPD), which can jam both GPS and mobile communication. These products are claimed to be intended for people who wish to prevent third parties from tracking them using location-based services but unfortunately such a device can just as easily be used by thieves to disable asset and vehicle tracking systems. PPDs could also unintentionally effect much more critical applications. A good example of this is a Federal Aviation Authority investigation into the reason why a GPS-based landing systems used at New Jersey's Newark Airport suffered from periodic breaks in reception. PPDs used by truckers on the nearby freeway were eventually identified as the cause [38].

11.5.2 Future Efforts on Securing GNSS

Redesigning GNSS to allow for secure civil location services will in all likelihood not happen anytime soon and securing civil location services does not appear to be an immediate concern. GPS III, which is scheduled for deployment in 2014, introduces a second civilian channel and a 'Safety of Life' channel, alongside a backward compatible civilian channel, but none of these channels provides for any security mechanisms. GPS III does include improved anti-jamming and security measures on the military channel. Only one of Galileo's five forthcoming services does allow for jamming resistance and encryption. This service, 'public regulated services' (PRS), could potentially alleviate secure location issues for some systems as it is intended not only for defence purposes, but also for law enforcement and emergency services in addition to selected critical telecommunication, energy and transport applications. Access to this channel will, however, be regulated and only authorised parties will be able to use a PRS-capable receiver, so it is likely that this service will not be implemented in most products intended for the civilian market [39].

Of course, if security mechanisms were ever to be implemented on civilian channels this might adversely affect their ubiquitous usability and their reliability. For example, a cryptographic solution would need a suitable key management system, which would allow for timely key distribution to both transmitters and receivers. This especially adds to the complexity and cost of the receiver, and some users with limited security needs might not be satisfied with a system where they could lose service out in the wilderness or on the ocean because their device was unable to receive a key update. This means that the responsibility rests on systems designers to find ways to improve existing receiver architectures, hardening these against simple attacks, to take GNSS security risks into account when designing location services, and to incorporate adequate fail-safe measures, such as back-up non-GNSS solutions.

11.6 Conclusion

As location information gathered by embedded systems become more significant, the embedded location systems used must be designed to be secure and reliable. A location system's underlying components must therefore be resistant to location fraud, whereby the system is made to generate incorrect location information by a malicious entity, and yet continue to function reliably. There are numerous proposals for implementing secure position estimation methods, which are meant to be resilient against intentional attacks and operational errors. Although some of these proposals are arguably useful in certain situations, all of these approaches exhibit some practical security weaknesses so there is still scope for research and development work in this area. GNSS is a major source of location information for embedded systems found in numerous scenarios where reliability is crucial, such as air traffic control and military applications. Despite its importance, GNSS has only limited security, a situation which has been recognised but is unlikely to be rectified anytime soon. This means that at this stage the onus is on the designer of GNSS receivers, and embedded designers, to ensure that their systems are made resistant to jamming, meaconing and spoofing attacks as much as is possible.

References

1. C.A. Patterson, R.R. Muntz and C.M. Pancake. 'Challenges in Location-Aware Computing'. IEEE Pervasive Computing, Vol. 2, No. 2, pp 80–89, June 2003.
2. A. Boukerche, H.A.B. Oliveira, E.F. Nakamura and A.A.F. Loureiro. 'Secure Localization Algorithms for Wireless Sensor Networks'. IEEE Communications Magazine, Vol. 46, No. 4, pp 96–101, April 2008.
3. A.I.G-T. Ferreres, B.R. Alvarez and A.R. Garnacho. 'Gauranteeing the Authenticity of Location Information'. *IEEE Pervasive Computing,* Vol. 7, No. 3, pp 72–79, July 2008.
4. Y. Zhou, Y. Fang and Y. Zhang. 'Securing Wireless Sensor Networks: a Survey'. *IEEE Communications Surveys & Tutorials,* Vol. 10, No. 3, pp 6–28, September 2008.
5. X. Chen, K. Makki, K. Yen and N. Pissinou. 'Sensor Network Security: A Survey'. *IEEE Communications Surveys & Tutorials,* Vol. 11, No. 2, pp 52–73, 2009.
6. ISO/IEC 24730. Information technology - Real-time locating systems (RTLS).
7. N.O. Tippenhauer, K.B. Rasmussen, C. Popper and S. Čapkun. 'Attacks on Public WLAN-based Positioning Systems'. *Proceedings of ACM/Usenix International Conference on Mobile Systems, Applications and Services (MobiSys),* 2009.
8. G.P. Hancke, K. Mayes, and K. Markantonakis. 'Confidence in Smart Token Proximity: Relay Attacks Revisited'. *Elsevier Computers& Security,* June 2009.
9. Y. Desmedt, C. Goutier and S. Bengio. 'Special Uses and Abuses of the Fiat-Shamir Passport Protocol'. *Advances in Cryptology (CRYPTO),* Springer-Verlag LNCS 293, pp. 21, 1987.
10. Y.C. Hu, A. Perrig and D.B. Johnson. 'Packet Leashes: A Defense Against Wormhole Attacks in Wireless Networks'. *Proceedings of INFOCOM,* pp. 1976–1986, April 2003.
11. G.P. Hancke and M.G. Kuhn. 'Attacks on "Time-of-Flight" Distance Bounding Channels'. *Proceedings of First ACM Conference on Wireless, Network Security (WISEC'08),* pp. 194–202, March 2008.
12. J. Clulow, G.P. Hancke, M.G. Kuhn, T. Moore. 'So Near and Yet So Far: Distance-Bounding Attacks in Wireless Networks'. *European Workshop on Security and Privacy in Ad-Hoc and Sensor Networks (ESAS),* Springer-Verlag LNCS 4357, pp. 83–97, September 2006.

13. S. Brands and D. Chaum. 'Distance Bounding Protocols'. *Proceedings of Advances in Cryptology (EUROCYPT '93)*, Springer-Verlag LNCS 765, pp 344–359, May 1993.
14. S. Čapkun and J-P. Hubaux. 'Secure Positioning in Wireless Networks'. *IEEE Journal of Selected Areas in Communications*, Vol. 24, No. 2, pp. 221–232, February 2006.
15. A. Harter, A. Hopper, P. Steggles, A. Ward and Paul Webster. 'The Anatomy of a Context-Aware Application'. *Proceedings of Fifth Annual ACM/IEEE International Conference on Mobile Computing and Networking, MOBICOM'99*, pp. 59–68, August 1999.
16. Z. Li, W. Trappe, Y. Zhang and B. Nath., 'Robust Statistical Methods for Securing Wireless Localization in Sensor Networks'. Proceedings of the International Symposium on Information Processing in Sensor Networks, 2005.
17. L. Lazos and R. Poovendran. 'SeRLoc: Secure Range-Independent Localization for Wireless Sensor Networks' *Proceedings of the 3rd ACM workshop on Wireless, Security*, pp. 21–30, 2004.
18. L. Lazos and R. Poovendran, 'Hirloc: High-Resolution Robust Localization for Wireless Sensor Networks'. IEEE Journal on Selected Areas of, Communication, Vol. 24, No., pp. 233246. February 2006.
19. A. Alkassar, C. Stuble and A. Sadeghi. 'Secure Object Identification: or Solving the Chess Grandmaster Problem'. *Proceedings of New Security Paradigms, Workshop*, pp. 77–85, 2003.
20. K.B. Rasmussen and S. Čapkun. 'Implications of Radio Fingerprinting on the Security of Sensor Networks'. *Proceedings of IEEE SecureComm*, 2007.
21. B. Danev, T.S. Heydt-Benjamin and Srdjan Capkun. 'Physical-layer Identification of RFID Devices', *Proceedings of USENIX Security Symposium*, 2009.
22. B. Danev and S. Capkun. 'Transient-based Identification of Wireless Sensor Nodes'. *Proceedings of the ACM/IEEE International Conference on Information Processing in Sensor Networks (IPSN)*, 2009.
23. G. DeJean and D. Kirovski. 'RF-DNA: Radio-Frequency Certificates of Authenticity'. *Proceedings of International Workshop Cryptographic Hardware and Embedded Systems (CHES 2007)*, Springer-Verlag LNCS 4727, pp. 346–363, September 2007.
24. M.G. Kuhn. 'An Asymmetric Security Mechanism for Navigation Signals'. *6th Information Hiding Workshop, Springer-Verlag LNCS 3200*, pp 239–252, May 2004.
25. P. Papadimitratos, M. Poturalski, P. Schaller, P. Lafourcade, D. Basin, S. Čapkun and J-P Hubaux. 'Secure Neighborhood Discovery: A Fundamental Element for Mobile Ad Hoc Networking'. *IEEE Communications Magazine*, February 2008.
26. N.O. Tippenhauer and S. Capkun. 'ID-based Secure Distance Bounding and Localization'. *Proceedings of European Symposium on Research in Computer Security (ESORICS)*, 2009.
27. G.P. Hancke and M.G. Kuhn. 'An RFID Distance Bounding Protocol'. *Proceedings of IEEE/CreateNet SecureComm*, pp. 67–73, September 2005.
28. J. Reid, J.M.G Nieto, T. Tang and B. Senadji. 'Detecting Relay Attacks with Timing-Based Protocols'. *Proceeding 2nd ACM Symposium on Information, Computer and Communications, Security*, pp. 204–213, March 2007.
29. D. Singelée, B. Preneel. 'Distance Bounding in Noisy Environments'. *European Workshop on Security and Privacy in Ad-Hoc and Sensor Networks (ESAS)*, Springer-Verlag LNCS 4572, pp. 101–115, 2007.
30. Drone Hijacking? Thats Just the Start of GPS Troubles. Wired.com, July 2012. http://www.wired.com/dangerroom/2012/07/drone-hijacking/all/
31. Vulneribility Assessment of the Transportation Infrastructure Relying on the Global Positioning System. John A. Volpe National Transportation Systems Center, 2001.
32. Global Navigation Space Systems: Reliance and Vulnerabilities, Royal Academy of Engineering, March 2011.
33. GNU Radio http://gnuradio.org
34. Ettus http://www.ettus.com
35. N.O. Tippenhauer, C. Popper, K.B. Rasmussen and S. Capkun. 'On the Requirements for Successful GPS Spoofing Attacks'. Proceedings of ACM Communication and Computer Security (CCS), October 2011.

36. T.E. Humphreys, B.M. Ledvina, M.L. Psiaki, B.W. O'Hanlon and P.M. Kintner. Assessing the Spoofing Threat: Development of a Portable GPS Civilian Spoofer. Proceedings of GNSS Conference, September 2008.
37. Military jamming of GPS in Scotland suspended. BBC, October 2011. http://www.bbc.co.uk/news/uk-scotland-highlands-islands-15242835
38. GPS Jamming: No Jam Tomorrow. The Economist, March 2011.
39. Galileo System Overview, European Space Agency. http://www.esa.int/esaNA/galileo.html

Chapter 12
Automotive Embedded Systems Applications and Platform Embedded Security Requirements

Jan Pelzl, Marko Wolf and Thomas Wollinger

Abstract Contemporary security solutions in the automotive domain usually have been implemented only in particular applications such as electronic immobilizers, access control, secure flashing, and secure activation of functions or protection of mileage counter. With cars, which become increasingly smart, automotive security will play a crucial role for the reliability and trustworthiness of modern automotive systems. In this chapter, we will introduce the topic of automotive security and provide motivation for security in embedded automotive platforms.

12.1 Introduction: Smart Embedded Platform Automotive

Modern vehicles, especially when considering vehicles of the premium segments, are equipped with more than 50 embedded microcontrollers, the so-called Electronic control units (ECU). The ECUs provide a growing diversity of system functions to control most processes in the car such as engine control, steering and braking systems, multimedia and entertainment systems, anonymous driving equipment and information systems (like navigation systems and traffic control). These applications are realised as embedded systems and range from simple control units to high-end processors whose computing power approaches that of current PCs. These ECUs are connected via various vehicular buses (e.g. CAN, MOST, LIN, etc.) forming a complex highly networked and distributed system. The Controller Area Network (CAN) bus is a synchronic serial bus connecting different ECUs [5]. The primary reason for

J. Pelzl (✉) · M. Wolf · T. Wollinger
ESCRYPT GmbH, Embedded Security, Bochum, Germany
e-mail: jan.pelzl@escrypt.com

M. Wolf
e-mail: marko.wolf@escrypt.com

T. Wollinger
e-mail: thomas.wollinger@escrypt.com

K. Markantonakis and K. Mayes (eds.), *Secure Smart Embedded Devices,*
Platforms and Applications, DOI: 10.1007/978-1-4614-7915-4_12,
© Springer Science+Business Media New York 2014

the CAN bus was to reduce the cable harness size and thereby, the weight as well as the cost. Media oriented systems transport (MOST) is a bus developed especially for the transfer of multimedia data in automobiles. MOST is a serial bus system for audio, video and data signals using fibre optic cable. The Local interconnect network (LIN) bus has been introduced for the inexpensive communication with intelligent sensors and actors in the vehicle. The LIN bus is typically used in connecting the vehicle doors or seats.

This heterogeneous, multi-rank communication in-vehicle network is in today's vehicles extended by connecting mobiles devices (like smart phones) to the multimedia ECU. Considering the fact, that today the cost for electronics and software is approaching 30 % of all manufacturing costs of the vehicles, the network and the ECUs will be even more important in the future. Figure 12.1 illustrates the possible communication connections a vehicle will have that connect all ECUs using different on-board automotive buses within a complex network. In addition to the internal communication, future vehicles will support more and more external communication. Local wireless communication like Wi-Fi or Dedicated short range communication (DSRC) will support communication with other vehicles (Vehicle-to-Vehicles communication - V2V) and to the infrastructure (Vehicle-to-Infrastructure communication - V2I). Infrastructure, in this case, can be any device the vehicles will communicate with in the future such as traffic lights, traffic control centre, gas stations, or traffic signs. The global wireless communication using Universal Mobile Telecommunications System (UMTS), GSM/GPRS, Tetra or others will allow direct vehicle communication such as automatic emergency calls (eCall) in case of an accident [16]. In addition, common computer interfaces will become more and more usual in vehicles as well. The communication to user specific devices will be for example via USB, Bluetooth or Ethernet. Global Positioning System (GPS) and Galileo reception will be supported for future positioning applications and services.

The massive innovation through electronic technology as well as the network communication in today's vehicles and the fact that safety improvements are expected using vehicle-to-vehicle and vehicle-to-roadside units communication (V2X communication), underlines the importance of functional integrity in this context. Critical examples include electronic safety aid systems, like local danger warnings, traffic

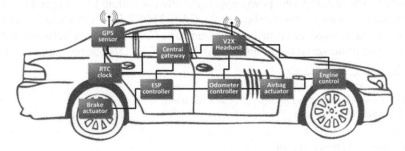

Fig. 12.1 Vehicular on-board network and external communications interfaces

light pre-emption, or electronic emergency brakes. While these functionalities should improve safety, new security requirements need to be considered in order to prevent malicious attacks. Attacks can be manifold and can have many different motivations, and could manifest themselves for example as apparent malfunctions or influencing traffic flow with faked messages. IT security is increasingly considered in the auto- motive domain; however, it is only introduced to niche applications like electronic immobilizers, access control, secure flashing, and secure activation of functions or protection of the mileage counter. Hence, the vast majority of software and hardware systems in current cars are not equipped with security functionality. The reason is that in the past, car IT systems did not need security functions, because there was very little incentive for malicious manipulation. Secondly, security tends has tended to be an afterthought in automotive IT system, because the achievement of the core function is the focus when designing a system. Future security designs should con- sider the whole communication links, e.g. from the sensor which records a physical variable to the controller that reacts to the data representing the variable value. Note that the sensor could be installed deep within the vehicle whereas the controller could be many kilometres away in a service facility. Hence, one needs to secure the internal system components (like the sensors and platforms/ECUs), critical applications (e.g. motor control or user applications), and communication links between internal and external communication devices as well as the infrastructure devices and support applications. Considering the V2X communication, the channel starts within one vehicle passing information to another vehicle and/or infrastructure via a wireless channel. The receiving party performs actions based on the incoming data, which could have critical effect such as slowing down because of a dangerous situation.

12.1.1 Smart Communication Platform

With increasingly powerful CPUs and the demand for information technology in the automotive domain, manifold smart communication platforms are likely to emerge in the next few years. Common concepts plan to route all communication of a vehicle with the outside world via a central smart platform. Such a platform provides all required communication capabilities as well as the necessary security mechanisms to protect critical assets such as valuable data and privacy. Typical future applications will include V2X communication, including communication from vehicle to vehicle and vehicle to infrastructure such as roadside units (traffic lights, toll collection and/or parking garage). Furthermore, the on board network will have the possibility to connect via the Internet to home networks and provide information about automated systems of the driver's home. Moreover, intelligent transport systems (ITS) will increasingly depend on smart communication platforms to provide, for example, intelligent traffic management for the connected car as well as location based services for the driver.

Integration of mobile phones and smart phones via different communication interfaces such as Bluetooth, wireless local area network (WLAN) or near field communication (NFC) is currently under investigation by automotive manufacturers. As an example, mobiles might replace the traditional key fob for accessing the car. Driver specific configurations of the car (e.g. favourite radio settings, navigation, etc.) will be provided by the link to such mobile user devices. Furthermore, the integration of Internet connectivity into cars and usage of dedicated services provided via the Internet such as infotainment applications (e.g. media streaming) or office applications (email and word processing) in cars will increase. Many future applications will require smart and secure platforms in order to protect assets such as valuable data and guarantee privacy.

12.1.2 Smart After-Market Platform

From a business perspective, modern cars offer the possibility for the manufacturer to make money on the road. Similar to the recent development in the mobile phone sector, smart platforms in cars additionally allow for interesting after-sales models to provide additional value and commercial services to the driver. Technically speaking, a communication platform in a car together with respective security mechanisms bears the potential to be used as a platform for after-market sales. Examples include on-demand business models, for example, for content such as video and music on demand or location based services such as ticketing applications for parking.

Furthermore, customisation of in-car electronics via software applications and configuration becomes possible. With applications on such a platform, the driver of a car might choose different skins for the head unit and the interaction with the car. As another example, linking home automation with the in-car network in order to provide information about the status of the driver's home at any time is currently being developed for in-car platforms. Combined with security, software application markets comparable to existing markets for applications for mobile phones are feasible for the automotive domain. For car manufacturers, feature activation and custom configuration of cars on the road, including consumer electronics and engine control are within reach. With such a system, a driver could for example buy more engine power for the weekend trip.

From a business point of view, after-sales will only work if security measures enforce the business model and prevent fraud. From a dependability point of view, security becomes inevitable in order to enforce safety.

12.1.3 Smart Future Platform

In future, many additional applications based on smart platforms are imaginable and extend current existing business-models and possibilities. As an example, smart traffic management systems will increasingly use the possibility to connect interactively to smart automotive platforms in order to obtain real-time information about the status of a car in terms of speed, destination, and other sensor information. On the other hand, traffic information will be evaluated and transmitted instantaneously by such platforms in order to provide optimal traffic routing and prevent traffic jams. Collaboration between platforms and integration with public transport systems extend the information exchange to other forms of transportation and allow for optimisation of traffic in a more cooperative manner. Consequently, reduction of $CO2$ emission can be optimised by optimal usage of cars. Saving energy or avoiding traffic jams might be achieved by creating specific incentives for drivers such as, for instance, free parking in particular areas and times, intelligent management of car sharing or car pools of companies

With the dawn of electric cars, new business models and requirements in conjunction with energy distribution arise. In this context, charging electric cars will become one of the logistic challenges of the next generation since the process of charging takes at a lot of time and when considering millions of electric cars. It also requires a highly sophisticated distribution of energy from the power plants to the batteries of cars. Secure measurement of energy, secure transmission of control data as well as security for new business models in the area of smart grid and smart metering all add demands for supporting security mechanisms in future automotive platforms. One basic requirement of all of these applications is security, in order to provide a trusted and reliable system and defend against fraud and manipulative attacks.

12.2 Security Aspects of Smart Embedded Automotive Platforms

In general, we have the same goals for automotive security as in most other IT security systems. Hence, we want to achieve the following security properties for the embedded automotive platforms:

- Confidentiality (e.g. for trade secrets and know-how).
- Authenticity/Integrity (e.g. for original OEM software).
- Availability (e.g. for vehicle safety functionality).
- Non-Repudiation (e.g. for error or driving logs).
- Privacy (e.g. for latest navigation destinations).

It is very important to note that all the measures implemented to provide and ensure the security properties must themselves be protected against malicious manipulations (also referenced as the "security of security mechanisms"). Thus, the following section gives an overview of potential attackers, attack paths and resulting security threats in the automotive domain together with some real-world examples and countermeasures.

12.2.1 Automotive Attackers

Automotive systems will need to defend against different types of potential attackers. These attackers may have different levels of access to the vehicle, different expertise and different financial resources. A summary of the different attacker types and the attack potential can be found in the Table 12.1. Table 12.1 classifies the different resources of the attacks in three levels, namely low, medium and high. Hence, considering the financial strength of an attacker, low would indicate that the attacker has some hundred dollars and he will be able to buy normal equipment like a computer or laptop. The medium attacker has however for example some thousand dollars and therefore can buy special equipment (like more expensive measuring equipment). The attacker rated high can essentially buy whatever is needed for the attack, like special purpose key search machines.

It should be noted that we have to protect the vehicles against third party attackers but also against the vehicle owner/driver. The latter case may be most relevant when the owner could benefit financially from a successful attack, e.g. saving some money using additional features without payment, or reducing tax or insurance fees as well as inflating the resale value of the vehicle. Even more dangerous are tuning activities that may not only causes illegitimate warranty claims, but also increase the safety risk to the driver and the other travellers. The garage worker can risk human lives by using counterfeit spare parts with or without the knowledge of the owner of the vehicle. Even more damage can be caused by competitors or suppliers who can try to gain economical advantage or cause bad publicity. Professional criminals have extensive knowledge and financial capability to carry out successful attacks and then to make money from the sale of stolen vehicles or parts. As with all IT systems, automotive platforms may be targeted by amateur hackers (or script kiddies) that attack systems for fun and reputation. Although these attackers may not have criminal intent, they can educate criminals and cause very damaging publicity.

12.2.2 Automotive Attack Paths

The introduction gave us a brief overview on how complex today's vehicles are, with many potential points of attack. In the following section, we present the different layers or attack perimeters a potential attacker can use to mount an attack. As shown in Fig. 12.2, automotive attack paths can be generally classified into in-vehicle (*interior*) *attacks and attacks that can be mounted externally* (exterior).

Table 12.1 Attackers in the automotive domain

Reference classification according [1]	Exemplary automotive role	Target knowledge	Time frame	Technical experience	Technical equipment	Financial resources
0: Script kiddies	Driver	Low	High	Low	Low	Low
1: Clever outsiders	Garage worker	Medium	Medium	Medium	Medium	Medium
2: Knowledgeable insiders	Professional thief	High	Medium	Medium	High	Medium
3: Funded organisations	Competitors, counterfeits, ac. research	High	High	High	High	High

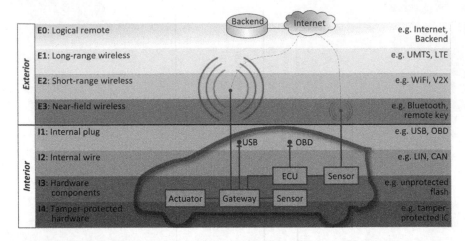

Fig. 12.2 Different attack perimeters of an attacker on a vehicle

12.2.2.1 In-Vehicle Attack Paths (Interior)

In-vehicle attack paths cover all attacks, where the attacker needs to be physically present and needs access to the vehicle interior by being able to open at least a door or the bonnet of the car.

Internal plugs (I1):
Most prominent attacks today (e.g. mileage manipulations or vehicle theft) use standardised on-board interfaces such as On-board diagnosis (OBD) or USB interfaces, which exist in all modern vehicles to easily access vehicular on-board electronics and on-board infotainment units, respectively. Thus, attackers can use physically standardised attack equipment and standardised protocols. The only individual adaptions needed are carmaker or car model individual parameters or commands, which can often be found on the Internet, while dedicated security mechanisms such as user authentication at OBD-level, are rather seldom implemented

On-Board Communication Attacks (I2):
For the on-board communications attacks, the attacker goes beyond obvious interfaces such as OBD or USB and directly connects to the internal vehicle bus system, which is normally a CAN bus system [5]. For attacks, using the internal bus system, the engine hood or some hatch in the interior has to be opened and some wires or bus connectors have to be attached directly with the attacking device (e.g. a laptop using a standard CAN card). Once successfully connected, this attack can use the generally standardised, publicly available protocols to generate commands and attempt parameter modification. Similar to today's I1 interfaces, dedicated security mechanisms such as user authentications are generally not implemented.

Simple Attacks on Hardware Level:
Simple attacks at the hardware level comprise all attacks, which can be executed with basic know-how and some simple equipment in a limited time frame, such as manipulating unprotected flash memories, completely deactivating or replacing certain devices, cutting wires or accessing proprietary debug interfaces.

Sophisticated Attacks on Hardware Level:
The attacks one can conduct directly on the hardware are usually very complex and can only be performed successfully by highly skilled experts. These attacks include bypassing dedicated hardware tamper protection measures such as special enclosures, tamper response mechanisms and/or carrying out sophisticated side-channel attacks [12].

12.2.2.2 External Attack Paths (Exterior)

External attack paths cover all attacks where the attacker does not need direct physical access to the car. Moreover, for some external attack paths, the attacker can be a great distance away, for instance using the Internet for access.

Near-Field Wireless Attacks (E3):
E3-level attacks misuse externally available near field wireless interfaces such as Bluetooth or the wireless car key interface to attack the vehicle without touching the car physically. However, the attacker usually has to be near the car.

Short-Range Wireless Attacks (E2):
E2-level attacks misuse externally available short-range wireless interfaces such as Wi-Fi or V2X communication interfaces [13] to attack the vehicle and without being directly noticeable around the car.

Long-Range Wireless Attacks (E1):
E3-level attacks misuse externally available long-range wireless interfaces such as UMTS cellular network interfaces or other long-range vehicular communication interfaces such as Radio Data System (RDS) or Global Positioning System (GPS), where the attacker is not in sighting distance, but can be kilometres away.

Logical Attacks (E0):
Logical attacks will be conducted by misusing an inherent logical or systematic fault of the overall IT (security) design or the corresponding (backend) implementation to compromise the vehicular IT system. Hence, logical attacks typically exploit vulnerabilities in the corresponding backend IT infrastructure or any indirect vehicle access, for instance, by any kind of malware, where attacks can be executed from anywhere in the Internet. Logical attacks also include social engineering, which exploits human weaknesses, typically by trying to trick a person to reveal sensitive and valuable information.

12.2.3 Automotive Security Threats and Risks

Not every potential attack path introduced in Sect. 12.2.2 implies a real security threat or risk. In fact, the actual security risk is the resulting product of two factors, the complexity and hence the likelihood of a certain attack path and the corresponding damage a successful attack would yield in worst case. For determining the likelihood of a certain attack path there exist some qualitative metrics as proposed for instance by the common criteria (CC) [14] which are linked to the individual attacker model (cf. Sect. II.1 12.2.1). The impact can in turn be divided into at least three damage sub categories. Thus, a successful attack can have safety (up to fatal casualties), financial (stolen car) or privacy (monitoring) implications.

However, security risk analyses of modern vehicles are not straightforward and there are unfortunately no standardised methodologies and procedures available. However, there are some best practice approaches such as [15]. Applying such an approach to a current vehicle, could yield the security risk evaluation as presented in Table 12.2.

12.2.4 Security of Automotive Safety Mechanisms

The most important security objective is the protection of automotive safety and hence all automotive-safety-related components, communications, functions and interfaces. This includes protection against any direct malicious encroachments to all dedicated passive (e.g. belts, airbags, [near] accident detection) and active safety mechanisms (e.g. Anti-lock braking system (ABS), Electronic stability program (ESP), lane assistant), but in particular all driving-related components and functions (e.g. vehicle steering, braking or shifting).

However, indirect safety encroachments, which do not directly target a dedicated safety component, can also lead to serious safety threats. This is true for instance, for security attacks, which enable the installation of potentially unsafe counterfeit components. Another example is unauthorised modification, for example, of the odometer, which may result in a safety problem if the unofficial manipulation could, accidentally change or overwrite critical safety functionality. Finally, it is important to realise that there are many sub-systems, such as windscreen wipers, infotainment units, or seat adjustments, which appear non-safety critical yet, could cause safety problems if they act without control while driving, due to a security attack.

Table 12.2 Examples of automotive security risk evaluations

Automotive system	Attack likelihood	Potential damage			Security risk
		Safety	Financial	Privacy	
Theft protection (electronic immobilizer, component identification)	High	Low	High	N/A	High
Drivetrain systems (e.g. ABS, EPS)	Low	Medium	Low	N/A	Medium
Active safety systems (e.g. airbag)	Low	Medium	Low	N/A	Medium
Data storage (e.g. odometer, error codes, event data recorder)	High	Low	High	High	High
Infotainment (e.g. radio, GPS navigation)	High	Low	Medium-High	Low	High
External communication systems (e.g. Internet access, proprietary auto maker systems, V2X)	Medium	Low	Low	Medium	Medium
Local wireless communication systems (e.g. Bluetooth and tyre pressure monitoring system)	Medium	Low	Low	Medium	Medium
Local wired connection (e.g. OBD-II)	High	Low	Low	Low	Medium

12.2.5 Security of Automotive Legal Applications

Another important security objective is the protection of all automotive components, communications, functions and interface, which enforce effective legal restrictions. Examples of attack targets could include:

- Legal restrictions regarding environmental laws (e.g. exhaust control, engine control, fuel detection. mandatory maintenance intervals). An attack might increase engine performance, but reduce efficiency.
- Legal restrictions regarding fiscal laws (e.g. odometer, identification, and restrictions related to tax reductions and payments). An attack might seek to avoid or reduce payments.
- Legal restrictions regarding traffic and labour laws (e.g. max speed, TV activation while driving, tachograph for maximum driving periods). An attack might seek to improve driver comfort and/or profit at the expense of (other's) safety.
- Legal restrictions regarding export laws (e.g. export-restricted equipment such as night vision equipment or export-restricted markets). An attacker might wish to circumvent political objectives.

12.2.6 Security of Automotive Business Models

The protection of automotive business models is a financially oriented security objective and covers all automotive components, communications, functions and interface, which enforce a vehicle-based (aftermarket) business model. Examples of some mainly financially relevant automotive applications are:

- Component or feature activation, where the Original Equipment Manufacturer (OEM) usually has already installed all necessary components (e.g. a fully equipped infotainment unit) in the car, but has it not yet completely activated it. This would be in order to sell customers enhanced features (e.g. special multimedia decoders, extended information) using an on-demand base business model, for instance.
- There are several pay-as-you-drive business cases such as pay-as-you-drive car insurances, pay-as-you-drive car taxes, pay-as-you-drive car leasing or car rental, or pay-as-you-drive warranty or maintenance plans.
- There are special discounts for renting or buying a car based on some special car usage restrictions (e.g. limited usage locations, limited usage periods, limited mileage, limited performance values etc.) or for some dedicated car "branding" (e.g. some mandatory advertisement at every car startup),
- There is an aftermarket software and information business (e.g. in-car app store).
- In future there may even more sophisticated automotive business applications such as third party hardware rentals (e.g. a third party pays to use in-car hardware

resources), smart grid integration use cases for electric vehicles, or the commercial use of the car as a sensor for local traffic, weather, or road conditions.

12.2.7 Automotive Privacy Aspects

The last but not least important security objective is privacy, which usually means the privacy of vehicle owners, drivers, and passengers, and their data. Since modern cars create, process, store and communicate more and more information, privacy becomes an increasingly important issue. Although driving a vehicle has never been fully anonymous, there are several (new) automotive privacy concerns, including:

- Location privacy with respect to vehicle external parties. For instance, location information may leak via unique identifiers for V2X applications, tolling applications, or, any wireless identification tokens (e.g. tyre pressure monitoring system, in-car radio-frequency identification (RFID) components, wireless access authorisation tokens for private garages etc.).
- Location privacy with respect to vehicle internal parties. An attack might proceed for instance by (manually) reading out the "last trips" navigation list or by internally logging the vehicle location with software applications for pay-as-you-drive services or by any other in-vehicle software application with access to the vehicle location data.
- General vehicle usage privacy, which includes usage time, speeds, or even logged vehicle control activities, for instance, from the automotive black box called "the event data recorder" or the digital tachograph for large commercially driven vehicles.
- Personal privacy, which means all personal information, which is or will be processed and/or stored in vehicles such as vehicular access permissions, personal address books, personal phone or Internet access histories or even personal health data; for instance, with regard to the new mandatory eCall system [16].
- Vehicle maintenance or service activity privacy (if feasible), which enforce some effective restrictions about vehicular maintenance (e.g. car inspections) or service activities (e.g. fueling/charging, parking activities).

12.2.8 Real-World Automotive Security Incidents

The "history" of automotive security virtually starts with the introduction of first electronic components into mass production vehicles in the early 1990s. The first attacks targets were electronic theft protection mechanisms, which always sent the same unprotected "password," which could easily be intercepted and replayed. Other attack targets, which still apply today, include unauthorised manipulations of Engine control unit (ECU) software (chip tuning) to circumvent effective legal or safety restrictions (cf. Sect. 12.2.3), or mileage manipulations.

The introduction of external wireless communication interfaces such as the cellular network interface for GM's OnStar system extended the attack surface again, even though real-world attacks remained rare and mostly affected third-party aftermarket components [17]. The most prominent academic proof-of-concept attack in the 2000s was probably the complete break of the Keeloq remote key entry system [18], which is used by many major OEMs. Keeloq is a proprietary cryptographic authentication protocol, which replaced the simple to hack plain password remote entry systems of the early 1990s. However, Keeloq had some inherent cryptographic weaknesses and, even worse, it was deployed using an insecure key management approach based on a few globally unique keys, which once discovered could be easily reused for other cars. Some current (2011) very prominent real-world attacks, even though they were executed by academics, were the comprehensive attacks done by the Center for Automotive embedded systems security (CAESS). The first [19] successfully demonstrated over a range of experiments, both in the lab and in road tests, the ability for unauthorised control of a wide range of automotive functions and completely ignore driver input including disabling the brakes based on prior internal physical access. The second attack [20] successfully demonstrated external remote attacks showing that "exploitation is feasible via a broad range of attack vectors (including mechanics tools, CD players, Bluetooth and cellular radio), and further, that wireless communications channels allow long distance vehicle control, location tracking, in-cabin audio exfiltration and theft". Another currently and prominent real-world attack, which is particularly related to vehicular privacy, successfully demonstrated how the complete lack of protection for the mandatory vehicle Tyre pressure monitoring system (TPMS) could lead to serious privacy and security vulnerabilities [21].

Furthermore, the future already promises some further important application challenges for automotive security. A short-term example is the (mandatory) introduction of eCall [16], which gives assistance (e.g. in case of car accident) based on an inherent cellular network connection for voice and emergency data communication. A midterm example (around 2015) is the introduction of vehicle-to-infrastructure (V2I) or vehicle-to-vehicle (V2V) communication systems, as foreseen by the corresponding automotive manufacturer car-2-car communication consortium (C2C-CC). Both examples will introduce the mass-market application of long-range wireless interfaces for remote car access, and potentially affect the vehicle driving behaviour based on information received from external parties.

12.2.9 Examples of Automotive Security Mechanisms

Nowadays, IT security is deployed in manifold ways in modern cars. Starting with the communications inside the cars and between cars and the outside world, up to hardware and software security measures, one can find many security implementations in the automotive domain.

12.2.9.1 Communication Security

One of the most prominent aspects for vehicle security is communication security. As already mentioned in the introduction, modern vehicles have many internal and external communication channels and interfaces that require adequate protection against malicious manipulations or encroachments into confidentiality or privacy. In general, we can distinguish between in-vehicle wired communications and vehicle-external wireless communications. Examples for in-vehicle wired communications are the following.

- Interior-based in-vehicle wired communication (e.g. USB, SD card).
- Basic in-vehicle wired communication (e.g. on-board diagnosis).
- Extended in-vehicle wired communication (e.g. CAN, LIN).
- Expert in-vehicle wired communication (e.g. test and debug interfaces, on-chip).

Examples for vehicle-external communications that required adequate protection are the following.

- Low-range wireless vehicle communication (e.g. Bluetooth, tyre sensors).
- Mid-range wireless vehicle communication (e.g. Wi-Fi, V2X systems).
- Long-range wireless vehicle communication (e.g. cellular network communications such as UMTS, GSM).
- One-way wide-range wireless vehicle communication (e.g. FM radio, GPS, traffic message channel).

12.2.9.2 Software Security

Another potential security target is automotive software. Due to its flexibility and cost-efficiency, software more and more replaces most electromechanical solutions. However, from the security perspective, software is also much easier to manipulate or to replace without authorisation than electro-mechanical solutions. Moreover, automotive software can currently be duplicated almost free of costs. This makes automotive software especially susceptible to malicious encroachments that may be difficult to detect even by proficient investigators. Hence, efficient security mechanisms for automotive software are required including:

- Software authenticity/integrity protection.
- Software runtime protection.
- IP and expertise theft prevention.
- Authenticated software updates.
- Data confidentiality and privacy.
- Data availability.
- Data access control.

12.2.9.3 Hardware Security

Finally, automotive hardware is one of the earliest security targets in the automotive domain. Even though hardware manipulations are often more costly and more difficult than software manipulations or communication encroachments, they can be very powerful and effective and are particularly difficult to protect against. Hence, efficient security mechanisms for automotive hardware components are required for instance:

- Hardware integrity and safety protection.
- Enforcing technical, legal, or environmental restrictions.
- Enforcing a dependable base for most upper-layer security mechanisms.
- Preventing fabrication and installation of counterfeits.

12.3 Smart and Secure Open Automotive Platforms Platform

The integration of communication interfaces into automotive networks as well as the combination of different applications on the same ECU result in many security requirements to be addressed in today's automotive platforms. This section introduces the concept of an Open vehicular secure platform (OVERSEE). OVERSEE [22] is realised through a European research project and describes an open vehicular IT platform that provides a protected standardised in-vehicle runtime environment and onboard access and communication point. Therefore, the main objectives of the OVERSEE platform are the IT security and dependability. Hence, OVERSEE enforces a strong level of isolation between independent applications and seeks to ensure that vehicle functionality and safety cannot be harmed by any other application. Today, every new automotive project causes the development of a new and project specific Electronic Control Unit (ECU), which causes immense costs and project risks. Furthermore, currently no obtainable universal device is able to connect vehicle internal and external networks in a secure and common way. The idea of OVERSEE can be split in the following two main parts:

1. Open platform for the execution of OEM and non-OEM applications.
2. Secure single point of access to ITS communications.

In the following section, we describe the concept of virtualisation in OVERSEE, the underlying security services architecture and security implementation.

12.3.1 OVERSEE Virtualisation

The automotive applications running on an OVERSEE platform are executed in protected runtime environments for maximum dependability and security.

Fig. 12.3 OVERSEE vehicular security architecture

Applications are prevented from influencing each other. In addition, OVERSEE allows secure communication to connected networks. To achieve this goal, virtualisation is one of the main concepts. The applications are executed in runtime environments, which are abstracted from the physical hardware. The runtime environments for applications are called partitions. Partitions could be:

- Single application partitions with or without Operating System (OS) / Real Time Operating System (RTOS).
- Clusters (with or without OS/RTOS) serving more than one application.
- Clusters could be static (fixed set of applications).
- Dynamic (e.g. application store approach).

The different partitions with typical applications are shown in the figure below.

The partitions are controlled by the virtualisation system. Virtualisation offers a temporal and spatial partitioning platform to execute several execution environments on one physical Electronic code unit (ECU) with very low overhead, yet increasing the reliability of the applications. The virtualisation has the following benefits:

- Application independence.
- Uniform view of system resources (hardware and software).
- System resources are controlled in a deterministic way.
- Hardware independence.
- Uniform system view through a standardised programming interface.
- Increased portability by providing a virtual machine.

From a cost perspective, the OVERSEE platform reduces the initial costs for hardware and efforts as well as the follow-up costs and efforts for later system modification, certification, and validation. New applications can be added or modified without affecting the other partitions in the system. Only those parts of the system that change must be re-certified or re-validated. It facilitates parallel development and an easy integration process.

The OVERSEE approach uses the XtratuM hypervisor [11] to realise the virtualisation. The approach is beneficial because of the intensive consideration of security, dependability and reliability issues within the development. The programming interfaces for developing applications are publicly available. Hence, developers are able to quickly and efficiently develop new automotive applications and integrate security and dependability aspects right from the start. The communication interface is based on existing standards and therefore it is possible to connect most recent and new vehicle internal and external networks with only small effort. As security issues are an integral part, connecting new networks would be possible without the fear of creating new backdoors for attackers. The security of communication via OVERSEE and with the applications executed on OVERSEE will be based on a small and well-defined message and command set. The access of applications executed on OVERSEE to the communication interfaces as well as the incoming interfaces is protected by a message filtering firewall. The firewall is customisable by user policy rules.

12.3.2 OVERSEE Security Services Architecture

The virtualised architecture enables different runtime environments to run in parallel on the same platform with different levels of trustworthiness. This enables the creation of secure and isolated services, which can be reached over dedicated channels. The need for integrity and trustworthiness can be limited to a minimal number of modules in this way and evaluated separately from the user specific part of the OVERSEE platform. Based on such a trusted base, further enhanced security services can be added. In this section, the general architecture of the security concept will be explained.

As seen in Fig. 12.4, the architecture involves a hardware security module (HSM) developed by the EVITA project [3]. The EVITA HSM provides many services and features serving as a base for the security concept. The HSM is logically coupled to the security service partition of the OVERSEE platform which provides a secure and isolated runtime environment for security services in general. This security services can be requested by the other partitions through secured communication channels.

The EVITA HSM adheres to a large extent to the TPM specification [7] and therefore can assure the secure boot process of OVERSEE. The secure boot process starts with the authentication of the XtratuM hypervisor, the secure I/O partition and the security services partition. The hash values of the actual configuration of these partitions are compared with the known configuration values of the platform in the HSM. Based on the comparison, many actions can be taken and/or enforced by the architecture. One action would be to restrict access to cryptographic keys by coupling the correct hash value of the system configuration to the access rights of cryptographic keys stored in the HSM. Furthermore, the HSM can provide signed attestations of the system configuration, which can be used by external entities to remotely validate the integrity of the system.

Today's systems usually start from a stand-by (or hibernated) state which excludes the secure boot process. Furthermore, the fact that harmful changes can only be recognised in the next boot process already can cause severe security flaws.

The parallel execution of the security service partition can react in such a situation. The integrity of specific memory areas or stored data can be periodically validated on-the-fly or after specific actions like warm-start or software installation. The actions upon recognition of non-expected configuration can vary from just giving notifications up to stopping partitions. The security of stored data is assured by services for encrypting individual files. A possible candidate for an encrypting file system would be dm-crypt.[1] The key material for the encryption is stored in the HSM, providing a sealed storage. The file system of the user partitions are provided by the secure I/O partition using Virtio [8].

The OVERSEE architecture provides a central point for handling software installations, providing a mandatory verification procedure by means of authorisation, integrity, authenticity, compatibility and dependencies. The software package structure is based on a standardised Debian package [9]. The central handler serves as a proxy between the partitions and the repositories and validates the packages before forwarding to the destination and initiating an installation. It also provides functionality for updating whole partition images or the hypervisor. The direct services provided by the security service partition can be summarised as follows:

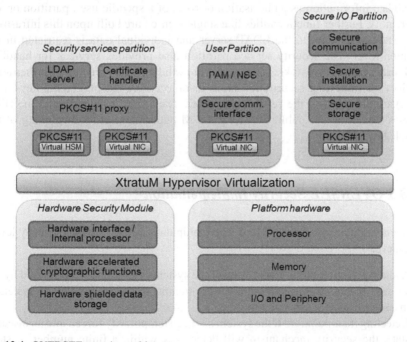

Fig. 12.4 OVERSEE security architecture

---

[1] dm-crypt is a Linux device-mapper target that provides transparent encryption of block devices using the new Linux 2.6 crypto API (cf. http://www.saout.de/misc/dm-crypt/).

- A controlled interface to cryptographic functions and key material hosted by the HSM.
- Central handling of authentication and authorisation information.
- Certificate handling.

The interface to the security module is based on the PKCS#11 [10] specification which is supported by most of the security modules (e.g. smart cards, hardware security modules, etc.). This enables the use of other security hardware instead of the EVITA HSM. The PKCS#11 interface is tunnelled to the other partitions by a proxy communicating with a PKCS#11 client driver hosted at the other partitions. The OVERSEE PKCS11# proxy design enables parallel access to the cryptographic functions and objects of the security hardware. It also provides a layer for restricting access of the individual services for each partition sending a request over the PKCS11# proxy.

The central handling of authentication and authorisation data is enabled by a Lightweight directory access protocol (LDAP) server (e.g. OpenLDAP) by the security service partition. This server can be invoked by the other partitions via NSS (Name Switch Service) or LDAP based PAMs (Pluggable Authentication Modules). Also direct usage of the LDAP server through look-up services can be used to retrieve data to validate information as authorisation or roles of a specific user, partition or any other entity. Further functionalities like single sign-on are built upon this infrastructure. The access right to the LDAP server and individual data is restricted in the partition level. The security service partition also provides services for handling certificates and security tokens like certificate validation and creation. Furthermore, services for importing security attributes or objects (e.g. cryptographic keys, new users) are provided by the certificate handler. The secure key storage in the HSM enables the storage of public keys (e.g. OEM public key) at the beginning of the vehicle lifecycle.

12.3.3 OVERSEE Security Implementation

Concerning data, software and hardware manipulation there are two security levels to achieve:

1. Avoiding manipulation: Every state of the system is continually controlled by the security mechanism in real time, which attempts to determine that the system is in a secure state at any point in time.
2. Detecting manipulation: If the system has been temporarily forced into an insecure state, the security mechanism will detect this within a finite number of steps after this event. Practically, we are often interested in real-time or near real-time detection.

While it is desirable to secure crucial vehicle components such as the complete engine, this seems to be out of scope today. Hence, we focus on securing the CUs connected to the internal buses. The overall security goals can easily be sum-marised, to avoid and detect manipulation, in order to maintain software and hard-ware integrity. If confidentiality is required, it should be provided for direct channels and side-channels (signal leakage). From an IT-security point of view, security con-trollers form a basis for secure platforms in automotive applications. The general term security controller refers to special microprocessors protected against active (tam-pering and other invasive attacks) and passive (timing attacks, simple differential power analysis, internal collision attacks, EM analysis, template attacks and many others) physical attacks. They offer a number of pre-implemented cryptographic ser-vices such as DES/3DES, AES, hash functions, long number arithmetic for public key operations, RSA, ECC, secure generation of random numbers,. These crypto-graphic functionalities are often implemented as co-processors. Security controllers are also able to store data in secured memory, e.g. the data can be written once, but can afterwards only be read out or only used by the security controller for crypto-graphic operations. Most of these controllers are smart card derivations delivered in traditional microcontroller packages (e.g. DIL, TSSOP, DSO, etc.). Such security controllers possess 8-, 16- or 32-bit central processing units with clock frequencies between 8 and 66 MHz, 2–16 kBs RAM, 16–256 kB ROM and up to 400 kB Electri-cally Erasable Programmable Read-Only Memory (EEPROM). There are, however, security controllers with larger (up to several MBs) EEPROM and ROM.

As an advantage, such controllers are special-purpose high-security solutions with well-understood hardware and firmware architectures and thoroughly evaluated and certified security by state certification bodies within formalised certification procedures (e.g. by FIPS 140-2 [6] or Common Criteria (CC) [14] Protection Profiles (PP) [2]). Security controllers are already available on the market and can be produced relatively quickly after developing the corresponding embedded software application, resulting in relatively low manufacturing costs (prices in the range of one to several Euros for high volumes).

At the same time, there are several technical drawbacks impeding their potential widespread adoption in automotive applications:

- Relatively low computational performance in view of the need of real-time reaction in safety-relevant applications.
- Relatively low data transmitting capabilities, which may prevent standard secu-rity controllers from controlling broadband in-car communication and external interfaces on-line.
- Relatively narrow range of operational conditions. For example, the allowed tem-perature range is as a rule from −20 to +85 degrees Celsius, which is to be compared to the required values for automotive applications from −40 to +105 degrees Celsius. For solving these problems, some modifications in the hardware core through semiconductor manufacturers may be necessary.
- Typical lifetime of a smart card chip is usually shorter than that of a car.

More recently, security controllers specific to the automotive domain have evolved. As an example, automotive ICs with Secure Hardware Extension (SHE) pro-vide a specific security extension tailored to typical use-cases in the automotive domain [4]. Security hardware for the next generation of automotive communication systems is a hot topic in industrial research. Secure and trustworthy intra-vehicular communication is the basis for trustworthy communication among cars or between cars and the infrastructure. Therefore, the objective of the E-safety Vehicle Intrusion Protected Applications (EVITA) project [3] is to design, verify, and prototype a security architecture for automotive on-board networks where security-relevant components are protected against tampering and sensitive data are protected against compromise when transferred inside a vehicle. By focusing on the protection of the intra-vehicle communication EVITA complements, other e-safety related projects that focus on the protection of the vehicle-to-X communication. Compared to SHE, EVITA offers additional security functionality for many (future) use-cases in the automotive domain.

12.4 Conclusions

In this chapter, we introduced the notion of automotive security. Contemporary security solutions in the automotive domain are usually applied to niche applications such as electronic immobilizers, access control, secure flashing, and secure activation of functions or protection of the mileage counter. The primary reason is that legacy car IT systems did not need security functions, because there was very little incentive for malicious manipulation. Secondly, security has tended to be an afterthought in automotive systems, because the achievement of the core function is the focus when designing a system.

In the last decade, automotive security has evolved from considering individual niche products to a more holistic view. New approaches in automotive security consider the system view of the car, and in future, networks of cars and infrastructure. Consequently, security is required in many stages of automotive systems including the internal system components, the applications, the communication links between internal and external communication devices as well as the infrastructure devices and applications. With cars, which become increasingly smart, automotive security will play a crucial role for the reliability of future automotive systems.

References

1. K. Finkenzeller, RFID Handbook: Radio-Frequency identification fundamentals and applications, Wiley, 1999.
2. European Technical Standards Institute (ETSI), http://www.etsi.org.
3. M. Mouly, M-B Pautet, The GSM System for Mobile Communications, Cell & Sys. Correspondence 1992.
4. Third Generation Partnership project (3GPP), http://www.3gpp.org

5. ETSI SAGE Group (originally), 3G Security; Specification of the MILENAGE algorithm set: An example algorithm set for the 3GPP authentication and key generation functions f1, f1*, f2, f3, f4, f5 and f5*; Document 1: General, 3GPP TS 35.205.
6. Security Algorithms Group of Experts (SAGE), www.portal.etsi.org/sage/.
7. NIST, Advanced Encryption Standard, FIPS 197, 2001, http://csrc.nist.gov/publications/fips/fips197/fips-197.pdf
8. 3GPP, Specification of the SIM Application Toolkit for the Subscriber Identity Module - Mobile Equipment (SIM - ME) interface (Release 1999) 3GPP TS 11.14 V8.18.0, 2007–06.
9. The Java Card Forum http://www.JavaCardforum.org/
10. 3GPP, Security mechanisms for the (U)SIM application toolkit; Stage 2 (Release 5) TS 23.048 V5.9.0, 2005–06.
11. GlobalPlatform, GlobalPlatform Card Specification, 2006.
12. 3GPP, Specification of the Subscriber Identity Module -Mobile Equipment (SIM - ME) interface (Release 1999) TS 11.11 V8.14.0 (2007–06).
13. International Standard Organisation, "ISO/IEC 7816, Information technology - Identification cards - Integrated circuit(s) cards with contacts- Part 4 Interindustry commands for interchange", http://www.iso.org, 1995
14. David Wagner and Ian Goldberg, "GSM Cloning", ISAAC Berkley, http://www.isaac.cs.berkeley.edu/isaac/gsm.html, 1998
15. Anderson Ross, Kuhn Markus, "Tamper resistance - a cautionary note ", second USENIX workshop on electronic Commerce Nov 1996.
16. Paul Kocher, "Timing Attacks on Implementations of Diffie-Hellman RSA DSS and Other Systems", Advances in Cryptology - CRYPTO '96, LNCS 1109, 104–113, 1996.
17. Paul Kocher, Joshua Jaffe and Benjamin Jun,"Differntial Power Analysis, Advances in Cryptology - CRYPTO '99, LNCS1666, 388–397, 1999.
18. E. Biham, A. Shamir, "Differential Cryptanalysis of DES-like Cryptosystems. Journal of Cryptology", Vol. 4 No. 1, 1991.
19. Kumar Sandeep et al, "How to break DES for €8,980 ", CHES 2006, http://www.crypto.ruhr-uni-bochum.de
20. Eli Biham, Adi Shamir, "Differential Fault Analusis of Secret Key Cryptosystems", Technicon Computer science dept - Technical report CS0910.revised, 1997.
21. Tiago Alves and Don Felton. TrustZone: Integrated hardware and software security: Enabling trusted computing in embedded systems. www.arm.com, July 2004.
22. Mayes Keith and Markantonakis Konstantinos, On the potential of high density smart cards, Elsevier, Information Security Technical Report Vol11 No3 2006.

Chapter 13
Analysis of Potential Vulnerabilities in Payment Terminals

Konstantinos Rantos and Konstantinos Markantonakis

Abstract Payment systems fraud is considered in the center of several types of criminal activities. The introduction of robust payment standards, practices and procedures has undoubtedly reduced criminals' profit, and significantly hardened their work. Still though, all payment systems' components are constantly scrutinised to identify vulnerabilities. This chapter focuses on the security of payment terminals, as a critical component in a payment system's infrastructure, providing an understanding on potential attacks identified in the literature. The attacks are not only limited to those aiming to insult terminals' tamper-resistance characteristics but also include those that target weak procedures and practices aiming to facilitate the design of better systems, solutions and deployments.

13.1 Introduction

Payment systems have always been high on the list of attractive targets due to their ubiquity and the ease of exploitation of compromised assets.[1] The introduction of smart cards has significantly reduced credit and debit card losses. However, figures reveal that fraudsters have been seeking alternative paths to attack payment systems and their actions constantly shift to adapt to new trends and practices in

---

[1] Figures reveal that payment card fraud is one of the most profitable attacks for fraudsters and costly for the card payments industry to defeat. In the U.S. alone, card fraud costs the card payments industry an estimated US$8.6 billion per year [1].

K. Rantos (✉)
Technological Educational Institute of Kavala, Kavala, Greece
e-mail: krantos@teikav.edu.gr

K. Markantonakis
Information Security Group, Smart Card Centre, Royal Holloway, University of London, London, United Kingdom
e-mail: k.markantonakis@rhul.ac.uk

K. Markantonakis and K. Mayes (eds.), *Secure Smart Embedded Devices, Platforms and Applications*, DOI: 10.1007/978-1-4614-7915-4_13,
© Springer Science+Business Media New York 2014

payments technology. As with any system's security, some could argue that it is the attackers that implicitly drive technology, which then has to face new challenges and exploitation methods.

EMV payment standards were introduced to fortify the highly vulnerable magnetic-stripe cards-based payments with the deployment of chip cards accompanied by an extensive list of security measures adopted to make the system robust. One of the key components that comprise this security chain is the payment terminal used to accomplish a transaction, whose security with an emphasis on attacks against them is examined in this chapter. Most of the attacks presented in this chapter, although being against the UK's EMV payment system Chip and PIN,[2] they do not directly attack Chip and PIN. They rather target vulnerabilities left in the card payment system, whilst magnetic stripe acceptance remains an option. The majority of the attacks rely on the work carried out by the Security Group of the Computer Laboratory in Cambridge University.

The security introduced by EMV in payments is complemented by another set of specifications defined and published by the Payment Card Industry (PCI— www.pcisecuritystandards.org), which is seeking high security levels for the data relating to payment transactions. Five major payment card brands,[3] who also founded the PCI Security Standards Council (PCI SSC) developed and incorporated in their data security compliance programs the PCI data security standard (PCI DSS) [5], a set of requirements for cardholder data protection.

Deployed EMV payment systems, such as the Chip and PIN has been successful in reducing the fraud types it was designed to address. Counterfeit fraud is at its lowest levels since records began. Fraud has shifted away from the face-to-face environment to card-not-present [23], where currently EMV Chip and PIN has no impact. However, there are numerous occurrences of proven fraudulent activity, mainly related to the Chip and PIN, which include but are not limited to the following:

1. May 2006: Skimming attack at Shell stations in the UK, although on a very limited scale, resulted in more than £1 m losses in customers' accounts.[4] Fraudsters were able, through tampered payment terminals, to record card numbers and PINs and create cloned magnetic-stripe on cards used for money withdrawal and payment transactions in more security relaxed environments. Following the skimming attack, Shell had temporarily stopped using PIN-based cardholder verification and reverted to the traditional handwritten signature.

2. February 2007: Cambridge security experts managed to relay valid account data between a genuine cards and a remote genuine terminal (via tampered card and terminal) to successfully complete a transaction to which the legitimate

---

[2] Chip and PIN (http://www.chipandpin.co.uk) is the UK's flavour of EMV introduced in 2004 and fully rolled-out in February 2006.

[3] Listed in alphabetical order: American Express, Discover Financial Services, JCB International, MasterCard Worldwide and Visa Inc.

[4] http://news.bbc.co.uk/2/hi/uk_news/england/4980190.stm

cardholder did not consent [17].[5] The attack is known as "relay attack" and is described in Sect. 13.3.7.

3. November 2007: The same team from Cambridge tapped the communications between the card and the PIN entry device (PED), for two popular PEDs, and managed to intercept all necessary information for producing counterfeit magnetic-stripes on cards capable of cash withdrawal from automated teller machines (ATMs) that rely only on magnetic stripe information.

4. August 2008: UK Police recovered a PIN pad tampering kit together with fake cards. Advertisements about selling the know-how to bypass terminals, together with guidance and a proprietary bluetooth transmitter and receiver for $4,000 appeared in forums in the same period. The hack was based on a vulnerability on communication interception, i.e. on tapping into an unencrypted communication channel on PEDs,[6] and involved stolen re-engineered PEDs installed into retail outlets.

5. October 2008: Tampered terminals from China and Pakistan used in shops and supermarkets,[7] as reported by Dr. Joel Brenner, allowed card details to be relayed over mobile networks to overseas fraudsters for producing cloned magnetic-stripe cards mainly used for card-not-present transactions and cash withdrawal. The fraudsters masterfully opened the terminals, perfectly resealed them so that tampering could not be spotted unless carefully examined by professionals, and put them back in the supply chain to be exported to Britain, Ireland, the Netherlands, Denmark and Belgium. The simplest way to spot a tampered device, was through weighing them instead of disassembling them. To avoid being immediately spotted, the criminals used the cloned cards two months after the information had been stolen.

To be able to perpetrate the aforementioned attacks, fraudsters have to compromise one of the components that comprise the payment system, including the terminal, and bypass security measures and/or procedures deployed to make the system robust. Although the majority of the terminals are evaluated and certified under high assurance levels of corresponding certification schemes, such as the requirement for all PEDs (PIN pads) deployed in UK to be evaluated to an Assurance Level of EAL4+ under the Common Criteria methodology [19], research in this area and the aforementioned incidents have shown that there are still ways to bypass those strong security mechanisms for the benefit of the attacker.

This chapter starts with an introduction to payment terminals and their security characteristics based on the applicable security standards. Logical, physical and procedural attacks that have managed or can be used to bypass their security features are described in the following sections.

---

[5] http://www.theregister.co.uk/2007/02/06/card_security_attack/

[6] http://www.theregister.co.uk/2008/08/13/pin_security_analysis/page2.html

[7] http://www.telegraph.co.uk/news/uknews/law-and-order/3173346/Chip-and-pin-scam-has-netted-millions-from-British-shoppers.html

13.1.1 EMV Standard

EMV specifications were developed to facilitate secure interoperable credit and debit applications and the required interaction between point of sale terminals and chip cards where the payment application is stored. Prior to the introduction of chip based cards, the payments card industry was fighting against high-motive criminals with magnetic stripe cards with additional security features such as cardholder details embossing, use of holograms, and the use of cardholder verification code (CVC).[8] Security evaluated Integrated Circuit or Chip Cards was the next step.

EMV introduced the following main security features for payment transactions:

- **Authentication Methods**. Used for authenticating the card and verifying the legitimacy of integrated circuit card (ICC)-resident data.

  - **Static data authentication (SDA)**: SDA is an off-line authentication method where the terminal verifies the issuer's signature on a set of card specific data, namely signed static authentication data (SSAD) [14]. Due to the static nature of the authentication data the same set is used in all transactions throughout a card's lifetime. Moreover, SSAD are provided to the terminal during authentication in the clear, hence easily intercepted by adversaries while in transit. These properties result in a weakness in the system in which SDA cards can be subject to attacks, such as replay based attacks or creation of skimmed and counterfeit EMV cards (SDA cloning) [13]. Although not the most secure method, the attacks are only relevant for off-line usage, which is usually limited to low value transactions.

  - **Dynamic data authentication (DDA)**: DDA goes beyond examining against card data alteration and checks that a card is genuine by verifying the existence of a valid card resident cryptographic key. This is done with the creation of a digital signature by the card, namely signed dynamic authentication data (SDAD), on ICC-resident and dynamically generated data including a terminal-provided nonce. DDA eliminates the threat of replay attacks, hence DDA is not subject to the attacks that SDA is, and counterfeit cards can be identified even for off-line transaction authorisation. However, as will be described later, due to lack of message binding, i.e. the VERIFY (PIN) command is not bound in some way to the generation of SDAD, DDA is subject to a different type of attack known as a man-in-the-middle or "wedge" attack.

  - **Combined data authentication (CDA)**: CDA is a method that combines DDA and Application Cryptogram Generation which is required on-line for verifying transaction details. This combination distinguishes CDA from DDA and is the method that ensures that the transaction cryptogram has not been corrupted.[9]

---

[8] Aka card verification value (CVV or CVV2), card validation code (CVC or CVC2) or Value, or Card Security Code

[9] As Professor Chris Mitchell points in his Lecture Slides (Available: http://www.isg.rhul.ac.uk/cjm/IY5601/IY5601_B_060205_83-156.pdf) CDA, if appropriately used, makes EMV robust against wedge attacks.

To achieve that, EMV requires the card's signature to also include the card's application cryptogram, which carries details about the transaction.

- **Cardholder verification method (CVM)**: EMV defines more robust verification methods for the cardholder to protect against lost or stolen card type of fraud [3]. Without removing the traditional hand-written signature based authentication and on-line PIN verification (available in some debit markets) EMV provides plaintext and enciphered (for DDA/CDA cards) off-line PIN verification.
- **Off-line and On-line transaction authorisation**: EMV offers the terminal and the card the ability to decide whether to approve transactions off-line, decline transactions off-line, or request on-line authorisation of the transaction based on a risk analysis made during the transaction and on pre-defined values.
- **Secure messaging between the card and the issuer**: It is used to ensure the integrity the commands and data and the confidentiality of some data (e.g. PIN update) sent by an issuer to the card.

The above security features protect EMV based payment transactions which typically consist of the following steps:

- **Card Initialisation**: the card returns the answer-to-reset (ATR) sequence.
- **Application Selection**: the terminal selects one of the payment applications available on the card.
- **Read Application Data**: from the chip: including primary account number, cardholder name, expiration date, CVM list, and issuer's public key and certificate.
- **Off-line Data Authentication**: card and terminal choose a data authentication method, one of SDA, DDA or CDA, based on their capabilities.
- **Cardholder Verification**: A method supported by the terminal and the card is chosen for verifying the cardholder including one of: on-line PIN, off-line encrypted PIN, or off-line plaintext PIN.
- **Terminal Action Analysis**: Following a risk analysis based on results from on-line data authentication, checks on the chip's authorisation to participate in the transaction, checks from the terminal's side to determine whether there is a need for on-line processing, and considering rules in the terminal and the chip, the terminal requests for on-line approval, off-line approval, or off-line decline.

  - If the terminal decides to proceed off-line, it asks the card for a transaction certificate (TC).
  - If the terminal decides to go on-line, it asks the card for an authorisation request cryptogram (ARQC)
  - If the terminal decides to reject the transaction, it asks the card for an application authentication cryptogram (AAC).

  Note that the terminal's decision on the transaction might be altered by the card.
- **Card Action Analysis**: On a similar action analysis performed by the terminal, the card also performs a risk management to protect the issuer from fraud and its decision could be to go on-line (sending to the issuer an ARQC), to decline the

transaction off-line (sending an AAC), or to approve it off-line (computing and sending a TC).

- **On-line Processing**: If terminal and card decide for on-line approval, a request is sent by the terminal to the issuer for authorisation and on-line authentication. The response might also carry an authentication response cryptogram (ARPC), an issuer authentication cryptogram which is sent to the card for verification [15].

13.2 Current Terminal Status

This section outlines existing terminal technology and the current status of security requirements as these are dictated by standards and commonly accepted practices.

13.2.1 Types of Terminals

EMV specifications [20] define a payment terminal as "the device used in conjunction with the ICC at the point of transaction to perform a financial transaction. The terminal incorporates the interface device and may also include other components and interfaces such as host communications." Based on this definition, many terminal derivations are specified based on attendance, connectivity, capabilities and configurations [20].

A terminal typically comprises of at least the following functional components which are not necessarily standalone parts and can be combined and/or integrated into others:

- Interface device (IFD) [aka card reader or card acceptance device (CAD)]: it is the unit that can communicate with the card and in the case of an ICC, exchange data with it. It is a very appealing component as its direct contact with the card allows tampering for intercepting communications that take place with the card during a transaction.
- PIN entry device (PED) (aka PIN pad): in markets that use PIN at Point of Sale it is the unit used by the cardholder to enter the card's PIN.
- Point of sale (POS): it is the unit that controls the other components and the payment application.

An additional component that becomes more essential when the POS comprises of a set of units that work together to provide the required terminal functionality, is the terminal management system (TMS). The TMS has the overall control of the payment process and can be used to parameterise the system. Networking devices (such as switches and routers) are also necessary for the communications with the acquirer to complete on-line transactions but these are typically not part of the terminal.

Configurations marketed by vendors include the following:

- **Standalone terminals**: they do not require any additional components for completing the transaction, as they contain a card reader (magnetic stripe or chip-based), payment application, PED, network connectivity (Ethernet, wireless, GSM, GPRS) and a receipt printer.
- **POS integrated**: the payment terminal is integrated with the merchant's point-of-sale which provides much of the required functionality, except for the PED which might be a separate device connected via USB or RS-232 and typically with built-in logical and physical security.
- **Unattended self-payment terminals**: they have all the necessary interfaces (reader, PIN pad, network connectivity) and functionality (payment application) for completing the transaction and (optionally) providing a receipt. They are typically used by gas/petrol stations and as parking, ticketing and vending machines.
- **Wireless payment terminals**: standalone or POS integrated ones which have a wireless interface to the networking infrastructure. This is a crucial difference for these terminals as the deployment of wireless communications in the cardholder data environment opens the door to many additional attacks related to intercepting and manipulating over the air communications, or to the wireless infrastructure [4].

The evolution of terminals and the inclusion of additional functionality has allowed vendors to market multiple configurations to satisfy the diversified needs of various environments. Terminals are now connected to open public networks giving adversaries remote access to them and accommodate all sorts of connectivity interfaces, such as wireless. They are using common operating systems and share resources with other applications hosted on the same equipment. All the above, although useful in terms of functionality, offer intruders alternative paths to attack and make payment terminals subject to common operating systems vulnerabilities and malicious code attacks.

13.2.2 Where does Security Apply?

Terminal security should be an integral part of the overall payment system, similarly to any other computing device, and each of the components that comprise it should adopt the appropriate security mechanisms to mitigate the risk associated with their functionality. The requirements set by commonly adopted standards, internationally accepted evaluation criteria, common practices and guidelines, will reduce the risk without eliminating it. Flaws might be introduced by components that are not part of the target of evaluation or by inadequate security policies and procedures.

Security breaches can occur in any layer of the payment terminal stack as this is depicted in Fig. 13.1.

- At the application layer, logical separation of the applications running at the terminal and use of dedicated working space (sandboxing and memory protection

Fig. 13.1 Terminal layers

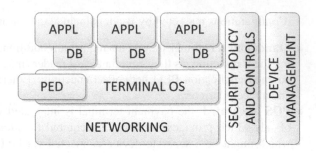

through segmentation and virtualization) allows data to be used exclusively by the application and prevent unauthorised access during processing.

- Storage of cardholder data, even in encrypted form, in a merchant's database might be an appealing target for the attacker who will attempt to exploit alternatives and gain unauthorised access through these, most likely, less secure systems.
- Use of common PC operating systems makes terminals subject to the same types of attacks as all other computing equipment with the same platform. Malware infection including the most dangerous rootkits, remote command shells, packet sniffers and Trojan-driven key loggers are some of these threats these terminals have to face. VISA has identified in [31] the three high-level areas of vulnerabilities for POS environments, which are *Remote Access Security, Host Security* and *Network Security*. This highlights the need to deal with a merchant's environment similarly to any other sensitive computing environment.
- The deployment environment and personnel's security awareness, training and practices also significantly affect system's security. These diversities impose different requirements and the corresponding operating environments may be enhanced by additional controls and practices to further mitigate identified risks.
- The shift towards a connected sophisticated device paves the way to new explorations by adversaries as the increased functionality is bound to introduce vulnerabilities. Attacks can now be mounted remotely since payment devices are now networking nodes. Therefore terminals' network security has to be an integral part of the overall security of the payment system and transaction. The wireless interface, where applicable, is also a very attractive target for fraudsters who explore the technological advances and the ease by which air transmitted data can be intercepted. The threats related to wireless payments are the following:

  – Unauthorised card read: Unauthorised reader used for reading and storing card information.
  – Eavesdropping: Listening into the communication between the card and reader during a point-of-sale transaction
  – Relay attack: Using unauthorised card read and relaying card responses from an intermediate device to another device at a distant point-of-sale.

Wired connections also have their flaws where lack of hard-wired connections between the different units, such as being able to "detach a PED from its terminal by simply pulling the wire out" [19] introduces similar vulnerabilities.

PCI has set the cornerstones regarding the assets that need protection and the types of measures that have to be deployed by stakeholders while it allows merchants to deploy their own "compensating controls" to satisfy DSS requirements if the suggested methods cannot be used. The core PCI specifications, issued as security standards against which vendors are encouraged to seek compliance, focus on the following three key components.

- **PCI data security standard** (PCI DSS): comprises a set of requirements for protecting cardholder data, applicable to all entities involved in a transaction, i.e. merchants, processors, acquirers, issuers, services providers and third parties authorised to manage cardholder data. It explicitly defines the types of transaction related data, namely Account Data, that need to be protected and categorizes them into Cardholder Data and Sensitive Authentication Data [5] (Table 13.1).
- **PCI PIN transaction security** (PCI PTS): aims to protect the PIN in PIN processing devices (PIN-PADs, hardware security modules (HSM)s, Unattended Payment Terminals).
- **PCI payment application-data security standard** (PA-DSS): aims to protect sensitive cardholder data in payment applications of software vendors.

PCI standards are only applicable if PANs are stored, processed or transmitted, while PCI DSS [5] explicitly prohibits storage of Sensitive Authentication Data after authorisation, even in encrypted form. It is only the PAN, expiration date, service code, and cardholder name that can be stored by merchants, so that transaction data are traceable for management purposes, in which case precautions for safe storage must be used [6]. Although these restrictions do not preclude that sensitive data can

Table 13.1 PCI data categorization

Cardholder data	Sensitive authentication data
Primary account number (PAN) (unique payment card number that identifies the issuer and the particular cardholder account. Typically embossed on the card but also contained on the magnetic stripe and the chip)	Full magnetic stripe data or equivalent on a chip
Cardholder name (the name of the cardholder as this is also printed on the card)	CAV2/CVC2/CVV2/CID
Expiration date (also printed on the card)	PINs/PIN blocks
Service code	

be leaked through alternative channels, as it is demonstrated by the attacks described here, additional measures that can be used to mitigate the risk include end-to-end (aka point-to-point) encryption [8, 27–29], tokenization [30], masking [5], truncation and data suppression [26].

13.3 Types of Attacks

Transaction security and terminal security are closely coupled and the latter can be considered a key component in the infrastructure deployed to protect payment transactions. As in any other type of attack, criminals seek for the weakest link in the transaction environment to exploit and this might be provided by the terminal. Terminals pose an attractive challenge for criminals to exploit, sometimes mainly due to their direct accessibility, such as the unattended ones in petrol stations, vending machines, or kiosks. According to [25] "Once a criminal has access, investigators familiar with this type of attack report that it only takes crooks about 30 s to remove the entire card device from a gas pump and replace it with an identical one fitted with electronic skimmers". Then the terminal becomes a valuable source of cardholders' data.

Due to chips' advanced security properties many attacks aim for the less secure features left on the card mainly for backward-compatibility as well as for cross-border interoperability, i.e. the card's magnetic stripe mechanism and data. The attacks described here mainly target hybrid systems that support both chip and magnetic stripe transactions. They seek to bypass the advanced security features of the chip and downgrade the system's security by "fooling" the system to accomplish a magnetic stripe based only transaction. They exploit vulnerabilities in applications, terminal design (hardware), deployment and operational policy (such as choosing to use low cost SDA only cards with the inherent vulnerabilities instead of the more advanced DDA cards).

13.3.1 Attacking the Supply Chain

The majority of security measures taken to protect terminals focus on their operational status. However, many attacks are mounted prior to installation at merchant's premises, mainly exploring weak procedures in the supply chain. This highlights the need to secure terminals throughout the device life cycle and the delivery process, i.e. during manufacturing, in transit, temporary storage, distribution, repair, disposal, and while being installed.

Supply chain attacks typically involve device (components or modules) tampering to enable unauthorised access to cardholder and sensitive data. For instance, prior to being provided to a merchant or acquirer, a PC-based terminal might be loaded with malicious code which intercepts account data and transfers them to an adversary.

The aforementioned allegations that hit the news in 2008[10] about Chip and PIN card readers in Europe, were stating that tampered readers were used to copy swiped card details and relay account data over the telephone network to remotely operating organised crime syndicates in Pakistan and China.

The devices were physically tampered with extra hardware either during manufacturing or supply. They were subsequently given to merchants, who unwittingly became part of the attacker's network, supplying their customers' card data to the fraudsters. The scam success was based on the sophisticated device tampering which left no external signs of device interference and therefore was very difficult even for the professionals to spot.[11] Moreover, the scam was not immediately identified, since the fraudsters were patient enough to wait for at least two months (after they had collected the relayed data and used them for producing cloned cards) before they started using these cards. It was only after the cloned cards were started being used that experts identified the attack through alerts on charges anomalies that triggered fraudulent activity. The scam has contributed significantly to the total number of compromised account details in 2008 (more than 280 million accounts [25]).

This is only one example of how a tampered terminal might reach the merchant's premises and become part of the infrastructure. One should not overlook the fact that the terminal might not reach the merchant through the legal predefined secure path but it might be a previously stolen one or one that has been obtained by a fraudster through legal means, manipulated by the fraudster and then put into operation at a merchant's location where security measures are relaxed enough to allow such a deployment [25]. Use of a non-certified or unauthorised terminal, poses an additional threat to the merchant.

Anti-tampering mechanisms should be deployed to protect against physical penetration attempts using means of drilling, lasers, chemical solvents substances, opening covers, splitting the casing (seams) and using ventilation openings [7]. Payments brands use state-of-the-art equipment for an initial assessment of the device's integrity using X-rays, which allow the assessor to identify any additional unauthorised modules in the terminal [25]. As with physical tampering where extra hardware is installed in the device, tampering can also have the form of malicious code injection and execution.

Assuming that devices leave the manufacturing stage intact, according to aforementioned standardised specifications and other security requirements, there are several measures that can be taken regarding the post-manufacturing secure handling and management of terminals, a set of which is issued by the Secure POS Vendor Alliance. This set highlights the need to define and manage rigorous procedures to securely handle "storage and transport, transfer of accountability from

---

[10] Details were given by the US National Counterintelligence Executive, Dr Joel Brenner in a Daily Telegraph interview, http://www.telegraph.co.uk/news/uknews/law-and-order/3173346/Chip-and-pin-scam-has-netted-millions-from-British-shoppers.html.

[11] Johnston et al. [21] demonstrated that bypassing tamper-indicating security, aka security seals, can sometimes be quite trivial.

manufacturer to the entity performing the initial key loading, device authentication, key management, incident response, outsourcing, and auditing" [24].

13.3.2 Exploiting Inadequate Security Measures

Although the aforementioned measures refer to protecting the device prior to its installation at merchant's location, there are also several issues that have to be considered regarding proper device management during its operation lifecycle. Negligent and insecure deployment and usage or management of the terminal (such as service and replacement), non-compliant to vendor's recommendations or the security policy of the retail store poses a significant threat to the payment system's security.

Several incidents that can lead to a security breach originate from the weakest link in an otherwise adequately protected system, that is human interaction. Security unaware personnel or lack of expertise can help adversaries bypass strict security measures and gain unauthorised access to the terminal and its components. Examples include but are by no means limited to the following incidents:

• Ad-hoc use of a payment terminal at conferences, forums, fairs or events, by inexperienced staff.
• Leaving authorised technicians unattended during any repairs.
• Leaving builders unattended or not inspecting their work (allowing them, for instance, to install a microcamera).
• A bogus engineer claiming to be a terminal technician and walks into the merchant's premises for onsite maintenance purposes or for fixing an allegedly malfunctioning terminal.

The first line of defence for merchants is to safeguard unauthorised physical access to the terminal. Typical access restriction measures can be taken, including monitoring through CCTV cameras to detect terminal replacement and tampering. Such measures however, should not contradict the need for protecting cardholder data, especially sensitive ones, such as PIN numbers entered on the pad or data printed on the card, such as CVV. PINs entered by cardholders not acting in due diligence, must not be subject to unauthorised disclosure due to a misplaced and/or misconfigured CCTV monitoring system to which fraudsters might easily gain unauthorised access and intercept video footage.

There should be a balance between merchant's assets protection and user's privacy. For instance, due to security reasons, the merchant might have to look away while the card is being used so that he/she will not be accused of shoulder surfing. This practice however, constitutes a vulnerability to the system as is the case with the relay attack described in Sect. 13.3.7 where the adversary might try to use his/her own unauthorised components during transaction processing.

13.3.2.1 Wireless Communications

A (potentially unprotected) wireless interface is also a candidate point of failure through which adversaries can intrude to the protected networking perimeter and bypass other efficient security measures deployed according to the merchant's policy. Fraudsters can intercept the required communications by listening to the air interface in order to capture all transmitted data (mainly targeting unencrypted traffic), or by deploying fake/rogue access points in order to route legitimate traffic through them.

Even if wireless infrastructure is not deployed as part of the payment system, e.g. is used for the rest of communications at the retailer's environment or is part of neighbouring networking, measures should be taken so that traffic is not accidentally or deliberately routed via these unauthorised networks. Strong wireless authentication and encryption, access control and network segmentation are some of the mechanisms that should be deployed to prevent information leakage and unauthorised access to network resources. PCI has issued corresponding recommendations for wireless networks used in a merchant's environment, being part of the cardholder data environment (CDE) or not [4].

Exploiting insecure wireless networks was one of the simplest, yet most profitable attacks carried out in the USA, which brought to the criminals information of more than 40 million credit cards stolen by sniffing data from vulnerable wireless networks to capture credit card numbers, PINs and other account information.[12]

13.3.2.2 Merchants' IT Systems

Merging payment terminals and systems with the rest of the computing equipment and IT infrastructure introduces new threats to the infrastructure and creates alternative paths to be exploited. The old, yet still effective, SQL injection attacks, which allow hackers to gain unauthorised access to the system and get cardholder data stored in it are only one example. Using this technique, a group of hackers managed to steal more than 130 million debit and credit card details from the systems of five retailers in the US.[13] Although such an attack is not directly related to a terminal it demonstrates the new threats that merchants are likely to face, emerging from the use of new generation terminals with advanced functionality and features in terms of applications support.

Another method of credit card fraud which targets high-volumes of cardholder data, is through breaching payment companies' and large retailers' systems where these data are held. If they are not adequately protected, e.g. unencrypted, or encrypted using vulnerable key management procedures, or unmasked, it is quite likely that the attackers will find a way to eliminate or bypass the network security perimeter and gain unauthorised access to them. Data logging, eavesdropping, electronic monitoring or even compromising emanation from components of a point of

---

[12] http://www.theregister.co.uk/2008/08/06/id_fraud_hacking_case/

[13] http://www.theregister.co.uk/2009/08/17/heartland_payment_suspect/

sale device, wiretapping, pinhole cameras and redundant equipment are some other methods that fraudsters are likely to deploy in order to reach the illegitimate profit they are seeking.

13.3.3 Skimming

Skimming is the act of illegitimately copying account related data on a card's magnetic stripe in order to produce a fraudulent copy of the victim's card that can successfully participate in payment transactions. It is a common way of attacking payment systems[14] as it can be performed on unattended terminals to gather a high volume of account data without alerting the cardholder or the stakeholders, as opposed to targeting a specific card, e.g. through card stealing. In 2008, in Europe alone, there were 10,302 attacks reported [2].

Although the source of the intercepted data can be any component of the payment system that handles them insecurely, terminals provide the necessary physical interface to the fraudster for mounting such an attack and therefore can be used as the vehicle for creating skimmed cards. Data sent to acquiring bank are commonly unencrypted [12] and therefore easily intercepted, while those exchanged between the chip card and the terminal included all the information needed to make a fake magnetic stripe card. The industry responded to this threat and since January 2008 the international payments schemes have mandated the use of unique security codes in chip data, namely icard verification value (CVV for ICCs) to mitigate the risk of magnetic stripe clones being created from data intercepted from chip-based transactions.[15] iCVV enables issuers to identify fraudulent use of chip data in magnetic-stripe read transaction processing. Magnetic-stripe based transactions are still subject to skimming. Physically and/or logically tampered terminals, dedicated card readers hidden under the counter desk, or terminals that do not adequately protect data transmission, such as wireless data transmission in the clear can provide an adversary all the necessary data for card skimming.

A common skimming practice is the use of small electronic devices, known as skimmers, that are masterfully positioned within the card-reader interface, e.g. a card slot of an ATM, in such a way that they cannot be easily detected, especially by an unaware cardholder, and have the capability to copy the information stored in the card's magnetic stripe. They are typically combined with microcameras pointing to the PIN pad, by transparent PIN pad overlays that record cardholders' keystrokes, or by even more sophisticated devices, such as thermal cameras [22], used for capturing the corresponding PIN.

Although skimming has traditionally been targeting magnetic stripe cards, where card copying is feasible with a skimming device, more sophisticated attacks for stealing data from chip based cards also emerge. Chip cloning, as opposed to mag-

---

[14] Skimming devices are even sold on Internet forums for about 8,000€.

[15] According to [23], at the end of 2011, more than 134 million UK cards had unique iCVV.

netic stripe skimming, is not practically feasible especially for sensitive data such as private and symmetric keys. Therefore, attacks on chip-cards typically aim for stealing static card data and creating legitimate appearing cards with limited/reduced functionality. Such an attack is even more attractive in countries that still use the traditional magnetic-stripe based payment schemes. This allows attackers to successfully use the fraudulent cards abroad for less-secure magnetic-stripe based transactions.

EMV SDA cards are more vulnerable to card skimming due to the following:

- PIN transfer from PED to an SDA-enabled card is only in the clear within the confines of a tamper responsive unit, hence the PIN can be intercepted from an undetected tampered device. Note that if the IFD is separate from the PED the transmission should be encrypted.
- Use of static data across multiple authentication instances allows an attacker to simply mount replay attacks using a seemingly valid card.

If an SDA card transaction is accomplished off-line, the terminal and the merchant are not able to detect the fraudulent transaction. The card will produce for the terminal an invalid Transaction Authorisation in the form of TC which cannot be verified by the terminal (since it does not have the keys). The attack will only be detected if the transaction goes on-line. In that case, since the rogue card does not have the required set of keys it cannot compute a valid ARQC (for on-line authorisation) and therefore will be easily detected by the issuer.

13.3.4 Covert Channels to PINs

One of the most critical data to protect in a transaction is the PIN entered by the cardholder, typically on a secure PED. Once an adversary gets hold of a lost or stolen card or manages to create a counterfeit card (magnetic stripe or SDA card) through skimming (see Sect. 13.3.3), the data necessary for performing ATM cash withdrawals or even using the SDA card for buying goods on merchant's site, is the PIN. Although intercepting communications between card and PED, and device tampering (see Sect. 13.3.5) are two sophisticated methods that a fraudster can use to get access to account data and the PIN, they are not the only ones.

Apart from the common and typical ones, where the PIN is held with the card in the stolen wallet, or stored as a phone number on a stolen mobile, there are many other PIN interception techniques such as the ones that utilise merchant's insecure infrastructure. An easier way to access the PIN is to get it from the merchant's environment not as data entered and transmitted over the merchant's network but indirectly, through other methods which can be exploited by both insiders and outsiders. As with any environment, staff can pose a threat either by introducing vulnerabilities, accidentally or deliberately, or due to their knowledge and potentially expertise.

Covert channels that can be used to harvest cardholder's data, such as the PIN, include the following:

- **Using the misplaced CCTV camera**: a security camera pointing to the till where the PED is also installed and used by customers for entering their cards' PINs can become the source for the PIN in question.
- **Wrongly positioned PEDs**: PEDs placed in locations where PIN entry is not obscured or not protected by additional measures such as a shroud, can give third parties access to a PIN at the time this is entered, through shoulder surfing techniques where an unaware, negligent or distracted user can enter the PIN with no precautions, such as covering the pad with his/her other hand.
- **Hidden microcameras**: positioned close to the PED and pointing towards it, such as a charity box placed on the counter desk or a pinhole camera placed by unauthorised personnel or technicians.
- **PIN pad overlays**: they are placed over the original keypad and allow PIN data capturing. They form a very popular way for unauthorised access to PINs during cash withdrawal from ATM machines and typically range from the very simple ones where keystrokes leave a mark on the overlay or are temporarily stored into its memory, up to the very sophisticated ones where a GSM module is also attached to it and allows the real-time transmission of captured PINs to the fraudster (possibly with some other cardholder data if the overlay is combined with a skimming device deployed on the same terminal or ATM).
- **Noted PINs**: usually found in stolen or lost wallets.
- **Thermal cameras**: used shortly after the transaction on the PIN pad can reveal which keys were pressed based on the keys' temperature [22].

Unattended terminals are even more susceptible to these types of attacks where the fraudsters who have physical access to the device can more easily manipulate the attacked device or the surrounding environment.

13.3.5 PIN/PIN Block Interception and Cracking

Covert channels to PINs are one option for an attacker to obtain these sensitive data while the other is to attack the points that they are stored or communicated, such as an insecure communication channel between the card and the IFD. Even if the PED has all the necessary functionality to protect the PIN by encrypting it before passing it to the card for off-line verification, this might not be supported by the card, i.e. the card cannot handle encrypted PINs because it only supports "Plaintext PIN verification". SDA cards do not support PIN encryption and therefore PINs are provided only in plaintext when off-line PIN processing is the selected CVM, hence PIN interception might be possible through tampering.

Intercepting communications between a PED positioned at an ATM and the financial institution on the other side of the communication channel would also give attackers access to encrypted PINs.[16] The encrypted PIN typically travels through multiple nodes and HSMs where it is decrypted and re-encrypted with the appropriate keys shared between pairs of HSMs. This process continues until the encrypted PIN reaches the issuer who has to perform the necessary check for PIN matching. Attackers may exploit known HSM vulnerabilities, like weaknesses on the security API [18], sometimes caused by misconfigurations, in order to get access to the decryption keys of the HSM and/or gain access to the decrypted PIN.

Key material might also be accessible through reverse-engineering, which gives the attackers the opportunity to bypass the security measures and disassemble the code in order to gain unauthorised access to the terminal's internals and functionality. An example of such an attack is found in [32] where the authors demonstrated for one specific device that access to private key material is feasible, although with many prerequisites. Such attacks require physical access to the device which, although viable for insiders or for inappropriately disposed or terminated HSMs, creates another obstacle for fraudsters as opposed to the remotely mounted attacks that aim for security breaches through HSMs' API, or even emissions analysis.

13.3.6 Manipulating the Terminal-Card Interface

Skimming attacks described in Sect. 13.3.3, aim for recording the account data of a valid card during a transaction. Two other attacks aim for manipulating the messages exchanged between the terminal and the card, exploring communication protocol vulnerabilities, without being detected by the participating genuine terminal. More specifically, with the "wedge" [9] and the "YES card" attacks described in this section the fraudster intervenes between the terminal and the card to alter the exchanged messages and deviate from the normal protocol execution. Both take advantage of the fact that the response to a VERIFY (PIN) command is not authenticated.

An example of a simple, yet allegedly effective, manipulation of this interface was recently demonstrated for a number of cards in Germany: because of a Y2K + 10 flaw around 30 million cards became inoperable in 1st Jan of 2010 because the chip was unable to recognise the year 2010.[17] Apparently, a workaround to this problem, until properly patching the terminals, was to seal the chip with a small piece of Scotch tape so to fool the terminal that there is no chip on the card, or the chip is malfunctioning and therefore "force" it to use the magnetic stripe instead.[18]

---

[16] http://www.wired.com/threatlevel/2009/04/pins/

[17] http://www.google.com/hostednews/afp/article/ALeqM5isP_cJaxnqSGaPVgUy0P3tSvpqrA

[18] http://www.spiegel.de/wirtschaft/soziales/0,1518,670433,00.html

13.3.6.1 Wedge Attack

A wedge attack is one where a malicious dedicated device, typically under the attacker's control, is placed between the merchant terminal and a genuine card to intercept and manipulate data exchanged between those two (the details of such an intermediary device are given in [9]). It typically allows a criminal "to use a genuine card to make a payment without knowing the card's PIN" [9, 16]. The attack, as the authors claim, although cannot be used on ATMs, is proven to work not only for card-not-present attacks but also for use at point of sale, and differs to the "YES card" attack in that it can be used even if the merchant goes on-line for transaction authorisation.

The attack is a man-in-the-middle one where the adversary tries to bypass PIN verification by convincing the terminal that PIN verification was successful and the card that a different cardholder authentication method is used, i.e. handwritten signature. To achieve that, the adversary suppresses the VERIFY (PIN) command, formats and sends to the terminal a response (the value 0x9000), which now considers that PIN verification was successful. The VERIFY PIN command never reaches the card which now believes that the cardholder was verified using another method, such as a handwritten signature. Fooling the terminal is feasible because the card's response on a PIN check request is not properly authenticated. The same method can be successfully deployed even on a DDA capable card.[19]

Cardholder verification method results are not communicated to the card (typically during an application cryptogram generation request) and therefore the legitimate card has no means to identify the type of method actually used and therefore cross-check the result with its own transaction records. Terminal verification results (TVR) [15] focus on failure codes, and in the case of a successful authentication, they do not carry any information about the details of it or the method used to authenticate the cardholder.

With the wedge attack, the authorisation cryptogram is created by the genuine card and therefore the attack cannot be detected even if it goes on-line. Although some card generated data sent to the issuer during on-line authorisation in an issuer proprietary field, called issuer application data (IAD), can reveal the type of authentication used, not all acquirers' terminals can parse this information correctly and therefore compare it with their own transaction record. Such a comparison would reveal the inconsistencies between the card's and the terminal's records and would result in declining the transaction.

Similarly, a solution proposed in [9] suggests inclusion of the cardholder verification method result (CVMR) in the transaction authorisation computed by the card. The CVMR contains the cardholder verification result together with the method used for this (e.g. PIN or handwritten signature) and is stored by the terminal. Upon passing the CVMR to the card, it can check against its own records regarding the verification method and verify the accuracy of it. If another man-in-the-middle attack

---

[19] http://www.lightbluetouchpaper.org/2009/08/25/defending-against-wedge-attacks/

occurs manipulating CVMR, e.g. replacing "PIN OK" message with "Signature", it will be the terminal's responsibility to perform a similar check.

As previously mentioned, given current EMV specifications, CDA capable cards are robust against wedge attacks. The attack would also be prevented if the messages exchanged during a transaction were linked together instead of being independent, a property that allows an adversary to inject malicious messages in between and manipulate the legitimate ones.

13.3.6.2 YES Card

A YES card attack is fairly similar to a wedge attack and executed using a counterfeit card that has account data from a stolen card copied on it,[20] but is modified so that it can accept any PIN. Although a wedge attack can work with any genuine card and will succeed even if the transaction goes on-line, a YES card on-line transaction will be immediately spotted as fraudulent, as the fake card does not have the necessary keys to create a valid cryptogram for issuer's checks. However, it benefits compared to the wedge attack in that it does not require the use of additional malicious hardware.

In this attack, fraudsters exploit two vulnerabilities of the static nature of an SDA card:

1. The card's response to PIN verification is not supported by an authentication mechanism and can be replayed or generated by an unauthorised entity.
2. The transaction authorisation is checked only if the terminal goes on-line and forwards it to the card issuer. If the risk analysis made for the transaction does not require on-line verification, the message authentication code (MAC) generated by the card is wrongly accepted by the terminal and successfully completes a transaction.

A "YES card" attack is accomplished with a card which typically has the following characteristics:

- The card contains an authorisation cryptogram that might have been intercepted during a past successful transaction.
- The card responds to a CHECK PIN command with the value 0x9000, i.e. successful, for any argument (PIN) passed with the command.
- The card will typically generate an error message and terminate the transaction if the terminal seeks on-line authorisation to avoid being detected.

The attack scenario is as follows:

- The attacker using a cloned/forged SDA card which carries all static data of a genuine EMV SDA card (certificates and application data), aims for merchants

---

[20] The attack is only successful with SDA cards used off-line and not with DDA or CDA cards, or on-line transactions as the fraudster cannot have access to the keys necessary for card data authentication.

and transactions that most likely will complete off-line[21] constituting the attack undetectable at the time that it is performed. Since an SDA card is authorised using only static data, card details and digital signature can be intercepted and replayed in a subsequent transaction.

- The next step is cardholder authorisation where the cardholder has to enter the card PIN which is in turn forwarded to the card for a validity check. Given that the card is a "YES card" it will respond with a 0x9000 value, meaning that PIN check is successful, for any PIN value.
- The card generates an (invalid) TC on transaction data as a response to a terminal's authorisation request. As the transaction terminates without going on-line, there are no checks being performed on the presented TC as the terminal does not have the necessary key set, and therefore the forged data go undetected. It is only when these data are sent to the card issuer that the fraud is identified but by that time the fraudster has already walked out with the goods in hand. Note that the forged card cannot produce a valid TC as it does not have the valid set of keys for the specific card and therefore if the transaction goes on-line for authorisation purposes the issuer's checks will fail and the transaction will be declined.

13.3.7 Relay Attacks

In a relay attack, a genuine card and the cardholder participate in a remote transaction, essentially paying for goods purchased by a fraudster and for which the cardholder never consented. There are a lot of prerequisites for this attack to succeed with the most demanding one being the use of a tampered terminal, which has to communicate with a bogus card during a genuine card transaction. Tampering, as with the other attacks, can be done either by the merchant or to a bunch of terminals during manufacturing or prior to being delivered to an unsuspected merchant. In the former case, the risk is very high for a merchant to accept since a few bogus transactions combined with cardholders' disputes can reveal the attacks' origins, hence the source is easily identified.

Moreover, given that the details of the attack are known, the attack can be very easily detected, even after the very first fraudulent transaction. However, it cannot be underestimated mainly because tampering might be simply the result of malicious software being injected into the terminal, hence multiple terminals controlled by a fraudster can be used for this purpose in a manner unpredictable by fraud detection tools.

The attack described in [10, 17] requires collusion of two or more persons and a tampered terminal. All communications between a genuine card and the tampered

---

[21]   According to http://www.dailymail.co.uk/news/article-389084/Millions-danger-chip-pin-fraudsters.html: "Of the 6.2billion transactions on a credit, debit or charge card carried out every year in this country, one in five happens 'off-line', meaning the chip and pin terminal does not connect to the cardholder's bank."

terminal are relayed to a bogus/manipulated card which communicates with another remote genuine terminal. Tampered terminal and card create a transparent communication bridge between legitimate card and terminal. Therefore goods are purchased by the legitimate cardholder, who unwittingly participates in a bogus transaction, that is never dispatched to the genuine cardholder. The authors used the mafia-fraud example introduced in [11] to demonstrate how a card used for paying for lunch at a restaurant ends up being charged for buying a diamond from an adjacent jewellery shop.

The tampered terminal, which as previously mentioned accepts a genuine card for a legitimate transaction, has the following characteristics:

• It relays all communications with the genuine card to the bogus card and is able to receive data from it. This can be accomplished using either wireless interface controlled by the fraudsters or wired communications considering though that the delays introduced by the use of this channel are not significantly expanding the card-terminal interface latency.
• It appears to and interfaces with cardholders as a genuine one.
• It prohibits communications with the issuer.
• It synchronises genuine and counterfeit card and controls delays that would result in transaction termination due to excessive delays in card-terminal (genuine one) interface, even though "existing smart card systems are tolerant to very high latencies" [17]. Even in that case though some responses expected by the genuine terminal can be requested in advance.

Countermeasures against this attack, described in [17], are based on distance bounding protocols and include setting upper limits on the acceptable delay between a request and a response, based on the acceptable distance between card and terminal. Although, a viable solution for dedicated devices, it might not be acceptable for computing platforms that incorporate POS functionality to which a terminal is attached or integrated, possibly consisting of mobile components, and therefore delays might be affected by a number of factors, such as computational and network load at the time the transaction takes place.

13.4 Conclusions and Future Considerations

This chapter focused on issues related to terminal security and on methods that can be used to bypass their security features. The majority of the attacks presented here require physical access to the terminal and, in most cases, infringement of its integrity.

However, as technology shifts towards terminals' integration with the rest of the computing platform and the terminal becomes part of the IT infrastructure of merchants, the possibility of exploiting other vulnerabilities to launch the same or similar attacks, cannot be excluded. Therefore, future deployments will also have to consider the following:

- Vulnerabilities introduced by the use of common operating systems typically designed without considering the requirements of a payment system and its peculiarities, not to mention, lack of a necessary security certification
- Terminals connected to open networks (IP-enabled devices) with widely used and standardised communication protocols that paves the way for remote attacks. If this is to be used in conjunction with networking capable cards (e.g. new generation Java Cards), there will be more security issues to resolve, as new exploitation paths are bound to be created for attackers.

It is worth mentioning that some of the attacks described here can be avoided by using DDA[22] or CDA-enabled cards and by using on-line transactions. Although this increases the cost of the infrastructure, it provides the means for the stakeholders to reduce losses resulting from the use of SDA cards. Moreover, financial institutions do not only depend on terminals' security to counteract these attacks but continuously invest on, deploy and enhance back end systems with advanced fraud engines that can effectively detect abnormal transactions indicating possible attacks. As more sophisticated attacks enter the scene suitable countermeasures provided by EMV and adopted by stakeholders to mitigate the associated risks, will prove a very valuable and effective asset in their constant effort to reduce loss figures.

References

1. Aite Group: Card Fraud in the United States: The Case for Encryption. January 2010. Available: http://www.aitegroup.com
2. ENISA, ATM crime: Overview of the European situation and golden rules on how to avoid it. August 2009. Available: www.enisa.europa.eu
3. EMVCo. A Guide to EMV. Version 1.0. May 2011. http://www.emvco.com
4. PCI, SSC Wireless Special Interest Group Implementation Team - Information Supplement: PCI DSS Wireless Guideline. Available: https://www.pcisecuritystandards.org/pdfs/PCI_DSS_Wireless_Guidelines.pdf
5. Payment card Industry (PCI) Data Security Standard: Requirements and Security Assessment Procedures. Version 2.0. October 2010. Available: https://www.pcisecuritystandards.org
6. PCI, SSC: PCI Data Storage Do's and Dont's. Available: https://www.pcisecuritystandards.org/pdfs/pci_fs_data_storage.pdf
7. PCI Encrypting PIN Pad (EPP) - Security Requirements, v2.1. January 2009. Available: https://www.pcisecuritystandards.org/documents/epp_security_requirements.pdf
8. Payment Card Industry (PCI) Point-to-Point Encryption. September 2011, Available: https://www.pcisecuritystandards.org
9. Murdoch, S. J., Drimer, S., Anderson, R., and Bond, M.: Chip and PIN is Broken. IEEE Symposium on Security and Privacy (2010) pp 433–444.
10. Anderson, R., Bond, M., and Murdoch, S. J.: Chip and SPIN. Computer Security Journal v 22 no 2 (2006) pp 1–6.
11. Desmedt, Y., Goutier, C., and Bengio, S. Special uses and abuses of the Fiat-Shamir passport protocol. In Advances in Cryptology CRYPTO 87: Proceedings (1987), vol. 293 of LNCS, Springer, p. 21.

---

[22] From 1st January 2011 schemes mandated that all new and replacement cards support DDA. At the end of 2011, 98 million DDA cards were in issue in the UK [23].

12. Murdoch, S.J., EMV flaws and fixes: vulnerabilities in smart card payment systems. Available: http://www.cl.cam.ac.uk/sjm217/talks/leuven07emv.pdf
13. Everett D. Chip and PIN Security. Available: http://www.smartcard.co.uk/Chip and PIN Security.pdf
14. EMV Iintegrated Circuit Card Specifications for Payment Systems - Book 2: Security and Key Management. Available: https://www.emvco.com
15. EMV Iintegrated Circuit Card Specifications for Payment Systems - Book 3: Application Specification. Available: https://www.emvco.com
16. Murdoch, S. J., Drimer, S., Anderson, R., and Bond, M.: EMV PIN verification "wedge" vulnerability, February 2010. Available: http://www.cl.cam.ac.uk/research/security/banking/nopin/
17. Drimer, S., and Murdoch, S. J.: Keep your enemies close: Distance bounding against smartcard relay attacks. In USENIX Security Symposium, August 2007. Available: http://www.usenix.org/events/sec07/tech/drimer/drimer.pdf
18. Centenaro, M., Focardi, R., Luccio, F., Steel, G.: Type-based analysis of PIN processing APIs. In: Backes, M., Ning, P. (eds.) ESORICS 2009. LNCS, vol. 5789, pp. 5368. Springer, Heidelberg (2009).
19. The UKCARDS Association: Security guidance for card acceptance devices - Deployed in the face-to-face environment.
20. EMV Integrated Circuit Card Specifications for Payment Systems: Book 4 - Cardholder, Attendant, and Acquirer Interface Requirements, June 2008. Available: www.emvco.com.
21. Johnston, R. G., Garcia, A. R., and Pacheco, A. N.: Efficacy of tamper-indicating devices. Journal of Homeland Security (April 2002).
22. Mowery, K., Meiklejohn, S., Savage, S.: Heat of the Moment: Characterizing the Efficacy of Thermal Camera-Based Attacks. In 5th USENIX Workshop on Offensive Technologies, August 2011. Available: http://www.usenix.org/events/woot11/tech/final_files/Mowery.pdf
23. Financial Fraud Action UK: Fraud - The Facts 2012. Available: http://www.financialfraudaction.org.uk
24. SPVA Lifecycle of a Secure Payment Device: Post Manufacturing Stage, June 2011, Available: www.spva.org.
25. Mastercard, Understanding Terminal Manipulation at the Point of Sale. Available: http://www.mastercard.com/us/company/en/docs/Terminal_Manipulation_At_POS.pdf
26. Visa Best Practices for Primary Account Number Storage and Truncation. Available: http://usa.visa.com/download/merchants/PAN_truncation_best_practices.pdf
27. European Association of Payment Service Providers for Merchants. Point-to-Point Encryption and Terminal Requirements in Europe. May 2011. Available: http://www.epsm.eu
28. VISA, Guide to Data Field Encryption. Available: http://www.visacemea.com/ac/ais/uploads/AIS_Guide_0610_Data_Field_Encryption.pdf
29. Mastercard Worldwide, An Analysis of End-to-end Encryption as a Viable Solution for Securing Payment Card Data. Available: http://www.mastercardacquirernews.com/pdfs/encryptionAnalysis.PDF
30. Visa Best Practices for Tokenization Version 1.0. Available: http://usa.visa.com/download/merchants/tokenization_best_practices.pdf
31. CISP Bulletin, Top three POS system vulnerabilities identified to promote data security awareness. November 2006. Available: http://usa.visa.com/download/merchants/top_three_pos_system_vulnerabilities_112106.pdf
32. Bond, M., Cvrcek, D., and Murdoch S.J.: Unwrapping the Chrysalis, In: Technical report, No. 592, 2004, Cambridge, GB, p. 15, ISSN 1476–2986.

Chapter 14
Wireless Sensor Nodes

Serge Chaumette and Damien Sauveron

Abstract This chapter addresses the key points of wireless sensor nodes: applications, constraints, architecture, operating systems, and security concerns. It does not pretend to be exhaustive but to provide the major references on these topics.

14.1 Introduction

Huge advances in Microelectromechanical systems (MEMS) and Wireless communications in the last decade of the twentieth century gave birth to new paradigms, where cheap, small size communicating sensors have been developed and integrated in many devices and large hardware/software environments. In this chapter, we target standalone wireless sensing devices, so-called wireless sensor nodes, which means that we focus on the device itself and not on the way it can be integrated within a global wireless sensor network. These small hardware pieces are becoming a key component of the digitization of the real world, thanks to their ease of deployment and the benefits that they can bring to human life in general, like infrastructure management (such as power grids) and environmental protection for instance. They are quite different from expensive isolated sensors (i.e., not intended to be a part of a whole swarm) and achieve complex measurements (related to a given phenomenon), and subsequent computation operations. Indeed, the strength of small sensor nodes is their ability to self-organize as a large network which enables measurement very close to a possibly dangerous phenomenon. They can cover a wide area and so are able to observe the evolution and spreading of complex events. In Sect. 14.2 we

S. Chaumette (✉)
LaBRI,University of Bordeaux, Bordeaux, France
e-mail: serge.chaumette@labri.fr

D. Sauveron
University of Limoges, Limoges, France
e-mail: damien.sauveron@unilim.fr

K. Markantonakis and K. Mayes (eds.), *Secure Smart Embedded Devices,*
Platforms and Applications, DOI: 10.1007/978-1-4614-7915-4_14,
© Springer Science+Business Media New York 2014

will show that wireless sensor nodes are used in both military and civil applications, health care, and power grid monitoring, among many other domains. In Sect. 14.3 we address constraints, like cost, energy-consumption, and network management (even though, as explained above, this last point is not a central topic of this chapter). In Sect. 14.4 the generic architecture of a wireless sensor node and the major features of the associated operating systems are described. We eventually present the security concerns in Sect. 14.5.

14.2 Applications

It is widely acknowledged that the first uses of advanced sensing technologies were military. In 1949, the US Navy announced its intention to exploit passive sonars for anti-submarine warfare (ASW) purposes using hydrophonic sensors deployed on the seafloor. In 1961, the sound surveillance system (SOSUS) [4] provided deep-water long-range detection capabilities, which were successfully used during the Cold War. It is now used by national oceanographic and atmospheric administration (NOAA) to observe a number of natural phenomena like submarine earthquakes [51] or the activities of some animals [22]. However, the concept of Distributed Sensor Networks only appeared in the early 1980's, driven by defense advanced research projects agency (DARPA) through a first program which was followed in 2001, due to evolution in the domain of MEMS, by the sensor information technology (SensIT) program [43]. The primary goals of SensIT were the creation of a new class of software for distributed microsensors, the development of novel methods for ad hoc networking of deployable microsensors and the extraction of right and timely information from a sensor field. Typical military applications of sensors are related to the collection of information that can be used to enforce situation management:

- monitoring friendly forces, equipment, and ammunition;
- battlefield surveillance;
- reconnaissance of opposing forces and terrain;
- targeting (e.g., for missiles);
- battle damage assessment;
- nuclear, biological, and chemical attack detection and reconnaissance.

Since early 2000, the number of civil applications has increased. Sensors now permit the creation of low-cost monitoring systems. Many domains, for instance biology, climatology, geology, etc., benefit from integrated sensors and sensor network technology. These sensors can for instance be used:

- to alert the civilian population when a volcanic eruption is going to occur [67] and to collect data without any risk to compromise the physical integrity of the persons in charge of taking the samples [65];
- to detect forest fire [37] or floodings [15];
- to forecast environmental pollution;

- to observe animals in their living environment without human presence [42, 49];
- to help farmers collect accurate measures of phenomena, which have a potential impact on their agricultural production [16].

As mentioned in the introduction, sensors can also be used in:

- health care, like telemonitoring of human physiological data [41], tracking and monitoring doctors and patients inside a hospital, environmental control in office buildings [54], etc.;
- infrastructure protection by monitoring bridges, tunnels, pipelines, power grids, etc.;
- home automation and smart environments [32], interactive museums [54], vehicle tracking and detection [60], etc.

Beyond these well-known applications, some other are much more unusual. For example, Simon et al. [61] propose to use a large number of cheap sensors communicating through an ad hoc wireless network, to detect and accurately locate shooters in urban environments. Their system supports multiple sensors failures, provides good coverage and high accuracy, and is capable of overcoming multipath effects.

14.3 Constraints

As illustrated in the above section, Wireless Sensor Nodes can be used for an extremely wide range of applications. However, they suffer from several constraints that need to be known and understood in order to properly target their possible domains of use.

14.3.1 Costs: Production Versus Performance

When a phenomenon needs to be observed, the decision to use wireless sensor nodes or classical sensors is mostly based on technical reasons. For instance, when considering the ability of the system to observe the spreading and evolution of a phenomenon to measure, it appears that a wireless sensor network is probably a good candidate because it allows wider and more accurate geographic coverage. Nevertheless, the financial cost of building the sensing system is also a prominent parameter. A wireless sensor network can be composed of a few to several thousands[1] devices and the overall price of the sensing system may thus be huge. To be viable, their production

---

[1] In practice very few applications use thousands of nodes. However theoretically, for instance, in military applications like battlefield surveillance, a huge number of nodes may be required. The biggest wireless sensor networks publicly known that have ever been built are: a system based on MyriaNed and composed of more than 1000 nodes [14, 38]; an 800 nodes network called the "Largest Tiny Network Yet" deployed at Berkeley [5] in 2001; WISEBED [21] which is made of

cost should remain very low (for example less than 1$/node). Indeed, even if not all networks are composed of thousands of nodes, there can be several networks operating in parallel for different purposes and thus the total number of nodes can still be huge. At the same time, they should offer good performance to sense phenomena and if required enough computing power to handle data locally, so as to overcome the network-related constraints and usage consequences such as energy-consumption (see below). Thus, their design is always the result of a trade-off between performance and cost. Obviously there is no hidden cost due to the deployment of wires (and associated devices) as can be found in classical networks, since by definition wireless sensor nodes do not require any infrastructure, however, in some cases there may be a cost for a wireless sensors spectrum licence.

14.3.2 Energy

To enable the sensors to operate during their whole mission time, energy saving is a major concern of both the manufacturers and the users, even though it is widely acknowledged that designing energy efficient communication components is a challenging task. Furthermore, it still remains that even with a power-efficient radio frequency (RF) technology, the energy consumption due to the communication between two nodes follows at best an inverse-square law of their distances. For this reason, to save power, it is needed to 'split' long distances between two communicating nodes by using multi-hop communication and routing [24]. Software is thus also largely involved in the energy-saving process. However, this way to communicate implies more cooperation between the nodes of the network in order to relay messages using power-efficient routing algorithms. It is thus a challenge to design low duty cycle radio circuits to relay messages between neighbors without losing any message. Strategies based on "wakeup on demand" (using two radios) [59] or "adaptive duty cycles" [68] are being studied to circumvent this problem. Another solution is to setup a network backbone with a subset of nodes that remain active.

In addition, it is possible to minimize [8] power consumption in a multi-hop sensor network, by processing sensed data locally to minimize transmission. The economy comes from the fact that the cost to transmit a large quantity of raw data coming from the sensor is more important than the cost of locally processing them, extracting the useful information, and eventually sending these information over the wireless network. Even if it is true that communication consumes a large quantity of energy in a sensor architecture, it still remains that the rest of the hardware and especially the processing unit must also be designed to be efficient. At software level, the operating system, protocols, algorithms, etc., must also be power-efficient and thus power-aware.

(Footnote 1 continued)
around 750 nodes but split in several subnetworks; the architecture described in [7] that uses 273 sensors and 47 wireless nodes to monitor nectarine orchard.

When battery capacity cannot be sufficient, solar cells [20] can be added to supply additional power. But they increase the manufacturing cost and are not suited to all deployments. However, other power scavenging methods, which enable wireless nodes to be completely self-sustained exist [55], at least as prototypes. Among these methods are those using temperature gradients, human power, wind or air flow, vibrations [56], etc.

14.3.3 Management: Self and Decentralized

When large-scale wireless sensor networks are deployed, relying on a centralized base station that would manage the topology and the routing is by nature not possible. This is due, among other parameters, to the cost in terms of energy spent to communicate over large distances. However, even though it is true that lower cost (in terms of energy consumption) multi-hop communication can be used to reach a base station, it still remains that in the particular case where nodes are mobile, the base station can become physically unreachable. Thus, wireless sensor networks often rely on decentralized management. For instance the decisions related to routing are computed locally based on information collected from neighbors using algorithms that once again optimize the energy consumption of the system.

In addition, it must be remembered that the nodes are deployed in a possibly adverse environment where they must thus operate without any human intervention. Therefore, configuration, adaptation, maintenance, and repair should be performed in an autonomous manner [24]. The nodes then need to have self-management and context-awareness capabilities [50], like self-organization and self healing. These require the ability to adapt configuration parameters according to changes in the environment, such as network disruption (the goal here being to maintain a given network topology) and to support self-protection (the ability to detect and protect against attacks). These are two fundamental features.

14.4 Architecture and Operating System

There are a plethora of readily available wireless sensor nodes which combine diverse hardware architectures with various operating systems. To try and make things clear, this section describes a tentative generic architecture of a node and the main features that are usually supported by the associated operating systems. Figure 14.1 presents the architecture. The sensing unit is the aim for which the sensor node has been designed. The rest of the unit is needed: to process the acquired data items or those received from neighbor nodes; to communicate with neighbor nodes; to supply power to all the boards and components. As can be seen, the sensing, power, and communication units exchange/communicate over buses such as general purpose input/output (GPIO), secure data input/output (SDIO), universal serial bus (USB),

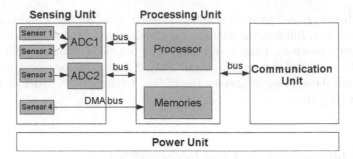

Fig. 14.1 Generic architecture of a wireless sensor node

serial peripheral interface (SPI), Inter-Integrated Circuit (I2C), the two latter being the most commonly used. As illustrated in Fig. 14.1, a direct memory access (DMA) bus can even be used so as to save the computing cycles of the processor.

As aforementioned, energy supply and management are crucial in sensors. The power unit may be a simple battery or it can generate part (or all) of the necessary power using solar cells, or even MEMS-based advanced technologies.

14.4.1 Sensing Unit

Figure 14.1 shows that a sensing unit can integrate several sensors and analog to digital converters (ADCs). The crucial hardware element of a sensor node are obviously the sensors themselves (even though the other components are also of major importance). Their goal is to monitor changes of some external physical phenomenon and to output a continuous analog signal as a function of time, which then needs to be converted to a digital value. The choice of the embedded sensors is done by the manufacturer depending on the physical phenomena to observe. It should also be noted that many architectures now support the notion of shields that are purpose-specific sensor boards, that can be plugged in depending on the target application. Thanks to the advances in MEMS, there exist sensors for most physical values, from heart rate to temperature (an illustrative list of sensors is given Table 14.1). The readers interested in building their own node with specific sensors are referred to a later chapter where the Arduino platform (a Do It Yourself open hardware platform) is described. A sensor that requires power to measure a physical information is called active while one that derives power from the energy provided by the sensed phenomenon is called passive.

To convert the signal produced by a sensor to information that can be handled by the processing unit, an ADC must be used to digitize the analog values. A bus is then used to feed the processing unit with the resulting digital information. As can be seen in Fig. 14.1, an ADC can be shared by different sensors. The decision of having one single ADC for several sensor units or one ADC per sensor unit or even not to embed

Type of sensed phenomena	Example of sensor
Acoustic	Microphone
Chemical	pH sensor
Electromagnetic	Magnetometer
Flow	Anemometer
Humidity	Hygrometer
Mechanical	Tactile sensor
Motion	Accelerometer
Optical	Photodiode
Position	Gps
Pressure	Barometer
Radiation	Radioactivity counter
Temperature	Thermocouple

Table 14.1 Examples of sensed phenomena and associated sensors

any ADC is done by the manufacturer based on the nature of the information (for instance, in terms of sampling frequencies and in terms of resolution) that the sensor is designed to measure. Multiplexing at ADC level can be used for sensors whose monitored information does not change with high frequency (or at least remains under a given threshold).

Although this will not be detailed here, sensor nodes can also integrate actuators such as LEDs, motors, etc.

14.4.2 Processing Unit

The processing unit provides the computing power of the system and connects the sensing and the communication modules. It consists in a processor and different kinds of memory banks: a nonvolatile memory (for instance Flash memory or EEPROM) to store the programs (application(s) and OS) and a volatile fast memory (different brands of RAM are possible) to store the objects used by the operating system (like the application stack), the sensed data items, etc. The processor can be of different types: field programmable gate array (FPGA), digital signal processor (DSP), application-specific integrated circuit (ASIC), or microcontroller. The latter embeds memories and several additional hardware features such as a clock generator.

However, if DSPs and FPGAs are special purpose energy-efficient processors that can be used for some well-defined and simple sensing tasks, they are not suitable for setting up a modular architecture. In the sensor network context, the sensing tasks can vary a lot, the overall hardware configuration can be modified by adding a new sensor, and the software layers must then be able to redefine the sensing operations. It is acknowledged that a microcontroller is the best solution when dynamic code loading and updating are required. For instance, it makes it possible to load post-issuance new energy-efficient software components and protocols. It is less powerful than a DSP or a FPGA, but easier to program (in C or assembly language). The main drawback of DSPs is their lack of flexibility when reprogramming is needed, whereas the drawback of an FPGA is that its production and design costs remain high. An ASIC provides the best performance, but again reconfiguration is difficult to achieve. Most of the time ASICs are used to complement microcontrollers or DSPs in low level tasks, such as wireless communication.

From an architectural point of view, the processing unit can be organized according to the Von Neumann (for FPGAs and ASICs) or Harvard (or even Super Harvard) architectures (for DSPs or microcontrollers).

14.4.3 Communication Unit

The communication unit, i.e., the transceiver, may be an optical device like in the Smart Dust motes project [50], but most of the time, it is an RF device. However, as aforementioned in Sect. 14.3.2, hardware is not sufficient and software multi-hopping is the most efficient and usual way to communicate in sensor networks. Therefore, different types of short-range and energy-efficient radio technologies are used such as those based on IEEE 802.15.4 like ZigBee [29], WirelessHART [62], ISA100.11a [64]; IEEE 802.11 [23] like WiFi; IEEE 1451 [45]; or other proprietary technologies. Because of its high-power consumption when switching the RF module on and off, Bluetooth is not used very often. Although there are not many benchmarks available, the reader interested in a comparison between these technologies can refer to the three following papers [46, 48, 63].

14.4.4 Major Features of Operating Systems

The operating system of a WSN is a small software layer linking the application layer to the hardware layer. It enables applications to interact with the hardware resources through services or libraries and is in charge of scheduling and prioritizing tasks. It can manage memory, power, etc. In addition, it provides application developers with a few basic abstractions and paradigms to express their algorithms. For example, to support concurrent tasks, they often provide either a multithreaded or an event-based programming environment [24].

Nevertheless, there is most of the time no clear separation between the operating system and the applications running on top of it. WSN operating systems generally consist of a number of selectable lightweight modules which are linked together at compilation time to create a monolithic program code which is in charge of sensing, processing, and managing communications. Some other operating systems provide an indivisible system kernel along with a set of library components for building applications [24]. More details on some common operating systems like TinyOS [33], SOS [36], Contiki [26] or LiteOS [17] can be found in [24].

14.5 Security Concerns

Until recently, most attacks on wireless sensor nodes have targeted the protocols used to communicate between nodes integrated in a network, therefore the intrinsic security of the nodes themselves is a major concern. For instance, nodes can be captured and then tampered with since they often operate in an adversary environment. We first address security of the nodes before presenting the main network-related concerns.

14.5.1 Security of Wireless Sensor Nodes

In our opinion, tamper resistance will be an important feature of sensing nodes operating in adverse environments (which is the most common case). In 2000, the authors of [18] mentioned the need for tamper resistance; however, because of the (low) cost constraint, they assumed that tamper protection for sensor nodes was limited. They even recommended not to use unattended sensors for some military missions where classified algorithms have to be used. It is true that developing tamper resistant nodes will be more expensive at the beginning of the production due to all the stages needed to be able to provide a high trust regarding the added security mechanisms. However, based on the smart card experience, one can think that the price will certainly quickly decrease when the production process will integrate these constraints.

Until 2006, node-capture has been considered for instance in [28, 53] that only the resilience of the network has been addressed. The proposed approaches were based on algorithmic solutions, e.g., through routing protocol [25], instead of a tamper-resistant mechanism, or based on redundancy to cross-check measures for consistency. The first studies [10, 11, 27] which have targeted the security of nodes themselves were done quite recently. Benenson et al. [10, 11] have developed a *"design space for physical attacks"* on nodes and have provided a framework for realistic security analysis in wireless sensor networks. Attacks (e.g., via a JTAG Test Access Port, via the Bootstrap Loader) against unattended sensor nodes in the field were discovered and countermeasures were proposed. For example, detection by its neighbors of the removal of a node through the communication protocol or by the node itself using for example an acceleration sensor, and if the removal is detected then the node is revoked of the network in the first case or it erases its confidential data in the second case. Side-channel attacks were only mentioned in the conclusions by Z. Benenson et al.; however, these threats have been considered by K. Eagles et al. [27]. In this chapter, the authors have developed a comparative threat analysis framework and a methodology to catalog threats, vulnerabilities, attacks, and countermeasures for smart cards (contact and contactless) and wireless sensor network nodes. One of the goals of their research was to determine security lessons learned from the world of smart cards that could be applied to the nodes of a wireless sensor network.

They clearly conclude that the nodes are subject to many attacks that exist on smart cards, and that tamper resistance features that are implemented within smart cards should also be considered for nodes.

Despite the very interesting conclusions of these studies, it is surprising to note that very few efforts have been made in this direction to improve the security of nodes. However, in 2010, Bialas published two papers [12, 13] in which he considered security issues of sensors used for high-risk applications. Among his contributions, he proposed Common Criteria-related security design patterns for the development of sensors' security features. This is an important step. However, applications with crucial security and reliability requirements, such as the studied sensors in charge of detecting methane in a mine [12], have not been subject to real Common Criteria evaluation and certification processes. A high level of trust for sensor nodes will only exist when Common Criteria certificates will be issued. However, the company named "Ultra Electronics 3eTI" seems close to deliver the first commercial sensing product that can reach such a level of assurance [1]. They manufacture a product called "EnergyGuard Appliance 3e-723" [2] which is a real-time energy monitoring and control system with built-in security mechanisms that allows energy managers to analyze usage at the building and base/campus level. The webpage of the product [3] claims it has been validated FIPS 140-2, Level 2, that Common Criteria EAL2 and EAL4 are pending and that it complies with DoDD 8500.1 and DoDD 8100.2.

In parallel and regardless of the conclusions of the paper by Eagles et al., three papers [30, 34, 35] have been published which exploit the network to attack the nodes themselves. Here is what the authors have been able to achieve. In [34], they compromised a Von Neumann architecture-based sensor with a classical stack attack. In [35], it is explained how a mal-packet carrying only specially crafted data can exploit memory-related vulnerabilities and utilize existing application code in a Harvard-based architecture sensor to propagate itself without disrupting sensor's functionalities. In [30], the authors succeeded in achieving a remote code injection attack on a Harvard-based architecture sensor, which was considered impossible before. This attack enables adversary to gain full control of the target sensor and for example to inject a worm that can then propagate through the wireless sensor network and possibly create a sensor botnet, or eavesdrop the network, etc.

To counter these recent attacks, Hu et al. present in [39] the design and implementation of a trusted sensor node, called trustedFleck. It uses a commodity trusted platform module (TPM) chip to extend the capabilities of a standard wireless sensor node (Fleck) to provide security services such as message integrity, confidentiality, authenticity, and system integrity. In addition, they provide services like secure software update and remote attestation.

The reader interested in software remote attestation approaches like SWATT [58] and ICE-based [57] schemes must be aware that attacks [19] exist against them and that a hardware root of trust is certainly the best reliable solution. Recently a debate [31, 52] concluded that while software-based code attestation is a useful security primitive, its design principles are not yet fully understood. The last advance [47] on this topic at the time of writing this chapter does not target sensor nodes.

14.5.2 Security in Networks of Wireless Sensor Nodes

As mentioned in the introduction of this section, most of the classical attacks that have been studied until now are related to the network protocols and not to the nodes themselves. Classical attacks like replay, packet injection, or corruption are well known and will thus not be described here. It should also be noted that operations like data aggregation and clock synchronization [44] will not be presented here in spite of the security issues that they raise, because they are application specific.

14.5.2.1 Denial-of-Service Attacks

A first class of security concerns in a sensors network is related to availability. Indeed, it is very easy for an adversary to stop the network operation with Denial-of-Service (DoS) attacks [66].

For example at physical layer, a jamming attack can be achieved by interfering on radio frequencies used by the sensor network. This kind of attack is very efficient since in most of the topologies, it does not require an important number of "attacking nodes" to perform it. However, some countermeasures to make it more complex have been developed, like:

- using spread-spectrum communication [e.g., Frequency-Hopping Spread Spectrum (FHSS)] which forces attackers to jam on a wide frequency band;
- jamming detection which enables the nodes to be switched in low consumption mode and awakened periodically to check if the jamming attack is still in progress.

Tampering with a node is also considered a DoS attack, since once captured, it is possible to destroy it or try to compromise it. Currently suggested countermeasures consist in:

- disabling the node and deleting its information when it believes that it has been compromised. It should be noted, however, that this is not very satisfactory, since the node is no longer available, which is the goal of DoS.
- camouflaging or hiding nodes in their physical environment [9], but this is not an information technology-based countermeasure!

At link-layer level, a DoS can be achieved with collision attacks which may consist in exploiting the medium access control backoff and retransmission procedures. In addition, even if collisions occur for only a few bits, the sending node should send its packet again, thus consuming more energy (this is an exhaustion attack). Fortunately, this can be partially addressed using error correcting codes and best-effort delivery protocols.

At network-layer level, a DoS can be achieved using routing loop attacks [44] to create loops in message routes so that messages are constantly forwarded around this loop, draining batteries of the nodes involved in the loop, and preventing the message from reaching its final destination.

At transport-layer level, a flooding attack can be performed by repeatedly requesting new connections to a given node which then undergoes a memory exhaustion and thus refuses further connections from legitimate nodes. A desynchronization attack can also be achieved to elicit resource-costly retransmissions [24].

14.5.2.2 Routing Attacks

There are a large number of possible attacks on routing protocols [6]. Here are a few:

- the blackhole attack in which an attacker attempts to be on the path of one or more routes to drop all the traffic so that the transmitted data never reach their destination. A variant is the selective forwarding attack, in which the attacker drops only the data matching certain criteria. This latter is more difficult to detect than a blackhole attack.
- the sinkhole attack is another variant of the blackhole attack, in which the traffic is not dropped, but disrupted or tampered.
- the rushing attack [40] in which an attacker exploits the nature of the route discovery procedure of on-demand routing protocols to increase its probability to be chosen as an intermediary node between a source and a destination.
- the sybil attack consists in an attacker presenting multiple (impersonated or false) identities to the other nodes of the network to influence the decision taken in cooperation inside the network (for instance to become a part of the route). Impersonation can be done with node or base station identities [44]. A variant of the sybil attack exists for geographic routing protocols where an attacker claims to be at several locations (instead of impersonating an identity) simultaneously with the hope to be chosen as a forwarding node.
- the wormhole attack in which two colluding attackers using an out-of-band channel between them, divert most of the traffic from the rest of the network. This then enables attacks like blackhole, sinkhole, etc.

While using wireless link encryption and authentication mechanisms can prevent most attacks from an outsider of the network, it cannot be effective against an inside attack, coming from a compromised node. As seen in this section, there are many possibilities to compromise a node and thus an intrusion detection system (IDS) is clearly an additional security mechanism that can help protecting the system.

14.5.2.3 Conclusion on Security

To the best of our knowledge, at the time of writing these lines, there is not yet any publicly available cheap wireless sensor node which would implement tamper-resistant features so as to provide a high level of trust and that could be deployed in a real network to defeat all aforementioned attacks.

Acknowledgments The authors want to thank the reviewers for their constructive comments which were helpful to improve this chapter.

References

1. 3eti: Company overview http://www.ultra-3eti.com/assets/1/7/3eTI_-_Company_Overview_ (07--26-2011).pdf
2. Datasheet of energyguard appliance 3e–723 http://www.ultra-3eti.com/assets/1/7/3e-723_ EnergyGuard.pdf
3. Energyguard appliance 3e–723 webpage product http://www.ultra-3eti.com/products/sensor_ networks/energyguard_appliance/
4. Globalsecurity.org. sound surveillance system (sosus). http://www.globalsecurity.org/intell/ systems/sosus.htm
5. Largest tiny network yet (2001). http://webs.cs.berkeley.edu/800demo/
6. Secure routing in wireless sensor networks: attacks and countermeasures (2003). doi: 10.1109/SNPA.2003.1203362. http://dx.doi.org/10.1109/SNPA.2003.1203362
7. Integrated smart sensing systems (2007). http://dpi.projectforum.com/isss/11
8. Akyildiz, I., Su, W., Sankarasubramaniam, Y., Cayirci, E.: Wireless sensor networks: a survey. Computer Networks 38(4), 393–422 (2002). doi: 10.1016/S1389-1286(01)00302-4. http:// www.sciencedirect.com/science/article/pii/S1389128601003024
9. Anjum, F., Sarkar, S.: Security in sensor networks. In: Mobile, Wireless and Sensor Networks: Technology, Applications and Future Directions. John Wiley & Sons.
10. Becher, E., Benenson, Z., Dornseif, M.: Tampering with motes: Real-world physical attacks on wireless sensor networks. In: in 3rd International Conference on Security in Pervasive Computing (SPC (2006).
11. Benenson, Z., Cholewinski, P.M., Freiling, F.C.: Vulnerabilities and attacks in wireless sensor networks. Wireless Sensors Networks Security pp. 22–43 (2007). http://www1.informatik.uni-erlangen.de/filepool/publications/zina/attacker-models-bookchapterIOS_Press.pdf
12. Bialas, A.: Common criteria related security design patterns alidation on the intelligent sensor example designed for mine environment. Sensors 10(5), 4456–4496 (2010). doi: 10.3390/s100504456. http://www.mdpi.com/1424-8220/10/5/4456/
13. Bialas, A.: Intelligent sensors security. Sensors 10(1), 822–859 (2010). doi: 10.3390/s100100822. http://www.mdpi.com/1424-8220/10/1/822/
14. Bisscheroux, M.: Largest deployment of myrianed wireless nodes (2010). http://wsn.chess.nl/? p=50
15. Bonnet, P., Gehrke, J.E., Seshadri, P.: Querying the physical world. IEEE Journal of Selected Areas in Communications 7(5),10–15 (2000).
16. Burrell, J., Brooke, T., Beckwith, R.: Sensor and actuator networks- Vineyard computing: sensor networks in agricultural production. IEEE Pervasive Computing 3(1), 38–45 (2004). doi: http://dx.doi.org/10.1109/MPRV.2004.1269130
17. Cao, Q., Abdelzaher, T.: liteos: a lightweight operating system for c++ software development in sensor networks. In: Proceedings of the 4th international conference on Embedded networked sensor systems, SenSys '06, pp. 361–362. ACM, New York, NY, USA (2006). doi: http://doi. acm.org/10.1145/1182807.1182855.
18. Carman, D.W., Kruus, P.S., Matt, B.J.: Constraints and approaches for distributed sensor network security. Tech. Rep. 010, NAI Labs, The Security Research Division Network Associates, Inc. (2000). http://www.cs.umbc.edu/courses/graduate/CMSC691A/Spring04/papers/ nailabs_report_00-010_final.pdf
19. Castelluccia, C., Francillon, A., Perito, D., Soriente, C.: On the difficulty of software-based attestation of embedded devices. In: Proceedings of the 16th ACM conference on Computer and communications security, CCS '09, pp. 400–409. ACM, New York, NY, USA (2009). doi: http://doi.acm.org/10.1145/1653662.1653711.

20. Chandrakasan, A., Amirtharajah, R., Cho, S., Goodman, J., Konduri, G., Kulik, J., Rabiner, W., Wang, A.: Design considerations for distributed micro-sensor systems. In: Proceedings of the IEEE 1999 Custom Integrated Circuits Conference, pp. 279-286 (1999).
21. Chatzigiannakis, I., Fischer, S., Koninis, C., Mylonas, G., Pfisterer, D.: Wisebed: An open large-scale wireless sensor network testbed (2010). http://dx.doi.org/10.1007/978-3-642-11870-8_6.10.1007/978-3-642-11870-8_6
22. Clark, C.W., Mellinger, D.K.: Application of navy iuss for whale research. The Journal of the Acoustical Society of America 96(5), 3315–3315 (1994). doi: 10.1121/1.410808. http://link.aip.org/link/?JAS/96/3315/1
23. Crow, B., Widjaja, I., Kim, J., Sakai, P.: IEEE 802.11 Wireless Local Area Networks. IEEE Communications Magazine pp. 116–126 (1997).
24. Dargie, W., Poellabauer, C.: Fundamentals of Wireless Sensor Networks: Theory and Practice. Wireless Communications and Mobile Computing. Wiley (2010). http://books.google.fr/books?id=8c6k0EVr6rMC
25. Deng, J., Han, R., Mishra, S.: A performance evaluation of intrusion-tolerant routing in wireless sensor networks. In: Proceedings of the 2nd international conference on Information processing in sensor networks, IPSN'03, pp. 349–364. Springer-Verlag, Berlin, Heidelberg (2003). http://dl.acm.org/citation.cfm?id=1765991.1766015
26. Dunkels, A., Gronvall, B., Voigt, T.: Contiki - a lightweight and flexible operating system for tiny networked sensors. In: Proceedings of the 29th Annual IEEE International Conference on Local Computer Networks, LCN '04, pp. 455–462. IEEE Computer Society, Washington, DC, USA (2004). doi: http://dx.doi.org/10.1109/LCN.2004.38.
27. Eagles, K., Markantonakis, K., Mayes, K.: A comparative analysis of common threats, vulnerabilities, attacks and countermeasures within smart card and wireless sensor network node technologies. In: Proceedings of the 1st IFIP TC6 /WG8.8 /WG11.2 international conference on Information security theory and practices: smart cards, mobile and ubiquitous computing systems, WISTP'07, pp. 161–174. Springer-Verlag, Berlin, Heidelberg (2007). http://dl.acm.org/citation.cfm?id=1763190.1763209
28. Eschenauer, L., Gligor, V.D.: A key-management scheme for distributed sensor networks. In: In Proceedings of the 9th ACM Conference on Computer and Communications Security, pp. 41–47. ACM Press (2002).
29. Farahani, S.: ZigBee Wireless Networks and Transceivers. Newnes, Newton, MA, USA (2008).
30. Francillon, A., Castelluccia, C.: Code injection attacks on harvard-architecture devices. In: Proceedings of the 15th ACM conference on Computer and communications security, CCS '08, pp. 15–26. ACM, New York, NY, USA (2008). doi: http://doi.acm.org/10.1145/1455770.1455775. http://doi.acm.org/10.1145/1455770.1455775
31. Francillon, A., Castelluccia, C., Perito, D., Soriente, C.: Comments on efutation of on the difficulty of software-based attestation of embedded devices (2010).
32. Frank, R.: Understanding Smart Sensors. Measurement Science and Technology 11(12), 1830 (2000). doi: http://dx.doi.org/10.1088/0957-0233/11/12/711
33. Gay, D., Levis, P., Culler, D.: Software design patterns for tinyos. ACM Trans. Embed. Comput. Syst. 6 (2007). doi: http://doi.acm.org/10.1145/1274858.1274860.
34. Goodspeed, T.: Exploiting wireless sensor networks over 802.15.4. In: ToorCon 9 (2007).
35. Gu, Q., Noorani, R.: Towards self-propagate mal-packets in sensor networks. In: Proceedings of the first ACM conference on Wireless network security, WiSec '08, pp. 172–182. ACM, New York, NY, USA (2008). doi: http://doi.acm.org/10.1145/1352533.1352563. http://doi.acm.org/10.1145/1352533.1352563
36. Han, C.C., Kumar, R., Shea, R., Kohler, E., Srivastava, M.: A dynamic operating system for sensor nodes. In: Proceedings of the 3rd international conference on Mobile systems, applications, and services, MobiSys '05, pp. 163–176. ACM, New York, NY, USA (2005). doi: http://doi.acm.org/10.1145/1067170.1067188.
37. Hefeeda, M., Bagheri, M.: Wireless sensor networks for early detection of forest fires. In: IEEE 4th International Conference on Mobile Adhoc and Sensor Systems, MASS 2007, 8–11 October 2007, Pisa, Italy, pp. 1–6. IEEE (2007). doi: http://dx.doi.org/10.1109/MOBHOC.2007.4428702

38. Heukoop, C.V.: Alwen 1000 node experiment. In: Elektronica (2010). http://wsn.chess.nl/wp-content/uploads/2010/02/AlwEN-1000node-exp-Elektronica-janfeb2010.pdf
39. Hu, W., Tan, H., Corke, P., Shih, W.C., Jha, S.: Toward trusted wireless sensor networks. ACM Trans. Sen. Netw. 7, 5:1–5:25 (2010). doi: http://doi.acm.org/10.1145/1806895.1806900.
40. Hu, Y.C., Perrig, A., Johnson, D.B.: Rushing attacks and defense in wireless ad hoc network routing protocols. In: Proceedings of the 2nd ACM workshop on Wireless security, WiSe '03, pp. 30–40. ACM, New York, NY, USA (2003). doi: http://doi.acm.org/10.1145/941311. 941317. http://doi.acm.org/10.1145/941311.941317
41. Johnson, P., Andrews, D.C.: Remote continuous physiological monitoring in the home. Journal of Telemedicine and Telecare 2(2), 107–113 (1996).
42. Juang, P., Oki, H., Wang, Y., Martonosi, M., Peh, L.S., Rubenstein, D.: Energy-efficient computing for wildlife tracking: design tradeoffs and early experiences with zebranet. SIGPLAN Not. 37, 96–107 (2002). doi: http://doi.acm.org/10.1145/605432.605408. http://doi.acm.org/10.1145/605432.605408
43. Kumar, S., Shepherd, D.: SensIT: Sensor Information Technology for the warfighter. In: Proceedings of the 4th Conference on Information Fusion, pp. 3–9. Montreal, Canada (2001).
44. Larsson, A.: Report on the state of the art of security in sensor, networks (2011).
45. Lee, K.: Ieee 1451: A standard in support of smart transducer networking. In: Proceedings of IEEE Instrumentation and Measurement, vol. 2, pp. 525–528. IEEE (2000). http://ieeexplore. ieee.org/xpl/freeabs_all.jsp?arnumber=848791
46. Lennvall, T., Svensson, S., Hekland, F.: A comparison of WirelessHART and ZigBee for industrial applications. In: Factory Communication Systems, 2008. WFCS 2008. IEEE International Workshop on, pp. 85–88 (2008). doi: http://dx.doi.org/10.1109/WFCS.2008.4638746
47. Li, Y., McCune, J.M., Perrig, A.: Viper: verifying the integrity of peripherals' firmware. In: Proceedings of the 18th ACM conference on Computer and communications security, CCS '11, pp. 3–16. ACM, New York, NY, USA (2011). doi: http://doi.acm.org/10.1145/2046707. 2046711
48. Lopez, J., Roman, R., Alcaraz, C.: Analysis of security threats, requirements, technologies and standards in wireless sensor networks. In: A. Aldini, G. Barthe, R. Gorrieri (eds.) Foundations of Security Analysis and Design V, Lecture Notes in Computer Science, vol. 5705, pp. 289–338. Springer Berlin/Heidelberg (2009). http://dx.doi.org/10.1007/978-3-642-03829-7_10.
49. Mainwaring, A., Culler, D., Polastre, J., Szewczyk, R., Anderson, J.: Wireless sensor networks for habitat monitoring. In: Proceedings of the 1st ACM international workshop on Wireless sensor networks and applications, WSNA '02, pp. 88–97. ACM, New York, NY, USA (2002). doi: http://doi.acm.org/10.1145/570738.570751.
50. Mills, K.: A brief survey of self-organization in wireless sensor networks. Wireless Communications and Mobile Computing 7(7), 823–834 (2007).
51. Nishimura, C.E., Conlon, D.M.: Iuss dual use: Monitoring whales and earthquakes using sosus. Marine Technology Society Journal 27(4), 13–21 (1994).
52. Perrig, A., van Doorn, L.: Refutation of n the Difficulty of Software-Based Attestation of Embedded Devices (2010). http://sparrow.ece.cmu.edu/group/pub/perrig-vandoorn-refutation.pdf
53. Perrig, A., Stankovic, J., Wagner, D.: Security in wireless sensor networks. COMMUNICATIONS OF THE ACM 47(6), 53–57 (2004).
54. Rabaey, J.M., Ammer, M.J., da Silva, J.L., Patel, D., Roundy, S.: Picoradio supports ad hoc ultra-low power wireless networking. Computer 33, 42–48 (2000). doi: 10.1109/2.869369. http://dl.acm.org/citation.cfm?id=619053.621512
55. Roundy, S., Steingart, D., Frechette, L., Wright, P., Rabaey, J.: Power Sources for Wireless Sensor Networks. pp. 1–17 (2004). http://www.springerlink.com/content/b0utgm8ahnphll3l
56. Roundy, S., Wright, P.K., Rabaey, J.: A study of low level vibrations as a power source for wireless sensor nodes. Computer Communications 26(11), 1131–1144 (2003). doi: http://www. sciencedirect.com/science/article/pii/S0140366402002487
57. Seshadri, A., Luk, M., Perrig, A., Doorn, L., Khosla, P.: Scuba: Secure code update by attestation in sensor networks. In: in Proceedings of ACM Workshop on Wireless Security (WiSe6). ACM, pp. 85–94. Press (2006).

58. Seshadri, A., Perrig, A., Doorn, L.V., Khosla, P.: Swatt: Software-based attestation for embedded devices. In: In Proceedings of the IEEE Symposium on Security and Privacy (2004).
59. Shih, E., Bahl, P., Sinclair, M.J.: Wake on wireless: an event driven energy saving strategy for battery operated devices. In: MobiCom '02: Proceedings of the 8th annual international conference on Mobile computing and networking, pp. 160–171. ACM, New York, NY, USA (2002). doi: http://dx.doi.org/10.1145/570645.570666
60. Shih, E., Cho, S.H., Ickes, N., Min, R., Sinha, A., Wang, A., Chandrakasan, A.: Physical layer driven protocol and algorithm design for energy-efficient wireless sensor networks. In: Proceedings of the 7th annual international conference on Mobile computing and networking, MobiCom '01, pp. 272–287. ACM, New York, NY, USA (2001). doi: http://doi.acm.org/10.1145/381677.381703.
61. Simon, G., Maróti, M., Lédeczi, A., Balogh, G., Kusy, B., Nádas, A., Pap, G., Sallai, J., Frampton, K.: Sensor network-based countersniper system. In: Proceedings of the 2nd international conference on Embedded networked sensor systems, SenSys '04, pp. 1–12. ACM, New York, NY, USA (2004). doi: http://doi.acm.org/10.1145/1031495.1031497.
62. Song, J., Han, S., Mok, A., Chen, D., Lucas, M., Nixon, M., Pratt, W.: Wirelesshart: Applying wireless technology in real-time industrial process control. In: Proceedings of the 2008 IEEE Real-Time and Embedded Technology and Applications Symposium, pp. 377–386. IEEE Computer Society, Washington, DC, USA (2008). doi: 10.1109/RTAS.2008.15. http://dl.acm.org/citation.cfm?id=1440456.1440604
63. Song, J., Han, S., Mok, A.K., Chen, D., Lucas, M., Nixon, M.: WirelessHART: Applying Wireless Technology in Real-Time Industrial Process Control. In: Real-Time and Embedded Technology and Applications Symposium, 2008. RTAS '08. IEEE, vol. 0, pp. 377–386. IEEE, Los Alamitos, CA, USA (2008). doi: http://dx.doi.org/10.1109/RTAS.2008.15
64. Surhone, L., Tennoe, M., Henssonow, S.: Isa100.11a. VDM Verlag Dr. Mueller AG & Co. Kg (2010). http://books.google.fr/books?id=F_BMYgEACAAJ
65. Werner-Allen, G., Johnson, J., Ruiz, M., Lees, J., Welsh, M.: Monitoring volcanic eruptions with a wireless sensor network. In: Wireless Sensor Networks, 2005. Proceeedings of the Second European Workshop on, pp. 108–120. IEEE (2005). doi: http://dx.doi.org/10.1109/EWSN.2005.1462003
66. Wood, A.D., Stankovic, J.A.: Denial of service in sensor networks. Computer 35, 54–62 (2002). doi: http://dx.doi.org/10.1109/MC.2002.1039518.
67. Wright, R., Flynn, L., Garbeil, H., Harris, A., Pilger, E.: Automated volcanic eruption detection using MODIS. Remote Sensing of Environment 82(1), 135–155 (2002). doi: http://dx.doi.org/10.1016/S0034-4257(02)00030-5
68. Ye, W., Heidemann, J., Estrin, D.: Medium access control with coordinated adaptive sleeping for wireless sensor networks. IEEE/ACM Trans. Netw. 12, 493–506 (2004). doi: http://dx.doi.org/10.1109/TNET.2004.828953.

Chapter 15
Near Field Communication

Gerald Madlmayr, Christian Kantner and Thomas Grechenig

Abstract Near field communication (NFC) is a radio frequency (RF) based proximity coupling technology allowing transactions within a range up to 10 cm. With NFC, a key technology is on its way into the consumer's most personal device, allowing the customer to use his devices for secure services such as payment or ticketing but also for service initiation or data exchange. Interoperability is one of the most important goals to be achieved prior to the roll out of devices and services, in order to satisfy the consumer's expectations. This chapter deals with different operating modes and use cases that can be implemented with NFC technology with the main focus on mobile phones. This high level description is backed up with a look into the hardware architecture for NFC as well as the software stack in mobile phones. The chapter ends with a description of tags and tag formats for the NFC ecosystem.

15.1 Introduction

Radio frequency identification (RFID) technology is used in many daily applications. For the consumer, RFIDs are unnoticed and simple to use, they offer a popular alternative to conventional communication channels. Starting with simple access control possibilities up to complex data memories, quite different applications can be realized. A further development represents NFC [16], a technology for the fast and uncomplicated exchange of small amounts of data. It opens new perspectives

G. Madlmayr (✉) · C. Kantner · T. Grechenig
Research Group for Industrial Software (INSO), Vienna University of Technology,
Vienna, Austria
e-mail: gerald.madlmayr@inso.tuwien.ac.at

C. Kantner
e-mail: christian.kantner@inso.tuwien.ac.at

T. Grechenig
e-mail: thomas.grechenig@inso.tuwien.ac.at

K. Markantonakis and K. Mayes (eds.), *Secure Smart Embedded Devices,*   351
Platforms and Applications, DOI: 10.1007/978-1-4614-7915-4_15,
© Springer Science+Business Media New York 2014

regarding the application development on all kind of consumer devices. Nowadays, NFC is being introduced to mobile phones.

NFC is an amendment to the existing contactless smart card systems, but still compatible to them. It is presented in ISO 18092 (NFCIP-1), supporting cards compliant to ISO 14443 [9] and Sony's proprietary FeLiCa system as well as NFC's own communication method. NFC allows wireless transactions over a distance of up to 10 centimetres. This is part of the *Touch and Go* philosophy giving the user a new dimension of usability. Hence, NFC-enabled handsets allow the consumer to interactively participate in the *Internet of Things* in a way like never before.

Consumers can use their handsets to retrieve further information by touching tags integrated within posters, products, or shelves. Alternatively, the handset itself can be used as a transponder, and therefore provides additional functionality in terms of applications and identification. This vision requires interoperability on different layers and a common agreement of industry players integrating technology and applications.

15.2 NFC Technology

15.2.1 Physical Layer

On the physical layer, NFC data are exchanged by two inductively coupled coils, one per appliance, generating a magnetic field with a frequency of 13.56 MHz. The field is modulated to facilitate data transfers. For communication, one device acts as the initiator (starting the communication), whereas the other device operates in target mode (waiting for the initiator). Typically, two devices are involved in the communication [3].

The roles of the devices, initiator and target, are assigned automatically during the listen-before-talk concept, which is part of the mode switching of NFC. In general, each NFC device acts in target mode. Periodically, the device switches into initiator mode in order to scan the environment for NFC targets (= polling) and then falls back into target mode. If the initiator finds a target an initiation sequence is used to establish communication before exchanging data.

NFC distinguishes two operation modes for communication: passive and active modes (Fig. 15.1).

15.2.1.1 Passive Mode

In passive mode only the device that starts the communication (the initiator) produces the 13.56 MHz carrier field. A target introduced to this field may use it to draw energy, but must not generate a carrier field of its own. The initiator transfers data by directly modulating the field, the target by load modulating it. In both directions, the

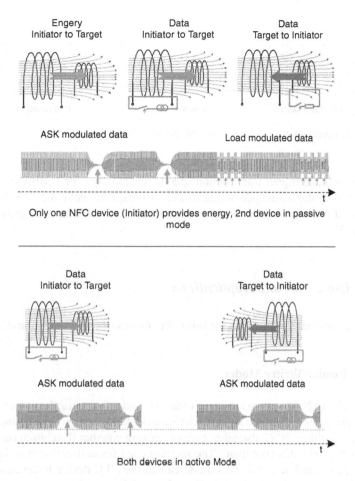

Fig. 15.1 Active and passive operation mode of an NFC device [13]

coding complies with ISO14443 or FeLiCa, respectively. This enables NFC devices to communicate with existing contactless smart cards. The term load modulation describes the influence of load changes on the initiator's carrier field's amplitude. These changes can be perceived as information by the initiator. Depending on the size of the coils, ranges up to 10 cm and data rates of 106, 212, and 424 kbits/s are possible.

15.2.1.2 Active Mode

When in active mode, both appliances generate an RF field. Each side transmits data by modifying its own field, using an amplitude shift keying (ASK) modulation scheme. Advantages compared to passive mode include a larger operating distance

NFC Phone   NFCPhone   NFC Phone   RFID   RFID   NFC Phone
Initiator   Target   Reader/Writer   Transponder   Terminal   CardEmulation

NFC peer-to-peer-Mode   *Reader/Writer-Mode*   *Card Emulation-Mode*

Fig. 15.2 Different operating modes of an NFC device [13]

(up to 20 cm) and higher transmission speeds (eventually over 1 MBit/s). To avoid collisions, only the sending device emits a electromagnetic field; the receiving entity switches off its field while listening. If necessary, these roles can change as often as needed [13].

15.2.2 Use Cases and Applications

An NFC compliant device offers the following modes of communication (Fig. 15.2):

15.2.2.1 Reader/Writer Mode:

In this mode, an NFC system acts as an ordinary reader for contactless smart cards. If two or more cards are present in the reader's carrier field one is selected using an anti-collision algorithm. NFC also takes care of sensing whether the chosen card is ISO 14443-A/B or FeLiCa compliant. The method used for anti-collision is dependent on the type of card detected. This mode causes the NFC device to act as an active device. From an application's view, there is no difference between a conventional and an emulated terminal, accesses to the contactless token proceed equally [3].

Operating in this mode, the NFC device can read and alter data stored in NFC compliant passive (without battery) transponders. Such tags can be found on e.g. a *SmartPoster* allowing the user to retrieve additional information by reading the tag with an NFC device. Depending on the data stored on the tag, the NFC device takes an appropriate action. If e.g. a URI was found on the tag the handset could open a Web browser.

15.2.2.2 Card Emulation Mode:

Card emulation mode is the reverse of reader/writer mode; a contactless token is emulated in passive mode. Due to the fact that the card is only emulated, it is possible to use one NFC device in place of several *real* smart cards. Which card is presented to the reader depends on the situation and can be influenced by software. Additionally, an NFC device can contain a secure element to store the information for the emulated card in a secure way [3].

In this case, an external reader cannot distinguish between a smart card and an NFC device in card emulation mode. This mode is useful for contactless payment and ticketing applications, and a single NFC-enabled handset is capable of emulating multiple contactless smart card applications.

15.2.2.3 Peer-to-Peer Mode:

This mode is specific to NFC. After having established a link between the two participants (the method is equal to ISO 14443-A), a transparent protocol for data exchange can be started. The data block size can be chosen freely, with an maximum transmission unit (MTU) limited to 256 bytes. The main purpose of this protocol is to enable the user to send his/her own data as soon as possible (i.e. after a few milliseconds). In a peer-to-peer session, the initiator is always in active mode, whereas the target may be active or passive. This helps the target to reduce its energy consumption and is therefore especially useful if the initiator is a stationary terminal (e.g. a ticket machine) and the target a mobile device (e.g. a mobile phone) [3].

The NFC peer-to-peer mode (ISO 18092) allows two NFC-enabled devices to establish a bidirectional connection to exchange contacts, bluetooth pairing information, or any other kind of data [10]. Cumbersome pairing processes are a thing of the past thanks to NFC technology. To establish a connection, a client (NFC peer-to-peer initiator) is searching for a host (NFC peer-to-peer target) to setup a connection; then the near field communcation data exchange format (NDEF) is used to transmit the data.

15.3 Hardware Integration

In order to provide all three application modes, a mobile has to include an NFC chip, a secure element, and a host controller. Their tasks and functionality will be outlined in the following sections.

15.3.1 NFC Chip

The NFC hardware in the mobile phone has to take over various tasks. One core functionality of the NFC chip is the analogue front end which is responsible for modulating and demodulating the 13.56 MHz carrier signals from the antenna to digital signals. Additionally, the communication between the host controller as well as the secure element (if present) is part of the chip. The NFC chip is managed by the host controller and an appropriate software stack with an API which is available to the developer. The functions of the NFC chip will most likely be an integrated part of future baseband processors in order to reduce costs and save valuable space in mobile devices.

15.3.2 Secure Element

The secure element is typically a smart card controller which is capable of emulating a *real* smart card. For security purposes, it is typically constructed in a tamper proof way. The software architecture of a typical secure element provides an environment, where applications are downloaded, personalized, managed, and removed independently with varying life cycles (e.g. using Global Platform functionality). It is also possible that the secure element is completely emulated in software. There are already general purpose CPUs in the market that support a trusted execution environment (TEE) where data can be processed in a secure way (e.g. ARM with its TrustZone architecture).

Payment or ticketing applications place high demands on security of the secure element. The chip must be highly reliable and robust to withstand different kinds of attacks. Also the manageability of the secure element is an important property [2]. Actually, a secure element function is not mandatory in an NFC device, although such devices cannot be used for card emulation mode.

- Security: Security implies confidentiality, integrity, availability, and authentication. Security as defined in ISO 14980 is: 'The purpose of information security is to ensure business continuity and minimize business damage by preventing and minimizing the impact of security incidents.'
- Manageability: A secure element can contain applications and data from different sources. Hence, means for application and data management need to be available. There should also be the possibility to remotely lock the secure element in case of loss or theft.

A secure element has to provide the following functionality [11]:

- Secure memory: A secure memory is necessary to store sensitive data like private keys, root certificates, and personal data.
- Cryptographic functions: Protocols for secure data exchange usually rely on cryptographic functions to provide security for the sensitive data, as this information must not leave the secure element without encryption.
- Secure environment for code execution: A secure element has to contain a unit to execute code in a secure way which cannot be monitored.

The secure element can be implemented in different ways in a mobile device. Smart cards or secure smart card chips are possible options for a secure element, whereas the following implementations are common for an NFC enabled handset:

- Embedded secure element as a chip

- Subscriber identity module (SIM) Card
- Secure digital card (SD-Card)

For the communication with the secure element application protocol data units (APDU) are used. The communication interface is standardized in the ISO standard for smart cards, ISO 7816. In order to link the secure element to the NFC Chip and the host controller, different interfaces are considered:

- Single wire protocol (SWP) to link a Universal Integrated Circuit Card (UICC) to the NFC chip
- SigIn-SigOut-Communication (S2C) for linking the NFC chip with embedded secure elements or SD-Cards [19]
- ISO 7816 interface for the communication between the host and the secure element.

Java Card OS is a widespread platform for the secure element which is already an industry standard for chip cards and is getting more and more popular for SIM cards as well.

15.3.3 Host Controller

The host controller has to manage the NFC chip on the one side (e.g., reading/writing data to a tag, switching between application modes) as well as to communicate with the secure element on the other side (e.g., to view the content of a card application stored on the secure element). The communication between the NFC chip and the host controller is handled by the host controller interface (HCI). The communication between the host controller and the secure element is not specifically standardized, but is based on APDUs as defined in ISO 7816-4. In case the secure element is the UICC, the communication is routed through the cell phone baseband chipset via the radio interface layer (RIL) which holds control over the UICC. In case of the embedded secure element or an SD-Card, it depends on the integration of the hardware within the phone.

Figure 15.3 shows the logical parts of each component and how they communicate with each other.

To save space, the embedded secure element and the NFC chip can be packaged into one chip. The chipset manufacturer NXP for example offers such a product under its PN65 series. This type of chip has a so-called *SmartConnect* architecture (see Fig. 15.4).

Thus the secure element, in this case NXP's SmartMX, can operate in the different modes [13]:

- Wired: In wired mode, the embedded secure element can be accessed by the host controller through the NFC chip. Therefore, an application running on the phone is able to fetch data stored on the secure element. The secure element is not visible to an external reader and therefore card emulation is disabled.

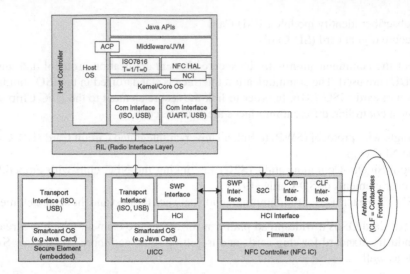

Fig. 15.3 Architecture for the integration of NFC into a mobile phone

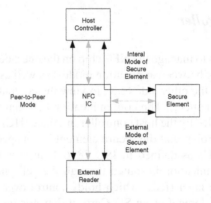

Fig. 15.4 SmartConnect architecture for NFC chips with an embedded secure element [13]

- Virtual: In virtual mode, the embedded secure element represents a virtual card and the NFC chip is in card emulation mode. The secure element cannot be accessed from an application running on the phone.
- Off: In this case the secure element is turned off, the communication with the secure element is not possible at all and the NFC chip can be used in reader/writer mode or for peer-to-peer (P2P) communication.

This is a special form of integrating both, NFC chip and an embedded secure element into a mobile phone. Handset manufactures like Nokia (3220, 6131, 6212) or Samsung (X700n, Nexus S, and Nexus Galaxy) use variants (one chip and two chip solutions) of this architecture in their phones.

15.4 NFC and Linux

The Linux kernel is used in an increasing number of computer platforms and Android is one of the most famous members of the Linux family. Besides Desktop and Server environments, Linux is used in many embedded devices such as Internet routers and connected consumer electronics. NFC has the potential to greatly enhance the user experience for such devices and so far, several initiatives have been started to integrate NFC into Linux environments:

- Native Android support: Google has integrated NXP's FRI library with a kernel driver for the PN544 chip. The whole SW architecture focuses on NXP NFC hardware and is maintained by NXP and Google. The software supports all main NFC features: reader/writer mode, secure element support for card emulation, basic SWP support, and peer-to-peer mode.
- Open NFC for Android: This is similar to the native approach, but with the focus on Inside Secure hardware. Android phone manufacturer would be responsible for the integration of Open NFC on their own products [8].
- libnfc: libnfc implements NFC functionality completely in the user space and thus depends on the existing drivers. Furthermore, the library is platform independent. It supports reader/writer mode and peer-to-peer functionality [22].

All the above-mentioned initiatives are incomplete and either miss out particular features or are focused on certain hardware. This adds justification to a new approach which is starting to implement NFC support into the linux kernel: Linux NFC Subsystem as of kernel 3.1 [14]. The NFC Linux Subsystem follows the principles of Linux open source projects:

- Vendor independent (drivers are needed for new hardware)
- Portable Operating System Interface (POSIX) compliant
- Sockets and Netlink for data exchange and device control

The first included drivers are for devices based on NXP chips PN533 (via USB) and PN544 (via I2C).

At the moment, the Linux Subsystem provides only limited functionality, but work is ongoing to eventually support all NFC features, including full support for card emulation. Currently, Nokia and OpenBossa [7] are working on this implementation goal.

15.5 NFC Integration in Android

In the following section, the software architecture for the integration of the NFC hardware architecture into an operating system will be described. This is shown for Google's Android operating system and the NXP NFC hardware platform (native Android support). To have a sound understanding on what the NFC chips exactly do, an overview of the components is given first.

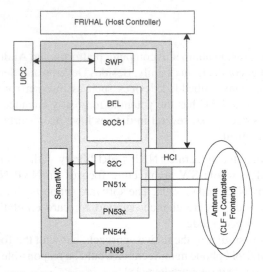

Fig. 15.5 Different chip variants of NXP's PN-family

15.5.1 NFC Chip

The Nexus Galaxy ships with a PN65 chip (see Fig. 15.5) which contains different hardware and software components, such as [20]:

- A PN512 NFC transmission module for contactless communication at 13.56 MHz.
- A micro controller (80C51 core with 32 kbyte of ROM and 1 kbyte of RAM) running the firmware for the PN512 transmission module. The combination of the micro controller and the PN512 is also called PN531.
- An additional interface and software stack to use a SIM card as the secure element. Therefore, the chip needs to implement the so-called SWP.
- A secure smart card chip. In this case, a P5CN072 Secure Dual Interface PKI Smart Card Controller, SmartMX, which can be used as the embedded secure element. This secure element is running a Java Card OS.

The chip used in this phone supports both an embedded secure element as well as a secure element implemented within the SIM card. The software running on the host is thus able to send commands to the NFC chip through the host controller interface in order to user either the embedded or SIM card secure element to emulate a virtual smart card.

From an integration point of view, it does not make any difference if the handset manufacture uses the PN544 or the PN65 as both chips have the same interfaces and use the same pin layout. The only difference is the SmartMX with is included in the PN65 which cannot be directly contacted from outside the chip.

15.5.2 API for the NFC Chip

The NFC software stack running on the host of an Android OS-based device is called the forum reference implementation (FRI) and is already part of Android since Version 2.3 (Gingerbread). The stack is implemented in pure ANSI C and communicates with the */dev/pn544* device of the Android variant of the Linux kernel. On top of the native NFC software stack, there is a Java native interface (JNI) layer that builds the bridge to the Android Java development environment for the Android developer. Finally, the android system development kit (SDK) provides Java APIs which can be used by any app running on the device in order to communicate with the NFC chip in the phone. This API can be used for reader/writer mode, P2P mode, detecting external fields, or targets as well as switching on and off the card emulation mode (Fig. 15.6).

On the J2ME platform, there is already an API standardized for this purpose: Contactless Communications API (JSR257). This JSR was released in 2006 and describes the necessary interfaces in order to allow contactless transactions with a J2ME application running on the handset. Thus, this API makes use of the read-er/writer mode as well as the NFC peer-to-peer mode. The JSR257 already implements the near field communication data exchange format (NDEF) and the basic record type definitions (RTD) published so far by the NFC-Forum [12]. Unfortunately, the Android implementation and the J2ME implementation are not compatible.

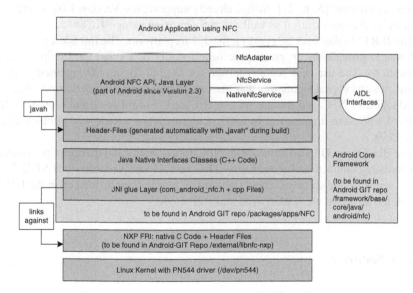

Fig. 15.6 Software stack for the integration of an NFC chip into a handset; by the example of the Google Nexus S/Nexus Galaxy

15.5.3 API for the Secure Element Access

Additionally, an API is required to access the secure elements in the phone. Accessing the embedded secure element within the PN65 can be done through the *SmartConnect* architecture and the FRI. The embedded secure element first needs to be switched into wired mode. Then a communication channel has to be established. After that APDUs can be sent to the smart card chip to read and write information from the secure memory.

Accessing the SIM card involves more software (SW) layers. The host controller of an Android OS-based device cannot directly talk to the SIM card, but must use functions from the radio baseband controller which finally connects to SIM Card. Thus, the host controller needs to send commands to the radio interface layer (RIL) which then talks to the SIM Card. As the RIL is a proprietary implementation and full control of the UICC is not mandatory for Android, it is up to the phone manufacturer to support the necessary RIL functions. The open source project secure element evaluation kit (SEEK) from G&D is for example providing tutorials and an open source stack for accessing the SIM card from an Android application. Phone manufacturers can use this module and integrate it into their Android variant. Google investigates the integration of full UICC access into the official Android code. So far the official implementation is not available.

The GSM association (GSMA) as well as the SIM Alliance agreed on an API for accessing the secure element as well as the security mechanisms using the UICC as the secure element [5, 6, 21]. SEEK already supports the Version 1.01 of the SIM Alliance's API specification as well the authentcation using PKCS#15.

The JSR177 takes over this part on the J2ME platform. The intended goal of this API was to provide the cryptographic functionality of a smart card chip to J2ME applications. The use of a secure storage for Digital Rights Management (DRM) certificates and digital signatures was also a use case during the definition. With the introduction of NFC and the use of a smart card chip for tag emulation, this API received a boost in importance. In 2007, a maintenance release was published [11] (Fig. 15.7).

The Blackberry OS also comes with NFC functionality and therefore provides an API for using contactless functionality. This new API is available since SDK 7 and allows access to the secure element (which can be either embedded or in the UICC) as well as the use of the reader/write and P2P mode.

15.5.4 Security

When looking at security of NFC different aspects are relevant. There are different threat models and attack scenarios for NFC usecases [15]. The most valuable information is stored in the secure element. Hence, this component is implemented as a separate hardware chip in the mobile device. Access to the secure element is possible

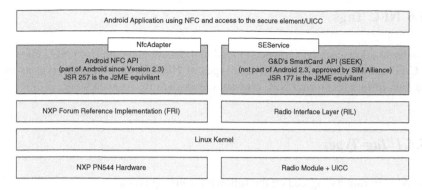

Fig. 15.7 APIs for accessing the different functionalities in an NFC-enabled handset

through the contactless interface of the NFC chip or through an application running on the host controller.

Accessing data in the secure element usually requires the appropriate keys. The most common authentication between an external reader and a secure element is a three pass mutual authentication using symmetric keys. After the authentication, a secure channel is established which allows the two parties to exchange data in a secure way. Although the data stream is routed through the NFC chip, eavesdropping information at this point is useless as the communication is encrypted.

Accessing data in the secure element from an application running on the device is the big advantage of NFC in comparison to usual smart cards. The communication is possible through SEEK (Android) or the JSR177 (J2ME). As these APIs provide access to the secure element of the device special care must be taken in order to restrict the access to those APIs.

All applications using these restricted APIs must be signed with an appropriate certified key. This mechanism is called access condition policys (ACP) enforcement. The ACP is part of the operating system and validates the signature of the application running on the host. In this case, there are certificates (PKCS#15) on either the SIM card or on the phone that are used to validate the signatures of applications that wish to access the secure element.

As the ACP is part of the operating system and therefore implemented in software it can be modified. Especially for systems which are available as open source (e.g., Android), the ACP is easy to disable . Nevertheless, it provides at least an additional barrier to accessing and hacking the UICC from malware and the attacker/customer would also have to root his device and flash a custom ROM into it.

For J2ME, there is an attack method which abuses the fact that there is no byte code verification of an application installed on the device. Thus through modifications in the byte code an application is able to access resources which normally are not available (e.g., accessing the filesystem) [4].

15.6 NFC Tags

In order to allow each NFC device to read and decode the data from NFC Tags, the NFC Forum has defined four different types of NFC Forum compliant tags as well as a data format for storing NFC relevant data structures on such a tag.

15.6.1 Tag-Types

The NFC Forum has agreed on the following four tag types.

Type 1: Type 1 Tag is based on ISO/IEC 14443A. This tag type is read and re-write capable. The memory of the tags can be write protected. Memory size can be between 96 bytes and 2 kbytes. Communication Speed with the tag is 106 kbit/s. Example: Innovision Topaz

Type 2: Type 2 Tag is based on ISO/IEC 14443A. This tag type is read and re-write capable. The memory of the tags can be write protected. Memory size can be between 48 bytes and 2 kbytes. Communication Speed with the tag is 106 kbit/s. Example: NXP Mifare Ultralight, NXP Mifare Ultralight C

Type 3: Type 3 Tag is based on the Japanese Industrial Standard (JIS) X 6319-4. This tag type is pre-configured at manufacture to be either read and re-writable, or read-only. Memory size can be up to 1 Mbyte. Communication Speed with the tag is 212 kbit/s. Example: Sony Felica

Type 4: Type 4 is fully compatible with the ISO/IEC 14443 (A & B) standard series. This tag type is pre-configured at manufacture to be either read and re-writable, or read-only. Memory size can be up to 32 kbytes; For the communication with tags APDUs according to ISO 7816-4 can be used. Communication speed with the tag is 106 kbit/s. Example: NXP DESfire, NXP SmartMX with JCOP.

The specifications for the tag types are available for free from the NFC-Forum website [1]. Note that Mifare Classic is not an NFC forum compliant tag, although reading and writing of the tag is supported by most of the NFC devices as they ship with an NXP chip. Due to its reported security weaknesses, the NXP Mifare Classic should be regarded as obsolete and not recommended for new systems [18].

15.6.2 NFC Data Exchange Format (NDEF)

The NFC forum has defined a structure for writing data to tags or exchanging it between two NFC devices. The format is called NDEF. A so-called NDEF message can contain multiple different NDEF records also referred to as record type definitions (RTD). An NDEF message has to contain at least one RTD. An RTD is an information set for a single application, as an RTD may only contain isolated information such as text, a uniform resource indicator (URI), Multipurpose Internet Mail Extensions (MIME) media type, a business card or pairing information for other technologies. The different RTD specifications are available from the NFC Forum website. NDEF is a binary data format with a TLV (tag/length/value) structure. The maximum size of a standard NDEF record is 2^{32}-1 Bytes. As lots of NFC applications do not need so much data, the NDEF specification defines a so-called short record with a maximum length of 255 Bytes. Payloads of NDEF records can include nested NDEF messages or chains of linked chunks.

An NDEF record includes three parameters to describe its payload [17]:

- The payload length: The payload length indicates the number of bytes in the payload.
- The payload type: The NDEF payload type identifier indicates the type of the payload. NDEF supports URIs, MIME media type constructs, and an NFC-specific type format as type identifiers. By indicating the type of a payload, it is possible to hand over the payload of the records to the appropriate application on the NFC device.
- The payload identifier: The payload may contain an absolute or relative URI as the payload identifier. The use of an identifier enables payloads that support URI linking technologies to cross-reference other payloads.

The structure of an NDEF record is shown in Fig. 15.8. The header additionally includes the following parameters in the first byte:

- Message begin (MB): Indicates whether this is the first NDEF records of the NDEF message or not.
- Message end (ME): Indicates whether this is the last NDEF records of the NDEF message or not.
- Chunk flag (CF): The chunk flag bit can be set to segment the payload into multiple record with are serialized with one message.

Fig. 15.8 Structure of an NDEF record

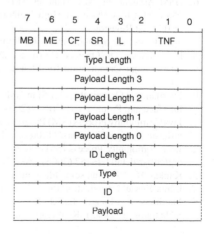

- Short record (SR): The short record bit is set to '1' in case the record size is not longer than 255 bytes.
- ID length (IL): This bit is set the record contains the payload identifier and payload identifier length.

15.7 Conclusion

NFC integrates sophisticated RFID and smart card technology into mobile devices. Although the industry has been pushing the technology through the NFC Forum since 2003, it seems to be the integration of NFC into Google's Android platform (in 2010) that has finally pushed the technology into the consumer market. Thus, NFC is on the verge of becoming a ubiquitous technology like bluetooth and WiFi.

The combination of existing contactless applications such as credit card payment and the upcoming NFC capabilities like P2P provides the basis for complete new interaction models between the virtual and physical worlds.

References

1. Nfc-forum. http://www.nfc-forum.org/
2. Bishwajit, C., Juha, R.: Mobile Device Security Element. Mobey Forum, Satamaradankatu 3 B, 3rd floor 00020 Nordea, Helsinki/Finland (2005)
3. Dillinger, O., Langer, J., Madlmayr, G., Muehlberger, A.: Near field communication in embedded systems. In: Proceedings of the Embedded World Conference 2006, vol. 01, p. 7 (2006)
4. Gowdiak, A.: Java 2 micro edition (j2me) security vulnerabilities. In: Proceedings of the Hack in the Box Security Conference (2004)
5. GSM Association: GSMA NFC UICC Requirement Specification Version 2.0. GSMA London Office, 1st Floor, Mid City Place, 71 High Holborn, London WC1V 6EA, United Kingdom, 2.0 edn. (2011). 2st Revision
6. GSM Association: NFC Handset APIs and Requirements v2.0. GSMA London Office, 1st Floor, Mid City Place, 71 High Holborn, London WC1V 6EA, United Kingdom, 2.0 edn. (2011). 2st Revision
7. openBossa Inc.: openbossa website. http://www.openbossa.org/ (2011)
8. InsideSecure: The Open NFC Project, (2011). http://www.open-nfc.org/
9. International Organization for Standardization: ISO/IEC 14443 Part 1–4: Proximity cards (2003)
10. International Organization for Standardization: ISO/IEC 18092: Near Field Communication - Interface and Protocol (NFCIP-1) (2004)
11. Java Community Process (SM) Program: Java Security and Trust Services API (SATSA). http://java.sun.com/products/satsa/ (2004). JSR177 Final Release
12. Java Community Process (SM) Program: Java Contactless Communications API. http://jcp.org/en/jsr/detail?id=257 (2006). JSR257 Final Release
13. Kunkat, H.: NFC und seine Pluspunkte. Electronic Wireless 01, 4–8 (2005)
14. Linux Kernel Organization Inc.: The Linux Kernel Archives, (2011). http://www.kernel.org/
15. Madlmayr, G., Langer, J., Schaffer, C., Scharinger, J.: Nfc devices: Security and privacy. In: S. Jakoubi, S. Tjoa, E.R. Weippl (eds.) Proceedings of the 3rd International Conference on

Availability, Reliability and Security, vol. 03, p. 6. DEXA Society, IEEE Computer Society (2008)

16. Michahelles, F., Thiesse, F., Schmidt, A., Williams, J.R.: Pervasive RFID and Near Field Communication Technology. IEEE Pervasive Computing 6(3), 94–96, c3 (2007). doi:http://doi.ieeecomputersociety.org/10.1109/MPRV.2007.64

17. NFC Forum: Nfc data exchange format (ndef). www.nfc-forum.org/resources/ (2007). Letzter Zugriff am 10.3.2008

18. Nohl, K.: Cryptanalysis of crypto-1. http://www.cs. virginia.edu/ kn5f/Mifare. Cryptanalysis.htm (2008). Letzter Zugriff am 12.12.2008

19. NXP: S2C Interface for NFC (2005). http://www.nxp.com

20. NXP: PN65 — Near Field Communication (NFC) SmartConnect Module in a single package (2006). http://www.nxp.com

21. SIMalliance Limited: Open Mobile API specification V2.02. SIMalliance Limited, 29/30 Fitzroy Square, London W1T 6LQ, 2.02 edn. (2011). 2st Revision

22. Verdult, R., Conty, R.: libnfc.org - Public platform independent Near Field Communication (NFC) library, (2011). http://www.libnfc.org/

Chapter 16
The BIOS and Rootkits

Graham Hili, Keith Mayes and Konstantinos Markantonakis

Abstract There exist many documents, guidelines and application-level programs attempting to secure various operating systems (OS), but there is much less documentation and software for protecting lower levels subsystems such as the Basic Input Output System (BIOS). Security professionals are well aware that the security on any system is as strong as its weakest link as an attacker will seek to break into a system with the least amount of effort. In this chapter we will focus on the BIOS, and describe its main functions as well as the potential for attacks and countermeasures. After discussing the BIOS and analysing how it might be compromised, we will go on to consider rootkits. Installing a rootkit is often the next stage of an attack once the BIOS has been compromised, allowing the attack to take full control of the target system. We will discuss what rootkits actually are, how to identify that a system has been infected with a rootkit, and how to try and prevent such attacks in the first place. It should be note that the issues raised in this chapter have also provided justification for specialist hardware security measures such as the Trusted Platform Module (TPM) [1–3] described in Chap. 4.

16.1 The BIOS

The Basic Input/Output System, sometimes referred to as the "boot loader" or "system loader", has a very important function. This piece of software, which is

G. Hili (✉) · K. Mayes · K. Markantonakis
Information Security Group, Smart Card Centre, Royal Holloway,
University of London, London, United Kingdom
e-mail: hili.graham@gmail.com

K. Mayes
e-mail: keith.mayes@rhul.ac.uk

K. Markantonakis
e-mail: k.markantonakis@rhul.ac.uk

K. Markantonakis and K. Mayes (eds.), *Secure Smart Embedded Devices,*
Platforms and Applications, DOI: 10.1007/978-1-4614-7915-4_16,
© Springer Science+Business Media New York 2014

located within an integrated circuit embedded on the motherboard of personal computers or servers, is the first piece of software that is executed when the system starts up. The initial function of the BIOS is to perform a power-up test on hardware components. After these tests have been performed, it will initialise the hardware to a known state. The last step performed by the BIOS is to search and locate a valid OS on a secondary storage device and boot it up.

16.1.1 The BIOS Subsystem Functionality

The BIOS performs various power-up tests on the system hardware. For examples, the Random Access Memory (RAM) presence, capacity and parity of stored data may be checked. The BIOS may also perform tests to identify that certain important hardware components such as a keyboard are connected to the system. If the entire set of tests is successful, the BIOS will move on to the next step and initialise all hardware to a default level.

After the hardware has been properly checked and initialised, the BIOS will try to locate the Operating System (OS). The BIOS needs to keep information about the connected devices to find the OS. This information is stored on the BIOS integrated circuit in a non-volatile memory, often called the BIOS RAM. If a valid OS is located, it will be transferred into the main computer RAM and eventually booted up. From this point forward the OS has full control of the system. The OS will carry out further system customisation, identification and initialisation of connected devices, such as network cards or smart card readers. At this stage, OS services and third-party applications start running.

The BIOS typically provides a basic User Interface (UI). This UI is usually menu driven, but on newer BIOS versions an enhanced windowed graphical user interface (GUI) is provided. The UI/GUI application can be used to modify the BIOS settings. Such settings can include, but are not limited to, general BIOS information, devices that are eligible for booting the OS, and various BIOS parameters such as system time and BIOS password.

In older systems, such as those using MS-DOS [4] the OS was loaded in real mode. In real mode the OS and applications (See Fig. 16.1) could access any part of the memory as there was no enforced memory protection and segregation. In this mode BIOS fulfilled another very important function, being the only component used to perform input and output operations to and from secondary storage. More recent OS implementations are more secure as they work in protected mode. In this mode the memory is segmented into different areas so that the memories used by applications and OS are isolated from one another. Furthermore, in protected mode the input and output to secondary storage is done via dedicated drivers and subsystem components instead of the BIOS. Therefore on newer systems, the BIOS lie idle once the OS has been booted, as indicated in Fig. 16.2.

The BIOS has continued to evolve over the past few years with the introduction of virtual machines and faster network connectivity. Newer and more advanced BIOS

Fig. 16.1 BIOS in real mode

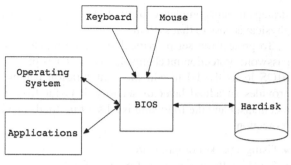

Fig. 16.2 BIOS not used in newer systems after OS is loaded

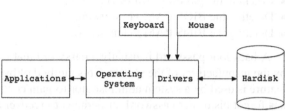

systems can be configured to load virtual machines instead of just one local OS. This allows for different OS to coexist on the same computer.

On the newest systems the conventional BIOS is effectively replaced with a boot process called Unified Extensible Firmware Interface BIOS (UEFI) [5–7], which provides much more flexibility for virtual machine integration and hardware auto-detection, as well as faster boot time. Despite this advance, the term BIOS is still widely used in the computer industry to refer to the piece of software that boots up a particular system or server.

16.2 Attacks on the BIOS Subsystem

There is no shortage of individuals and organisations that would attack all aspects of IT systems so what actually gets attacked depends on the likely benefit to the attacker. Therefore, the obvious question is why would an attacker attempt to compromise the BIOS subsystem? The BIOS provides limited functionality and does not hold important user data, however, it controls which OS is loaded on a particular system. This is a very interesting and powerful target for an attacker, who may seek to load an alternative and perhaps malicious OS instead of the original.

As previously mentioned, the BIOS have the start-up system configuration parameters stored in the BIOS non-volatile memory. One important parameter used by the BIOS during boot time is the location of the original OS. Changing this parameter alone would result in a different OS being loaded. An attacker could use such a strategy to gain complete control of a system, especially if the attacker has temporary physical access to that system.

To protect the set of critical stored parameters, a traditional BIOS includes a password protection mechanism, so the user is prompted for a password before the BIOS loads the UI to edit the configuration parameters. Although the password provides an added layer of security, it is far from being an adequate protection mechanism and the following is a list of methods that have been used to bypass this protection:

- Using a backdoor password.
- Cracking the password BIOS via software.
- Deleting the BIOS RAM via software.
- Deleting the BIOS RAM via hardware.

A backdoor password is usually a password that is built into the BIOS software by the manufacturer of the system. This method of bypassing the password security feature is used by a system administrator to gain control over the BIOS after a user has forgot his or her password, or perhaps to recover a machine after a disgruntled employee has changed the BIOS password before leaving a company. Developers often provide such a method to recover the BIOS in user computers and the passwords can usually be found in the BIOS or motherboard manuals. For more advanced systems, such as servers, the administrator typically had to call the manufacturer and give the machine ID or serial number in order to get the password.

Another method for legitimately recovering the BIOS password is to use a software application that can either be downloaded from the BIOS or motherboard vendor websites. The password has to match with the locally stored copy and in older BIOS versions this was stored in clear text within the BIOS RAM. Newer BIOS versions store only the hash of the password, which has added some security, but perhaps not to the extent that one might think. The BIOS has limited memory for storing information and so only part of the hash value may be used, which increases the potential for collisions (multiple data sets that have the same hash value). Cracking a password using a specialist software utility, typically takes anything from few seconds to few minutes, depending on the processor being used. A search on the Internet can find several utilities including software used on the Microsoft Windows OS e.g. CMOSPWD [8] and !BIOS [9].

Some of the software used to crack passwords can also be used to clear the BIOS configuration parameters completely, as is the case with CMOSPWD, which alone could wreck the normal operation of the system. Unfortunately, a malicious attacker will not think twice about clearing the BIOS configuration data of a target machine. Another common way to clear the BIOS configuration is to remove the on-board system battery. The BIOS parameters are stored in a small RAM and this only represents a non-volatile memory while a continuous source of power is supplied. When the system is switched off, the RAM relies on a small back-up battery and so if this is removed or disconnected the stored configuration data will be lost. Note that this strategy would not work if the BIOS non-volatile storage was based on EEPROM or Flash memory.

The most popular means of clearing the BIOS configuration data is via software, as this method is less invasive and means the attacker does not need to tamper with the hardware or necessarily be physically close to the target machine. It is also worth noting that certain motherboard vendors provide specific proprietary ways to clear the BIOS data. For example, on older Toshiba laptop systems, connecting a special dongle to the parallel port could clear the BIOS configuration parameters.

16.2.1 Countermeasures to BIOS Attacks

It is difficult to protect against physical BIOS attacks without a trusted hardware module [10, 11] of some kind (see Sect. 3.6). If an attacker has physical access to the system, he could probably clear the BIOS data in one way or another. The first protection mechanism that is simple yet effective is to enclose sensitive systems in special metal enclosures that discourage physical tampering. These enclosures will make a physical attack more difficult and time consuming, but cannot prevent a determined attacker from succeeding. Obviously drawbacks to such protection include the added cost, space and weight requirements.

Another control that could be in place is to make the BIOS boot only from a certain secondary storage device. This can eliminate the fact that compromised OS can be loaded on a system. Unfortunately, this is a BIOS specific functionality and not supported by all systems.

16.3 Rootkits

In this section we will begin to explore rootkits, although in these few pages we can only skim over what has become a major topic in information security research, with attack methods evolving all the time.

16.3.1 Introduction to Rootkits

The term rootkit is actually made up of two words; "root" and "kit". The former has it origins in UNIX-like OS, where logging in as root on gives the user full administrative control over the system. This is useful for administration purposes, such as installing a new system wide component or adding a new user account or user group. The term "kit" is commonly used by system administrators to refer to a set of applications that carry out a certain task or administrative process. Therefore, the term rootkit implies a set of applications that potentially give a user unlimited control over a system.

A quick search on the Internet will find various rootkits, both in source code and executable formats, for various OS. A reliable search for such code can be conducted on the SANS website [12]. There are also rootkit-creation applications; a common one called Mpak [13] (the site always changes due to the nature of the software). If you look deeper, you will even find service companies offering to develop custom rootkits for specific OS. In fact, there is a thriving black market on the Internet for the sale of either full rootkits or zero day attacks that can be converted into rootkits. The trading sites are difficult to find and access is usually only granted to people who are formally introduced by some existing member of the "community".

The choice of rootkit will depend on the system OS. For example, a famous rootkit called t0rn [14] was targeted for Linux [15] platforms and effected common system administration applications. A particular rootkit may have several components to carry out its intended purpose. For example, a rootkit will need a process to install the kit (installation phase), which will usually be an application layer process. After a successful installation, the rootkit will delete any temporary files that where used or downloaded in order to cover its tracks (clean-up phase). The installed rootkits are usually complicated pieces of software with many interconnected sub components, ranging from OS to application layer components, and are designed to remain undetected. One such example of how a rootkit manipulates the OS or applications to remain undetected is by modifying the application used to list the directories and files on a system. When a user invokes the compromised application to list files and directories, the listing will not show files related to the rootkit itself.

Some newer rootkits can also work at the kernel level. Attacking the kernel level would ensure that the rootkit will have complete control over the OS. The rootkit would then be able to load and delete OS modules as necessary. An example would be to show the user that an antivirus module has been loaded in main memory, but in fact the rootkit loaded a fake antivirus.

Rootkits are now beginning to target hypervisors [16–19]. A hypervisor is a virtual OS manager, this is used to host several OS on a single system. This technique is becoming very popular on high-end servers as it allows more resources of a system to be used concurrently. Targeting the hypervisor is a very powerful use of a rootkit and an attacker can then present a completely different system to a user than what is actually going on in the background. Rootkit attacks on hypervisors is another major research topic and a detailed discussion is beyond the scope of this chapter. In the remainder of the chapter we will focus on rootkit infection of traditional systems.

16.4 Rootkit Infections

In this section, we will introduce some concepts related to the threat of rootkit infection. There is a misconception by some users and administrators that because they have an antivirus application and/or firewall they are secure from such threats. Such applications will detect many attacks, but they are far from fool proof and can still get "rooted" (rootkit infected).

For machines that are connect to the Internet, an attacker will likely prefer to mount a remote attack rather than trying to physically compromise the machine. There are

Fig. 16.3 Example of a drive by download attack

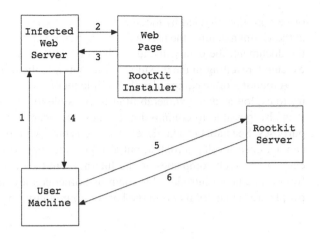

several strategies for remote attacks such as "drive by download", "phishing", or even directly attacking the network. Here we will just cover some common attack methods and the reader is referred to [20, 21] for more detailed coverage.

One common remote attack starts by modifying an existing and legitimate web-server accessible via the Internet. For example assume that website http://ABCXYZ.com is a very popular site. The attacker will first attack the webserver where this site is hosted. Once this webserver is under the control of the attacker (gaining control usually involves putting a rootkit on the server), he will insert some malicious scripts that will run when a user lands on the target website.

When a user's web browser accesses the familiar/legitimate site http://ABCXYZ.com (Steps 1 and 2 in Fig. 16.3), the infected server will download a small payload script (Steps 3 and 4) that will run on the user machine to download the full rootkit to the user machine (Steps 5 and 6).

Once the rootkit is on the user machine it will start executing. Usually, the first objective is for the rootkit to install itself in a safe area of the hard disc. After this is achieved, the rootkit will start running and modifying the components available on the local machine. Modifications include, but are but not limited to, removing the anti virus protection, opening particular ports on the local machine and modifying the OS so that the user will not notice such changes. Once the rootkit is fully installed, the attacker will have full control of the machine and will usually also install spyware software on the victim machine. The spyware will harvest the user's sensitive and private information, such as passwords, PINs and account details etc.

Another approach is to exploit vulnerabilities found on common applications that are installed on target machines. One such attack was to exploit a vulnerability found in Adobe Acrobat Reader that executed a harmful script contained in PDF documents [22]. The attacker had to craft a special portable document format file (PDF). This file was meant to appear legitimate, for example it could be a document containing information regarding popular topics, a game tutorial or a business document such as in invoice. The document could be sent to the victim via an email message or perhaps downloaded by the user from a web server. When this document arrives at the victim machine, the user will try to open it via Acrobat Reader. When opening the document, the data will cause a buffer overflow and a malicious script will be executed, resulting in the download of a rootkit to the target machine.

A rootkit is often equipped with multiple methods to infect a particular OS and will try to exploit as many vulnerabilities as possible to increase its chances of success. It is also possible to commission the development of custom rootkits for specific purposes. Specialist rootkit developers are typically motivated by financial gain and can not only develop a rootkit, but also provide support for their software! It is not easy to find such companies on the Internet, but they can be found by looking into forums and being introduced to them by someone they trust. Usually, these trusted people are called "runners" in the hacker community lingo.

16.4.1 Detection of Rootkits

Detecting a rootkit on an infected machine is not an easy process[23–25]. This is due to the fact that if the OS is compromised, it and the applications that depend on it, will not behave as intended. Output from commonly used system administration commands, will be altered. For example, when compiling the list of processes on a compromised system, the rootkit will not report all running processes. Another example would be when listing all opened ports; the compromised ports used to control the rootkit will not be listed.

The big question is of course is how can we identify something that is designed to hide its existence? There have been applications written to help system administrators to identify rootkits, but there is no sure-fire solution for all scenarios. Here we will describe some software tools that can be used and briefly mention how to identify a rootkit infection via other methods such as behavioural analysis. Again, the methods listed here are not the only ones that can be used, and the list is by no means exhaustive.

The most important thing to keep in mind if you suspect your OS has been infected by a rootkit is not to trust your loaded OS *at all*. Having an OS on a CD-ROM is very handy as it should be read-only and serves the purpose of providing a method to load an OS that is fully under our control and trustworthy. A common live CD (a full OS on a CD-ROM) used for such a purpose is BackTrack [26], which uses a powerful Linux-based OS.

Searching the Internet for rootkit removers will yield a large number of results. Some of these results are from well-known companies such as Sophos [27] and Symantec, although there is no software that will discover all rootkits. This is due to the fact that rootkit developers also know about these tools and they write their rootkits to be as stealthy as possible. Another problem is that rootkit removers from free or not well-known suppliers might themselves be malicious.

The most common way to check for rootkits on a system is by using integrity checking methods, which in principle can be reliable and generate few false positives. To carry out such a check we need to have a clean bootable and trusted OS, be it on a USB stick or preferably on CD-ROM. We boot from this media to have a clear and trusted OS before checking the integrity of files on the local hard disc. Of course these days we are accustomed to loading OS and application patches on an almost daily basis and so the difficulty of maintaining the latest reference integrity check values should not be underestimated.

An integrity check will typically make use of a simple cryptographic hash algorithm, a common hash algorithm that has been used for such a purpose is MD5 [28] due fast performance even on large files. The MD5 result from a file can be cross checked with a value that we know is correct and perhaps obtained via a trusted reference site. It is important to note that the MD5 algorithm is no longer recommended for use and so there is potential for attackers to target the hash values that are meant to help safeguard the system security.

Two common software applications used for integrity checking are called Tripwire [29] and AIDE [30]. They are both widely used in corporate environments and reasonably well regarded by information security experts. The tools compute the file integrity checks for the target machine and then cross reference them with hashes from clean systems or from the website of the OS vendor. However, rootkits might also modify such software applications to redirect them to malicious sources of hash results making them report that everything is OK even though the rootkit is installed on the system.

The common integrity check method, although very easy to use and very reliable, has its limitations. One of the limiting factors is that this method is not intended to be used on the user's data files. This is due to the fact that these types of files may change very frequently from normal usage, whereas executable and application files should only change when there is an upgrade. Therefore, the rootkit checking tools may simply ignore the directories where users store their data, which can lead to another problem. If the rootkit is hidden somewhere in a user data file, then even if we manage to clean the system after reloading the OS, the users will start working again with their infected files and the payload for the rootkit may be downloaded yet again.

Another method to detect rootkits is to dump the main memory of a system while it is running the compromised OS. The rootkit will have to be loaded like any other process when it infects an OS. The fact that it is not shown by normal OS functions does not mean it is not there. If a memory dump is done, we can then analyse this dump file in a debugger on a trusted system. If the memory dump can be captured correctly then the method is effective and any rootkit should show up. The problem with such a method is that it is very specialised, requires low-level knowledge of the OS, and is time consuming. Analysing a memory dump is not at all trivial, as it will contain thousands of lines in machine code. Modern OS implementations have applications to capture dumps; for example, Windows has the sysdm.cpl application, but again, this method should not be trusted if you suspect that the system has been compromised. The best way is to use an external application and run it from a CD-ROM. If you are trying to capture a Windows memory dump, a tool to use is DebugDiag [31] from Microsoft. There exist other third party tools that can be found by doing a search on the Internet, although caution is advisable when considering tools from little known suppliers.

A further method that can be used to detect the presence of rootkits is the behavioural analysis of applications. Again, this is a specialised and very time-consuming process involving analysis of running applications perhaps within a debug environment. For example, if we are using a local standalone application, the network transmissions could be a strong indication that the system has been compromised. Unfortunately modern applications are increasingly complex with many OS API calls and links to external servers, so rootkit analysis is difficult and can lead to a large number of false positives. A free Windows debugger is called OllyDBG [32], but many alternatives are available.

16.4.2 Removal of Rootkits

The removal of a rootkit from a system is not at all an easy process and it can consume a significant amount of expert IT admin support. There is no shortage of tools, both free and commercial; however, none of them work on all rootkits. A starting point could be Sophos Anti-Rootkit [33], which has received some good reports and it is free for personal use.

The complexity of today's OS implementations and applications make it extremely difficult to completely understand if a system call is legitimate or not. Many experts in the field of rootkits believe that the most reliable solution to a rootkit compromised systems is simply to re-install the OS. However, it should be appreciated that reloading the OS alone does, not make the machine "clean" and it is necessary to sanitise all the user data to prevent against re-infection.

Most documentation on how to clean rootkits recommends disconnecting the system from the network, which might not always be feasible, especially if the system is supporting mission critical activities that are unaffected by the rootkit. Furthermore, disconnecting the system from the network might change how the rootkit behaves. For example, once the rootkit cannot detect an active network it may stop sending data to the network interface.

Of course the best way to defend against rootkits attacks is to try very hard to keep them out in the first place, by observing best practices and policies for IT/Information security. This may include making sure that the users of the system do not run applications in super user mode, which gives them full control of the system. While in normal user mode, a rootkit will not have full privileges to write to the protected area of the OS.

Common Internet security software such as antivirus and firewall provides a worthwhile protective barrier against rootkit installation and can remove some infections, but it cannot be expected to defend against zero day attacks. When using Internet security software it is important to use all the features that might help to protect the system against rootkits. For example, an email verification process could be used on a mail server to try to identify problems and suspicious behaviours in incoming mail messages. A secure proxy layer might also be used to identify potential malicious websites.

It should not be overlooked that the user is an important part of the security barrier. Education is important as an informed and motivated user is a security asset whereas a naive and/or careless user is a liability. If users are on their guard when opening suspicious emails from unknown senders and can refrain from clicking on odd web links and email attachments then there should be fewer rootkit infections. Furthermore, users can be encouraged to report any "odd" IT behaviour that might help with rootkit detection. Unfortunately, due to the sophisticated and evolving nature of the threat, user education will not eradicate the problem and even the best experts are caught out from time to time.

16.5 Conclusion

In this chapter, we started by introducing the BIOS as a low-level software component used to boot up computer systems. We emphasised the vital importance of the BIOS and that if it is compromised it could undermine the security protection provided by the OS and applications. We also looked at ways that the BIOS could be attacked and how we might protect against this.

We introduced the subject of rootkits and how they are used (effectively) by attackers to take control of a system, often as a follow on to a successful BIOS attack. We considered example methods (that were not totally satisfactory) for the detection, and removal of rootkits; not forgetting the fact that the user has a significant role to play in the prevention of malware infection.

There has been a lot of past research into attacks and countermeasures exploiting the BIOS and/or rootkits and this chapter provided just a brief introduction. However, it is worth noting that much of the past work has been based on computers in the conventional form of PCs and laptops; whereas the future may be more dominated by mobile phones. We are already seeing great interest from legitimate users in rooting their phones to bypass controls that they consider to be unwanted or unfair. Mobile phones are increasingly complex, rushed to market and in general have a range of OS types that are often based on proprietary and secretive solutions. The prospect of mobile phones being targeted for more malicious and possibly fast spreading BIOS/rootkit attacks seems quite probable. Fortunately, the situation is not hopeless and the interested reader is referred to Chap. 4 which describes how specialised trusted hardware could be used to address the rootkit problem.

References

1. Mitchell, Chris, ed. "Trusted computing." Institution of Electrical Engineers, 2005.
2. Pearson, Siani, and Boris Balacheff. Trusted computing platforms: TCPA technology in context. Prentice Hall PTR, 2003.
3. Grawrock, David. "The Intel safer, computing initiative." ISBN-10976483262 (2005).

4. An Inside Look at MS-DOS, Tim Paterson, http://www.patersontech.com/Dos/Byte/InsideDos.htm.
5. Intel Web Site, Defining the interface between the operating system and platform firmware, http://www.intel.com/technology/efi/.
6. Hu, Yin, and Haoyong Lv. "Design of Trusted BIOS in UEFI Base on USBKEY." Intelligence Science and Information Engineering (ISIE), 2011 International Conference on. IEEE, 2011.
7. ZHOU, Zhen-liu, et al. "Research and Implementation of Trusted BIOS Based on UEFI." Computer Engineering 8 (2008): 062.
8. CmosPwd Website, http://www.cgsecurity.org/wiki/CmosPwd.
9. Bios320 download site, http://www.technibble.com/downloads/misc/BIOS320.exe.
10. Ghaleh, Hossein Rezaei, and Shahin Norouzi. "A new approach to protect the OS from off-line attacks using the smart card." Emerging Security Information, Systems and Technologies, 2009. SECURWARE'09. Third International Conference on. IEEE, 2009.
11. Hendricks, James, and Leendert Van Doorn. "Secure bootstrap is not enough: Shoring up the trusted computing base." Proceedings of the 11th workshop on ACM SIGOPS European workshop. ACM, 2004.
12. System Administration, Networking, and Security Institute, http://www.sans.org/.
13. MPac Article, By Robert Lemos, SecurityFocus, http://www.theregister.co.uk/2007/07/23/mpack_developer_interview/.
14. System Administration, Networking, and Security Institute, What is t0rn rootkit?, Paolo Craviero, http://www.sans.org/security-resources/malwarefaq/t0rn_rootkit.php.
15. What Is Linux: Overview of the Linux Operating System, http://www.linux.com/learn/new-user-guides/376-linux-is-everywhere-an-overview-of-the-linux-operating-system.
16. Rutkowska, Joanna, and Rafa Wojtczuk. "Preventing and detecting Xen hypervisor subversions." Blackhat Briefings USA (2008).
17. Gavrilovska, Ada, et al. "High-performance hypervisor architectures: Virtualization in hpc systems." Workshop on System-level Virtualization for HPC (HPCVirt). 2007.
18. Leinenbach, Dirk, and Thomas Santen. "Verifying the microsoft hyper-v hypervisor with vcc." FM 2009: Formal Methods (2009): 806–809.
19. Microsoft, Introduction to the Hypervisor in Windows Server 2008, http://www.microsoft.com/en-us/server-cloud/hyper-v-server/overview.aspx.
20. Stone-Gross, Brett, et al. "Your botnet is my botnet: analysis of a botnet takeover." Proceedings of the 16th ACM conference on Computer and communications security. ACM, 2009.
21. Mavrommatis, Niels Provos Panayiotis, and Moheeb Abu Rajab Fabian Monrose. "All your iframes point to us." (2008).
22. Adobe Systems, Adobe Security Bulletin, http://www.adobe.com/support/security/advisories/apsa09-01.html.
23. Levine, John, Julian Grizzard, and Henry Owen. "A methodology to detect and character-ize kernel level rootkit exploits involving redirection of the system call table." Information Assurance Workshop, 2004. Proceedings. Second IEEE International. IEEE, 2004.
24. Kruegel, Christopher, William Robertson, and Giovanni Vigna. "Detecting kernel-level rootkits through binary analysis." Computer Security Applications Conference, 2004. 20th Annual. IEEE, 2004.
25. System Administration, Networking, and Security Institute, RootKit Investigation Procedures, Sans Reading Room, http://www.sans.org/score/checklists/rootkits_investigation_procedures.pdf.
26. BackTrack, Linux Security Distribution, Offensive Security, http://www.backtrack-linux.org/.
27. Sophos Ltd official website, http://www.sophos.com/.
28. A study of MD5 Attacks: Insight and Improvements, J. Black, M. Cochran, T. Highland, http://www.cs.colorado.edu/~jrblack/papers/md5e-full.pdf.
29. TripWire (Community Version) official website: http://www.tripwire.org/.
30. AIDE official website: http://aide.sourceforge.net/.
31. Microsoft, Debug Diagnostic Tools version 1.1, http://www.microsoft.com/download/en/details.aspx?displaylang=en&id=24370.

32. OllyDbg Debugger, Official Website http://www.ollydbg.de/.
33. Sophos Anti root kit personal edition, http://www.sophos.com/en-us/products/free-tools/sophos-anti-rootkit.aspx.

Chapter 17
Hardware Security Modules

Stathis Mavrovouniotis and Mick Ganley

Abstract Hardware Security Modules/(HSMs), also known as Tamper Resistant Security Modules (TRSMs), are devices dedicated to performing cryptographic functions such as data encryption/decryption, certificate management and calculation of specific values such as card verification values (CVVs) or Personal Identification Numbers (PINs). What these devices offer is tamper response, the capability to detect any attacks on their surface and securely delete the sensitive content stored in their memory. Such devices are manufactured to meet specific criteria [e.g. Federal Information Processing Standard (FIPS)] and must be appropriately managed throughout their whole lifecycle. Together with encryption algorithms, cryptographic functions and vendor provided functionalities, they host one or more cryptographic keys that respond to automated or manual commands. Physical security and key management are essential in order to protect the confidentiality and integrity of the keys and these requirements are properly described in various standards. Due to the specific functionality of HSMs, there have been many published attacks via the command interface, which reinforces the need for adequate controls, both physical and logical, around these devices.

17.1 Introduction

The first question that needs to be addressed is what is meant by a Hardware Security Module (HSM)? In order for a device to be classified as an HSM, it must belong to

S. Mavrovouniotis (✉)
20 Pandrosou Str, P. Faliro, 17564 Athens, Greece
e-mail: mavrovouniotis@gmail.com

M. Ganley
Information Security Group, Smart Card Centre, Royal Holloway, University of London, London, United Kingdom
e-mail: mick.ganley@rhul.ac.uk

K. Markantonakis and K. Mayes (eds.), *Secure Smart Embedded Devices,
Platforms and Applications*, DOI: 10.1007/978-1-4614-7915-4_17,
© Springer Science+Business Media New York 2014

the family of Tamper Resistant Security Modules (TRSM) or Secure Cryptographic Devices (SCD), which are physically secure devices and/or tamper responsive, meaning that any attempt at penetration of the device will cause immediate erasure of all secret information stored in the memory of that device [1]. An HSM is any hardware device, with some level of tamper-resistance,[1] which is used for cryptographic processing. Of course, this rather broad definition would include smart cards and other devices that are discussed elsewhere in this book. It would also include, for example, devices used at the network level to provide high-speed encryption, devices for issuing/signing certificates for a Certification Authority (CA), devices using for time stamping, etc. Another good example is that of retail Point of Sales terminals (POS terminals) used for processing "Chip and PIN" transactions, which have a security core that is frequently referred to as an HSM.

The HSM itself is either a peripheral device to the host computer or bus-connected. Nowadays, most peripherally connected HSMs communicate with the host machine via Ethernet or fibre cable, but in the past a variety of communication protocols were usually supported; customers would choose the protocol that best suited the required transaction throughput and available budget. As well as having a port to allow communication with the host computer, HSMs usually support a variety of other input/output methods, for example smart card reader, key pad, a dedicated management port, printer port, a CD/DVD drive to allow software loading, or a console cable in order to perform the HSM management or the key ceremonies. In simple terms, therefore, an HSM has many of the characteristics of a PC, the main differences being the limited functionality and the physical and logical security of the device, which will be described later in the chapter.

17.2 HSM Usage

Clearly an HSM can be used in any situation where high-grade cryptographic security is required. An HSM is most commonly a hardware device or a Payment card industry (PCI) card that responds to commands sent to it by an application, via a vendor-specified application programming interface (API). It is generally straightforward to modify the API to meet customer requirements; this may be done by the customer or, more usually, by the HSM vendor. HSM software/firmware is digitally signed, either directly or indirectly, using a vendor private key and verified using the corresponding public key installed in the HSM as part of the manufacturing process. Examples of the use of HSMs include the protection of personal data (e.g. health records, databases, etc), bulk encryption (e.g. satellite broadcasting) and trusted third-party services (certificate authorities, signature authorities, etc).

A typical 3-tier architecture containing an HSM is depicted in 17.1.

---

[1] The term "tamper resistant", in this context includes "tamper-evident" and "tamper-detective" that will often appear in this chapter and which can be used interchangeably, as well as "tamper responsive" which refers to the reaction of the device in a tamper attack.

Fig. 17.1 A typical 3-tier
architecture including an
HSM

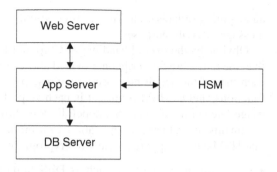

A message for the application server would be sent to HSM for cryptographic processing, prior to being sent to the next tier of the process. In a case of encryption of financial record, the data would be received by the web server and would then be passed to the application server. The application server would then, with the aid of the HSM, encrypt the data to pass on to the DataBase server to be stored.

In a case of a financial transaction, the application server would pass the financial data to be verified to the HSM, which would calculate, on the fly, a cryptographic value [(such as a Card Verification Value (CVV)] with the corresponding key Card Verification Key (CVK) and compare it with the value provided in the transaction, in order to approve or decline the transaction. As stated earlier, we will focus on HSMs used in the financial sector. Such an example is their use in "Chip and PIN" payment cards, such as a debit or credit cards. The two principal areas where HSMs are used with such cards are:

- data preparation, card personalisation and Personal Identification Number (PIN) mailer printing, as part of the issuing process;
- transaction processing.

In terms of card personalisation, a variety of secret or sensitive values need to be generated and loaded onto the card; these include a number of cryptographic keys (symmetric keys used during transaction processing and asymmetric keys, together with certificates, used for authentication purposes), as well as cryptographic values, such as a PIN and CVVs (CVV / CVV2 / iCVV). These values are typically generated with the use of specific cryptographic keys PIN Verification Key (PVK) for PIN, CVK for CVV values—we will call them functional keys in general) during the data preparation process (the data generation creates the demographic data) and then transmitted to the card personalisation system for loading onto the card. The PIN itself may need to be transmitted to another organisation for printing on some form of PIN mailer. The protection of such data, during generation, transmission and loading/printing, is provided by HSMs.

During transaction processing, at least in the online case, an HSM is used by the card issuer to ensure the integrity of transaction messages. In the particular case of the card being used at an Automatic Teller Machine (ATM) to withdraw cash, the PIN is encrypted by the ATM and sent to the issuer for verification, possibly via an acquiring organisation. The acquirer would use an HSM for PIN translation and message integrity purposes.

Obviously, the overall handling of Chip and PIN cards is far more complicated than the rather brief description provided above. Each part of the issuing process and each part of the acquiring process requires specific keys for specific cryptographic functions, and an HSM is present in each step of this process. We will discuss key usage later in the chapter, but hopefully the above serves to illustrate HSM usage within this single financial application. Continuing with the above example, it is clear that HSMs must support a variety of cryptographic mechanisms, such as:

- cryptographic algorithms, such as DES (although no longer accepted), 3-DES, AES, RSA, SHA-1, SHA-256;
- cryptographic key management, including key generation, key derivation, key distribution, key storage, etc;
- data encryption, in particular PIN encryption techniques;
- data integrity, including Message Authentication Code (MAC) generation and verification, digital signature generation and verification;
- CVV generation and verification;
- PIN generation and verification, PIN translation;
- chip cryptographic keys generation and transmission, cryptographic values generation and verification.

A typical HSM used solely in financial applications will support all of the above functionality including many PIN generation/verification, PIN block formats, key management techniques, MAC algorithms and encryption modes. The sheer number of commands that are supported by an HSM, including many that are not actually used by the application(s), raises a number of security concerns that we will investigate later in this chapter.

Whilst it may be tempting simply to provide a range of "primitive" functions for an HSM and allow application developers to build more complex functions from these primitives, this would have an adverse effect on performance and could have serious implications for the security of command processing. Security evaluation of such a solution would also be difficult, as discussed later in this chapter. Instead, an HSM API will typically include many rather complicated functions, often with a range of options. For example, in the Chip and PIN application discussed earlier, a single HSM command used by an acquirer could involve PIN translation, including a format change, data decryption using one key and then re-encryption using a different key (even possibly a different mode of encryption) and MAC verification followed by the generation of a new MAC.

The typical format of a command for an HSM used in a payment transaction would be:

- Command header

  - Command code
  - Command data

Table 17.1 Sample HSM command and responses messages

Command message (with comments)	
Header	Typically for use by the calling application
Command code	Unique command identifier
Source PIN encryption key	Encrypted using a local key
Destination PIN encryption key	Encrypted using a local key
Source MAC key	Encrypted using a local key
Destination MAC key	Encrypted using a local key
Source PIN block format	e.g. ISO 9564 format 0 [2]
Destination PIN block format	May be the same as the source format
Source MAC mode	e.g. ISO 9797 algorithm 3 [3]
Destination MAC mode	May be the same as the source format
PIN block data	(Optional) depends on PIN block formats
PIN block	Encrypted with the source PIN key
Message data	Transaction data
Source MAC	MAC on message data, using source MAC key
Command trailer	(Optional)
Response message (with comments)	
Header	Typically for use by the calling application
Response code	Unique response identifier
Error code	e.g. 00 = no errors
PIN block	Encrypted with the destination PIN key
Destination MAC	MAC on message data, using destination MAC key
Response trailer	(Optional)

- Command trailer

The corresponding response message is:

- Response header
  - Response code
  - Error code
  - Response data

- Response trailer

So, for example, an acquirer command that involves a PIN translation and a MAC verification and new MAC generation is described in Table 17.1:

In the above command, errors may occur in a variety of ways, for example:

- key parity errors, if using DES or 3-DES (stored keys may have become corrupted);
- an invalid PIN block format or MAC mode;
- a PIN block error (e.g. format of the plaintext PIN block does not match the specified format);
- MAC verification failure;
- command syntax error.

In the event of an error being detected the HSM should return an appropriate error code and continue to the next transaction. As already mentioned, a crucial function of an HSM is the protection of secret data, in particular cryptographic keys and PINs. In a following section, we consider HSM key management in more detail.

17.3 HSM Physical Security

The purpose of an HSM is to provide high-grade cryptographic security and a crucial aspect of this security is the physical security of the device. It must be emphasised, however, that this is only one aspect of HSM security—attacks via the HSM's API and procedural aspects of HSM security are equally important. Indeed, it could be argued that an HSM's physical security is the easy part; a physical attack is likely to be detected quickly, whereas a logical or procedural attack might never be detected!

An HSM's primary defence against physical attack is based around the concept of a tamper-resistant core, which is an HSM sub-system that contains all the sensitive components. As the most common approach, the security core will provide battery-backed volatile memory for the storage of plaintext cryptographic keys (such as the Host Master Key (HMK), discussed later) and all cryptographic processing will be performed within the core system. The tamper-resistant features of the core sub-system mean that should attack on the core be detected then the contents of the secure memory will be immediately deleted ("zeroised"). Typically, HSM software/firmware is stored in a combination of ROM and E2PROM and so will not be deleted if the device is tampered. The ANSI X9.24-1 standard [4] mandates the following:

> An HSM must have features that resist successful tampering, which includes penetration without zeroisation of security parameters, unauthorised modification of the HSM's internal operation or insertion of tapping mechanisms or non-intrusive eavesdropping methods to determine, record or modify secret data; such features must include one or more of the following:

- tamper-detection mechanisms, which must be active regardless of the HSM's power state;
- physical barriers to make successful tampering infeasible;
- sufficient resistance to tampering, so that successful tampering requires an extended period of time (absence of an HSM from its authorised location should be noticed before the tampered device is returned to resume cryptographic operations);
- the HSM's construction is such that successful tampering will cause visible damage to the device that is likely to be noticed after the device has been returned to its authorised location but before cryptographic operations are resumed—i.e. a tamper-evident feature.

Immaterial of the use of an HSM, because of its nature, the physical controls around it are very strict. This would most commonly mean the HSM is located within a high security area, probably locked inside a secure cabinet, under dual physical controls (each cabinet door would require two controls—for example, two keys, a key and a combination or a key and a biometric) or a similar and equally effective approach. Consequently, an attacker would find it very difficult to remove an HSM from its normal location without detection—unless he is an employee with physical access and rights to do so anyway—this is why the dual control should be enforced and actually the second person should not by default be entitled to be around the specific cabinet. However, the same may not be true for other types of HSM (e.g. the security core of a retail PIN pad). In general, therefore, the primary defence of an HSM against physical attack is the tamper-detection circuitry, which must zeroise secure memory as soon as an attack is detected.

Attacks that must be defended against include:

- drilling or otherwise penetrating the security core;
- low temperature attacks;
- attacks involving variations in voltage or current;
- power analysis or timing attacks (members of a class of attacks known as "side channel" attacks).

Typical defences against such attacks include wrapping the entire security core in some form of fine-grained electronic mesh and then encasing the core in epoxy resin. An attacker attempting to penetrate the resin is likely to break the mesh. If the mesh is broken or damaged in some way then the zeroisation circuitry is immediately triggered.

Other HSM defences include the use of physical locks, micro-switches, light-sensitive diodes, mercury tilt-switches, temperature sensors and sensors to detect variations in voltage and current. Side channel attacks are unlikely to be successful unless the attacker is able to penetrate the core (in which case a side channel attack is probably unnecessary!) and in any case HSM vendors usually build in defences against such attacks. Important note: Some of these controls can be enabled or disabled, so it must be stressed that they can protect the HSM only when activated. An HSM is considered as an HSM only if it has these controls activated!

Many early HSMs used by the financial industry had only rudimentary tamper-detection mechanisms, often no more than a couple of micro-switches to detect when an HSM's casing was opened. This could be easily by-passed by an attacker and so in such circumstances the physical and access security of the computer centre environment became the principal defence against an attacker. Nowadays, HSM security is usually evaluated against standard requirements and we now move on to consider such evaluations.

17.4 HSM Security Evaluation and Approvals

Although there are a number of standards detailing security requirements for cryptographic modules, for example ISO 13491 [5, 6], most HSMs used in the financial sector are evaluated against the requirements of the Federal Information Processing Standard (FIPS) 140-2 [7]. Devices used in some government applications may also need to be evaluated against the Common Criteria requirements (e.g. [8]).

More recently, a PCI standard for HSM security has been published (PCI-HSM, see [9]) and we will briefly consider this standard at the end of this section. For the time being, however, we will concentrate on FIPS 140-2. The FIPS 140-2 standard ("Security Requirements for Cryptographic Modules") specifies security requirements in 11 different areas and covers 4 different security levels, with level 1 being the lowest and level 4 being the highest. Each level builds on the previous level. The following table, copied from the FIPS 140-2 standard summarises the requirements for the different levels:

The term Operational environment refers to the management of the software, firmware and/or hardware components required for the module to operate. The abbreviations PP and EALx refer to the Common Criteria Protection Profile and Evaluation Assurance Level x, respectively (see [8]).

There is little purpose to be served in a detailed discussion of FIPS 140-2 in this chapter, but the following points are noted:

- devices approved to FIPS 140-2 level 1 or level 2 provide limited protection for cryptographic keys and other sensitive data; such devices are not appropriate for many financial applications; indeed, the example given in the standard of a device that could achieve level 1 approval is a PC encryption board;
- HSMs used in financial applications are typically approved to level 3 or 4; note however that for particularly sensitive applications the physical security requirements of level 3 may not be acceptable except in secure environments;
- there is a large "gap" between the level 3 and level 4 requirements, in particular the requirement for a formal model for design assurance; some HSMs meet the level 4 requirements in many areas and yet only receive approval to level 3;
- FIPS 140-2 evaluation does not consider side-channel attacks, such as power analysis, nor does it include command manipulation attacks, based on the device's API; this latter topic has already been discussed briefly and will be considered further later in this chapter. Level 3 is the approved level by all payment schemes security requirements.

Level 4 approval is hard to achieve and currently (early 2012) very few products have been approved to this level:

Those products with certificate #235 and lower were evaluated against the earlier FIPS 140-1 standard. A complete list of all FIPS 140 approved products can be found at [10]. The list gives an overall security level for each approved product, but also includes those areas where the overall level has been exceeded.

As previously mentioned, the PCI-HSM standard [9] has recently appeared and lists its security requirements in 4 categories (Tables 17.2 and 17.3):

- A: Physical security
- B: Logical Security
- C: Device Security during Manufacture

Table 17.2 Summary of FIPS 140-2 security requirements

	Level 1	Level 2	Level 3	Level 4
Cryptographic module specification	Specification of cryptographic module, cryptographic boundary, approved algorithms and approved modes of operation; description of cryptographic module, including all hardware, software and firmware components; statement of module security policy			
Cryptographic module ports and interfaces	Required and optional interfaces; specification of all interfaces and of all input and output paths		Data ports for unprotected critical security parameters logically or physically separated from other data ports	
Roles, services and authentication	Logical separation of required and optional roles and services	Role-based or identity-based operator authentication	Identity-based operator authentication	
Finite state model	Specification of finite state model; required states and optional states; state transition diagram and specification of state transitions			
Physical security	Production grade equipment	Locks or tamper-evidence	Tamper detection and response for covers and doors	Tamper detection and response envelope; EFP or EFT[a]
Operational environment	Single operator; executable code; approved integrity technique	Referenced PPs evaluated at EAL2 with specified discretionary access control mechanisms and auditing	Referenced PPs plus trusted path evaluated at EAL3 plus security policy modelling	Referenced PPs plus trusted path evaluated at EAL4
Cryptographic key management	Key management mechanisms: random number and key generation, key establishment, key distribution, key entry/output, key storage and key zercisation			
	Secret and private keys established using manual methods may be entered or output in plaintext form		Secret and private keys established using manual methods may be entered or output encrypted or with split knowledge procedures	
EMI/EMC[b]	47 CFR FCC[c] part 15, subpart B, class A (business use) applicable FCC requirements (for radio)		47 CFR FCC part 15, subpart B, class B (home use)	

Table 17.2 Summary of FIPS 140-2 security requirements

	Level 1	Level 2	Level 3	Level 4
Self-tests	Power-up tests: cryptographic algorithm tests, software/firmware integrity tests; critical function tests; conditional tests			
Design assurance	Configuration management (CM); secure installation and generation; design and policy correspondence; guidance documents	CM system; secure distribution; functional specification	High-level language implementation	Formal model; detailed explanations (informal proofs); preconditions and postconditions
Mitigation of other attacks	Specification of mitigation of attacks for which no testable requirements are currently available			

[a] Environmental failure protection and environmental failure testing
[b] Code of Federal Regulations (CFR) and Federal Communication Commission (FCC)
[c] Electromagnetic Interference/Electromagnetic Compatibility

Table 17.3 Products approved to FIPS 140-2 overall level 4

Certificate #	Vendor	Product
1505	IBM	IBM 4765 cryptographic coprocessor security module
1340, 956, 235, 146, 123, 112	AEP	Networks advanced configurable cryptographic environment (ACCE) various versions
1174, 930	Hewlett Packard	Atalla cryptographic sub-system (ACS)
661, 524	IBM	IBM eServer cryptographic coprocessor security module
541	AEP	Networks AEP enterprise CM
118	IBM	IBM eServer zSeries 900 CMOS cryptographic coprocessor
116	IBM	IBM 4758-002 PCI cryptographic coprocessor (miniboot layers 0 and 1)
115	Thales	Secure generic sub-system (SGSS)
40	IBM	IBM S/390 CMOS cryptographic coprocessor
5	IBM	IBM 4758 PCI Cryptographic coprocessor (miniboot layers 0 and 1)

• D: Device Security between Manufacture and Initial Key Loading

Many of the requirements for physical security are derived from the level 3 FIPS 140-2 requirements, although requirement A2 includes some side-channel attacks, such as power analysis. The logical requirements are generally more strict than the corresponding FIPS 140-2 requirements, in particular the key management requirements. Of particular interest, however, is requirement B9, which states:

"The HSM's functionality shall not be influenced by logical anomalies such as (but not limited to) unexpected command sequences, unknown commands, commands in a wrong device mode and supplying wrong parameters or data which could result in the HSM outputting the clear-text PIN or other sensitive information."

We will return to this topic later in this chapter.

Currently, only three products are listed on the PCI web site [11] as having been approved against the PCI-HSM requirements, namely the, HP Atalla Ax160, Thales payShield 9000 and Tokheim Crypto HSM+ devices.

17.5 HSM Management

Under normal operating conditions, HSMs are intended to work without any manual intervention. However, there are many HSM activities that require some form of human input, for example:

• HSM installation and initialisation, including generation of the highest level local key (the HMK, if used);
• define users and corresponding authorisations;

- generation, import/export and installation of other keys;
- configuration; for example, communication parameters and security policy;
- state changes, such as putting the HSM off-line to the host or requiring special authorisation for sensitive functions (e.g. key loading);
- enabling and disabling of commands;
- enabling and disabling of PIN block formats;
- audit functions;
- diagnostics and problem solving;
- other tasks, such as those relating to real-time clock management (e.g. "set time");
- firmware and other software (e.g. licence files) updates.

Such activities have in the past required direct access to the HSM, via a dedicated management port, while some HSM vendors now offer a "remote" solution for managing HSMs. This has many advantages in terms of personnel and time, especially when trying to manage a large number of geographically dispersed HSMs. Remote access requires additional security mechanisms to be in place, in particular mutual authentication between operators and HSMs and confidentiality of transmitted data.

Regardless of the actual mechanism employed, all management activities must be governed by detailed and rigorously enforced procedures. Security incidents are far more likely to occur because of poor management than an attacker somehow compromising a physically secure HSM located in a data centre.

An HSM's security policy can be configured to cover items such as:

- number of HMK components—as described later in the chapter;
- minimum number of components for plaintext key entry;
- enabled commands and PIN block formats;
- denying the use of single-length DES;
- minimum PIN length—for the incoming PIN from a transaction;
- various key export/import options (e.g. ANSI X9.17 not permitted);
- types of keys that can be exported or imported;
- permitted number of key check value characters;
- permitting clear PINs to be input or output, for example when PIN translation is performed;
- data encryption and decryption options;
- audit options;
- authorisation options.

The above configuration options are reasonably standard across most HSMs used for the processing of financial transactions, but vendors will typically offer a range of other configuration possibilities, for example:

- preventing a single-length DES key masquerading as a double- or triple-length key;
- encrypted decimalisation tables, i.e. tables used to map hexadecimal characters to decimal digits;
- weak PIN checking;

- minimum HMAC length for verification;
- specific restrictions on individual commands.

One area of concern for HSM management relates to the entry of plaintext key components which are combined to form a secret value, such as a cryptographic key. Such components are typically received from a partner organisation, are in paper form and must be entered by (trusted) security officers into the HSM, inside which they are combined to form the clear key, which is then output encrypted under the HMK. The issue is the entry of the components, which is usually done via some form of terminal, such as a PC, which leads to the possibility of the components being captured during entry (e.g. via a key logger or some form of device connected to the communication line to the HSM). In the past, such concerns have been mitigated by strict procedural controls but nowadays the payment industry is demanding that key components be entered via a more secure mechanism. For instance, requirement 13 of the PCI-PIN security requirements [1] states:

"The mechanisms used to load keys, such as terminals, external PIN pads, key guns, or similar devices and methods are protected to prevent any type of monitoring that could result in the unauthorized disclosure of any component."

One possibility would be the use of a secure retail PIN Entry Device (PED), approved against the PCI-PED security requirements [12]. HSM vendors are now actively seeking ways to meet this particular requirement. We will discuss the key management procedures in more detail in the next chapter.

17.6 Key Management

As mentioned before, different keys are used for different cryptographic processes and the whole of the proper functioning of the cryptographic model relies on the protection and proper use of the keys, which is the main principle of cryptography, as mentioned in all cryptography related publications. An HSM is essentially a cryptographic engine and it serves no useful purpose if secret or private cryptographic keys are exposed to an attacker during command processing. Hence, such keys must never appear in plain form outside the secure confines of the HSM. There is one exception to this rule, namely that if a key is required to appear in plain form outside the HSM then it must be in the form of two or more components and strict procedures must be followed to enforce the principles of dual control and split knowledge—ensure that the components cannot exist in the hands of one individual at any point in time. PCI-PIN Security Requirements [1] together with payment schemes standards, provide specific requirements about how the safety of the participating keys is preserved, during all phases of a key management lifecycle. In order to address this part of the chapter, we split the keys into three main categories as below:

1. Storage keys: keys such as the HMK, which is used to encrypt other keys when stored.

2. Transport keys: keys that are used to encrypt keys during key exchange, e.g. a Key Encrypting Key (KEK).
3. Functional keys: keys used to perform specific cryptographic functions and generate respective cryptographic values, such as PVKs, CVKs, chip authentication keys, PIN block encryption keys, etc.

There are two principal methods for protecting keys used by an HSM:

Method 1. Store all keys inside the secure memory of the HSM; in this case, when sending a command to the HSM a pointer to the key to be used must be included in the command message. This technique has one significant drawback, namely that if the HSM is tampered and loses its keys then all the keys must be reloaded into the HSM. In addition, if multiple HSMs are used (for reasons of throughput and/or redundancy) then all the keys must be loaded into each HSM.

Method 2. A single key, which we have already termed a "HMK" is loaded into the HSM and all other keys are encrypted with the HMK and stored in some form of key database accessible to host applications; this database can exist either on the database server of a 3-tier model, or as a file within a mainframe system.

Of course, if the HSM loses its HMK then the key must be reloaded into the HSM, but unlike method 1 this is the only key that needs to be reloaded. The drawback in method 2 is that should the HMK be somehow compromised then all the other keys in the system are potentially compromised as well. For this reason, strict procedures must be in place to protect the HMK, including plaintext storage of the key in component form under dual control and split knowledge. The HMK must be a "strong" key, for example a triple-length 3-DES key or an AES-256 key. It goes without saying that, regardless of the HSM key management technique, all keys that are stored inside the HSM should be backed-up. Interestingly enough, there is no mandate for backing up keys, only that if the keys are to be backed-up the same controls used for the production keys should be also enforced on the backup keys (e.g. requirement 27 of the PCI-PIN standard [1]).

Remark on terminology: Method 2 is the most commonly used HSM key management technique. This means that in general:

- function specific keys must be transferred encrypted under a transport key, such as a KEK;
- function specific keys and transport keys must be stored encrypted under a HMK, summarised in the Table 17.4:

Protecting the confidentiality of keys is one issue that is addressed by the methods described above, but it is equally important to protect the integrity of such keys. In particular, it must not be possible for an attacker to modify a key or to use a key for a purpose for which it is not intended. The requirements that "keys must be used only for their sole intended purpose" and that "cryptographic keys ever present and used for any function must be unique (except by chance) to that device" are basic

Table 17.4 Matrix of different types of keys and their storage/exchange

Function/key	Storage keys	Transport keys	Functional keys
Storage	In components	Under HMK	Under HMK
Exchange	Not applicable	In components	Under KEK

principles for protecting the keys and are two very important requirements of PCI standard for PIN Security requirements [1, requirements 19 and 20].

The key management is performed by a team of custodians, chosen and managed in a way that the principles of dual control and split knowledge are met. The role of custodian is crucial—they have access, although controlled, to all cryptographic material, together with physical access to the HSMs. Thus, the custodians must be appropriately trained, the key management procedures must be very well documented, and audit trails must exist and be maintained for every activity relating to key management: key generation, import/export, key storage/retrieval, key back up, key replacement or destruction and arguably most importantly key compromise. Once these basic principles are met and the HSMs as well as the keys are appropriately protected, the chances of key compromise are minimal.

If an attacker were to try and attack the keys themselves he would be looking for the following [13]:

- production keys used in the test environment, allowing the technical support staff to attack the key structure;
- PINs not protected by a secure PIN block, allowing "dictionary" attacks;
- failure to use approved cryptographic devices for PIN processing;
- cryptographic keys non-random, non-unique and never change;
- hard copy keys in the clear or in clear-text halves;
- few, if any, procedures documented; and,
- no audit trails or logs maintained.

We give two simple examples to illustrate the importance of these requirements:

Example 1 Suppose a double-length 3-DES key is encrypted using (some variant of) the HMK in Electronic Codebook (ECB) mode—this is a very common encryption mode and it is analysed in relevant bibliography. The attacker could replace the second half of the encrypted key with the left half of the key, so that the modified key is really a single-length DES key masquerading as a double-length key. The HSM could be used with the modified key to generate sufficient data to allow a brute-force attack on the left half of the key. This could then be used to expose the right half of the original key. This attack can be prevented by a variety of techniques. For example, the HSM could be configured to prevent such a key masquerade, by checking that all parts of a double or triple-length key are different. We will discuss a more generic technique shortly.   □

Example 2 Suppose a key is designated as a PIN encryption key (so, in particular, there is no HSM function that allows the key to be used to decrypt a PIN block).

Header (16 ASCII characters)	Optional Header Blocks (ASCII characters, variable length	Encrypted Key Data (variable length, ASCII encoded)	Key Block Authenticator (8 ASCII characters)

Fig. 17.2 TR-31 key block

If the attacker can change the key usage so that it appears to the HSM as a (generic) data encryption/decryption key then the key could be used to decrypt PIN blocks. One method, whereby, it may be possible to change key usage is via a combination of key export and key import. Until recently, HSM vendors used proprietary methods for local key management but were generally forced to use a "lowest common denominator" approach for exchanging keys among HSMs of different vendors. This approach usually involved exporting a key, encrypted under some higher-level KEK, using the ANSI X9.17 standard [14, now withdrawn), and in most cases when the key was encrypted under the KEK the original key usage could no longer be determined. Consequently, the attacker could easily import the key back into the HSM system as a different key type. HSM vendors have long recognised the importance of key usage and have employed a variety of techniques to ensure that a key is only used for its intended purpose. For example, different key types can be encrypted under different variants of the HMK, in some cases different variants are also used for different key parts. IBM HSMs use Control Vectors to define exactly how keys can be used by the HSM. Whilst such techniques can provide some level of protection for keys, the two examples above illustrate their limitations. □

The ANSI X9.24-1 standard [4] for retail financial key management mandates that keys must, amongst other things:

- be protected against disclosure and misuse;

and that 3-DES keys must:

- exist in a "key bundle";
- be secret and randomly or pseudo-randomly generated;
- have integrity, so that each key in the bundle cannot be altered in an unauthorised manner;
- be used as specified by the particular mode;
- be considered as a fixed quantity, in that it is not possible to manipulate part of the key, and that the key cannot be "unbundled".

Although the standard relates primarily to 3-DES keys, clearly the same requirements make sense for any secret or private key. The ANSI TR-31 standard [15] specifies a technique for meeting the requirements of X9.24-1 via the use of "key blocks". Although TR-31 is specifically for key distribution, it has been adopted and refined by some HSM vendors as a method of protecting local keys when encrypted under the HMK (Fig. 17.2).

The Key Data is encrypted using a variant of the KEK used to protect the key block, in Cipher Block Chaining (CBC) mode, with bytes 0–7 of the Header as the Initial Vector (IV). The Key Block Authenticator is a MAC over the rest of the key

Table 17.5 TR-31 key block header

Byte(s)	Field	Comments
0	Version ID	Value = A; current version
1–4	Key block length	Total length of key block
5–6	Key usage	e.g. key encryption, data encryption
7	Algorithm	e.g. DES, 3-DES, AES
8	Mode of use	e.g. encrypt only
9–10	Key version number	e.g. version of key in the key block or used to indicate that the key is a key component
11	Exportability	e.g. no export permitted
12–13	Number of optional blocks	Number of optional header blocks
14–15	Reserved for future use	Value 00

block data, calculated using a different variant of the KEK. The key block Header governs the use of the key contained within the key block and has the following structure (Table 17.5):

As mentioned, the TR-31 key block mechanism has been adopted and refined by some HSM vendors for local protection of keys using the HMK. Whilst the local use of key blocks will greatly improve the security of HSMs against the type of "key manipulation" attack described earlier, vendors are constantly battling against the need to maintain backwards compatibility for legacy systems and so the benefits of key blocks are nullified to a certain extent. There will still be potential problems involving key manipulation until all HSM vendors have introduced key blocks and legacy systems have been upgraded. This will be hopefully enforced in the next versions of PCI standards and payment schemes security requirements.

In conclusion, if the keys are administered in a proper way, and the HSMs are physically protected, an attacker, as the last resort, will focus on the attacks for command manipulation, which are addressed in the next section.

17.7 Command Manipulation Attacks

The final topic discussed in this chapter is that of HSM attacks based on the HSM's API, which we will designate "command manipulation attacks". This subject has already been mentioned a number of times and it is worth recalling requirement B9 of the PCI-HSM standard: "The HSM's functionality shall not be influenced by logical anomalies such as (but not limited to) unexpected command sequences, unknown commands, commands in a wrong device mode and supplying wrong parameters or data which could result in the HSM outputting the clear-text PIN or other sensitive information."

Two rather simple attacks have already been outlined, namely the use of a single-length DES key masquerading as a double-length key and changing key usage via a combination of key export and key import.

In the latter example, a PIN encryption key had its use changed to that of a data encryption/decryption key to allow PIN blocks to be decrypted. This same attack could also be used to decrypt keys in some earlier models of HSMs. A recent paper by Bortolozzo et al. [16] details a variant of this attack on a number of commercially available devices that support the PKCS#11 API [17].

Before describing some more sophisticated command manipulation attacks, it is worth asking the question as to whether such attacks are feasible in real-life. For example, some assumptions about the attacker need to be made:

- the attacker has detailed knowledge of the HSM and its command structure;
- the attacker is in a position to send commands to a "live" HSM;
- the attacker has access to live data, including keys (encrypted under the HMK) and transaction data;
- the attacker has knowledge of "standard" algorithms but does not generally have knowledge of proprietary techniques used by the HSM;
- the attacker may have physical access to the HSM but is not in a position to carry out sensitive management functions.

Financial institutions argue that these are not realistic assumptions and that physical restrictions, procedural arrangements, host configuration settings and comprehensive audit trails make impossible such types of attack. Whilst this may be true for some organisations, there is absolutely no guarantee that all organisations using an HSM have such stringent security regimes. It is probably the case that an "outsider" would find it extremely difficult to attack the system via command manipulation, hence the likely attacker would almost certainly be an "insider", probably with a number of system privileges.

The logging of HSM transactions in audit trails is potentially a major deterrent to an attacker, but somebody with detailed system knowledge may be able to get round that. If the attacker can directly access the HSM's host port then there may be no audit trail anyway. Finally, one problem with most audit trails is that they contain so much information that nobody bothers to look at them, at least not until it is too late!

So, the above assumptions are probably "not unreasonable" and a number of command manipulation attacks could perhaps be carried out by a privileged "insider". If the reader thinks this is an unduly pessimistic view of the security of HSM systems in financial institutions then he or she should read the article "Why Cryptosystems Fail" by Ross Anderson [18] or Anderson's book "Security Engineering" [19]. Many papers on command manipulation attacks have been written in recent years, with the most comprehensive treatment being given in Jolyon Clulow's MSc thesis [20]. One of the techniques described in this thesis is that of finding a single-length DES key via a "parallel search" technique, initially proposed in [21]. Here, by obtaining the same plaintext encrypted under many different keys, an exhaustive search to compromise one of the keys (but you cannot specify which one) can be speeded up significantly. For example, if 2k single-length DES keys all encrypt the same block of data then an exhaustive key search would expect to find one of the keys after an average of 255-k attempts. This technique forms the basis for a well-publicised attack [22] on the IBM 4758 HSM, summarised below.

Remark 1 The IBM 4758 HSM has been approved to FIPS 140-1 level 4 (see Table 17.3). The attack does not invalidate this approval, as the FIPS 140 security requirements do not cover API attacks. Although the published attack was specifically against the IBM 4758 HSM, which has long since been withdrawn, the basic idea is still applicable to possible attacks against modern devices.   □

Step 1. Use the parallel search technique to obtain the value of a single-length key, KDATA. This requires the use of the Encrypt function to generate the known plaintext/ciphertext pairs.

Step 2. Use the parallel search technique again, to obtain the value of a double-length KEK, which allows key export (KEKEXPORTER). The trick that allows this step is to make both halves of the KEK the same. This time, the corresponding plaintext/ciphertext pairs are obtained by exporting the known key, KDATA.

Step 3. Export all keys using the known key KEKEXPORTER and decrypt them at leisure.

One interesting aspect of this attack was the development of a DES search engine, based on an FPGA at a cost of less than $1,000, to carry out the searches in steps 1 and 2 in, approximately, 24 h for each stage.

A number of papers have been written on attacks on the IBM 3624 PIN verification technique (for example, [20, 23, 24]). This is a standard PIN generation algorithm and is implemented in most HSMs used in card payment systems.

One major benefit of this technique is that there is no need for the card issuer to maintain a database of (encrypted) PINs. Instead, a Customer PIN can always be regenerated or verified from input values, namely Account Related Data (ARD, for example the card's Primary Account Number), a PVK, a Decimalisation Table and an Offset. Specifically, the ARD, PVK and Decimalisation Tables are used to generate a Derived PIN, which is then combined with the Offset, to form the Customer PIN. This algorithm is described in Fig. 17.3. Note that the Derived PIN is a transitory value only—it never appears outside the HSM. In general, HSMs that support this PIN generation method have two specific commands:

- verify a Customer PIN;
- generate an Offset for a given Customer PIN.

The second command is to allow a customer to change his or her PIN. Note that the Derived PIN does not change, so that if a customer changes a PIN then the Offset must change to compensate.

One possible attack on this method (using the second command) is to run the command whilst making successive changes to the Decimalisation Table. Whenever the generated Offset is different from the correct value, the attacker can deduce one or more digits of the Derived PIN, from which the corresponding digits of the Customer

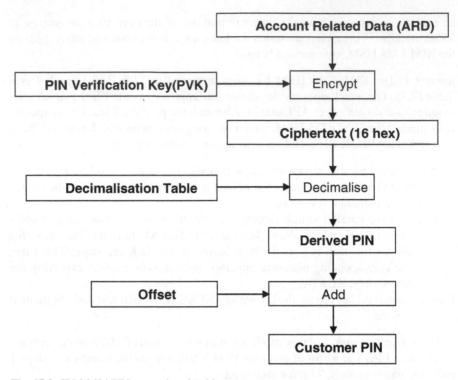

Fig. 17.3 IBM 3624 PIN generation algorithm

PIN can be calculated. This attack requires a maximum of 15 calls to the HSM. A slightly less efficient attack, which uses only the PIN verify command, involves modifying the Decimalisation Table, as above, to ascertain the PIN digits (but not their positions), via returned error codes, and then repeat the process this time also modifying the Offset. On average, this attack will reveal a Customer PIN in about 20 calls to the HSM.

In a third attack, it may be possible to compromise PINs for other customers via the use of insecure PIN blocks, when just one Customer PIN is known. For example, if the PIN for Customer X is known then the PIN for Customer Y could be translated to a PIN block format that does not involve an account number (e.g. ISO 9564, format 1) and then translated back to a format that involves the ARD of Customer X. By using the command to generate an Offset the PIN for Customer Y can be calculated. This attack requires only 3 calls to the HSM. The first two attacks described above can be easily defeated by using an encrypted Decimalisation Table, but this will not defend against the third attack.

One especially clever attack first described in Clulow's thesis (not involving the IBM 3624 algorithm) uses a combination of PIN verify and PIN translate commands to compromise PINs; essentially PINs are "flip-flopped" between different PIN block formats and error codes returned by the PIN verify command can be used to determine the PIN digits. The details of the attack are rather complicated and are not given here, but the interested reader should consult [20].

The above gives only a flavour of the types of command manipulation attacks that are possible. The crucial point is that many of these attacks are quite ingenious and use only "standard" HSM commands, so HSM vendors must be very careful when trying to satisfy requirement B9 of the PCI-HSM standard [9].

What, then, can be done to mitigate such attacks? The following suggestions would at least be a good starting point in addressing the problem:

Enabling and disabling functions: The golden rule should be that only those HSM features that are actually required should be enabled; this includes HSM commands, PIN block formats, PIN algorithms, etc; all other features should be disabled. In particular, enabling the generation of plain text PINs is a major risk, and is the only HSM related risk that is mentioned in a preventive measures paper, published by USSS and FBI [25].

Security policy: The HSM's security policy should be configured as "tightly" as possible, subject to the requirements of applications calling the HSM.

Key blocks: HSM vendors should introduce key blocks as soon as possible and customers should ensure that they are using this feature.

Formal methods of analysis: Some formal approaches to the analysis of HSM command sets have been defined (for example, the previously mentioned paper by Bortolozzo et al. [16]), although the results have been rather "patchy". HSM vendors should think about collaborating in the development of some sort of tool to enable formal analysis.

Vigilance: HSM vendors and HSM users should monitor academic papers describing new command manipulation attacks and (if necessary) modify HSMs as soon as possible to defend against such attacks. In addition, regular analysis of HSM command sets should be conducted by vendors, especially following major new releases or significant customisations.

Procedures: HSM users must ensure that all HSM-relevant procedures are strictly enforced, in particular that no unauthorised access to an HSM's host port is possible and that audit logs of HSM transactions are regularly monitored.

Whilst the above defences cannot guarantee immunity to command manipulation attacks, they would certainly make the attacker's life a lot more difficult.

17.8 Conclusions

In this chapter we have explained what we mean by an HSM, given some usages for HSMs (focusing primarily on the financial sector) and described the key management regime supported by many HSMs. Here we have also described two very simple attacks on the HSM's API that exploit weaknesses in the key management structure.

We then moved on to discuss the physical security of HSMs and HSM security evaluation, in particular against the requirements of FIPS 140-2. One specific requirement of the PCI-HSM standard was also highlighted, essentially that an HSM API should be immune to command manipulation attacks. Following a short discussion on HSM management, we moved back to the topic of API attacks and outlined a variety of (known) attacks that demonstrate the difficulty of meeting the PCI-HSM requirement. We concluded the discussion on command manipulation attacks by suggesting a variety of defences that could be used to reduce the likelihood of such attacks being successful.

References

1. "Payment card industry PIN Security Requirements", version 1.0, September 2011.
2. ISO 9564–1, "Financial services - Personal Identification Number (PIN) management and security - Part 1: Basic principles and requirements for PINs in card-based systems", 2011.
3. ISO 9797–1, "Information technology - Security techniques - Message Authentication Codes (MACs) - Part 1: Mechanisms using a block cipher", 2011.
4. ANSI X9.24-1, "Retail Financial Services Symmetric Key management, Part 1: Using Symmetric Techniques", 2009.
5. ISO 13491–1, "Banking - Secure cryptographic devices (retail), Part 1: Concepts, requirements and evaluation methods", 2007.
6. ISO 13491–2, "Banking - Secure cryptographic devices (retail), Part 2: Security compliance checklists for devices used in financial transactions", 2005.
7. FIPS 140–2, "Security Requirements for Cryptographic Modules", 2001, with some updates in December 2002.
8. "Common Criteria for Information Technology Security Evaluation", see http://www. commoncriteriaportal.org/.
9. "Payment card industry (PCI) Hardware Security Module (HSM) Security Requirements", version 1.0, April 2009.
10. http://csrc.nist.gov/groups/STM/cmvp/documents/140-1/140val-all.htm.
11. https://www.PCIsecuritystandards.org/approved_companies_providers/approved_pin_transaction_security.php.
12. "Payment card industry (PCI): POS PIN Entry Device, Security Requirements", version 2.1, January 2009.
13. "PIN Security Program: Auditor's Guide", version 2, January 2008, see http://usa.visa.com/download/merchants/visa_pin_security_program_auditors_guide.pdf.
14. ANSI X9.17, "Financial institution key management (wholesale)", 1985.
15. ANSI X9 TR-31, "Interoperable Secure Key Exchange Key Block Specification for Symmetric Algorithms", 2010.
16. M. Bartolozzo, R. Focardi, M. Centenaro & G. Steel, "Attacking and Fixing PKCS#11 Security Tokens", ACM Conference on Computer and Communications, Security, 2010, pp. 260–269.
17. PKCS#11, "Cryptographic Token Interface Standard", version 2.20, RSA Laboratories, June 2004.
18. R. Anderson, "Why cryptosystems fail", Proceedings of the 1993 ACM Conference in Computer and Communications Security, pp. 215–227. See also, http://www.cl.cam.ac.uk/users/rja14/wcf.html.
19. R. Anderson, "Security Engineering", (2nd Edition), Wiley, 2008.
20. J. Clulow, "The Design and Analysis of Cryptographic Application Programming Interfaces for Security Devices", version 4.0, M.Sc. Thesis at University of Natal, Durban, South Africa, dated 17 January 2003.
21. Y. Desmedt, F. Hoornaert & J.J. Quisquater, "Several Exhaustive Key Search Machines and DES", EUROCRYPT 86, 1986, pp 17–19.

22. R. Clayton & M. Bond, "Experience Using a Low-Cost FPGA Design to Crack DES Keys", presented at the CHES 2002 Workshop Francisco, 1st August. (http://www.cl.cam.ac.uk/rnc1/descrack/DEScracker.pdf).
23. M. Bond & P. Zieliński, "Decimalisation Table Attacks for PIN Cracking", University of Cambridge Computer Laboratory, Technical Report 560, dated February 2003. (http://www.cl.cam.ac.uk/TechReports/UCAM-CL-TR-560.pdf).
24. R. Anderson & M. Bond, "Protocol Analysis, Composability and Computation"; see http://www.cl.cam.ac.uk/rja14/Papers/bond-anderson.pdf.
25. Joint USSS/FBI Advisory February 2009, see http://usa.visa.com/download/merchants/20090212-usss_fbi_advisory.pdf.

Chapter 18
Security Evaluation and Common Criteria

Tony Boswell

Abstract Security evaluation of embedded devices presents a number of challenges, primarily because the relevant attacks for a particular device are determined by the software application that ultimately runs on or uses services from the embedded device, but the device is often designed and evaluated before details of this application context are known. This chapter examines how the common criteria (CC) security evaluation scheme can be used for embedded devices, and how current directions in the evolution of CC provide a particular opportunity to deal effectively with embedded device security.

18.1 Introduction

This chapter is concerned with security evaluation and certification of embedded devices, and focuses mainly on use of the common criteria (CC) because of its combination of generality (i.e. the ability to deal with a wide range of types of security functionality and security products), and its history of application and evolution in fields relevant to embedded devices. However, comparisons with two other evaluation schemes applicable to some types of embedded device (FIPS 140 and Payment Card Industry, PIN Transaction Security: PCI-PTS) are also given, demonstrating ways in which CC has learnt from other schemes, and how other schemes have adopted aspects of CC.

The sections below first discuss the basics of the embedded device security problem as they relate to security evaluation, then look at a basic model for evaluating embedded devices and then introduce the CC, discussing examples of their application to embedded devices. It is the continuing evolution of CC that is probably of most importance for embedded devices, since it seems likely that they will need

T. Boswell (✉)
SiVenture, Unit 6, Cordwallis Park, Clivemont Road, Maidenhead, Berkshire SL6 7BU, UK
e-mail: tony.boswell@siventure.com

K. Markantonakis and K. Mayes (eds.), *Secure Smart Embedded Devices,*   407
Platforms and Applications, DOI: 10.1007/978-1-4614-7915-4_18,
© Springer Science+Business Media New York 2014

Fig. 18.1 Layering model for
embedded devices

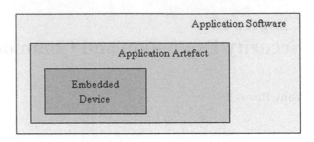

specific interpretation in the same way as for smart cards, building on and extending
work that is already under way in some other areas (such as payment terminals).

For the purposes of this chapter, the basic model of an embedded device and its
use in a larger IT system is illustrated in Fig. 18.1.

The model simply envisages that an embedded device is destined to be built into
some larger user-facing device, referred to here as an application artefact (i.e. a
physical thing such as a media player or mobile phone handset). The artefact in turn
makes application software visible to a user, and enables them to interact with the
application. Ultimately, the user is focused on using one or more applications (rather
than the artefact, or the embedded device). The application software needs to be
delivered to the user in some way, and therefore depends on the application artefact
to do this. Similarly, the application artefact depends on the embedded device to
supply some of its functionality. The security dependencies that arise from this are
illustrated in Fig. 18.2.

This means that the security requirements of the embedded device must also
consider its future context of use. If certification of the embedded device is to be
really useful then the security evaluation must support the needs of the dependent
layers above it.

18.2 Security Evaluation Issues

Because in general an embedded device has its security evaluated independently
of the application context a fundamental difficulty arises: how are the evaluators to
take into account the way in which "the device provides environment" and resources
for an arbitrary unknown artefact? The same uncertainty exists for the evaluation
of the application artefact, which has a definite purpose [such as a secure signature
device, or an integrated circuit (IC) for playing media] but where the particular
application software and its precise assets, peer entities, protocols and surrounding
system security features will typically be unknown at the time of the artefact security
evaluation.

Of course a number of functions provided by the device will be obviously security-
related, and thus likely to be used to implement critical security mechanisms for the

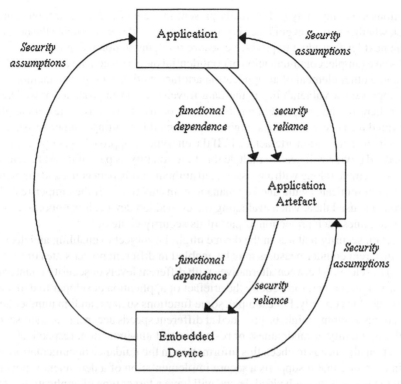

Fig. 18.2 Dependencies between layers in the embedded device model

application—these reflect the functional dependencies in Fig. 18.2. Cryptographic functions, a random number generator and authentication functions in the hardware would be typical examples of these. Not only must these functions be implemented correctly, but they must also operate in a generally secure manner, reflecting the "security reliance" dependencies in Fig. 18.2. The exact use of these functions by the application artefact, and the critical dependencies of the application software may be unknown when the embedded device is evaluated, and yet are clearly important in determining the assurance level that the device can reach, and the scope of functionality that should be analysed.

In evaluating the security of the embedded device, it is therefore necessary to make, and write down, assumptions about how it is to be used. This makes explicit the security assumption dependencies in Fig. 18.2, and also clarifies what the "security reliance" means for a particular device—for example, it should clarify whether the embedded device provides resistance to fault induction or whether it is a responsibility of its environment (i.e. the artefact and application). Making the assumptions visible is important both for use during the evaluation and to enable the creator of an application artefact or application software to select a suitable embedded device. Visibility means that an evaluator can judge not only whether the security functions are completely and correctly present in the embedded device, but can also check whether sufficient guidance is provided with the device to enable the developer of the next level product to produce a secure implementation using it.

As an example: communication of confidential data between the embedded device and some other element of an application artefact (perhaps a separate memory chip) may represent a vulnerability if the data travel over an unprotected bus. The bus might therefore be protected in one (or indeed both) of two ways: the data might be encrypted, or access to the bus might be prevented by putting tamper-resistant elements in the application artefact itself. If the encryption approach is adopted, then the embedded device will need to include this functionality as part of its evaluated security requirement (along with any associated authentication between it and the element that it communicates with, and key management functions). If the tamper-resistance approach is used then, when evaluating the embedded device, it is important to note that it assumes such protection as part of its security context.

Certain security features in the device might be subject to enabling and disabling (e.g. different countermeasures may be enabled in different power states or lifecycle states) or selection between alternatives with different levels of security countermeasures, and might therefore require the artefact or application developer to understand how to do this correctly. For example, some functions such as random number generation or data signing might be provided at different speeds according to their security level (e.g. quality of randomness or resistance to fault-induction, respectively). The evaluation also needs to check this information in the guidance documentation, once again to ensure that it supports a secure implementation of a dependent product.

In many cases the embedded device will have a target type of application artefact that is expected to use it (e.g. a media player or mobile phone handset), and a type of application software that is expected to run on it. Thus, some design assumptions can be made about the way in which it will be used, and about a range of identifiable security features related to these usages, generally having different levels of security assurance. In some cases the context may be determined by third-party specifications; an example of an emerging approach that provides a rich functionality "medium assurance" environment is the trusted execution environment specified by GlobalPlatform in [21–23]. In cases such as this, where a sufficiently definite and ubiquitous set of security assumptions, requirements and functions can be defined for an embedded device (or artefact), a common security evaluation requirement can be defined—in CC this is known as a Protection Profile (PP): the characteristics of a PP and some example PPs are discussed later.

In evaluation terms, combining separately evaluated products, or incorporating one previously certified product into another evaluated product (as the application artefact incorporates the embedded device) is referred to as composition. Historically, it has proven difficult to evaluate composite products in a satisfactory way, but the most successful approach has been the application of CC to smart card products, where there are typically two separate evaluations: first of the IC alone and then of the IC together with its operating system and application software. The main difficulty arises from the different states of knowledge of the evaluators of each separate part: in general it is assumed that these separate evaluations will be performed by different evaluation laboratories, and possibly under different national schemes. This is discussed later, in the section "CC Interpretation and Supporting Documents".

An important part of the security evaluation is of course to determine the types of attacks that must be considered for an embedded device. Of course the evaluation will need to include logical checks of the correct implementation of the security functions and their resistance to bypass (alternative methods of achieving an objective without passing through the intended controls) or tampering attacks (e.g. buffer overflows, or unexpected signal combinations), but the more difficult and time-consuming aspects are often the attacks on the hardware. Once again it is necessary to look ahead to the attacks on an application artefact, and from this the embedded device's security responsibilities can be deduced. In general there are three particular types of hardware-related attack that need to be considered for the application artefact[1]:

- Physical attacks, based on accessing and/or modifying device features such as printed circuit board (PCB) tracks or buses on ICs, sensors or filters, random number generator or device test interfaces.[2]
- Side-channel attacks (e.g. using statistical analysis of variations in power or electromagnetic emanations to determine a cryptographic key).
- Fault-induction attacks (e.g. inducing processing faults by applying voltage or frequency glitches, operating at an out-of-specification temperature, or by applying a laser to critical components on an IC).[3]

This classification of attacks focuses on the type of underlying vulnerability that an attacker searches for. However, there are several other ways of characterising types of attacks, and one more is noted here, because it focuses on the need to make use of the attacked device when exploiting a vulnerability:

- Destructive attacks: these render the embedded device (or application artefact) non-functional as a side effect of the attack. However, the attacker gains knowledge that typically enables construction of a fake, cloned or simulated device. These attacks may also be used to gather information about a target device in order to carry out one of the other types of attacks below.
- Disfiguring attacks: these damage the appearance of the device (or application artefact) so that there is no reasonable possibility of making use of the original artefact in any role where it is inspected: the visible changes would arouse suspicion. Such attacks may be of use to an attacker if the artefact does not need to be inspected when it is used, or perhaps where it is not inspected closely (e.g. a contactless application might allow the disfigured artefact to be concealed but still held close enough to an interface device to carry out a transaction while a fake artefact is presented and seems to the payment attendant to be in use).

---

[1] In some cases, different types of attacks may be combined into a more complex attack scenario (e.g. some attacks may be used to gather information or otherwise enable a further attack).

[2] For further discussion of physical attacks (and indeed other attack examples in the context of CC and smart cards), see [12].

[3] See Chap. 9 of [5], or [12] for discussion of fault-induction attacks.

- Non-invasive attacks: these leave the device (or application artefact) undamaged and essentially the same as before the attack (at least in physical terms: it may have been changed logically).

This exploitation-oriented view also draws attention to the number and types of samples of the artefact needed to prepare and carry out an attack. Some attacks may need only a physical sample of the artefact (e.g. to locate target components in relation to the tamper-resistance mechanisms), while others may need a functional device. There may be a further distinction between a functional device that holds operational cryptographic keys (and therefore can carry out real transactions), and one that holds expired or test keys: in some application contexts (e.g. payment terminals), it may be much harder for an attacker to obtain samples with operational keys.

The developer's main challenge is of course to find an appropriate balance between the cost of the device and the security that it provides, and this cost will include the financial cost to evaluate it, and the amount of time taken to complete a satisfactory evaluation. A simple partitioning of the security requirements between the components of an application system will tend to make evaluations shorter and cheaper, because the attack paths to be considered are more straightforward and can therefore be attempted (and hopefully dismissed as infeasible) relatively quickly. In the earlier example where the embedded device was exchanging confidential data with another component over a bus, an appropriate encryption method (using well-tried techniques and algorithms) is likely to be quicker and cheaper to evaluate than a complex tamper-resistance mechanism (which must protect both ends of the communication). A tamper-resistance technique will generally require a number of experimental attempts to defeat the mechanism, perhaps requiring relatively sophisticated test tools and a number of attempts to improve the practical aspects of attack technique. Recognising a good encryption technique that would render attacks on the bus futile will usually be quicker than the tamper-resistance experiments.

The embedded device may have further security considerations if it makes use of third-party intellectual property (IP), or reuses elements of another device from the developer. This may mean that there is a reusable "module" for which a suitable evaluation result has already been obtained (although of course this will be subject to some assumptions about how the reusable part is used, which must be checked as part of an evaluation). However, it can also mean that the device is vulnerable to reuse of attack methods from another device (cf. the attack described in [11]). Where a high security device reuses protection features present in a lower security device, this attack reuse becomes especially important.

18.2.1 The Security Evaluation Model

Based on the discussion above, the basic model of security evaluation of an embedded device used here reflects the three layers in Fig. 18.2: embedded device, application artefact and application software. It assumes three corresponding evaluations.

First the embedded device is evaluated and certified against its low-level security requirements, resulting in a certificate and a "composition interface" which describes the security information needed to develop and evaluate the next level product and to meet the assumptions of the device evaluation. Next the application artefact is evaluated against its own security requirements, demonstrating that the embedded device correctly meets the relevant security requirements of the artefact, and that the artefact satisfies the assumptions placed on it by the embedded device. This results in a certificate and composition interface for the artefact. Finally the application software is evaluated, adding further security requirements, showing that the obligations placed on it by the artefact have been met, and resulting in the top level security certification.

In reality, the same evaluation method may not always be applied at each layer. However, the general principle that some formal evidence-based process is required to check that each lower layer object fulfils security requirements of the higher layer, and that the higher layer object satisfies the assumptions of its lower layer component, seems necessary. This model is reflected in a number of different real-world schemes (including FIPS 140 and PCI PTS (PIN Transaction Security), as discussed below).

18.2.2 Structure and Use of the Common Criteria

The CC provide a way to perform independent, standardised, internationally recognised security evaluations and certifications of software and hardware products. CC is fully described in [1–3] (and, in terms of the evaluators" tasks, in [4]).[4] Further discussion of the evaluation process can be found in [5].

The essence of a CC evaluation is that a product is submitted as a target of evaluation (TOE) by its developer to an evaluation laboratory that has been accredited by the relevant national certification scheme [an evaluation laboratory is also known as an information technology Security Evaluation Facility (ITSEF) or CommerciaL Evaluation Facility (CLEF)], along with a set of evaluation deliverables including the security requirements set out in a specific CC form known as a security target (ST). The requirements for these evaluation deliverables are determined by the assurance level claimed in the ST, and are defined in part 3 of the CC. The evaluation is then carried out by the evaluation lab with reference to the three parts of the CC ([1–3]), using the evaluation methodology defined in the common evaluation methodology (CEM) [4].[5] Once it has completed evaluation and achieved certification, a product is listed on the international CC website at [26].

---

[4] The three parts of CC have also been adopted as international standard ISO 15408, and the common evaluation methodology (CEM) as ISO 18045.

[5] For description of a more specific "vulnerability-centric" approach to evaluation, involving a more specific model of interaction with the developer, see the UK scheme document SIN 092 "Vulnerability-centric Evaluation: Improving Evaluations by Putting Vulnerabilities First" at http://www.cesg.gov.uk/servicecatalogue/CCITSEC/Pages/Formal-Documentation.aspx.

A CC certificate applies to a specific version of a product, and therefore when the developer changes the product the resulting version is no longer certified. Depending on the nature of the changes involved, the developer may be able to undertake a process known as "maintenance" in which the changes are described in the form of an impact analysis, and (subject to certain conditions) the certification may be updated with a Maintenance Report (this is then attached to the certified product entry on the CC website). If the changes are more significant [e.g. if they directly affect the implementation of the security functional requirements (SFRs)], then a re-evaluation is required; however, even in this case the evaluators will generally reuse as much of the evidence and conclusions from the previous evaluation as possible. More details of the CC maintenance process are given in [15].

Common criteria is implemented as a national scheme in each country that chooses to use it.[6] To avoid the need to separately evaluate and certify a product in each country, however, most countries that use CC are signatories to the common criteria recognition arrangement (CCRA).[7] The CCRA provides for two types of signatories: certificate authorising participants, who both produce their own certificates (via a national certification body: CB) and recognise those produced by other certificate authorising participants, and certificate consuming participants, who do not issue their own certificates but undertake to recognise those issued by the certificate authorising participants. It is important to note that international recognition under the CCRA only applies up to an assurance level of EAL4 (as defined in CC part 3 [3]).

At the time of writing, there are 26 countries in the CCRA, of which 16 are certificate authorising. These countries provide representatives to the bodies that manage and maintain the CC (e.g. producing new versions of CC and CEM, and issuing supporting guidance documents)—primarily the common criteria management committee (CCMC) and common criteria development board (CCDB).

Within Europe, there is an additional grouping of CBs under the SOGIS mutual recognition agreement [25]. Members of this group recognise certificates issued by other certificate authorising SOGIS members at assurance levels of up to EAL7.

The main source of reference for CC (including the criteria themselves, certified products, the CCRA, and the international evaluation infrastructure) is the international website at [26].

---

[6] In fact there are a number of other applications of CC below the national level, with Certification Bodies that are not national government organisations. However, since these organisations cannot produce internationally recognised certificates as described later, they are ignored for the purposes of this chapter.

[7] Both CCRA and senior officials group information system security (SOGIS) discussed later allow a participant not to recognise a certificate from another member where national security issues are involved, or where the specific case would conflict with other national law.

18.2.3 Structure of Common Criteria

The CC consists of three parts:

- Part 1. Introduction and General Model [1]: this part defines the terminology used in CC, the basic actors, principles and objects involved, and how the criteria are applied to achieve the evaluation of a product. It includes the requirements for writing ST and Protection Profiles.
- Part 2. Security functional components [2]: this part is effectively a catalogue of templates for specifying common security functionality, and for identifying potential dependencies between security functions (e.g. between cryptographic operations and management of the keys that they use, or between auditing and identification of users instigating a recorded action). It is the source of most of the SFRs included in a product's ST.
- Part 3. Security assurance components [3]: this part is a catalogue of assurance requirements, and defined combinations of these requirements that are identified as evaluation assurance levels (EALs). Assurance levels and their use in PPs are discussed in more detail below.

In addition to the criteria themselves, there is also a definition of the activities required to evaluate some of the assurance requirements in CC part 3; this is found in the Common Evaluation Methodology (CEM) [4]. The CEM is followed by evaluators, and represents the common definition of evaluation activities that underlie the CCRA.

18.2.3.1 Protection Profiles and Security Targets

In a CC evaluation, the baseline document defining the scope of the evaluation and the assurance level is the ST. The scope of the evaluation includes identification of the evaluated version(s) of the product, the configurations of the product that are considered, and the security functionality covered by the evaluation. The detailed content of an ST is described in Annex A of CC part 1 and includes the SFRs that the product implements. Usually SFRs are drawn from CC part 2, but CC part 1 also describes how to define new extended SFRs.

The ST is a critical document for the evaluators, because it defines the scope of the evaluation work that needs to be done, but it is also a critical document for potential users of the TOE, and for developers who intend to build products that rely on the TOE. In the context of embedded devices, it is therefore important that the developer of an application artefact that will use a certain embedded device examines the ST for that device to confirm that the functions needed by the artefact have been included in the scope of the embedded device evaluation. Similarly, the developer of application software to run on the application artefact should confirm that the application artefact ST includes the necessary supporting functions for the application software (and that the evaluation scope includes the configuration(s) of the artefact that the application software needs). As a particular example, at both of these levels the specific cryptographic functions included in the evaluation scope will be important to the user of an evaluated product.

A PP is essentially a product-independent ST: it includes all of the content of an ST except for the TOE Summary Specification. A PP can be used to set requirements for a type of TOE, and therefore to achieve a degree of standardisation of the security expectations for that type of device. The intention of a PP is that STs will be written to claim compliance with it (the process for evaluating STs and PPs, including compliance of an ST to a PP, is included in the CEM [4]), and that this thereby allows comparison of similar evaluated products (because they are all evaluated against a common base requirement), and assists in building composite TOEs.

18.2.4 Assurance Requirements and Assurance Levels

CC includes the definition of seven standard hierarchic assurance levels in part 3 [3], with the lowest being EAL1 and the highest EAL7. Each assurance level is defined in terms of an increasingly demanding and wide-ranging set of assurance requirements that define evidence and evaluation activities in the areas of: design information (ADV), developer testing (ATE), guidance documentation (AGD), development lifecycle and delivery (ALC), and vulnerability analysis (AVA).

However, an ST does not have to adhere rigidly to the pre-defined assurance levels: it may also take a basic assurance level and augment it by adding requirements from a higher assurance level, or requirements that are not included in any assurance level. For example, CC part 3 includes the definition of a set of requirements for a developer's approach to flaw remediation (ALC_FLR).

Smart cards have typically used assurance levels based on EAL4 or EAL5 for both hardware and software aspects, but usually with augmentations to higher level requirements for development environment security (ALC_DVS) and vulnerability analysis (AVA_VAN). In this context it is worth recalling the earlier note that international recognition of CC certificates under the CCRA is limited to EAL4, although between SOGIS participants recognition is up to EAL7.

18.2.5 CC Interpretation and Supporting Documents

Although the CC documents provide a standard for specifying and evaluating assurance in a set of functional requirements, the CC definitions (of both SFRs and assurance) are at quite a high level. Furthermore, the origins of these requirements were in software, rather than hardware, and this is reflected in the way some of the requirements and evaluation activities are defined. It has therefore proven necessary to provide additional interpretation of what CC means in certain areas and for certain types of TOE. Such interpretation can be described in PPs or in CC supporting documents, both of which have been used extensively to provide interpretation for evaluating smart card hardware and software.

For smart cards, the main supporting documents are [6–9]. These cover the general ways in which CC requirements need to be applied to the types of deliverables and lifecycles applicable to smart card hardware and software, the requirements on evaluation labs that undertake smart card evaluations, the ways of calculating attack potentials applicable to smart cards (see the separate discussion of attack potentials below), and the ways in which the problem of composition (discussed above) is dealt with in the case of smart cards. In particular, the approach to composition requires that the evaluators of the dependent TOE (e.g. a smart card application) will check that the dependencies and assumptions between the TOEs match in terms of the ST content,[8] and that the developer's guidance for use of the lower level TOE (the smart card IC) and the assumptions it makes about its security environment, have been appropriately followed. Because it is unlikely that a lower level TOE will be able to implement its security functions perfectly (at least not at a reasonable cost), it has also proven necessary for a summary of the evaluation results for the lower level TOE to be passed on to the evaluators of the dependent device, and a specific mechanism has been defined to support this (see, in particular, [9]). This transfer of evaluator information (e.g. about intrinsic susceptibilities of an IC to side-channel or fault induction attacks) enables efficient and effective analysis of attacks that are carried out via the IC, but that ultimately affect the operating system or application software. For example: the evaluator of the composite TOE can check sensitive fault induction scenarios identified for the IC so that they can confirm not only that IC guidance has been followed, but that it is effective for protecting the specific functions used in the dependent TOE.

The main PP used for smart card ICs [10] is discussed as an example PP below.

18.2.6 Attack Potential Calculations

Probably the most important part of a CC evaluation (at least for the higher assurance levels such as those used by smart cards) is the vulnerability analysis activity described under the AVA_VAN family in CC part 3. The main requirement of AVA_VAN states:

> The evaluator shall conduct penetration testing, based on the identified potential vulnerabilities, to determine that the TOE is resistant to attacks performed by an attacker possessing <L> attack potential.

where <L> is one of: Basic, Enhanced-Basic, Moderate or High (which of these applies is determined by the assurance level). The general method for calculating attack potentials is given in Annex B of the CEM, but for smart cards a different

---

[8] In terms of assurance levels (EALs) in particular: at present the CC requirement for composition of a lower level product into a higher level one generally means certifying the lower level product at the same or higher assurance level as the higher one.

method is used as described in [8] (in fact this method is based on that used in the CEM for CC version 2.3). The smart card method in [8] divides the analysis of an attack into two stages: Identification and Exploitation. It allocates points for each of these stages in terms of: elapsed time for the attack, expertise required, specific knowledge of the TOE required by the attacker, number of TOE samples required, type of equipment required and all types of open samples (i.e. TOE samples with reduced security) required in order to identify the attack.[9]

The Identification and Exploitation stages separately assess the difficulty of discovering an attack, and the difficulty of putting it into practice. This reflects the fact that complex technical work leading to discovery and publication of a potential vulnerability may be carried out by different individuals and groups, with different motivations, as compared to those that would carry out real-world attacks. The risk-owner (i.e. in general the organisation that owns, or is liable for, the asset) may put into place a defence-in-depth strategy that tries to make practical exploitation of any attack more difficult (perhaps using mechanisms in the wider application system, such as accountability and audit, detection of fraudulent use and device/user blocking).[10]

Thresholds required to achieve each of the attack potential levels <L> above are listed as part of [8], and it is important to note therefore that potentially successful attacks may exist above this threshold of difficulty, even for a certified product. This situation emphasises how important it is for risk owners to understand the types of attacks that are considered in the evaluation of a device (or application artefact, or application software) that they are using, and the sort of level of difficulty that would be involved in attacks above that threshold. Gaining this sort of understanding, and participating in the setting of requirements in PPs for a technology type is one of the benefits and activities undertaken by a CC technical community (discussed below).

18.3 Evolution of Common Criteria

An important aspect of CC has been its ability to evolve, whilst retaining the international recognition of certificates within a growing set of CCRA participants. One dimension of evolution is the updating of the criteria (and the CEM), in terms of new versions and releases, which encompass editorial changes, clarifications, changes to requirements (SFRs and assurance levels) and even major changes in structure and philosophy of parts of the criteria.

---

[9] For an example in this vein see [11], which notes that in developing this attack (which was based on physically probing tracks on an IC), the attacker "began by buying chips in bulk for pennies apiece to experiment with".

[10] Even where risk mitigation of this sort is carried out, the impact of reputation damage needs to be considered. Although a full discussion of this aspect is outside the scope of this chapter, independent evaluation and certification under a recognised scheme would be one part of demonstrating that appropriate care has been taken in implementing security at each component level.

The other main dimension for evolving CC is the creation of interpretation in supporting documents and PPs, as discussed above. In recent years, this aspect of CC evolution has increasingly taken place through the efforts of technical communities. Several such communities are now working on a new generation of PPs that adopt an identifiably different style with the aim of making evaluation much more specific to separate product types and their associated technologies.

18.3.1 CC Technical Communities

When smart cards were first evaluated under CC, a number of significant problems were found in using the criteria. Many of these arose from the software origins of the CC, and the specialised hardware-related attacks that were vitally important for the vulnerability analysis and penetration testing of ICs. Initial attempts to solve these problems led to different national approaches and therefore a difficulty in maintaining international recognition of smart card certificates. Frustration with these early difficulties led to the formation of a CC technical community for smart cards, which in time produced the PPs and supporting documents that have enabled the smart card market to become, and remain, one of the major users of CC evaluation and certification. The smart card technical community is discussed in more detail in [12], but a critical feature is that it includes representatives from all the main stakeholder groups and from all the main countries involved in smart card evaluations and certifications, and thus benefits from technical expertise and experience of the day-to-day problems of CC evaluations.

The technical community is an essentially collaborative entity, and reduces the impact of creating new interpretation in the context of a single evaluation (and hence a single TOE, single developer, single evaluation lab and single CB). The community adds the wider reflection of the full range of stakeholders and of multiple participants in each stakeholder role. Furthermore, the community enables the interpretation to better take into account what is possible and achievable in real-world products, and the needs and expectations of risk owners.

Technical communities meet regularly and can respond relatively quickly to new developments (e.g. new publications of attack examples, such as in [11]), and also acts as a forum for discussing and resolving day-to-day issues arising in evaluations. The community continues to update the CC supporting documents, and the experience that underlies their consistent use across different national schemes and evaluation labs (e.g. by discussion of example attacks and their attack potential calculations).

After the initial creation of a technical community for smart cards, a further community was created to address the needs of evaluation of payment terminals (including the harmonisation of different national schemes and the PCI PTS scheme, as discussed below). This community produced the PP in [18], which includes community-specific needs such as a bespoke assurance level reflecting different evaluation needs and priorities for different parts of a payment terminal, and is producing similar supporting documents to those for smart cards (e.g. to help standardise attack potential calculations for payment terminals).

The main area of current evolution in CC is in extending this approach to many other areas (e.g. network devices, protected USB memory sticks, disc encryption and mobile devices), and to produce a new generation of PPs for each of these TOE types, as discussed in the following section.

18.3.2 New Generation Protection Profiles

Although the smart card community has produced both Protection Profiles and CC supporting documents, the starting point was to set down a PP that includes interpretations that match CC assurance requirements to the detail of the technologies, development and production environments, and attack types that apply to smart cards. This was then used to establish a base of common international experience in evaluation. The success of this approach has led to the creation of new technical communities, each aiming to produce a new style of PP that addresses the problems and concerns of its technology and usage. The PPs to emerge from this work are referred to here as "new generation PPs".

Probably the most important characteristics of new generation PPs are:

- Avoiding reliance on the traditional generic assurance levels in CC part 3. Instead they define assurance activities that describe specific ways to evaluate the SFRs and individual assurance requirements to match the technology and the application domain. An example of such a new generation PP is that for USB devices in [19].
- They are specific in their definition of the tasks of evaluators, enabling consistency in the conduct of evaluations. The assurance activities are deliberately specific in their detail, as well as matched to the technology and TOE type, so that it should be easy to recognise when the evaluation requirement has been met and the assurance demonstrated.

Embedded devices seem likely to be a part of a number of technical communities (for example, the existing communities for payment terminals and mobile devices), and the formation and growth of appropriate technical communities along with creation of new generation PPs seem likely to be the most appropriate path for dealing with the security problems of embedded devices, their application artefacts, and their application contexts.

18.4 Other Security Evaluation Schemes

Although this chapter is mainly concerned with security evaluation using the CC, it is interesting to compare this approach with other security evaluation schemes that are also in use for embedded devices. Two such schemes are examined briefly below.

18.4.1 FIPS 140

The cryptographic module validation program (CMVP) [24] was created in the US [in fact it is a joint effort from National Institute of Standards and Technology (NIST) and the Canadian communications security establishment (CSE)], and it is still demanded primarily for North American markets. The evaluation process is essentially similar to that of CC: a product is submitted to an independent evaluation lab, with deliverables as required in [13]. The lab carries out an evaluation against the criteria, produces a report and sends this to NIST (or CSE), who acts as the "validation authority" (equivalent to the CB in the CC process). On successful completion of the evaluation, the product is placed on the list of validated modules on the CMVP website [24].

FIPS 140 is specific to the evaluation of cryptographic modules, in contrast to the more general scope of CC, and this is reflected in the way the FIPS 140 requirements are specified in the standard [13]. Included are requirements for the use of only approved algorithms (meaning that they are approved under a separate FIPS process) and a number of other approved security functions. This specificity of algorithms and functions is at a lower level of detail than CC, and is one of the main differences between the schemes.

FIPS 140 specifies four hierarchic levels, numbered 1–4. Each of the levels includes requirements on module design (including the flows that can take place over ports and interfaces), operator roles and associated authentication, tamper protection, support from the underlying operating system, key management, self-test and development lifecycle. Level 1 works mainly at the level of the algorithm implementation to ensure that the module is basically well-designed and follows sound principles of cryptographic implementation—this is the only level that can be achieved by software alone. Level 4 has strong requirements for physical tamper protection, role-based authentication, controls on manual operations with keys and other requirements (requiring EAL4 for the underlying operating system), such that at Level 4 the module can operate without additional physical protection in the environment.

As noted above, there is a specific relationship between FIPS 140–2 and CC in the sense that the higher FIPS levels invoke CC requirements for the underlying operating system as part of protecting the cryptographic module. The underlying difference between CC and FIPS 140 is exemplified in the existence of the Derived Test Requirements [14] (DTRs) for each of the FIPS 140 levels. This document takes advantage of the specific type of module to which FIPS 140 is applied, and its requirements for approved algorithms and functions, to give relatively detailed instructions to evaluators (and hence to define in a relatively fixed way the activities that the evaluators undertake).

Perhaps the most appropriate comparison can be made by considering CC's use of PPs and supporting documents to specify requirements for specific types of TOE (such as smart cards). In the new generation CC PPs, there is also a movement towards providing more detail both about the technology, and about the evaluation methods (as contained in the defined assurance activities), and this aspect of CC's evolution could therefore be seen as incorporating some of the observed advantages of FIPS 140, whilst retaining the wider scope of product types, the greater degree of international recognition, and international ownership and management. With this in mind it is noted that FIPS 140-2 is much less specific with regard to side-channel and fault induction attacks (which are required to be analysed and tested under the smart card supporting documents in CC). The DTR states that this aspect of FIPS 140–2 is based only on analysis of developer documents; it does not require evaluator testing.[11]

18.4.2 PCI PIN Transaction Security Requirements

PCI PIN transaction security (PTS) is a set of modular security requirements for payment terminals incorporating PIN entry devices. The scheme is based on the requirements in [16] and an associated developer questionnaire [17]. As with CC and FIPS 140, the evaluation is carried out by an evaluation lab that is specifically approved for this evaluation scheme, with the lab performing analysis and testing, resulting in a report that is then submitted for acceptance, and recording of successful products on the list of approved devices on the PCI website [27].

The requirements cover: physical and logical security of the device and the PIN entry process; integration with the point-of-sale (POS) terminal; secure use of protocols; secure handling of data such as private and public keys, and account data; and device management (i.e. manufacturing and the process leading to initial key load). The evaluation approach is similar in many ways to FIPS 140, in that it lists very specific requirements based on its limited scope of application. Evaluation activities are also based on following a set of derived test requirements. Although PCI PTS does cover lifecycle requirements ("device management security requirements"), these are not validated by the evaluators.

The requirements have adopted the CC idea of attack potential calculations based on separate Identification and Exploitation phases as in the smart card interpretation of CC in [8], but with some differences. A CC evaluation is based on a single attack potential requirement for the entire TOE (and hence all the SFRs that the TOE implements). However in PCI PTS there is no requirement for the TOE as a whole, while some (but not all) of the individual requirements make reference to resisting a certain attack potential. Furthermore, some PCI PTS requirements specify minimum points required for the Exploitation phase alone. Finally, where minimum points thresholds are stated, these have different values to the thresholds for CC in [4, 8].

---

[11] It is interesting to note here that side-channel and fault induction attacks carried out by evaluators during CC evaluations of smart cards have historically led to the discovery of a variety of potential vulnerabilities, requiring either changes in the TOE itself, or additional guidance to enable software developers to mitigate the risks from these attacks. However, smart cards are generally used in unprotected environments, whereas FIPS 140-2 assumes some degree of physical protection in the environment except at level 4.

The PCI PTS requirements are also being incorporated into a CC PP for payment terminals in [18][12] (discussed as an example PP below), as a result of the desire to replace the large number of separate national acceptance schemes that a payment terminal currently has to undergo.

18.5 Example Protection Profiles

This section briefly surveys some PPs relevant to embedded devices. Naturally, much more could be said about each PP but such a survey is inevitably a hostage to the time that it is carried out and, given the move towards a new generation of PPs discussed earlier, it seems most relevant here to indicate features of interest in some of the PPs, which may give pointers to relevant features for future PPs for embedded devices.

18.5.1 Security IC PP

This PP [10] is the main one used for evaluations of smart card ICs (an older version of this PP, written for CCv2.3, was also extensively used; both versions have effectively the same requirements). Its assurance level is EAL4+,[13] with the augmentations increasing the assurance requirement for the development environment and requiring the TOE to resist an attack potential of High (instead of the default Enhanced-Basic level at EAL4).

The PP is notable for a number of reasons:

- It defines several extended SFRs (i.e. additional SFRs that are not present in CC part 2), to cover random number generation (FCS_RNG.1), test mode protection (FMT_LIM.1 & 2) and support for injection of data into each IC instance (including a unique identifier) during manufacturing (FAU_SAS.1).
- It contains extensive refinement of the assurance requirements to map them onto particular aspects of IC development and manufacturing.
- It requires resistance to High attack potential—this required extensive work within a technical community to create supporting documents [including [8], but also additional reference materials to support consistent rating of the range of applicable attacks at this level].
- The PP does not specify any cryptographic operations, even though all of its original TOEs implemented some. The scope of cryptographic functions covered by an evaluation is left to be added by the ST author.

---

[12] At the time of writing the PP [18] is based on an earlier set of PCI requirements that were superseded by PCI PTS. However, the requirements are generally similar and it should therefore be possible to update [18] to capture the requirements in [16].

[13] This shorthand notation means "EAL4 augmented with ALC_DVS.2 and AVA_VAN.5)".

- The PP allows flexibility about the inclusion of several categories of software (including test mode software) to be evaluated along with the hardware.
- The SFRs are described according to the structure of the TOE and the functionality it provides. This improves vastly on the approach taken in some PPs, where SFRs are listed alphabetically, which tends to leave the mapping to specific aspects of TOE functionality unclear and therefore liable to inconsistent interpretation by developers, users, evaluators and CBs.

18.5.2 Payment Terminal (Point of Interaction) PP set

This "PP" is in fact a set of three PPs that are described in a single document [18], and was mentioned in the earlier discussions of PPs from CC technical communities, and, in the discussion of the PCI PTS Requirements scheme, as a PP that brings together a range of separate national and international requirements into a single CC framework. The three PPs in this set have much of their functionality in common, and are all subject to the same sort of approval requirements—hence the presentation of the PPs as a core set of requirements with specified differences (e.g. only two of the three PPs include magnetic stripe transactions in scope). The assurance level in this PP is based on EAL2, but applies some requirements from EAL4 to certain parts of the TOE (typically the parts that handle the main payment system keys), defines extended assurance requirements,[14] and makes a number of refinements to the assurance requirements. This results in a set of assurance requirements that is not well-described by reference to the levels in CC part 3 (but which better reflects the interests and priorities of risk owners), and is therefore given a new name in the PP: "EAL POI".

The main points of interest in this PP set are:

- The definition of a bespoke assurance level was a significant divergence from previous CC practice. The adoption of this level reflects several aspects of the work of a technical community:

  - The community itself was competent to define its true assurance requirement because it included representatives of the relevant risk owners.
  - Writing the assurance requirements in a way that was consistent with existing CC methodology (and therefore compatible with CCRA), and defining additional supporting guidance to fill any gaps, was also within the competence of the community because it included evaluation labs (with experience of evaluation of such devices) and CBs.

- The assurance level definition is unusual in several ways: it combines elements of both EAL2 and EAL4; it uses extensive refinement of the requirements (e.g.

---

[14] These are analogous to extended SFRs: they are assurance requirements defined in the style of CC part 3, but not taken directly from CC part 3.

to explain what is required in terms of design deliverables); and it applies four different attack potential thresholds to different parts of the TOE, replacing the normal CC part 3 vulnerability analysis family (AVA_VAN) with a new family AVA_POI. The use of the bespoke assurance level is similar in some ways to the definition of assurance activities in the new generation PPs.

- The PP set has to remain generic in most areas, because it intends to cover a large variety of different scales and architectures in the implementation of compliant terminals. This raises the danger that it becomes difficult to map the PP requirements (especially at individual SFR level) onto a real TOE—hence there would be dangers of inconsistent application of the PP. To address this potential problem the PP includes significant amount of abstract design (e.g. in terms of TOE architecture and its application context) to structure the elements of the PP (from threats down to SFRs and assurance requirements).
- As part of its goal of bringing together other security evaluation scheme requirements, the PP includes application notes against many of the SFRs to note their correspondence to these other sources of requirements.
- The PP uses three extended SFRs to deal with random numbers (FCS_RND.1), proof of identity by the TOE to an external entity (FIA_API.1) and protection against emanations that might enable side-channel attacks (FPT_EMSEC.1). All three of these extended SFRs are reused from smart card PPs.

18.5.3 Trusted Platform Module PP

This PP [20] maps a large and relatively complex set of specifications into a CC framework. The PP therefore demands a much greater investment of effort by the reader, but benefits (e.g. in reduced ambiguity) from close correspondence to the main trusted platform module (TPM) specifications. The assurance level in this PP is EAL4+(ALC_FLR.1, AVA_VAN.4), where the augmentations add a requirement for a flaw remediation approach and require the TOE to resist an attack potential of Moderate. Points of interest include:

- The PP applies to hardware, firmware and/or software, depending on the combination used to implement the TPM—this is similar to the situation that may arise for a number of embedded devices, where implementation choices of visible security aspects may need to be left flexible at the time of the PP.
- The PP includes copious references to the main TPM specifications—these help to establish precision in the meaning of the PP requirements, and ultimately in the SFRs, which supports consistent use of the PP.
- The PP includes instructions to the ST author regarding parts of the SFRs that need to be completed in the ST (i.e. where the PP leaves some or all of an SFR for the ST author). Although such flexibility is a normal part of CC, it is generally helpful for an ST author to be guided as to the expected and allowed ways to complete the SFRs, especially in terms of the wider TPM specifications.

- An optional package for revocation functionality is included in the PP. This means that an ST does not need to implement this package in order to comply with the PP, but if it chooses to do so then the package can simply be included in the ST.
- The PP leaves some artefact-level implementation items unspecified, but highlights them in its objectives for the environment (see OE.Locality and OE.Physical_Presence in [20]).
- The PP has a strong focus on management operations, which is perhaps a likely characteristic of embedded security devices; they are likely to have specific initialisation and state management aspects that are reflected in SFRs. In several cases, the SFRs are refined,[15] adding a set of specific rules for management operations. Once again, this links the PP closely to the main TPM specifications.
- An SFR dealing with physical tamper resistance is included and specifies a minimum set of tamper resistance scenarios, allowing the flexibility for other specific scenarios to be added in the ST, if applicable.

References

1. Common Criteria for Information Technology Security Evaluation, Part 1: Introduction and General Model, Version 3.1 Revision 3, CCMB-2009-07-001, July 2009.
2. Common Criteria for Information Technology Security Evaluation, Part 2: Security Functional Components, Version 3.1 Revision 3, CCMB-2009-07-002, July 2009.
3. Common Criteria for Information Technology Security Evaluation, Part 3: Security Assurance Components, Version 3.1 Revision 3, CCMB-2009-07-003, July 2009.
4. Common Methodology for Information Technology Security Evaluation: Evaluation Methodology, v3.1 Release 3, CCMB-2009-07-004, July 2009.
5. Mayes K, Markantonakis K (eds) (2008), Smart Cards, Tokens, Security and Applications, Springer.
6. The Application of CC to Integrated Circuits, version 3.0 revision 1, CCDB-2009-03-002, March 2009 [Online] http://www.commoncriteriaportal.org/supporting/.
7. Requirements to perform Integrated Circuit Evaluations, version 1.0 revision 1, CCDB-2009-09-001, September 2009 [Online] http://www.commoncriteriaportal.org/supporting/.
8. Application of Attack Potential to Smart Cards, version 2.7 revision 1, CCDB-2009-03-001, March 2009 [Online] http://www.commoncriteriaportal.org/supporting/.
9. Composite product evaluation for Smart Cards and similar devices, version 1.0 revision 1, CCDB-2007-09-001, September 2007 [Online] http://www.commoncriteriaportal.org/supporting/
10. Eurosmart, Security IC Platform Protection Profile, version 1.0, BSI-PP-0035, 15 June 2007, [Online]http://www.commoncriteriaportal.org/files/ppfiles/pp0035b.pdf
11. Government Computer News, Engineer shows how to crack a "secure" TPM chip, [Online] http://gcn.com/articles/2010/02/02/black-hat-chip-crack-020210.aspx, (accessed 2 June 2011)
12. Boswell T, Smart card security evaluation: Community solutions to intractable problems, Information Security Technical Report, Volume 14 issue 2, May 2009, pp57-69.
13. NIST, Security Requirements For Cryptographic Modules, FIPS PUB 140–2, issued 25 May 2001, with change notices as at 12 March 2002 (The original FIPS 140 standard was FIPS

---

[15] This is an allowed operation on SFRs, subject to certain rules, as described in Sect. 8.1.4 of [1].

140–1; this was superseded by FIPS140-2, and FIPS 140–3 is in draft at the time of writing. In this chapter, FIPS 140 is used as a general name for the scheme.).

14. NIST, Derived Test Requirements for FIPS PUB 140–2, draft of 4 January 2011.
15. Assurance Continuity: CCRA Requirements, CCIMB-2004-02-009, version 1.0, February 2004, [Online] http://www.commoncriteriaportal.org/files/supplements/2004-02-009.pdf.
16. Payment Card Industry (PCI), PIN Transaction Security (PTS) Point of Interaction (POI) Modular Security Requirements, version 3.0, April 2010, [Online] https://www.pcisecuritystandards.org/documents/pci_pts_poi_sr.pdf.
17. Payment Card Industry (PCI), PIN Transaction Security (PTS) Point of Interaction (POI) Modular Evaluation Vendor Questionnaire, version 3.0, April 2010, [Online] https://www.pcisecuritystandards.org/documents/pci_pts_poi_vq.pdf.
18. Common Approval Scheme, Point of Interaction Protection Profile, version 2.0, 26 November 2010, [Online] http://www.ssi.gouv.fr/IMG/certificat/ANSSI-CC-cible_PP-2010-10en.pdf.
19. Information Assurance Directorate, Protection Profile for USB Flash Drives, version 1.0, 1 December 2011, [Online] http://www.niap-ccevs.org/pp/pp_usb_fd_v1.0.pdf.
20. Trusted Computing Group, Trusted Computing Group Protection Profile PC Client specific Trusted Platform Module TPM Family 1.2; Level 2, version 1.1, 10 July 2008, [Online] http://www.commoncriteriaportal.org/files/ppfiles/pp0030b.pdf
21. GlobalPlatform, The Trusted Execution Environment: Delivering Enhanced Security at a Lower Cost to the Mobile Market, February 2011, [Online] http://www.globalplatform.org/documents/GlobalPlatform_TEE_White_Paper_Feb2011.pdf.
22. Global Platform, GlobalPlatform Device Technology TEE Client API Specification, version 0.17, 27 April 2010, [Online] http://www.globalplatform.org/specificationsdevice.asp.
23. ARM, Building a Secure System using TrustZone Technology, issue C, April 2009, [Online] http://infocenter.arm.com/help/topic/com.arm.doc.prd29-genc-009492c/PRD29-GENC-009492C_trustzone_security_whitepaper.pdf.
24. Description of the CMVP, and list of CMVP validated modules, [Online] http://csrc.nist.gov/groups/STM/cmvp/index.html.
25. Mutual Recognition Agreement of Information Technology Security Evaluation Certificates, version 3.0, January 2010, [Online] http://sogisportal.org/.
26. International Common Criteria website, [Online] http://www.commoncriteriaportal.org.
27. PCI Security Standards council website, [Online] https://www.pcisecuritystandards.org.

Chapter 19
Physical Security Primitives

A Survey on Physically Unclonable Functions and PUF-Based Security Solutions

Ahmad-Reza Sadeghi, SteffenSchulz and ChristianWachsmann

Abstract Physically unclonable functions (PUFs) are an emerging technology and have been proposed as central building blocks in a variety of cryptographic protocols and security architectures. Among others, PUFs enable unique device identification and authentication, binding software to hardware platforms and secure storage of cryptographic secrets. Furthermore, they can be directly integrated into cryptographic algorithms and remote attestation protocols. In this chapter, we give an overview of the concept, properties, and types of intrinsic electronic PUFs, discuss potential attack surfaces and advanced PUF concepts as well as the most common applications of electronic PUFs. Further, we show new directions on logically reconfigurable PUFs (LR-PUFs) and PUF-based remote attestation and discuss open challenges.

19.1 Introduction

Physically unclonable functions (PUFs) are increasingly proposed as central building blocks in cryptographic protocols and higher level security architectures. Among others, PUFs enable unique device identification and authentication [44, 47, 54, 62], binding software to hardware platforms [12, 16, 18, 29], and secure storage of cryptographic secrets [34, 70]. Furthermore, they can be integrated into cryptographic

A.-R. Sadeghi (✉)
TU Darmstadt (CASED) and Fraunhofer SIT, Mornewegstraße 32, 64293 Darmstadt, Germany
e-mail: ahmad.sadeghi@trust.cased.de

S. Schulz
TU Darmstadt (CASED) and Macquarie University (INSS), Mornewegstraße 32,
64293 Darmstadt, Germany
e-mail: steffen.schulz@trust.cased.de

C. Wachsmann
TU Darmstadt (CASED), Mornewegstraße 32, 64293 Darmstadt, Germany
e-mail: christian.wachsmann@trust.cased.de

K. Markantonakis and K. Mayes (eds.), *Secure Smart Embedded Devices,*
Platforms and Applications, DOI: 10.1007/978-1-4614-7915-4_19,
© Springer Science+Business Media New York 2014

algorithms [2] and remote attestation protocols [55]. Today, there are already some PUF-based security products aimed for the market, mainly targeting IP-protection and anti-counterfeiting applications but also RFID systems [24, 67].

PUFs typically exhibit a challenge/response behavior: when queried with a specific *challenge*, the PUF generates a random-looking *response* that is stable over time. The security of PUFs depends on intrinsic manufacturing variations making PUFs physically unclonable and unpredictable. Even the manufacturer of the PUF should be unable to produce two PUFs with a similar challenge/response behavior. Furthermore, knowledge of a certain number of challenge/response pairs should not allow an adversary to predict PUF responses to unknown challenges.

There is a variety of PUF implementations [37]. The most appealing ones for integration into electronic circuits are *electronic PUFs*, which come in different flavors. *Delay-based PUFs*, such as arbiter PUFs [31, 35, 44] and ring oscillator PUFs [15, 38, 61] are based on race conditions or frequency variations in integrated circuits. *Memory-based PUFs* exploit the instability of volatile memory cells, such as SRAM cells [16, 21], flip-flops [32, 36], and latches [29, 60]. Finally, *coating PUFs* [46, 63, 64] use capacitances of a special dielectric coating applied to the chip housing the PUF.

In contrast to most cryptographic primitives, whose security can be related to well-established (albeit unproven) assumptions, the security of PUFs is assumed to rely on physical properties and is still under investigation. Existing PUF-based security solutions typically rely on assumptions that have not been confirmed for all PUF types. For instance, most delay-based PUFs have been shown to be susceptible to model building attacks that allow emulating the PUF in software [31, 35, 44, 51], which contradicts the unpredictability and unclonability properties. To counter this problem, additional primitives must be used: controlled PUFs [14] use cryptography in hardware to hide the responses of the underlying PUF from an adversary.

Since PUF responses are inherently noisy, they must be combined with error-correction mechanisms, such as fuzzy extractors [11] that remove the effects of noise before the PUF response can be processed in a (cryptographic) algorithm. Typically, the cryptographic and error correcting components and the connecting wires between them and the PUF must be protected against invasive and side-channel attacks.

Outline. This chapter gives an overview of the concept, properties, and types of intrinsic electronic PUFs (Sect. 19.2), discusses potential attack surfaces (Sect. 19.3), and advanced PUF concepts (Sect. 19.4) as well as the most common applications of PUFs (Sect. 19.5). Further, we show new directions on reconfigurable PUFs and PUF-based remote attestation (Sect. 19.6) and discuss open challenges (Sect. 19.7).

19.2 Physically Unclonable Functions

19.2.1 PUF Concept and Properties

A physically unclonable function (PUF) is a noisy function that is embedded into a physical object, such as an integrated circuit [1, 37, 45]. When queried with a *challenge c*, a PUF generates a *response r* ← PUF(c) that depends on both c and the unique device-specific intrinsic physical properties of the object containing the PUF. Since PUFs are subject to noise induced by environmental variations, such as supply voltage and ambient temperature variations, they return slightly different responses when queried with the same challenge multiple times.

PUFs are typically assumed to be *robust, physically unclonable, unpredictable* and *tamper-evident*, and several approaches to quantify and formally define their properties have been proposed (see the paper by Armknecht et al. [1] for an overview). Informally, robustness means that, when queried with the same challenge multiple times, the PUF returns a similar response with high probability. Physical unclonability demands that it is infeasible to produce two PUFs that cannot be distinguished based on their challenge/response behavior. Unpredictability requires that it is infeasible to predict the PUF response to an unknown challenge, even if the PUF can be adaptively queried for a certain number of times. Finally, a PUF is tamper-evident if any attempt to physically access the PUF irreversibly changes its challenge/response behavior significantly.

The properties of PUFs can either be evaluated theoretically, based on mathematical models of the underling physical processes [66, 68, 69], or experimentally by analyzing PUF instances built in hardware [19, 22, 23, 32, 65]. The first approach has the apparent drawback that mathematical models never capture physical reality in its full extent, which means that the conclusions on PUF security drawn by this approach are naturally debatable. The main drawback of the experimental approach is its limited reproducibility and openness: even though experimental results have been reported in literature for some PUF implementations, it is difficult to compare them due to varying test conditions and different analysis methods. Furthermore, raw PUF data is rarely available for subsequent research, which greatly hinders a fair comparison.

The security analysis of PUFs is further complicated by the drawbacks of existing approaches to formalize their security properties. Most PUF security models are not general enough and exclude certain PUF types (such as in [15, 45]), do not reflect all properties of real PUF implementations (for example in [2, 15, 16, 45, 52]), or include security parameters that cannot be determined for real PUF implementations in practice (such as in [2, 8, 52]). Recently, Armknecht et al. [1] proposed a PUF security framework that aims at providing security definitions that are compliant to standard game-based cryptographic security models and that allow engineers to evaluate and quantify the properties of PUF implementations.

19.2.2 PUF Types

There is a broad variety of PUF implementations that are based on very different physical characteristics, including optical, magnetic, and electrical effects. We focus on *electronic PUFs*, which can be easily integrated into electronic circuits without significant overhead using standard manufacturing processes. These PUFs are of particular interest since they enable the tight integration of PUFs into cryptographic primitives and higher level security architectures. Known electronic PUFs can be categorized as *delay-based PUFs* and *memory-based PUFs*, which will be explained in the following. A more detailed overview of various PUF types, including non-electronic PUFs, is given by Maes and Verbauwhede [37].

19.2.2.1 Delay-Based PUFs

Delay-based PUFs are based on race conditions or frequency variations in integrated circuits. The most popular PUFs of this type are arbiter PUFs [31, 35, 44] and ring oscillator PUFs [15, 38, 61].

Arbiter PUFs. The arbiter PUF is based on race conditions within integrated circuits. The basic arbiter PUF design has been presented by Lim et al. [31, 33] and consists of two identically designed signal paths consisting of wires and switching components and an arbiter at the end of both paths (Fig. 19.1).

The switching components allow the signal paths to be modified according to an external input, i.e., the PUF challenge. To evaluate the arbiter PUF, both paths are simultaneously excited with the same impulse signal. Depending on which of the two signals arrives first at the arbiter, one output bit is generated and used as PUF response. The delay caused by each signal path depends on the device-specific manufacturing variations of the transistors in the switching components and their connecting wires.

In addition, the delays of the signal paths are affected by environmental noise, such as temperature and supply voltage variations. The impact of noise is reduced by comparing the delays of both signal paths. Note that, in case the delay difference

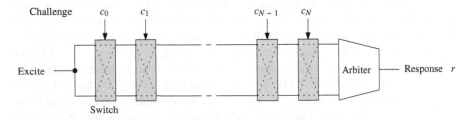

Fig. 19.1 Basic arbiter PUF design (for N bit challenge and 1 bit response)

between both signal paths is lower than the setup time of the arbiter, the response bit is independent of the delays and determined only by random noise (metastability).

Arbiter PUFs can be efficiently implemented on ASIC [31, 33], while implementations on FPGA seem to be difficult due to placing and routing constraints [37]. Moreover, the delays of the individual components of the signal paths are additive, which facilitates emulating the challenge/response behavior of the arbiter PUF in software (Sect. 19.3.1). To thwart these attacks, several variations of the basic arbiter PUF design have been proposed. The feed-forward arbiter PUF by Lim et al. [31, 33] uses the challenge and the output of intermediate arbiters to configure the signal paths. However, this design does not prevent emulation attacks [40, 41] and generates noisier responses due to increased metastability. As a countermeasure, Majzoobi et al. [39] propose lightweight secure PUFs, which are based on multiple interleaved arbiter PUFs. However, Rhrmair et al. [51, 52] show that lightweight secure PUFs of low complexity can be emulated using machine learning techniques (Sect. 19.3.1).

Ring Oscillator PUFs. Ring oscillator PUFs typically consist of several identically designed ring oscillators, which are loops of an odd number of inverters that, once stimulated, oscillate at a certain frequency. The oscillating frequency of each ring oscillator depends on the signal delays of its components, which are affected by manufacturing process variations and environmental noise. Basic ring oscillator PUF constructions typically do not support challenges. Challengeable ring oscillator PUFs can be implemented by integrating controllable delay elements into the ring oscillator circuits. Gassend et al. [13, 15] propose an alternative construction of a challengeable ring oscillator PUF that uses the challenge to select two out of a set of ring oscillators and derives a single-bit response based on the ratio of the oscillation frequencies of the selected ring oscillators. A similar approach by Suh et al. [61] derives the response bit based on which of the oscillation frequencies is higher (Fig. 19.2). While reducing the effect of noise on the ring oscillators, these constructions generate a high correlation between PUF responses [36], which reduces the unpredictability of their responses.

19.2.2.2 Memory-Based PUFs

Memory-based PUFs exploit the power-up behavior of volatile memory cells, such as SRAM cells [16, 21], flip-flops [32, 36] and latches [29, 60]. These memory cells are inherently instable circuits that, when an external data signal input is applied, enter one of two different stable states to store one bit of information. When no data signal is provided, most cells preferably enter the same state after each power-up, while some cells always enter a random state. The state the memory cell enters depends on the physical properties of the underlying transistors, which are affected by manufacturing process variations and environmental noise. Note that the amount of unique responses of a memory-based PUF is always limited by the number of its memory cells, i.e., the size of the underlying memory block.

SRAM PUFs. PUFs based on Static Random Access Memory (SRAM) have been proposed by Guajardo et al. [16] and Holcomb et al. [21]. The challenge to an SRAM

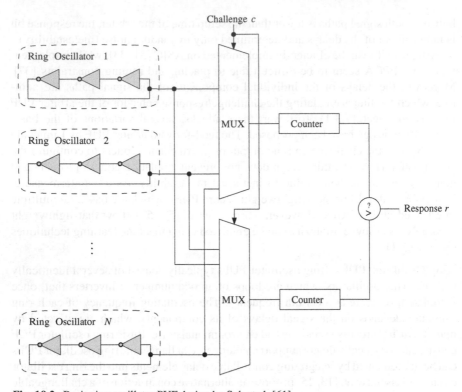

Fig. 19.2 Basic ring oscillator PUF design by Suh et al. [61]

PUF is a range of memory addresses, while the corresponding PUF response is the content of the uninitialized memory cells at those addresses.

An SRAM cell consists of two cross-coupled inverters that can store one bit of information and two additional transistors that are used to read and write data to the memory cell. Both inverters are typically designed to be identical in order to maximize write performance. When powered without applying a data signal, the SRAM cell will enter a state that depends on the threshold voltage mismatch of its transistors that is affected by manufacturing variations and environmental noise, in particular ambient temperature variations. SRAM cells with a large threshold mismatch always enter either the 0 or 1 state, while the state of cells with a small threshold mismatch is determined only by noise. In practice this means that some SRAM cells preferably enter the 0 state, others the 1 state, and some enter any of the two states with about the same probability. Cells that always enter the same state after power-up can be used as device-specific fingerprint since their behavior mainly depends on device-specific manufacturing process variations.

SRAM PUFs have been analyzed on FPGAs with dedicated SRAM [16, 17] and ASICs, including dedicated SRAM chips and SRAM embedded in micro-controllers [21, 22]. Note that each evaluation of an SRAM PUF requires the underlying SRAM to be powered down and up again, which can be problematic when the SRAM of the PUF is also used as random access memory by the device containing the PUF.

Butterfly PUFs. Butterfly PUFs have been proposed by Kumar et al. [29]. They emulate SRAM cells using cross-coupled data latches that, in contrast to SRAM PUFs, can be easily reset by triggering the set/reset input of the latches. Butterfly PUFs have been implemented and evaluated on FPGAs by Kumar et al. [29].

Flip-Flop PUFs. Maes et al. [36] propose flip-flops PUFs as an alternative to SRAM and butterfly PUFs that can be efficiently implemented on FPGAs. Flip-flop PUFs have been implemented and analyzed on ASIC by Van der Leest et al. [32]. In contrast to SRAM PUFs, flip-flop PUFs can be easily spread over the whole circuit to obfuscate the location of the individual flip-flops, which increases the difficulty of reverse-engineering and invasive attacks against the PUF.

Latch PUFs. Latch PUFs have been presented by Su et al. [59] in the context of device identification. These PUFs consist of an array of latches built from cross-coupled NOR gates. The threshold voltage differences of the underlying transistors, which are mainly caused by manufacturing process variations and affected by environmental noise, in particular ambient temperature variations, cause a mismatch in the latch. Hence, the state of the latches directly after power-up mainly depends on manufacturing process variations and can be used as device fingerprint. Su et al. [59] implemented and evaluated latch-based PUFs on ASIC.

19.2.3 Noise Compensation and Privacy Amplification

Many PUF-based applications require PUF responses to be reliably reproducible while at the same time being unpredictable [1, 2, 37]. However, since PUFs are inherently noisy and their responses are not uniformly random, they are typically combined with *fuzzy extractors* [11]. Fuzzy extractors consist of a *secure sketch* that maps similar PUF responses to the same value (noise compensation or error correction), and a *randomness extractor*, which extracts full-entropy bit-strings from a partially random source (privacy amplification).

Fuzzy extractors and secure sketches generally work in two phases (Fig. 19.3): in the *enrolment phase* some helper data h and a uniform bit string K (e.g., a cryptographic key) is derived from PUF response r. Helper data h is used later in the *reconstruction phase* to recover K from a distorted PUF response $r' = r + e$, where e is the error caused by noise. An important property of fuzzy extractors and secure sketches is that, after observing one single helper data value h, there is still some min-entropy left in r and K, which means that h can be stored and transferred publicly without disclosing the full PUF response r or secret K [11].

More detailed information on fuzzy extractors and a number of practical instantiations can be found in the work by Dodis et al. [11].

Fig. 19.3 Concept of fuzzy extractors

Enrolment Phase	Reconstruction Phase
Generate helper data h...	*...that is later used to recreate K.*
$r \leftarrow \mathsf{PUF}(c)$	$r' \leftarrow \mathsf{PUF}(c)$
$(K, h) \leftarrow \mathsf{Gen}(r)$	$K \leftarrow \mathsf{Rep}(r', h)$
Store (c, h)	

19.2.4 Characterizing the Unpredictability of PUFs

The unpredictability property of PUFs ensures that it is infeasible to efficiently compute the response of a PUF to an unknown challenge. This is an important property in PUF-based applications, such as authentication protocols, where the adversary could forge the authentication if he could predict the PUF response. Note that unpredictability should be independent of the operating conditions, such as ambient temperature and supply voltage variations, which could be exploited by the adversary.

Depending on the application, different degrees of unpredictability are required. For instance, most PUF-based authentication schemes (Sect. 19.5.1) require a strong notion of unpredictability, where the adversary can adaptively obtain a certain number of challenge/response pairs from the PUF of the device under attack and from similar PUFs on other devices [1]. In other applications, such as PUF-based key storage (Sect. 19.5.2), a weaker notion of unpredictability is sufficient, where the adversary is assumed to be unable to obtain challenge/response pairs of the attacked PUF.

The most basic evaluation method that gives a first indication of the unpredictability of a PUF is to compute the Hamming weight of its responses, which shows whether the distribution of the PUF response bits is biased toward '0' or '1'. Ideally, both values should be equiprobable and their fractional Hamming weight[1] should be 0.5 %. An indication of the uniqueness of a PUF can be given by computing the Hamming distance between responses form different PUFs to the same challenge. In the ideal case, responses from different PUFs should be independent and thus their fractional Hamming distance[2] should be 0.5 %.

A more precise assessment of the unpredictability and uniqueness of PUF responses can be done by leveraging statistical tests, such as the DIEHARD [42] or NIST [53] test suites. However, since these test suites are typically based on a series of stochastic tests, they can only give an indication about whether the PUF responses are random or not. Moreover, they often require more input data than typical memory-based PUF implementations can provide. Another approach to empirically assess the unpredictability and uniqueness of PUFs is estimating the entropy of their responses based on experimental data. In particular, *min-entropy* indicates how many bits of a PUF response are uniformly random. The entropy of PUFs can

---

[1] The fractional Hamming weight is the number of bits in a bitstring that are '1' divided by the length of the bitstring.

[2] The fractional Hamming distance is the number of bits that are different in two bitstrings divided by the length of the bitstrings.

be approximated using the context-tree weighting (CTW) method [71, 72], which is an algorithm related to data compression that allows estimating the redundancy of bitstrings [19, 23, 32, 65]. For certain PUF types, the entropy of responses can be computed under consideration of the physical structure and properties of the PUF. For instance, Holcomb et al. [22] compute the entropy of SRAM PUFs based on empirical data under the assumption that the individual bytes of an SRAM array are independent [22]. Alternatively, similar as in symmetric cryptography, the unpredictability of a PUF can be estimated based on the complexity of the best known attack against the unpredictability property [1, 37]. For instance, there are attacks [51] against delay-based PUFs that emulate the PUF in software and allow predicting PUF responses to arbitrary challenges (Sect. 19.3.1).

Evaluation results in literature are difficult to compare due to varying test conditions, different analysis methods and the fact that no representative data sets are publicly available. Hence, a fair comparison of the unpredictability property of different PUF instances based on the results in literature is hardly possible. The development of a common evaluation framework for PUFs and the analysis of PUFs implemented in the same technology is an important topic for future research.

19.3 Attacks Against PUFs and PUF-Based Systems

19.3.1 Emulation Attacks

Most delay-based PUFs are subject to emulation or model building attacks that allow emulating the PUF in software [31, 35, 44, 51]. These attacks collect a number of challenge/response pairs of the PUF and use them to derive a mathematical model, such as a formula that allows estimating the PUF response to a given PUF challenge.

A number of mitigations against these attacks have been proposed [31, 40, 41], which are all based on inserting nonlinearity into the delay circuit (Sect. 19.2.2.1). However, Rhrmair et al. [51] show that these approaches are vulnerable to emulation attacks based on machine-learning techniques, such as logistic regression and evolution strategies. One approach to counter emulation attacks are controlled PUFs [15] (Sect. 19.4.1).

19.3.2 Side-Channel Attacks

Side-channel attacks are hardware attacks that aim to extract secret data, such as cryptographic keys, from an electronic component. Hereby, the adversary observes the behavior (such as the power consumption, electromagnetic radiation, and/or timing behavior) of the component while it is using the secret data to be extracted. Since the behavior of the component is typically dependent on the data processed, it can leak information on this data. The fundamental underlying observation is that processing a data bit of value '1' typically consumes a different amount of power and/or time than processing a data bit of value '0'.

PUFs are typically used in combination with fuzzy extractors (Sect. 19.2.3) and most PUF-based applications (Sect. 19.5) require the plain PUF responses, i.e., before error correction and privacy amplification to be secret. Hence, side-channel attacks against PUF-based systems typically target the fuzzy extractor to gather challenge/response pairs and other information that eases emulation attacks on the underlying PUF (Sect. 19.3.1).

Research on side-channel analysis of PUFs and fuzzy extractors has been recently started and there are only a few published results. Karakoyunlu et al. [25] and Merli et al. [43] show side-channel attacks on implementations of common fuzzy extractors. Furthermore, Merli et al. [43] discuss potential side channel leakages of various PUF types. However, all known side channel attacks on PUF-based systems target the fuzzy extractor and are independent of the underlying PUF.

19.3.3 Fault Injection Attacks

Fault injection attacks aim to prompt erroneous behavior in a device by manipulating it in some way and, when combined with cryptanalysis, can lead to key recovery attacks. Faults may be injected in many ways, for instance by operating the device in extreme environmental conditions or by injecting transient faults into specific components of the device.

Attempts to operate the PUF outside its normal operating conditions, e.g., by varying its supply voltage or ambient temperature, will most likely affect the challenge/response behavior and thus the robustness and unpredictability of the PUF. Moreover, since implementations of fuzzy extractors and the underlying error correction algorithms are typically not resistant to fault injection attacks and exhibit data-dependent behavior, fault injection attacks can cause unintended leakage of PUF-related secret information, such as cryptographic keys bound to the PUF. In particular, most fuzzy extractors are not secure in case the helper data can be modified by the adversary [5]. Thus, robust fuzzy extractors have been proposed to prevent manipulations of helper data [10].

19.4 Advanced PUF Concepts

Several concepts have been proposed to enhance the security properties and functionality of standard PUFs.

19.4.1 Controlled PUFs

Most delay-based PUFs are subject to model building attacks that allow emulating the PUF in software (Sect. 19.3.1). One approach to counter this problem is controlled PUFs by Gassend et al. [15] that use cryptography in hardware to hide the actual PUF response from the adversary. Controlled PUFs typically apply a cryptographic hash function to the PUF challenges and/or responses, which introduces nonlinearity and breaks up the link between the actual PUF response and the output of the controlled PUF. Clearly, this does not address the fundamental weakness of delay-based PUFs. Moreover, to maintain verifiability of the controlled PUF, error correction must be applied before the noisy responses of the underlying PUF are processed by the cryptographic operation, which increases the complexity of the overall construction. Further, to protect against emulation attacks (Sect. 19.3.1), the cryptographic component and the error-correction mechanism as well as their connecting links must be protected against invasive and side-channel attacks, which may be hard to achieve in practice (Sects. 19.3.2 and 19.3.3).

19.4.2 Emulatable PUFs

The verification of PUF responses typically requires a database of reference challenge/response pairs (CRPs). This limits the scalability and efficiency of many PUF-based solutions and can be a serious drawback in many practical applications. One approach to counter this issue are emulatable PUFs, which, similar to public-key cryptography, allow the verification of PUF responses based on a publicly known mathematical model of the PUF. The concept of emulatable and publicly verifiable PUFs has been presented by Rhrmair et al. [48–50] as SIMPL systems (SIMulation Possible but Laborious). A similar concept known as *public PUFs* has been independently presented by Beckmann et al. [3]. The idea of both concepts is that the PUF can be emulated in software using a mathematical model of the physical properties of the PUF. However, this computation is assumed to take significantly more time than evaluating the actual PUF, which can be measured by a verifier in a PUF-based authentication protocol. This allows for the efficient verification of PUF responses by any entity with access to the mathematical model of the PUF, while preventing an algorithmic adversary from impersonating the PUF in the timeframe expected by the verifier. Concrete implementations of SIMPL systems have been presented by Rhrmair et al. [50].

Another approach to remove the need for a challenge/response pair database has been presented by Hammouri et al. [20, 44]. However, in contrast to SIMPL systems and public PUFs, their approach does not allow the public verification of PUF responses and requires the mathematical description of the PUF to be secret information that is only known to authorized entities, such as a verifier in an authentication protocol.

The security properties of practical instantiations of emulatable PUFs still need further evaluation.

19.5 Common Applications of PUFs

The most common applications of PUFs are identification, authentication, and secure key storage.

19.5.1 Device Identification and Authentication

The classical application of PUFs is the identification and authentication of physical objects, such as electronic devices. In fact, PUFs have been first proposed in the context of anti-counterfeiting solutions that prevent cloning (i.e., unauthorized copying) of products. There are many proposals to build identification and authentication schemes based on PUFs for various devices. We focus on solutions that are applicable to resource-constrained embedded devices, such as RFID systems.

One of the first proposals of using PUFs for RFID is by Ranasinghe et al. [47], who propose the manufacturer of a PUF-enabled RFID tag to store a set of challenge/response pairs (CRPs) in a database, which can later be used by RFID readers to identify the tag. The idea is that the reader chooses a challenge from the database, queries the tag and checks whether the database contains a tuple that matches the response received from the tag. One problem of this approach is that CRPs cannot be re-used since this would enable replay attacks. Hence, the number of tag authentications is limited by the number of CRPs in the database. This scheme has been implemented based on arbiter PUFs on RFID tags and its security and usability has been analyzed by Devadas et al. [9]. A similar approach based on the physical characteristics of SRAM cells has been proposed by Holcomb et al. [21].

A privacy-preserving PUF-based device authentication scheme has been presented by Gassend et al. [14]. They suggest to equip each tag with a PUF that is used to frequently derive new tag identifiers. Since readers cannot recompute these identifiers, the readers have access to a database that stores $(ID_0, ID_1, \ldots, ID_n)$ for each legitimate tag, where ID_0 is a random tag identifier and $ID_i = \text{PUF}(ID_{i-1})$ for $1 \le i \le n$. To authenticate to a reader, the tag first sends its current identifier ID_j and then updates its identity to $ID_{j+1} = \text{PUF}(ID_j)$. The reader then checks whether there is a tuple that contains ID_j in the database. In case the reader finds ID_j, it accepts the tag and invalidates all previous database entries ID_k, where $k \le j$ to prevent replay attacks. Another approach to PUF-based authentication by Bolotnyy and Robins [4] aims to prevent unauthorized tracing of tokens. A similar approach to PUF-based authentication has been proposed by Bolotnyy and Robins [4]. A major drawback of these schemes is that tokens can only be authenticated a limited number of times without being re-initialized, which enables denial-of-service attacks.

19.5.2 Secure Key Storage and Key Generation

PUFs can be used to securely bind secrets (such as cryptographic keys) to the physical characteristics of a device. The concept of PUF-based key storage has been presented by Gassend [13] and later generalized to physically obfuscated algorithms by Bringer et al. [7]. Instead of storing the key in nonvolatile memory that is vulnerable to invasive attacks, the key is extracted from the physical properties of the underlying hardware each time it is used. This protects the key against unauthorized readout by invasive attacks, such as probing attacks against nonvolatile memory. Moreover, in case a tamper-evident PUF is used, any attempt to physically extract the key from the PUF circuit changes the challenge/response behavior of the PUF and securely deletes the key bound to the PUF.

Since PUF responses are typically not uniformly random and subject to noise, they cannot be used directly as cryptographic keys. Hence, privacy amplification, which adds additional entropy to the PUF response, and error-correction techniques must be applied before PUF responses can be used as cryptographic keys. The most common approach to achieve this are fuzzy extractors [11] (Sect. 19.2.3).

Tuyls et al. [62] propose to use a PUF-based key storage for the secret authentication key of RFID tags. Since the key is inherently hidden within the physical structure of the PUF, obtaining this key by hardware-related attacks is supposed to be intractable for real-world adversaries [15]. According to Tuyls et al. [62], a PUF-based key storage can be implemented with less than $1,000$ gates, which is well within the capabilities of common RFID tags. Other authentication schemes for RFID exist that use PUF-based key storage to protect against unauthorized tracing of tokens [6, 54] and relay attacks [26].

19.6 Future Directions

19.6.1 Logically Reconfigurable PUFs

So far, most existing PUFs exhibit a static behavior, while a variety of applications would benefit from the availability of PUFs whose characteristics can be changed dynamically, i.e., reconfigured, after deployment. For instance, PUF-based key storage [34, 70] (Sect. 19.5.2) and PUF-based cryptographic primitives [2] may require that previous secrets derived from the PUF cannot be retrieved any more. Another examples are solutions to prevent downgrading of software [30] by binding the software to a certain hardware configuration, such as a PUF, which require the PUF behavior to be irreversibly altered upon installation of a software update.

Unfortunately, all known implementations of physically reconfigurable PUFs rely on optical mechanisms, reconfigurable hardware (such as FPGAs), or novel memory technologies [30], which all have serious drawbacks in practice. In particular, optical PUFs cannot be easily integrated into integrated circuits and require expensive and error-prone evaluation equipment, while FPGA-based solutions cannot be realized with non-reconfigurable hardware (such as ASICs) that is commonly used in practice [37].

In this context, several attempts to emulate physically reconfigurable PUFs have been made. One of the first proposals was integrating a floating gate transistor into the delay lines of an arbiter PUF, which allows physically changing the challenge/response behavior of the PUF based on some state maintained in nonvolatile memory [33, 34]. Other approaches restrict access to the interface of the PUF and use part of the PUF challenge as reconfiguration data [30, 31], which, however, works only for certain PUF types.

The concept and security properties of logically reconfigurable physical unclonable functions (LR-PUFs) have been recently formalized by Katzenbeisser et al. [27]. In contrast to classical, typically static PUFs, LR-PUFs can be dynamically reconfigured after deployment such that their challenge/response behavior changes in a random manner without replacing or physically modifying the PUF. The idea is amending a conventional PUF with stateful control logic that transforms challenges and responses of the PUF (Fig. 19.4). Katzenbeisser et al. [27] present and evaluate two different constructions for LR-PUFs that are simple, efficient, and can be easily implemented.

19.6.2 PUF-Based Remote Attestation

Remote attestation is a mechanism to report the software state of a remote computing platform (*prover*) to a verifying party (*verifier*). This generally requires trusted hardware to securely record and transmit the system state of the prover to the verifier. However, trusted hardware is often too expensive for resource-constrained embedded devices, such as wireless sensor nodes and RFIDs. Hence, software attestation was proposed as a lightweight alternative that exploits the computational constraints of a device to make statements about its internal software state [57, 58]. Specifically,

Fig. 19.4 LR-PUF concept

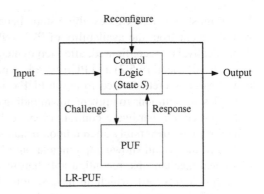

software attestation requires the prover to compute the response R to a given attestation challenge C within a given time frame. When receiving the correct response in the expected time, the verifier has assurance that only a specific attestation algorithm could have been executed within that time frame. The attestation algorithm is implemented as a checksum function that iteratively merges information gathered from the device, such as program memory samples, into the attestation response R. Hence, a timely and correctly computed attestation response provides assurance to the verifier that the prover is in the expected system state.

Standard software attestation makes two major assumptions: (1) the computational capabilities of the prover are known to the verifier and unmodified, and (2) the attestation algorithm is indeed computed by the prover and not delegated to another device. In most scenarios, software attestation is thus limited to attest only *local* provers such that their identity can be directly verified and undesired communication interfaces can be disabled. But even a local prover does not always guarantee that its identity is authentic, e.g., if multiple hardware revisions of apparently identical devices exist.

To overcome these problems, the attestation response R must be linked to the hardware it was computed on, which can be achieved by using PUFs [55, 56]. The idea is to include the responses of the prover's PUF into the computation of R while the software attestation is running. To assure that the attestation algorithm is not outsourced, the PUF is queried sufficiently often to overwhelm all external communication interfaces of the prover. Thus, the constraints of the communication interfaces of the prover are exploited, similar to the computational constraints exploited by standard software attestation. Due to the uniqueness of the PUF responses and their tight integration into the attestation algorithm, a correct and timely attestation response R provides assurance on the identity of a remote device as well as the integrity of its software state. A practical implementation PUF-based attestation has been presented by Schulz et al. [28, 56].

19.7 Open Questions and Challenges

Practical PUF Designs. Known electronic PUFs may be compromised since delay-based PUFs can be emulated using machine-learning techniques (Sect. 19.3.1) and memory-based PUFs can be read out completely since they have only a limited response space. While these PUFs can be used in many applications, such as PUF-based key storage (Sect. 19.5.2) and controlled PUFs (Sect. 19.4.1), that ensure that the adversary cannot access the challenge/response pairs of the PUF, the use of these PUFs in applications with strong unclonability and unpredictability requirements, such as device authentication schemes (Sect. 19.5.1) must be carefully considered. Moreover, in general PUF responses can be verified only when the verifier has access to a database of previously recorded challenge/response pairs (CRPs), which may lead to scalability problems in practice. Hence, one open challenge is the development and implementation of novel PUF designs that achieve the requirements of many existing theoretical PUF-based security solutions in literature, including resistance to emulation attacks, large (ideally exponential) challenge/response space to prevent complete readout of the PUF, public verifiability (i.e., no CRP database required to verify PUF response), tamper-evidence, physical reconfigurability, and small hardware footprint.

Common Evaluation Framework for PUFs. Currently, there is no common evaluation framework for PUFs that allows assessing and quantifying the security properties of real PUF implementations. The security properties of existing PUF-based security solutions in literature are proven in PUF security models that are typically not general enough and exclude certain PUF types, do not reflect all properties of real PUF implementations, or include security parameters that cannot be determined for real PUF implementations in practice. Hence, it is unclear whether these schemes can actually be implemented securely. Therefore, another open challenge is the development of a common evaluation framework for the analysis of PUF implementations that (1) captures the security properties of PUFs according to modern cryptographic standards and can be used to assess the security of PUF-based cryptographic schemes and security solutions, and (2) allows for empirically assessing and quantifying the most important properties of PUFs, including robustness, physical unclonability, unpredictability of responses to the same challenge to different PUF instances and to different challenges to the same PUF instance, and tamper-evidence. A promising first step in this direction has been presented by Armknecht et al. [1]. However, they do not consider all security properties of PUFs and do not show how their approach applies to other PUF implementations than SRAM PUFs.

Side-channel Analysis of PUFs. Many PUF-based applications such as PUF-based key storage require PUF responses to be inaccessible to the adversary, which is typically justified by the assumption of the PUF being tamper-evident so that any attempt to physically access the PUF response (such as an invasive attack) permanently changes the challenge/response behavior of the PUF. However, even when a tamper-evident PUF (such as a coating PUF) is used, it is currently unclear whether existing PUF implementations in integrated circuits leak information on their response over side channels, such as electromagnetic radiation or power consumption. Hence, the analysis of the side-channel leakage of known PUF implementations is an interesting open research problem.

19.8 Conclusion

Physically unclonable functions are a very interesting and promising approach to increase the security of embedded systems. They open new directions toward lightweight privacy-preserving protocols based on physical assumptions and cost-effective tamper-evident storage for cryptographic secrets that even cannot be learned or reproduced by the manufacturer of the corresponding PUF.

However, several aspects of PUFs and their deployment require further research. Since PUFs are bound to the device in which they are embedded, no other entity can verify the response r of a PUF to a given challenge c without knowing an authentic challenge/response pair (c, r) in advance. Current PUF-based protocols aim at circumventing this problem by providing the reader with a database that contains a set of challenge/response pairs that act as reference values for the responses of the interrogated PUF. However, this approach is not scalable and opens the possibility of replay-attacks. Furthermore, PUFs realizations require careful statistical testing before they can be safely deployed to real security-critical products, while, to our knowledge, there is no complete security evaluation framework for PUFs yet.

Acknowledgments This work has been supported in part by the European Commission under grant agreement ICT-2007-238811 UNIQUE.

References

1. Armknecht, F., Maes, R., Sadeghi, A.R., Standaert, F.X., Wachsmann, C.: A formal foundation for the security features of physical functions. In: IEEE Symposium on Security and Privacy (SSP), pp. 397–412. IEEE Computer Society (2011)
2. Armknecht, F., Maes, R., Sadeghi, A.R., Sunar, B., Tuyls, P.: Memory leakage-resilient encryption based on physically unclonable functions. In: M. Matsui (ed.) Advances in Cryptology (ASIACRYPT), *Lecture Notes in Computer Science (LNCS)*, vol. 5912, pp. 685–702. Springer Berlin/Heidelberg, Berlin, Heidelberg (2009)
3. Beckmann, N., Potkonjak, M.: Hardware-based public-key cryptography with public physically unclonable functions. In: S. Katzenbeisser, A.R. Sadeghi (eds.) Information Hiding (IH), *Lecture Notes in Computer Science (LNCS)*, vol. 5806, pp. 206–220. Springer Berlin/Heidelberg, Berlin, Heidelberg (2009)
4. Bolotnyy, L., Robins, G.: Physically unclonable function-based security and privacy in RFID systems. In: Conference on Pervasive Computing and Communications (PerCom), pp. 211–220. IEEE (2007)
5. Boyen, X.: Reusable cryptographic fuzzy extractors. In: ACM Conference on Computer and Communications Security (ACM CCS), pp. 82–91. ACM, New York, NY, USA (2004)
6. Bringer, J., Chabanne, H., Icart, T.: Improved privacy of the tree-based hash protocols using physically unclonable functions. In: R. Ostrovsky, R. De Prisco, I. Visconti (eds.) Security and Cryptography for Networks (SCN), *Lecture Notes in Computer Science (LNCS)*, vol. 5229, pp. 77–91. Springer Berlin/Heidelberg, Berlin, Heidelberg (2008)
7. Bringer, J., Chabanne, H., Icart, T.: On physical obfuscation of cryptographic algorithms. In: B. Roy, N. Sendrier (eds.) International Conference on Cryptology in India (INDOCRYPT), *Lecture Notes in Computer Science (LNCS)*, vol. 5922, pp. 88–103. Springer Berlin/Heidelberg, Berlin, Heidelberg (2009)
8. Brzuska, C., Fischlin, M., Schröder, H., Katzenbeisser, S.: Physically uncloneable functions in the universal composition framework. In: P. Rogaway (ed.) Advances in Cryptology (CRYPTO), *Lecture Notes in Computer Science (LNCS)*, vol. 6841, pp. 51–70. Springer Berlin/Heidelberg, Berlin, Heidelberg (2011)
9. Devadas, S., Suh, E., Paral, S., Sowell, R., Ziola, T., Khandelwal, V.: Design and implementation of PUF-based unclonable RFID ICs for anti-counterfeiting and security applications. RFID, 2008 IEEE International Conference on pp. 58–64 (2008)
10. Dodis, Y., Katz, J., Reyzin, L., Smith, A.: Robust fuzzy extractors and authenticated key agreement from close secrets. In: C. Dwork (ed.) Advances in Cryptology (CRYPTO), *Lecture Notes*

in Computer Science (LNCS), vol. 4117, pp. 232–250. Springer Berlin/Heidelberg, Berlin, Heidelberg (2006)

11. Dodis, Y., Reyzin, L., Smith, A.: Fuzzy extractors: How to generate strong keys from biometrics and other noisy data. In: C. Cachin, J. Camenisch (eds.) Advances in Cryptology (EUROCRYPT), *Lecture Notes in Computer Science (LNCS)*, vol. 3027, pp. 523–540. Springer Berlin/Heidelberg, Berlin, Heidelberg (2004)

12. Eichhorn, I., Koeberl, P., van der Leest, V.: Logically reconfigurable PUFs: Memory-based secure key storage. In: ACM Workshop on Scalable Trusted Computing (ACM STC), pp. 59–64. ACM, New York, NY, USA (2011)

13. Gassend, B.: Physical random functions. Master's thesis, Department of Electrical Engineering and Computer Science, Massachusetts Institute of Technology (MIT), The Stata Center, 32 Vassar Street, Cambridge, Massachusetts 02139 (2003)

14. Gassend, B., Clarke, D., van Dijk, M., Devadas, S.: Controlled physical random functions. In: Annual Computer Security Applications Conference (ACSAC), pp. 149–160. IEEE (2002)

15. Gassend, B., Clarke, D., van Dijk, M., Devadas, S.: Silicon physical random functions. In: ACM Conference on Computer and Communications Security (ACM CCS), pp. 148–160. ACM, New York, NY, USA (2002)

16. Guajardo, J., Kumar, S., Schrijen, G.J., Tuyls, P.: FPGA intrinsic PUFs and their use for IP protection. In: P. Paillier, I. Verbauwhede (eds.) Cryptographic Hardware and Embedded Systems (CHES), *Lecture Notes in Computer Science (LNCS)*, vol. 4727, pp. 63–80. Springer Berlin/Heidelberg, Berlin, Heidelberg (2007)

17. Guajardo, J., Kumar, S.S., Schrijen, G.J., Tuyls, P.: Physical unclonable functions and public-key crypto for FPGA IP protection. In: Field Programmable Logic and Applications (FPL), pp. 189–195. IEEE (2007)

18. Guajardo, J., Kumar, S.S., Schrijen, G.J., Tuyls, P.: Brand and IP protection with physical unclonable functions. In: IEEE International Symposium on Circuits and Systems (ISCAS), pp. 3186–3189. IEEE (2008)

19. Hammouri, G., Dana, A., Sunar, B.: CDs have fingerprints too. In: C. Clavier, K. Gaj (eds.) Cryptographic Hardware and Embedded Systems (CHES), *Lecture Notes in Computer Science (LNCS)*, vol. 5747, pp. 348–362. Springer Berlin/Heidelberg, Berlin, Heidelberg (2009)

20. Hammouri, G., Öztürk, E., Birand, B., Sunar, B.: Unclonable lightweight authentication scheme. In: L. Chen, M.D. Ryan, G. Wang (eds.) International Conference on Information and Communications Security (ICICS), *Lecture Notes in Computer Science (LNCS)*, vol. 5308, pp. 33–48. Springer Berlin/Heidelberg, Berlin, Heidelberg (2008)

21. Holcomb, D., Burleson, W., Fu, K.: Initial SRAM state as a fingerprint and source of true random numbers for RFID tags. In: Workshop on RFID Security (RFIDSec) (2007)

22. Holcomb, D., Burleson, W.P., Fu, K.: Power-up SRAM state as an identifying fingerprint and source of true random numbers. IEEE Transactions on Computers 58(9), 1198–1210 (2009)

23. Ignatenko, T., Schrijen, G.J., Škorić, B., Tuyls, P., Willems, F.: Estimating the secrecy-rate of physical unclonable functions with the context-tree weighting method. In: IEEE International Symposium on Information Theory (ISIT), pp. 499–503. IEEE (2006)

24. Intrinsic ID: Website. http://www.intrinsic-id.com/products.htm (2012)

25. Karakoyunlu, D., Sunar, B.: Differential template attacks on PUF enabled cryptographic devices. In: Workshop on Information Forensics and Security (WIFS), pp. 1–6. IEEE (2010)

26. Kardas, S., Kiraz, M.S., Bingol, M.A., Demirci, H.: A novel RFID distance bounding protocol based on physically unclonable functions. In: Radio Frequency Identification: Security and Privacy Issues (RFIDSec), *Lecture Notes in Computer Science (LNCS)*. Springer Berlin/Heidelberg, Berlin, Heidelberg (2011)

27. Katzenbeisser, S., Kocabaş, U., van der Leest, V., Sadeghi, A.R., Schrijen, G.J., Schröder, H., Wachsmann, C.: Recyclable PUFs: Logically reconfigurable PUFs. In: Workshop on Cryptographic Hardware and Embedded Systems (CHES), vol. 6917, pp. 374–389. Springer Berlin/Heidelberg, Berlin, Heidelberg (2011)

28. Kocabas, Ü., Sadeghi, A.R., Schulz, S., Wachsmann, C.: Poster: Practical embedded remote attestation using physically unclonable functions (2011)

29. Kumar, S.S., Guajardo, J., Maes, R., Schrijen, G.J., Tuyls, P.: Extended abstract: The butterfly PUF protecting IP on every FPGA. In: Workshop on Hardware-Oriented Security (HOST), pp. 67–70. IEEE (2008)
30. Kursawe, K., Sadeghi, A.R., Schellekens, D., Skoric, B., Tuyls, P.: Reconfigurable physical unclonable functions – Enabling technology for tamper-resistant storage. In: Workshop on Hardware-Oriented Security and Trust (HOST), pp. 22–29. IEEE (2009)
31. Lee, J.W., Lim, D., Gassend, B., Suh, E.G., van Dijk, M., Devadas, S.: A technique to build a secret key in integrated circuits for identification and authentication applications. In: Symposium on VLSI Circuits, pp. 176–179. IEEE (2004)
32. van der Leest, V., Schrijen, G.J., Handschuh, H., Tuyls, P.: Hardware intrinsic security from D flip-flops. In: ACM Workshop on Scalable Trusted Computing (ACM STC), pp. 53–62. ACM, New York, NY, USA (2010)
33. Lim, D.: Extracting secret keys from integrated circuits. Master's thesis, Department of Electrical Engineering and Computer Science, Massachusetts Institute of Technology (MIT), The Stata Center, 32 Vassar Street, Cambridge, Massachusetts 02139 (2004)
34. Lim, D., Lee, J.W., Gassend, B., Suh, E.G., van Dijk, M., Devadas, S.: Extracting secret keys from integrated circuits. IEEE Transactions on Very Large Scale Integration (VLSI) Systems 13(10), 1200–1205 (2005)
35. Lin, L., Holcomb, D., Krishnappa, D.K., Shabadi, P., Burleson, W.: Low-power sub-threshold design of secure physical unclonable functions. In: International Symposium on Low-Power Electronics and Design (ISLPED), pp. 43–48. IEEE (2010)
36. Maes, R., Tuyls, P., Verbauwhede, I.: Intrinsic PUFs from flip-flops on reconfigurable devices. In: Benelux Workshop on Information and System Security (2008)
37. Maes, R., Verbauwhede, I.: Physically unclonable functions: A study on the state of the art and future research directions. In: A.R. Sadeghi, D. Naccache (eds.) Towards Hardware-Intrinsic Security, Information Security and Cryptography, pp. 3–37. Springer Berlin/Heidelberg, Berlin, Heidelberg (2010)
38. Maiti, A., Casarona, J., McHale, L., Schaumont, P.: A large scale characterization of RO-PUF. In: Symposium on Hardware-Oriented Security and Trust (IIOST), pp. 94–99. IEEE (2010)
39. Majzoobi, M., Koushanfar, F., Potkonjak, M.: Lightweight secure PUFs. In: International Conference on Computer-Aided Design (ICCAD), pp. 670–673. IEEE (2008)
40. Majzoobi, M., Koushanfar, F., Potkonjak, M.: Testing techniques for hardware security. In: International Test Conference (ITC), pp. 1–10. IEEE (2008)
41. Majzoobi, M., Koushanfar, F., Potkonjak, M.: Techniques for design and implementation of secure reconfigurable PUFs. ACM Transactions on Reconfigurable Technology and Systems (TRETS) 2(1), 1–33 (2009)
42. Marsaglia, G.: The marsaglia random number CDROM including the Diehard battery of tests of randomness. http://www.stat.fsu.edu/pub/diehard/
43. Merli, D., Schuster, D., Stumpf, F., Sigl, G.: Side-channel analysis of PUFs and fuzzy extractors. In: J.M. McCune, B. Balacheff, A. Perrig, A.R. Sadeghi, A. Sasse, Y. Beres (eds.) Trust and Trustworthy Computing (TRUST), Lecture Notes in Computer Science (LNCS), vol. 6740, pp. 33–47. Springer Berlin/Heidelberg, Berlin, Heidelberg (2011)
44. Öztürk, E., Hammouri, G., Sunar, B.: Towards robust low cost authentication for pervasive devices. In: Conference on Pervasive Computing and Communications (PerCom), pp. 170–178. IEEE, Washington, DC, USA (2008)
45. Pappu, R., Recht, B., Taylor, J., Gershenfeld, N.: Physical one-way functions. Science 297(5589), 2026–2030 (2002)
46. Posch, R.: Protecting devices by active coating. Journal of Universal Computer Science 4(7), 652–668 (1998)
47. Ranasinghe, D.C., Engels, D.W., Cole, P.H.: Security and privacy: Modest proposals for low-cost RFID systems. In: Auto-ID Labs Research Workshop (2004)
48. Rührmair, U.: SIMPL systems: On a public key variant of physical unclonable functions. Cryptology ePrint Archive, Report 2009/255 (2009)

49. Rührmair, U.: SIMPL systems, or: Can we design cryptographic hardware without secret key information? In: I. Černá, T. Gyimóthy, J. Hromkovič, K. Jefferey, R. Královič, M. Vukolić, S. Wolf (eds.) Current Trends in Theory and Practice of Computer Science (SOFSEM), *Lecture Notes in Computer Science (LNCS)*, vol. 6543, pp. 26–45. Springer Berlin/Heidelberg, Berlin, Heidelberg (2011)

50. Rührmair, U., Chen, Q., Stutzmann, M., Lugli, P., Schlichtmann, U., Csaba, G.: Towards electrical, integrated implementations of SIMPL systems. In: P. Samarati, M. Tunstall, J. Posegga, K. Markantonakis, D. Sauveron (eds.) Workshop on Information Security Theory and Practices (WISTP), *Lecture Notes in Computer Science (LNCS)*, vol. 6033, pp. 277–292. Springer Berlin/Heidelberg, Berlin, Heidelberg (2010)

51. Rührmair, U., Sehnke, F., Sölter, J., Dror, G., Devadas, S., Schmidhuber, J.: Modeling attacks on physical unclonable functions. In: ACM Conference on Computer and Communications Security (ACM CCS), pp. 237–249. ACM, New York, NY, USA (2010)

52. Rührmair, U., Sölter, J., Sehnke, F.: On the foundations of physical unclonable functions. Cryptology ePrint Archive, Report 2009/277 (2009)

53. Rukhin, A., Soto, J., Nechvatal, J., Smid, M., Barker, E., Leigh, S., Levenson, M., Vangel, M., Banks, D., Heckert, A., Dray, J., Vo, S.: A statistical test suite for random and pseudorandom number generators for cryptographic applications. Special Publication 800–22 Revision 1a, NIST (2010)

54. Sadeghi, A.R., Visconti, I., Wachsmann, C.: Enhancing RFID security and privacy by physically unclonable functions. In: A.R. Sadeghi, D. Naccache (eds.) Towards Hardware-Intrinsic Security, Information Security and Cryptography, pp. 281–305. Springer Berlin/Heidelberg, Berlin, Heidelberg (2010)

55. Schulz, S., Sadeghi, A.R., Wachsmann, C.: Short paper: Lightweight remote attestation using physical functions. In: ACM Conference on Wireless Network Security (WiSec), pp. 109–114. ACM, New York, NY, USA (2011)

56. Schulz, S., Wachsmann, C., Sadeghi, A.R.: Lightweight remote attestation using physical functions. Tech. rep., Center for Advanced Security Research Darmstadt (CASED), Germany, Mornewegstraße 32, 64293 Darmstadt, Germany (2011)

57. Seshadri, A., Luk, M., Shi, E., Perrig, A., van Doorn, L., Khosla, P.: Pioneer: Verifying code integrity and enforcing untampered code execution on legacy systems. In: ACM Symposium on Operating Systems Principles (SOSP), vol. 39, pp. 1–16. ACM, New York, NY, USA (2005)

58. Seshadri, A., Perrig, A., van Doorn, L., Khosla, P.: SWATT: SoftWare-based ATTestation for embedded devices. In: IEEE Symposium on Security and Privacy (SSP), pp. 272–282. IEEE, Los Alamitos, CA, USA (2004)

59. Su, Y., Holleman, J., Otis, B.P.: A 1.6pJ/bit 96% stable chip-ID generating circuit using process variations. In: International Solid-State Circuits Conference (ISSCC), pp. 406–611. IEEE (2007)

60. Su, Y., Holleman, J., Otis, B.P.: A digital 1.6 pJ/bit chip identification circuit using process variations. IEEE Journal of Solid-State Circuits 43(1), 69–77 (2008)

61. Suh, E.G., Devadas, S.: Physical unclonable functions for device authentication and secret key generation. In: ACM/IEEE Design Automation Conference (DAC), pp. 9–14. IEEE (2007)

62. Tuyls, P., Batina, L.: RFID-tags for anti-counterfeiting. In: D. Pointcheval (ed.) Topics in Cryptology (CT-RSA), *Lecture Notes in Computer Science (LNCS)*, vol. 3860, pp. 115–131. Springer Berlin/Heidelberg, Berlin, Heidelberg (2006)

63. Tuyls, P., Schrijen, G.J., Škorić, B., van Geloven, J., Verhaegh, N., Wolters, R.: Read-proof hardware from protective coatings. In: L. Goubin, M. Matsui (eds.) Cryptographic Hardware and Embedded Systems (CHES), *Lecture Notes in Computer Science (LNCS)*, vol. 4249, pp. 369–383. Springer Berlin/Heidelberg, Berlin, Heidelberg (2006)

64. Tuyls, P., Škorić, B.: Secret key generation from classical physics: Physical uncloneable functions. In: S. Mukherjee, R.M. Aarts, R. Roovers, F. Widdershoven, M. Ouwerkerk (eds.) AmIware Hardware Technology Drivers of Ambient Intelligence, Philips Research Book Series, vol. 5, pp. 421–447. Springer Netherlands, Dordrecht (2006)

65. Tuyls, P., Škorić, B., Ignatenko, T., Willems, F., Schrijen, G.J.: Entropy estimation for optical PUFs based on context-tree weighting methods. In: P. Tuyls, B. Škorić, T. Kevenaar (eds.) Security with Noisy Data, pp. 217–233. Springer London, London (2007)
66. Tuyls, P., Škorić, B., Stallinga, S., Akkermans, A.H.M., Ophey, W.: Information-theoretic security analysis of physical uncloneable functions. In: A. Patrick, M. Yung (eds.) Financial Cryptography and Data Security (FC), *Lecture Notes in Computer Science (LNCS)*, vol. 3570, p. 578. Springer Berlin/Heidelberg, Berlin, Heidelberg (2005)
67. Verayo, Inc.: Website. http://www.verayo.com/product/products.html (2012)
68. Škorić, B., Maubach, S., Kevenaar, T., Tuyls, P.: Information-theoretic analysis of capacitive physical unclonable functions. Journal of Applied Physics **100**(2), 024,902–024,902–11 (2006)
69. Škorić, B., Maubach, S., Kevenaar, T., Tuyls, P.: Information-theoretic analysis of coating PUFs. Cryptology ePrint Archive, Report 2006/101 (2006)
70. Škorić, B., Tuyls, P., Ophey, W.: Robust key extraction from physical uncloneable functions. In: J. Ioannidis, A. Keromytis, M. Yung (eds.) Applied Cryptography and Network Security (ACNS), *Lecture Notes in Computer Science (LNCS)*, vol. 3531, pp. 99–135. Springer Berlin/Heidelberg, Berlin, Heidelberg (2005)
71. Willems, F.M.J.: CTW website. http://www.ele.tue.nl/ctw/
72. Willems, F.M.J., Shtarkov, Y.M., Tjalkens, T.J.: The context-tree weighting method: Basic properties. IEEE Transactions on Information Theory **41**(3), 653–664 (1995)

Chapter 20
SCADA System Cyber Security

Igor Nai Fovino

Abstract Modern industrial systems (e.g. power plants, water plants, chemical installation, etc.) make large use of information and communication technologies (ICT). In the past years, those systems started to use public networks (i.e. the Internet) for system-to-system interconnection, to provide new features and services. The migration from the traditional isolated system approach to an open system approach exposed these infrastructures to cyber-threats. The scope of this chapter is provide the reader with an overview of the cyber threats and vulnerabilities affecting the system control and data acquisition systems (SCADA), i.e. those systems in charge for monitoring and controlling the industrial processes, providing indications on possible mitigation techniques.

Keywords SCADA · DCS · Cyber security

20.1 Introduction

Modern industrial systems (e.g. power plants, water plants, chemical installation, etc.) make large use of ICT technologies. In the past years, those systems started to use public networks (i.e. the Internet) for system-to-system interconnection. As a result, thanks to this architectural advance, it was possible to provide new services and features. The technological shift from a virtually isolated architecture to an open (even if regulated) and interconnected system-of-system architecture has dramatically impacted the security of these installations. The case of Stuxnet [1] the first malware able to take the control of a field network, made the exposure to cyber-threats of system control and data acquisition systems (SCADA) extremely evident. SCADA systems are generally prone to different categories of vulnerabilities: protocol-based vulnerabilities, architectural vulnerabilities but also procedural

I. N. Fovino (✉)
Institute for the Protection and Security of the Citizen, Joint Research Centre, European Commission, via E.Fermi 1, Ispra 21027, Italy
e-mail: igor.nai@gmail.com

K. Markantonakis and K. Mayes (eds.), *Secure Smart Embedded Devices,*
Platforms and Applications, DOI: 10.1007/978-1-4614-7915-4_20,
© Springer Science+Business Media New York 2014

and conceptual weaknesses. The comprehension of these threats is the basis on which build more secure industrial infrastructures. The scientific community started to pay attention to the security of SCADA systems only few years ago. For an accessible survey, we point to [2]. Other surveys analyse more specific topics. Adam et al., in [13], presented an interesting high-level analysis of the possible threats to the SCADA system of power plants, a categorisation of the typical hardware devices involved and some high level discussion about the intrinsic vulnerabilities of the common power plant architectures. A more detailed work on the topic of SCADA security is presented by Chandia et al. [3]. In this work, the authors describe two possible strategies for securing SCADA networks, pointing out that several aspects have to be improved in order to secure these architectures. Some works discuss the security of SCADA communication protocols: for example, Majdalawieh et al. [19] presented an extension of the DNP3 protocol (we remind here that DNP3 stands for Distributed Network Protocol, and its a protocol widely used in SCADA systems), called DNPsec, which tries to address some of the known security problems of that Master-Slave control protocol (i.e. integrity of the commands, authentication, non repudiation, etc.). Similar approaches have been presented also by Heo et al. [7] while Mander et al. [6] presented a proxy-filtering solution aiming at identifying and avoiding anomalous control traffic. To understand the real exposure of SCADA systems to cyber-threats, testing campaigns have been conducted in recent years. In this context Masera et al. [15], presented the results of a field test campaign studying the real effects of a set of well-identified attack scenarios against the SCADA systems of an electric distribution station and a real power plant.

The goal of this chapter is to provide to the reader an overview of the main threats to which SCADA systems are exposed and to provide pointers to possible solutions. In the following, after providing an overview of a typical SCADA system, through concrete examples, the major classes of vulnerabilities affecting these systems will be described and possible mitigation actions will be presented and discussed. [1]

20.2 SCADA Architecture Overview

SCADA systems play a relevant role in the control and management of industrial installation. From a conceptual point of view, they are a "distributed control system" spread between two networks:

- *The Control (or Field) Network*: It hosts all the devices that, on one side, control the actuators and sensors of the physical layer and on the other side provide the *control interface* to the process network. A typical control network is composed of the following macro-elements: Programmable Logic Controller (PLC)-Remote Terminal Units (RTU) and Actuators. The PLCs receive data from the physical layer, elaborate on the basis of this data a "local strategy", and send back to the

[1] The content of this chapter summarises the results of the research activity conducted by the author within the context of the EU funded Escort Project (http://www.escortsproject.eu/)

actuators a set of appropriate commands. Moreover, the same PLCs provide, when requested, the data received from the physical layer to the SCADA servers in the process network, and execute the commands received by them.

- *The Process Network*: It contains the SCADA servers used to keep under control the physical layer. This network hosts the following elements:

  - *SCADA server/DCS*: In the literature of industrial control system, it is indeed made distinction between SCADA servers and Distributed Control Systems (DCS). The former more event driven and oriented to data gathering, the latter more process-oriented. In the modern world, the differences between SCADA systems and DCS are becoming more and more weak as the functionalities provided by these two systems are today almost overlapping. For that reason in the following they will be presented as if they are the same thing (even if they are in principle different). The goal of modern SCADA servers and DCS is to monitor and control processes and events which run in the industrial installation.
  - *Human Machine Interface (HMI)*: It provides a user friendly interface between the operator and the SCADA server.
  - *OPC Server*: OPC [8] stands for OLE for Process Control. It is a suite of standards defining a set of objects, interfaces and methods to facilitate interoperability among industrial control system (ICS) devices. An OPC Server is a software application that acts as an Application programming interface (API) or protocol converter. An OPC Server will connect to a device such as a PLC, DCS, RTU or a data source such as a database or user interface, and translate the data into a standard-based OPC format. OPC compliant applications such as an HMI, historian, spreadsheet, trending application, etc., can connect to the OPC Server and use it to read and write device data. OPC Servers are based on Server/Client architectures.
  - *Builder Server*: The Builder Server is the server usually used in order to "program" the PLCs. In order to work correctly it has to be directly connected with the control network, to be able to download the control code on the related PLC.

The glue allowing the system components to stay together is constituted by network devices (switches, transmitters, wireless devices). They are used to deliver data and commands across all the system. Being the most evident access point to a typical SCADA system, they are generally protected with traditional cyber-security appliances such as firewalls, intrusion detection systems, etc. Figure 20.1 provides a high level picture of the SCADA cyber layer, the Process Network and the control network.

20.2.1 SCADA Protocols Overview

The core of every SCADA system is constituted by the suite of communication protocols used to exchange commands/information between masters and slaves. SCADA protocols were designed in the 1970s for serial communication and only in the 1990s they have been "ported" on TCP/IP. Examples of SCADA protocols are

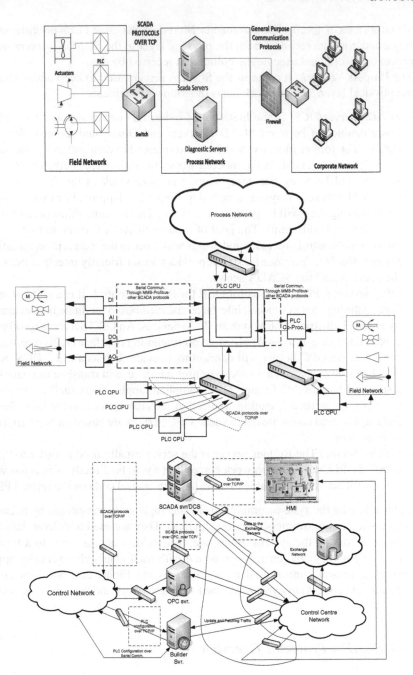

Fig. 20.1 **a** High level schema of the SCADA cyber layer, **b** Control network, **c** Process network

Modbus [18], DNP3 [19] (available for both serial and TCP/IP communication channels), IEC 60870-5 [20], Profibus [21] (serial communication), etc. Several vulnerabilities in the SCADA architectures are due to weaknesses of the SCADA protocols. To understand them, in the following we will provide an overview of one of the simplest and most used SCADA protocols, Modbus. We can consider this protocol as a good example to understand the tasks, the structure and the vulnerability classes associated with all the modern SCADA communication protocols.

20.2.1.1 Modbus

Modbus is an application layer messaging protocol, positioned at level 7 of the OSI model providing client/server communication between devices connected on different types of buses or networks. The Modbus transmission protocol was developed by Gould Modicon for process control systems. To get interconnected the devices can use either serial buses or TCP/IP-based communication channels. Almost all manufacturers produce equipment supporting the Modbus protocol and many systems using such protocols are involved in industrial process. It can be considered as a de facto industrial standard. The Modbus operates according to a classical master/slave principle; the protocol allows one master to manage up to 247 slaves. Only the master initiates a transaction. Such transactions can be: Query/response type (only a single slave is addressed) or Broadcast/no response type (all slaves are addressed). The Modbus protocol provides frames for the transmission of messages between master and slaves. The following steps compose the communication scheme:

1. The master sends a message containing the address of the intended receiver, what the receiver must do, the data needed to perform the action and a cyclic redundancy check (CRC) code.
2. The slave reads the message, and if the packet format is correct, it performs the required task and sends a response back to the master. Such reply usually contains the slave address, the action performed, the result of the action and a CRC code. If the initial message was of a broadcast type, there is no response from the slaves. The master can send another query as soon as it has received the response message. A time-out mechanism is used to avoid "deadlock" states.

An overview of the security weaknesses of Modbus is presented in Sect. 20.3.4.1.

20.3 SCADA Vulnerabilities and Attacks

Studies on the security of SCADA systems have showed how these systems are far from being considered secure and safe (we point the readers to the following works for a comprehensive survey in this context [2, 15]) To classify these vulnerabilities a *multi-layered* approach has been adopted: vulnerabilities can be caused by design errors (in architectures, in processes, in policies, in protocols) and implementation errors (again in protocols, software (SW), policies). Under this light in this chapter we adopt the following categorisation: *architectural vulnerabilities, security policy vulnerabilities, software vulnerabilities, communication protocol vulnerabilities*. In the following we provide a brief description of these categories.

20.3.1 Architectural Vulnerabilities

These vulnerabilities directly derive from weaknesses of the SCADA architecture adopted. Modern SCADA architectures are in general not so different in principle from the architectures used in the 1980s and 1990s for the following reasons:

1. These architectures have been tested for decades and then they are considered stable and well known.
2. The cost of the deployment of a completely new architecture is not negligible.
3. There is not a clear perception from the process engineers and the line managers of the benefit of investments in new architectures.

Migrating from an *isolated* to an *open* environment, from serial communication to TCP/IP communication, "traditional SCADA architectures" started to show all their limits. Process engineers tried to fix the new security problems, following the examples of the traditional ICT world, without considering that the peculiarities of the SCADA systems in several situations badly match with the constraints of traditional ICT security. Examples of architectural vulnerabilities common in typical SCADA systems are:

• The weak separation between the Process network and the Control Network. Since the field network is the inner part of the whole architecture, generally it is not strongly separated from the process network because, apparently, it is improbable for an attacker to be able to reach this area. However, several studies [12, 16] have showed how it is indeed relatively easy to reach this part of the network for someone determined at to damage the target system.
• The lack of authentication between the active components of the SCADA system (e.g. actuators-SCADA servers, actuators-RTUs, SCADA-Data exchange Servers). Since the control network has always been a closed environment, there was not a need for integrating authentication mechanisms among the different elements connected on this network. However, today the lack of authentication is an architectural vulnerability, which can be easily used in order to inflict any sort of damage (see next sections for details).
• The single point of failure represented by the process firewall (FW) and by remote access devices [(e.g. authentication servers (Radius server)].
• The scarce attention to network load balancing and redundancy.

Architectural vulnerabilities cannot easily be mitigated due to the impact architectures have on performance and specific needs of industrial systems. Moreover modifications of ICT architecture produce usually high costs for a company since their application often implies a stop of the production system.

20.3.2 Security Policy Vulnerabilities

An architecturally strong system still is vulnerable if it is badly maintained or badly used. In other words, a system in order to be considered robust has to be provided with both technical and policy level security layers. While in general the security policies related to the physical access to SCADA systems are quite well received and implemented in industrial systems, the security policies related to ICT in the process network are in general weak. Usually a core of policies is implemented, but it is the duplication of the traditional "ICT corporate" set of policies. In several situations these are not optimally suited for the process environment, and the operators may simply ignore those that could impact on their daily work. Example of security policy vulnerabilities can be:

- Lack of patching policies: Security patches are quite invasive (especially in the Windows world). On SCADA systems is common to find ad-hoc made software. Patches might interfere heavily with this software. Moreover, several patches require a system reboot in order to become effective, and the reboot of a SCADA server for example is an operation that might interfere with the production system. For that reason, security patching policies are not really popular within the process network.
- Rare or null antivirus update policies: also antivirus might interfere with the SCADA software. Moreover, the updates need to grant the access of the process system to the Internet or to insert into the architecture a parent server able to dispatch the new signatures. In reality it is generally preferred to update the signatures by hand, keeping the process network as isolated as possible.
- Nonrigorous access policies: Access policies are usually well codified, but badly implemented on the field (e.g. the classical post-it attached on a screen with password and login data). In that case what fails in the system is the process of security instruction of the operators.

In the past years international committees released several standards and best practices taking into consideration the security of industrial systems. Policies for SCADA systems should be rethought in the light of these standards and best practices. In the following, we provide a list of committees best practices and standards with indication about their content/subject of work.

API 1164 (Standard)
American petroleum institute (API) 1164 is an international standard providing guidance to the operators of oil and gas liquids pipeline systems for managing SCADA system integrity and security. This document embodies the API's Security Guidelines for the Petroleum Industry. This guideline is specifically designed to provide

the operators with a description of industry practices in SCADA security, and to provide the framework needed to develop sound security practices within the operator's individual companies.

IEEE 1686 (Standard)
IEEE 1686 is a technical standard defining security requirements for intelligent electronic devices (IED): This standard defines the general requirements to protect serial communications between master stations and remote terminal units from cyber attack, and to strengthen authenticated remote access to maintenance ports in RTUs.

ISO 27001 (Standard)
International organisation for standardisation (ISO) 27001-Information security management systems provides a model for an information security management system (ISMS). Even if quite general this standard can be adapted to the peculiar needs of SCADA systems.

NIST 800-53 (Best Practice)
NIST is a non-regulatory federal agency within the U.S. Department of Commerce. The *NIST 800-53-Recommended Security Controls for Federal Information Systems and Organisations* establishes an overall security program for industrial Information Systems by defining a series of security controls that embrace the whole life cycle of the system. The document contains a set of best practices and guidelines.

NERC CIP (Committee)
The North American Electric Reliability Corporation's (NERC) is the electric reliability organisation (ERO) certified by the Federal Energy Regulatory Commission to establish and enforce reliability standards for the bulk-power system. NERC CIP is a committee in charge for the definition of a suite of standards related to different aspects of the security of the electric world. Nevertheless, several of these standards might be taken as example also for more generic industrial systems and ICS.

ISA99 (Committee)
The International Society of Automation has created the ISA99 Committee to establish standards for implementing electronically secure manufacturing and control systems and security practices and assessing electronic security performance. The Committee's focus is to improve the confidentiality, integrity, and availability of components or systems used for manufacturing or control and provide criteria for procuring and implementing secure control systems. Of particular interest: ISA-99.01.01 describes concepts and models that form the basis for the ISA99. ISA-99.01.03 prescribes the requirements to establish quantitative system security compliance metrics for the system under consideration. ISA-99.02.01-2009 establishes an Industrial Automation and Control Systems Security Program. ISA-99.02.02 defines how to operate an Industrial Automation and Control Systems Security Program. ISA-99.03.03 defines standards for system security requirements and security assurance levels.

IEC 62351 (Committee)
International electrotechnical commission (IEC) 62351 is an international committee in charge for the definition of a standard for power systems management and associated information exchange—Data and communications security. Several of its indications find application in almost all the SCADA systems.

These are only few examples of international standards dealing with the security of industrial control systems. There are indeed a lot more, dedicated to different aspects and to different type of industrial architectures.

20.3.3 Software Vulnerabilities

These vulnerabilities are strongly related to the lack of rigorous security patching policies. Everything in a SCADA system is managed by software and it is impossible to guarantee that a piece of software is completely "bug free". Software vulnerabilities are extremely insidious since potentially they can allow an attacker to take full control of a target system. It is not possible in this chapter to list software vulnerabilities typical of SCADA systems, since considering the heterogeneity of these systems we would be probably obliged to list as "candidate vulnerability" every software vulnerability discovered till now. However, classical classes of vulnerabilities, which surely affect SCADA systems, can be the following: *Buffer overflows, SQL injection, Format String, Web-application vulnerabilities*. Moreover to these classes we need to also add all the vulnerabilities contained in the software deployed into the PLCs. These "software logics" are normally written without any attention to security issues. Considering their location (on board of PLCs) it is hardly possible to detect the presence of vulnerabilities in advance, but at the same time being so close to the actuators they can be one of the most dangerous sources of threat. For details about SCADA software vulnerabilities we point to [10].

20.3.4 Communication Protocol Vulnerabilities

Most of SCADA protocols, such as Modbus and DNP were designed several years ago, for control networks based on serial connections. When Ethernet connections became widely used as the physical connection layer for local networks, SCADA protocols were implemented over IP-based protocols, usually TCP. SCADA protocols usually do not have any protection mechanisms, such as authentication, authorisation, and encryption, due to their original design for serial cable. Since it is not possible to summarise in a single book chapter all the detailed vulnerabilities of all the known SCADA protocols (see for example [11]), in the following sections we highlight vulnerabilities and possible attacks exploiting intrinsic features of Modbus, using it as explicative example. Considerations on Modbus vulnerabilities can be easily extended to the majority of the other SCADA protocols.

20.3.4.1 Modbus Vulnerabilities

The delivery of Modbus messages using TCP introduces new levels of complexity with regard to the reliable dispatching of control packets in a process control environment with strong real-time constraints. In addition, it provides attackers with new means to target industrial systems. Modbus TCP lacks mechanisms for protecting confidentiality and for verifying the integrity of messages flowing between a master and slaves (i.e. it is not possible to discover if the original message content has been modified by an attacker). Modbus TCP does not authenticate the master and slaves (i.e. a compromised device could claim to be the master and send commands to the slaves). Moreover, the protocol does not incorporate any anti-repudiation or anti-replay mechanisms. The security limitations of Modbus can be exploited by attackers to damage or take the control of the field network of a SCADA system. Some key attacks examples:

- Unauthorised Command Execution: The lack of authentication of the master and slaves implies that an attacker can send forged Modbus messages to a pool of slaves. In order to execute this attack, the attacker must be able to access the network that hosts the SCADA servers or the field network. Carcano et al. [12] showed how it would be possible to create a malware able to perform a similar operation.
- Modbus Denial-of-Service Attacks: It would be possible to use the power of "maintenance commands" such as "forced stand-by", to disconnect the PLCs from the network and in this way perform a DoS.
- Man-in-the-Middle Attacks: The lack of integrity checks enables an attacker who has access to the production network to modify legitimate messages and send them to slave devices.
- Replay Attacks: The lack of security mechanisms enables an attacker to reuse legitimate Modbus messages sent to or from slave devices.
- Compromised Masters: Since anti-repudiation mechanisms are not implemented, it is hard to prove the trustworthiness of malicious masters, which could have been compromised.

20.4 SCADA Security Countermeasures

In the light of what presented in the previous section, is evident how SCADA systems need to be protected. In this section, we identified a set of security countermeasures. Such a list is not intended as exhaustive, but aims at providing to the reader a good starting point for more ad hoc tailored solutions.

20.4.1 Communication Protocol Countermeasures

It is possible to identify two types of communication protocols into an industrial installation network: traditional TCP/IP communication protocols and industrial communication protocols (Modbus, Profibus, DNP3, etc). With regard to TCP/IP countermeasures, the scientific and technical literature is quite well established. For that reason, in this section, attention will be paid only to the countermeasures related to SCADA protocols.

20.4.1.1 SCADA Protocols Countermeasures

The *Top ten vulnerabilities of control systems and their associated mitigation 2007* report by North American Electric Reliability Council (NERC) mentions the problem of SCADA protocols' vulnerability to cyber attacks, focusing the attention on the general lack of authentication and integrity preserving methods. In order to overcome this lack, several mechanisms have been proposed for different protocols. In the following, we briefly present some of these solutions:

Secure DNP3
This protocol adds both user and device authentication to the DNP3 protocol. It adds data integrity protection as well. The DNP3 user group started working on this secure variant in 2002. It released the first specification in 2007. Secure DNP3 provides protection between master stations and remote outstations. Authentication guarantees that messages arriving at a device comes from another previously legitimated device. Secure DNP3 is a modification of the standard DNP3 protocol, and modifies only the application layer. Thus, Secure DNP3 messages can be transmitted over TCP/IP, serial links, and all the other physical links suitable for DNP3. Nevertheless, this also means that it does not address issues specific to the pseudo-transport link and to the data link.

AGA 12
The American Gas Association (AGA) Cryptography Working Group developed a suite of open standards (AGA 12) to protect the data transmitted by SCADA systems, to authenticate the originators of messages on SCADA systems, and to ensure data integrity. AGA 12 requires that SCADA cyber security equipment can interoperate, independent of manufacturer or age. Initially, the AGA 12 working group addressed the problem of providing security features to serial communications of already installed SCADA systems. This first objective was set up due to the obvious reason that pipelines have a very long lifetime (7–20 years) and it is too expensive to replace them just for introducing security features. For more details about AGA 12, we point the readers to the AGA12 part 1 (background, policies and test plans), part 2 (retrofit link encryption for asynchronous serial communications) and part 3 (protection of networked systems) draft.

Secure Modbus

Just like the other SCADA protocols, Modbus is also vulnerable to a huge amount of attacks due to the lack of authentication and integrity mechanisms. The two possible solutions to secure Modbus are: (1) To embed the Modbus traffic into an SSL/IPV6 channel (2) to redesign the Modbus protocol to embed appropriate security mechanisms. The first solution can be deployed immediately with the actual technology, but raises reasonable doubts in term of usability and efficacy once one thinks about the need of creating a VPN comprising all the PLCs and RTUs of a field network. The second solution would be ideal in term of performance and manageability, but at the moment, a standard regarding a secure version of Modbus does not exist. The only work done in this direction, is a working protocol prototype of Secure Modbus, embedding authentication and anti-reply mechanisms, proposed by Nai et al. [17].

20.4.2 Filtering Coutermeasures

A quite common security suggestion for SCADA systems is to isolate the process and field network from the corporate network through firewalls. Modern firewalls are not able at the moment to understand and then to analyse in depth SCADA protocols. As described in [15] for maintenance purposes the process network might need to be accessed by external entities (e.g. remote operators, vendor support services etc.). In the real world the process network and the external network is in some way connected. A firewalling architecture is then needed to create a sort of air gap between the external network and the process network.

The high level objectives of a firewalling architecture for SCADA systems should be:

- Avoid direct connections between the process network and the Internet. This abstract rule should protect from (or mitigate the effect of) evident risks such as:

  - Direct bandwidth consumption DoS against the SCADA servers.
  - Direct injection of unauthorised control messages.
  - Malware infections.
  - Integrity and confidentiality breaches.

- Controlled access from the office intranet to the control network. Since the office intranet (that traditionally is part of the corporate network) is usually used by the operators, it is natural to grant them access to the process network. However, the office network is traditionally allowed to access the Internet, and, for that reason, a PC on this network might act as bridge for malicious external actors aiming at damage to the process network. The access from the intranet to the process network should be regulated through authentication, and mechanisms for enforcing the security of the communication between PC into the intranet and SCADA servers hosted into the process network should be put in place. A common solution can be for example the use of a firewall supporting VPN connections and Radius (Remote Authentication Dial In User Service) authentication. In this way a point-to-point encrypted connection could be established between the firewall and the PC into the intranet. The traffic flowing between this PC and the SCADA server beyond the firewall is protected against confidentiality and integrity threats while implementing also authentication mechanisms.

• Secure mechanisms for remote control and maintenance. A firewall for SCADA systems should allow remote access to the process and field network while ensuring the properties of security and controlled access listed before.

As said before, while modern firewalls are extremely advanced when called to analyse traditional ICT traffic, they are not able to analyse in depth the SCADA protocols. More precisely, while a firewall can easily perform a packet filtering, blocking unauthorised packets, not knowing the state of the system it is protecting, it cannot easily understand if a legitimate packet can be used to drive the system into a critical state. In this field Byres proposed a solution named Tofino [14] that aims at enforcing the SCADA architecture by filtering at low level each single packet sent to a target PLC/RTU. Such an approach provides a good protection for single packet attack scheme; however, still an open issue remains, related to more complex and subtle attacks. In order to better understand the problem, let us consider the following example: We have a system with a pipe in which flows high-pressure steam. The pressure is regulated by two valves (1 and 2). An attacker able to send packets on the process network sends a DNP3 packet to the PLC controlling the valve 1 in order to force its complete closure and a command to the PLC controlling the valve 2 in order to maximize the incoming steam. It is evident how such commands, when considered locally, are perfectly legitimate, while, altogether will bring the system to a critical state. In order to mitigate such a risk, it is necessary to provide the firewall with a detailed, explicit knowledge of the SCADA system under analysis (components, commands and critical states). Nai et al. [22], proposed a new type of firewall: when a malicious packet has been injected inside the process network the system could move from a secure state to a critical state (CS) that is defined by a set of configurations which might cause system stops, damages, etc. The firewall, while analysing the packets in search for known signatures, keeps updated a digital representation of the system physical state. In this way, every time a command brings the virtual image of the system into a critical state, an alert is raised. Complex SCADA attacks will be identified. The system virtual image, in particular, aims at representing a portion of the field system to be monitored. In other words, it is constituted by a collection of software objects which represent elements like valves, PLCs, actuators etc. The behaviour of these virtual elements is managed basically in three ways:

1. Analysing the command-response network traffic generated by the real system.
2. According to a behavioural profile.
3. By feeding periodically the virtual system through a synchronisation between the virtual system and the real system (basically the virtual system manager has the ability to emulate a SCADA server and, at periodical time intervals directly interrogates the field devices about their own actual state and configuration).

This kind of approach seems to promise good results in fighting against SCADA ICT attacks, since it has been proved to work also against zero-day attacks.

20.4.3 Monitoring Coutermeasures

Firewalls are powerful security bastions, however, the way in which they work is quite invasive (they have to stay physically in between the communication end-points to be effective). The jitters introduced by the presence of firewalls, especially in real-time networks, might create some problems. For that reason, in the past 10 years, firewall architectures have been often mixed with intrusion detection architectures (IDS). IDS techniques have the main characteristic to be passive, i.e. they analyse the behaviour of a network or of a PC silently without interfering too much with the environment under control. Traditionally, IDS techniques can be classified into two families on the basis of the source of information to be analysed:

- Network IDS: Sensors analysing network flows in search of attack proofs.
- Host IDS: Sensors installed on a target server, which analyse the operation it performs in search of malicious behaviours.

Host-based IDS are quite invasive since they need to be hosted by the same system one wants to monitor. For that reason, in SCADA systems Network Intrusion Detection System (NIDS) should be preferred. IDS can be classified also according to the techniques used to identify the threats:

- Signature-based IDS, which compare the information gathered with signatures which characterise a target attack.
- Anomaly based IDS, which compare the actual behaviour of the system with a "behavioural template" in search of deviations from the normal profile, i.e. in search of anomalies.

Both the techniques can be used in a SCADA system, the first in order to quickly identify known attacks, limiting the risk of false positives; the second in order to identify unknown attacks.

20.4.3.1 Limits of Intrusion Detection in SCADA systems

Modern Intrusion Detection systems are quite mature regarding the detection of traditional ICT threats and attacks; as in the case of firewalls, when speaking of SCADA traffic, while they are pretty efficient in performing single packet analysis, they generally fail in detecting complex attacks based on the use of legitimate commands to drive the system in a critical state. For example, if a malicious user, able in some way to have an access on the process network, starts to send legitimate Modbus packets to a pool of slaves (PLC) in order to change the state of the system, a traditional IDS will not be able to detect it since for it, the Modbus packets (contained in the payload of a TCP packet) are just *meaningless payload*. Only recently some extensions, for example for Snort (a well known IDS) have been developed in order to allow the IDS at least to analyse the single packets. In this context Carcano et al. [4] proposed an intrusion detection technique based on the concept of state analysis. However, while the technique per se is quite interesting, at the moment cannot be taken into consideration for use in production systems.

20.4.4 General Architectural Best Practices

Risk assessment has a relevant role in designing secure SCADA architectures, it enables selection of proper countermeasures during the design phase. A secure architecture is made by several technical solutions, processes, security procedures and measures. It is impossible to spot the perfect SCADA architecture for any possible use and environment. In order to properly design a secure control network, it is important to consider the specific setting of a SCADA system. Even if most networks are based on standard IT solutions, SCADA systems have specific characteristics differing from traditional IT environments. In particular, there are different risks and priorities. For that reason, when defining security countermeasures for SCADA installations, it is necessary to keep into consideration the impact that general purpose security countermeasures (e.g. firewalls) might have on the processes. Aspects such as introduced delays and computational constraints must be analysed and taken in high consideration.

20.4.4.1 Firewalls and Network Segregation

Firewalls are typically used to separate control networks from corporate networks. Separating networks highly restricts undesired access to the control system and can improve network performance by removing nonuseful network traffic. The configuration of firewalls needs specific solution for every organisation. In the following a set of general rules is described. In architectures without a demilitarised zone (DMZ) for shared servers:

- All rules should be stateful and specific both for IP address and port;
- Traffic from the corporate network to the control network should be limited only to packets coming from a controlled set of corporate addresses to a small portion of control devices, using the address portion of the rule;
- Access to servers should be allowed only from selected corporate IP addresses;
- Rules should forward packets only for ports of specific secure protocols, such as HTTPS. Allowing HTTP and FTP can be a security risk, even if sometimes they may be useful or necessary;
- Rules should forbid hosts outside the control network to initiate connections with hosts on the control network.

If a DMZ is used, then it is possible to configure the system avoiding any direct traffic between the corporate and the control network. All traffic from either side can be directed to the servers in the DMZ, providing more flexibility for protocols enabled through the firewall. In particular:

- Inbound traffic to control systems should not be allowed. Access to devices on the control network should be allowed through the DMZ.
- Outbound traffic through the control network firewall should be minimal, limited only to strictly necessary services and communications.
- All outbound traffic from the control network to the corporate network should be restricted using specific rules which restrict source addresses, destination addresses, services and ports.

A summary can be used as a guide for firewall configuration:

- The base rule set should be "deny all, permit none".
- Ports and services between the control network environment and the corporate network should be enabled and permissions granted on a specific case-by-case basis. There should be a documented business justification with risk analysis and a responsible person for each permitted incoming or outgoing data flow.
- All permit rules should be both IP address and TCP/UDP port specific, and stateful if appropriate.
- All rules should restrict traffic to a specific IP address or range of addresses.
- Traffic should be prevented from transiting directly from the control network to the corporate network. All traffic should terminate in the DMZ.
- Any protocol allowed between the control network and DMZ should explicitly not be allowed between the DMZ and corporate networks (and vice-versa).
- All outbound traffic from the control network to the corporate network should be source and destination-restricted by service and port.
- Outbound packets from the control network or DMZ should be allowed only if those packets have a correct source IP address that is assigned to the control network or DMZ devices.
- Control network devices should not be allowed to access the Internet.
- Control networks should not be directly connected to the Internet, even if protected via a firewall.
- All firewall management traffic should be carried on either a separate, secured management network (e.g. out of band) or over an encrypted network with multi-factor authentication. Traffic should also be restricted by IP address to specific management stations.

These should only be considered as guidelines. A careful assessment of each control environment is required before implementing any firewall rule set. Configuring firewalls need to take into account specific requirements and rules, beside the general rules mentioned before. Even if different organisations need different configurations, it is possible to use best practice documents also for specific requirements and settings. The industrial automation open networking association (IAONA) offers a template for analysing specific needs, assessing protocols, security risks, impacts and countermeasures [24].

20.4.4.2 Domain Name System

Domain name system (DNS) is primarily used to translate between domain names and IP addresses. Most Internet services rely heavily on DNS, but its use in control networks is relatively rare at this time. In most cases there is little reason to allow DNS requests out of the control network to the corporate network and no reason to allow DNS requests into the control network. DNS requests from the control network to DMZ should be addressed on a case-by-case basis. Local DNS or the use of host files is recommended, as well as the deployment of domain name system security extension (DNSSEC).

20.4.4.3 SCADA and Industrial Protocols

All SCADA and industrial protocols (for example MODBUS/TCP, EtherNet/IP, DNP317 and all the others used in ICS), are critical for communications to most control devices. Almost all of these protocols were designed without security built in and do not typically require any authentication to remotely execute commands on a control device. For that reason, they should only be allowed within the control network and not allowed to cross into the corporate network.

20.4.4.4 System Hardening

Some good practices may help harden SCADA systems.

- Removing or not installing software and functionalities that are not required. Removing or disabling unused services and ports reduces unauthorised accesses or use, ease configuring the system, and reduces chances to find an exploitable vulnerability.
- Physical and logical access to diagnostic and configuration ports should be protected, since they are usually related to critical services for the correct system functioning.
- All unused ports should be disabled, preventing unauthorised accesses. Thus, it is necessary to know exactly which ports and services are used by system components. This can be achieved also through the use of port scanners in a test environment.
- The use of removable devices, such as CDs, USB storage or laptop computers, should be restricted as much as possible. When it is necessary to use a removable media, anti-malware testing methods are needed.

- Organisations should test that the hardening of the system is robust, secure and functioning as supposed in the design phase.

20.4.4.5 Account Management

Account and user management is aimed to prevent or reduce unauthorised accesses.

- Organisations should implement a proper password policy. The policy should take into account password strength and expiration times. If a password cannot be changed frequently, an alternative protection mechanism may be appropriate. The policy should also cover the proper password encryption mechanisms.
- Unused default user accounts should be deleted.
- Network devices should support that passwords are encrypted within the device.
- Network devices should support role-based access mechanisms.
- Roles and access rights should be reviewed regularly.
- Default passwords should be changed.
- Critical functions should have stronger authentication methods.
- Each user should have an individual password that should not be communicated and divulged to any other person, nor stored in insecure places.
- Computer located in unattended places should have mechanisms for authentication, automatic locking and automatic disconnection.

20.4.4.6 Software Management and Update

Several attacks exploit known vulnerabilities and bugs of software. For this reason, software management and update procedures are necessary to avoid or recover this security issues. Actively managing vulnerabilities and related patches is aimed to reduce or prevent exploitations, and takes less time and efforts than recovering the system and responding after exploitation has been performed. Organisations should deploy documentation providing the software patching and hardening policy for their systems. The policies should be reviewed every year to address new threats and discovered vulnerabilities. The policy has to be consistent, for example software patching cannot reinstall software removed for hardening the system, or change security settings, etc. A typical model of software and patches is based on a pattern cycling through four phases:

1. Assessment and inventory, aimed to evaluate software components of the system, possible security threats and vulnerabilities, and whether the organisation can update software easily.
2. Patch identification, for locating software updates available, for understanding their relevance and effectiveness, and for determining whether an update responds to a security emergency or it is a normal software update.

3. Evaluation, planning and testing, aimed to decide which patched are to be deployed in the operational environment, to plan when and how to perform software updates, and to ensure that the software update conforms to the system, without compromising its business and operational aspects, using testing in a realistic operational environment.
4. Development, aimed to actually perform software updated in the operational environment, minimising the impact on the system.

20.5 Conclusion

Failures caused by cyber attacks against SCADA systems can directly impact on the operative capabilities of critical systems such as energy grids, water pipelines, but also air traffic control systems, train control systems etc. In other words, from their security depend the security and resilience of most of the so-called *critical infrastructures*. In this chapter we showed which are the main classes of vulnerabilities affecting them and which are possible high-level countermeasures. Trends on the evolution of ICT systems show that in the coming future technologies will converge more and more towards an integrated, interleaved system of systems. Smart Grids, Smart Cities, Clouds and the Internet of Things provide a clear indication of this trend. In a similar context, where industrial systems will be more and more open and accessible to the rest of the world, cyber-security will be even more important. Topics such as *digital identity* (in its broader meaning), *process integrity and security, secure and survivable communication, data provenance* will surely be in the top ten of the cyber-security research rank. The security evolution of SCADA systems (and more generally of industrial installations) needs to move on different levels: on the governance level corporate security and industrial security will need to be integrated in a more coherent vision. New protocols and architectures will be developed to face the challenges posed by open-integrated architectures. But on top of everything a strong cyber-security alphabetisation will be needed in the industrial sector. This evolution will require time and huge economical efforts but our society cannot ignore the problem anymore. The scope of this chapter was that of raising general attention to the problem to encourage the scientific, technical and policy-making communities to work together to make these systems more secure and indirectly make our society safer.

References

1. Karnouskos S., Stuxnet worm impact on industrial cyber-physical system security. IECON 2011—37th Annual Conference on IEEE Industrial Electronics Society, January 2012.
2. Igure V. M., Laughter S. A. and Williams R. D. "Security issues in SCADA networks". Computers & Security. 2006 V. 25, N.7, Pages 498–506 Month 10.

3. Chandia, R.; Gonzalez, J.; Kilpatrick, T.; Papa, M.; and Shenoi, S.; Security Strategies for Scada Networks. In *Critical Infrastructure Protection*, Eric Goetz and S. Shenoi (Eds.), Springer, Boston, Massachusetts, vol. 253, pp. 117–131, 2007.
4. Carcano, A.; Coletta, A.; Guglielmi, M.; Masera, M.; Nai Fovino, I.; Trombetta, A.; A Multidimensional Critical State Analysis for Detecting Intrusions in SCADA Systems. Industrial Informatics, IEEE Transactions on V. 7, I. 2, 2011, Page(s): 179–186.
5. Majdalawieh, M.; Parisi-Presicce, F. and Wijesekera, D.; DNPSec, Distributed Network Protocol Version 3 security framework. In Proceedings of the Twenty-First Annual Computer Security Applications Conference (Technology Blitz Session), Tucson, Arizona, USA, 2005.
6. Mander, T.; Nabhani, F.; Wang, L.; Cheung, R.; Data Object Based Security for DNP3 Over TCP/IP for Increased Utility Commercial Aspects Security. In Proceedings of the Power Engineering Society General Meeting, Tampa, FL, USA, June 24–28, pp. 18. IEEE, Los Alamitos (2007).
7. Hong, J. H. C. S.; Ho Ju, S.; Lim, Y. H.; Lee, B. S. and Hyun, D. H.; A Security Mechanism for Automation Control in PLC-based Networks. In Proceedings of the ISPLC '07. IEEE International Symposium on Power Line Communications and Its Applications 26–28 March 2007, pp 466–470, Pisa, Italy.
8. OPC: http://www.opcfoundation.org/ Last Access: 11/05/2012
9. Leszczyna, R.; Nai Fovino, I.; Masera, M.; Security Evaluation of IT Systems Underlying Critical Networked Infrastructures. In Proceeding of the 1st International Conference on Information Technology, Gdansk, Poland, 18–21 May 2008.
10. Cagalaban, G.; KIM, T; KIM, S; Improving SCADA Control Systems Security with Software Vulnerability Analysis. In Proceedings of the 12th WSEAS International Conference on Automatic Control, Modeling & Simulation. pp 409–414, 2010.
11. Edmonds, J.; Papa, M.; Shenoi, S.; Security Analysis of Multilayer SCADA Protocols. In Proceedings of the IFIP Critical Infrastructure Protection 2008. pp 205–221, 2008.
12. Carcano, A.; Nai Fovino, I; Masera, N. and Trombetta, A.; Scada Malware, a proof of Concept. In proceeding of the 3rd International Workshop on Critical Information Infrastructures Security, Rome, October 13–15, 2008.
13. Creery, A.; Byres, E.J.; Industrial Cybersecurity for power system and SCADA networks IEEE Industry Application Magazine, July-August 2007.
14. http://www.tofinosecurity.com. Last Access 02/12/2009
15. Dondossola, G.; Masera, M.; Nai Fovino, I.; Szanto, J.; Effects of intentional threats to power substation control systems. International Journal of Critical Infrastructure, (IJCIS), Vol. 4, No. 1/2, 2008.
16. East, S.; Butts, J.; Papa, M.; Shenoi, S.; A taxonomy of Attacks on the DNP3 Protocol. In proceedings of the third IFIP international conference on Critical Infrastructure Protection, Hannover, NH, 2009.
17. Nai Fovino, I.; Carcano, A. and Masera, M.; Secure Modbus Protocol, implementation, tests and analysis. In Proceeding of the Third Annual IFIP Working Group 11.10 International Conference on Critical Infrastructure Protection, Dartmouth College, Hanover, New Hampshire, USA, March 22–25, 2009.
18. http://www.modbus.org/
19. http://www.dnp.org/default.aspx; DNP consortium. Last access: 05/01/2012
20. IEC/TC 57 IEC 60870-5-104; http://www.iec.ch; International Electrotechnical Commission. Last access: 05/01/2012
21. http://www.profibus.com/nc/downloads/downloads/profibus-technology-and-application-system-description/display/; Profinet Foundation. Last Access: 05/01/2012
22. Nai Fovino, I.; Carcano, A.; Masera, M. and Trombetta, A.; A State Based Intrusion Detection System for Modbus Protocol. In Critical Information Infrastructures Security. Lecture Notes in Computer Science 2010. Springer Berlin / Heidelberg. Isbn: 978-3-642-14378-6 pp. 138-150. Vol. 6027.

23. Nai Fovino, I. and Masera, M.; A service oriented approach to the assessment of Infrastructure Security. In Proceeding of the First Annual IFIP Working Group 11.10 International Conference on Critical Infrastructure Protection, Dartmouth College, Hanover, New Hampshire, USA, March 19–21, 2007.
24. The IAONA Handbook for Network Security Draft/RFC v0.4, Industrial Automation Open.

Part IV
Practical Examples and Tools

Chapter 21
An Overview of PIC Microcontrollers and Their Suitability for Cryptographic Algorithms

Mehari G. Msgna and Colin D. Walter

Abstract The use of microcontrollers is widespread. They occur in most electronic devices, such as point of sale (POS) terminals, ATMs, printers and traffic signals. They can differ from each other in terms of architecture, processing capacity, storage capacity and supported hardware features. The purpose of this chapter is to present a brief introduction to one group of them which can be used for cryptography, the PIC microcontrollers, and give a detailed, practical account of how to investigate their strength against side channel analysis.

Keywords PIC controller · Oscilloscope · Side channel leakage · Trace · Differential power analysis.

21.1 Introduction

A *microcontroller* (μC) is a tiny computer in a single integrated circuit, dedicated to performing a small number of specific tasks, with limited processing capacity and limited memory compared to that of a personal computer. It is also known as a "computer-on-a-chip". At its heart is a microprocessor, or Central Processing Unit (CPU), which is a multi-purpose computation engine that is built as a single integrated circuit. It also has peripheral circuits to perform tasks such as reading and writing memory contents, converting digital signals to analogue and vice versa,

M. G. Msgna (✉)
Information Security Group, Smart Card Centre, Royal Holloway, University of London,
London, United Kindgom
e-mail: mehari.msgna.2011@rhul.ac.uk

M. G. Msgna
Information Network Security Agency, Addis Ababa, Ethiopia

C. D. Walter
Information Security Group, Royal Holloway, University of London, London, United Kindgom
e-mail: colin.walter@rhul.ac.uk

K. Markantonakis and K. Mayes (eds.), *Secure Smart Embedded Devices,*
Platforms and Applications, DOI: 10.1007/978-1-4614-7915-4_21,
© Springer Science+Business Media New York 2014

and communicating with external devices. Several basic characteristics differentiate microprocessors from each other and these include the instruction set, the processing capacity, the available memory and the clock speed. A typical microcontroller has bit manipulation instructions, access to input/output devices and efficient interrupt mechanisms.

Microcontrollers do not have an independent interface such as a keypad or screen display. Instead they are embedded inside larger devices which they control. Normally a microcontroller is designed to execute a small number of programs in order to perform a specific job; they cannot be transferred from a television to a printer, or from a microwave to a washing machine, for example. By including only the hardware features required for a specific task, their cost can be reduced. However, some are designed to be reprogrammed for any purpose. They are called *general purpose* microcontrollers.

21.2 Microcontroller Structure

Microcontrollers can be classified into different categories based on their family, architectural design, bus size, processing capacity, memory layout and instruction set. Figure 21.1 presents values for these criteria. The size of the arguments processed in a single instruction is generally small, typically 8 or 16 bits.

Microcontrollers with a reduced instruction set computer (RISC) architecture use simple, single clock cycle instructions. However, for specific tasks, the number of instructions per application can be reduced by having multiple operations within a single instruction which lasts for several clock cycles. This gives a complex instruction set computer (CISC) architecture. For example, a CISC microcontroller needs only one instruction to multiply two memory contents whereas a RISC microcontroller needs four to perform the same task. The details are in Table 21.1. A microcontroller is said to have a specific instruction set computer (SISC) architecture if its processing unit and instruction set are customised to do a specific type of job.

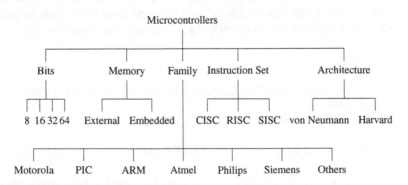

Fig. 21.1 Classification criteria of microcontrollers

Table 21.1   CISC, RISC and equivalent pseudo-code for multiplying two memory contents

mul mem-loc1, mem-loc2	load a, mem-loc1	a ← mem-loc1;
	load b, mem-loc2	b ← mem-loc2;
	mul a, b	a ← a*b;
	store mem-loc1, a	mem-loc1 ← a;

The two most common microcontroller architectures are *von Neumann* and *Harvard*. Those based on the von Neumann architecture [14, 37] have a single data bus for fetching program instructions and program data. Both the program instructions and data are stored in a common main memory. When the microcontroller has to perform a task, it fetches the instruction first and then the data associated with it. Harvard architecture microcontrollers use separate buses to access the program's instructions and data. Such a configuration allows parallel memory access to occur. The microcontroller can then fetch both the next instruction and its data simultaneously while executing the current instruction. This generally leads to improved performance.

21.3 Peripheral Interface Controllers

21.3.1 PIC Architecture

Peripheral Interface Controller (PIC) microcontrollers from Microchip Technology [27] belong to the Harvard architecture family [17, p. 74]. They have different modules working together to execute installed applications including the Microprocessor unit (MPU), Program Memory, Data Memory, Bus, Input/Output modules and other support devices (see Fig. 21.2).

The microprocessor unit (MPU) is the main module of the microcontroller chip. It performs arithmetic and logic operations, and controls the microcontroller's status. The MPU has three main components: the arithmetic logic unit (ALU), Registers and the control unit (CU). The ALU has three modules: the instruction decoder, the status register and the accumulator (working register). The instruction decoder interprets instructions fetched from the device's program memory. The accumulator is a register which holds the intermediate result of an arithmetic or logic operation before the final result is moved to the destination memory location or register. The final module is the status register, which is used to save or indicate the status of the processor. It flags exceptions and contains, among other things, the arithmetic status of the ALU, register bank selection bits and the reset status of the microcontroller.

The MPU also uses registers. These memory locations are used for storing information for certain tasks. There are three different types of register in the MPU, namely the program counter (PC), bank select registers (BSRs) and file select

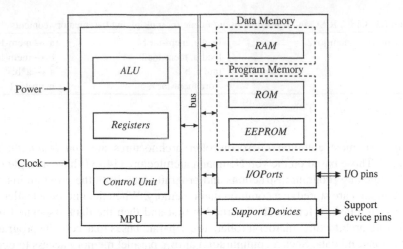

Fig. 21.2 General architecture of microcontrollers

registers (FSRs). The PC stores the memory address of the next instruction to be executed, BSRs are used to implement direct addressing of the data memory, and FSRs contain pointers for indirect addressing of the data memory.

The third component of the MPU is the Control Unit, which provides timing and control signals for all operations in the microcontroller, including both internal and external read and write operations.

21.3.2 Memory

In general there are two types of memory in microcontrollers: *volatile* and *non-volatile*. Volatile storage is memory where all the previous data are generally lost within at most a second or so if the power is turned off. Such storage is a form of random access memory (RAM). On the other hand, non-volatile storage retains information even when the power source is removed. Examples are read-only memory (ROM) and electrically erasable programmable read-only memory (EEPROM). Information stored in ROM cannot be modified. However, contents of EEPROM can be erased and reprogrammed repeatedly by applying a higher electrical voltage through the program pin of the microcontroller.

The memory of Harvard microcontrollers is organised into two separate blocks: the *program memory* and the *data memory*. They are on physically separate buses so that instructions cannot be used as data or *vice versa*. In von Neumann architectures the MPU uses a single main memory to store both program instructions and data.

Modern high performance microcontrollers incorporate aspects of both Harvard and von Neumann architectures. The on-chip cache memory is divided into an *instruction cache* and a *data cache* which store copies of values that are used frequently by the MPU. The MPU uses Harvard architecture when accessing the cache.

21.3.3 Other Components

The microprocessor unit (MPU) interacts with external peripheral devices through the I/O unit, and is connected to the external world through groups of pins known as *ports*. These ports can be input, output or bidirectional and each is associated with a unique register. Data can be read from, or written to, a port by reading from, or writing to, the register associated with it. The pins of the microcontroller can be configured to send or receive analogue or digital data.

The *Clock* is a timing signal which is used to synchronise the microcontroller's modules while they execute arithmetic and logic operations or to move data from one memory location to another. The clock signal can be generated inside the microcontroller chip or obtained externally via the clock input pin.

The MPU communicates with the program and data memory using a *bus*, which is a group of wires connecting the different modules. There are usually two types of bus: the *address bus* and the *data bus*. The former carries the address of a memory location and its size depends on the logarithm of the number of accessible memory locations. Once the memory location is accessed its content is transferred to the data bus. Where the same bus is shared for address and data fetch operations, an extra bit is added to specify the kind of information on the bus.

Other support modules include timers, analogue to digital and digital to analogue converters (ADCs and DACs respectively), serial and parallel ports, and universal synchronous/asynchronous receiver/transmitters (USARTs) (see Fig. 21.3). A *timer* allows the microcontroller to measure the precise execution time of selected tasks. A microcontroller can also have subsidiary MPUs, called *co-processors*, which generally help with heavy arithmetic and logic operations such as those used in cryptography [16]. The MPU interacts with external peripherals through serial and parallel ports [2, 30, 31].

21.3.4 Development Tools

An Integrated Development Environment (IDE) is used to develop software applications, e.g. the MPLAB from Microchip [24]. It normally has at least the following features:

- Program editing.
- Support for a range of microcontroller families.
- Program debugging.
- Program loading to the microcontroller.

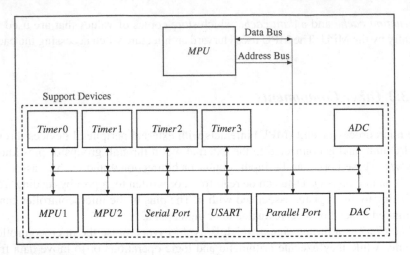

Fig. 21.3 Support modules in a PIC microcontroller

- Program simulation.

The IDE supports high level program development in, for example, C, Basic or Assembly, and provides a *debugger* for finding syntactic and semantic errors in the program code. The *simulator* [32] enables the program to be tested on a PC, thereby avoiding the time needed to load the code during prototype verification. The IDE takes the program file and compiles it into a HEX file of machine code which the microcontroller can run. Once completed and tested, the compiled program must be *loaded*, which is the process of writing the machine code into the internal memory of the microcontroller. For this, the IDE has to be connected to an external device in which the microcontroller is mounted, such as the MPLAB ICD 2 (see Fig. 21.4).

21.3.5 Summary

Microcontrollers are used in many aspects of our daily life, ranging from a room temperature controller to running complex algorithms. They have also been heavily used for implementing cryptographic algorithms such as AES, DES and RSA. We have reviewed mainly just the architecture of PIC microcontrollers, but most microcontrollers on the market share similar architectures and design principles.

21.4 AES on a PIC

Cryptography is the art of transforming understandable information into an unfathomable form called *ciphertext*. Only those who possess the secret key can, in

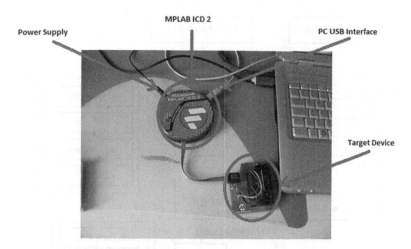

Fig. 21.4 A photo of the MPLAB ICD 2

theory, transform the ciphertext back into its original form. Recently, the security of IT infrastructures has become heavily reliant on embedded systems, such as smart cards and tokens, which use cryptographic algorithms to provide the security. In the remainder of the chapter we evaluate the security of two PIC microcontrollers against one of the main types of attack on embedded systems, differential power analysis, and illustrate the process in the context of the AES symmetric key algorithm.

21.4.1 Implementation of AES

In October 2001 after an open invitation for a new symmetric key algorithm, the National Institute of Standards and Technology (NIST) selected Rijndael [9] to replace the ageing DES. Rijndael is a block cipher algorithm designed by two Belgian cryptographers, Joan Deamen and Vincent Rijmen. NIST then published Rijndael as the advanced encryption standard (AES) [13] after fixing the block size to be 128 bits.

AES takes two parameters: the 128-bit plaintext/ciphertext message block and the secret key. The size of the secret key can be 128 bits, 192 bits or 256 bits. Each key length makes the algorithm behave in a slightly different way. Using a larger key size not only offers higher security but also makes the whole process of AES proportionately longer. AES performs a number of transformation *rounds* that convert the input plaintext into the final ciphertext output. The plaintext/intermediate data is represented as a 4 × 4, 4 × 6 or 4 × 8 matrix data block of byes (depending on the key length) when it is processed by the round functions. Each round consists of several operations, including one that uses a round key derived from the encryption/decryption key. These functions are: *add round key, substitute byte, shift*

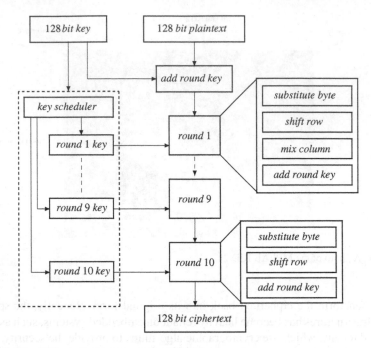

Fig. 21.5 The AES encryption process for 128-bit keys

row and *mix column*. Each function has forward and reverse versions corresponding to the encryption and decryption operations, respectively. Being a simple XOR, the function for the add round key is the same for both directions. In the other cases the reverse functions are the appropriately named inverse substitute byte, inverse shift row and inverse mix column functions. The 10-round encryption process for AES with a 128-bit key size is illustrated in Fig. 21.5. The larger key sizes require instead 12 and 14 rounds, respectively.

In an experiment AES was implemented in electronic code book (ECB) mode [11] on two different families of PIC microcontrollers: the PIC16F84A [25] and the PIC16F876 [26]. The implementation required 32 registers to store the byte values of the cipher key and input data. The first 16 hold the round key bytes and the last 16 store the input data or intermediate state vector bytes. The implemented algorithm was tested for correctness against the test vectors published on the NIST website [1].

21.5 Attack Example

Cryptanalysts often analyse the security of encryption systems by modelling cryptographic functions as *mathematical* objects. Conventional methods such as differential [4] and linear [21] cryptanalysis are used to find weaknesses in cryptographic algorithms. However, these techniques cannot be used to explore vulnerabilities in the *implementation* of cryptographic algorithms on a particular hardware configuration. In reality, implementations of cryptographic algorithms release information about the secret key. This information is known as *side channel* information and could be used to reveal the secret key of the cryptographic algorithm. *Side channel analysis* is the interpretation of the side channel information released by the target device while running the cryptographic algorithm using the set-up and inputs described in an attack. Various side channel attacks which use different types of side channel information have been proposed. Examples of side channel information are execution time [10, 18], power consumption [19], electromagnetic emission [36] and faulty computation output [5, 6] from the device under observation.

Power analysis attacks use minute variations in the power consumption of the device as the side channel information. The objective of an attack is to reveal the secret key information used in the execution of a cryptographic algorithm. It requires the attacker to record and analyse the instantaneous power consumption of the device by measuring variations in voltage or current. The recording produces *power traces* which are sequences of frequent, regular, digitised samples of these variations in the target device. The attacks are classified into various types depending on the number of power consumption traces that they need and the way they are analysed. The two most common power analysis attacks types are simple power analysis (SPA) [7, 19] and differential power analysis (DPA) [19]. DPA is discussed briefly in the following subsections.

21.5.1 Differential Power Analysis

DPA [19] is more powerful than SPA [7, 19] because the attacker does not need as much detailed knowledge about the implementation of the algorithm and the hardware under attack. On the other hand, more side channel information is required. Instead of trying to derive the secret key directly form the power traces as in SPA, DPA takes advantage of a statistical analysis methodology to facilitate the recovery of the secret information. It exploits the dependency of the power consumption of a cryptographic device on the intermediate data processed by it. The disadvantage of DPA when compared to SPA is that DPA needs a much larger number of power consumption traces (typically greater than 1,000).

For DPA to be used on a cryptographic algorithm, the intermediate data normally needs to be a function of known information, such as plaintext or ciphertext, and of secret information, which is derived directly from the cipher key in a known way. This intermediate data can be represented by a function $f(p, k)$, where p is the known information and k is the secret key. This makes DPA suitable for attacking the most commonly deployed cryptographic algorithms including the data encryption standard (DES) [35] and the advanced encryption standard (AES) [13]. In the case of public key algorithms, it often suffices for the intermediate data to be a function of random, uniformly distributed data, rather than known data, and part of the unknown private key such as an individual bit.

To help understand how DPA works against 128-bit AES, consider the following example. Let P be the plaintext input, K the first round key (which, in this case, is the full secret key), M the intermediate value and S the substitute byte function. P, K and M are 128 bits long, which means they are all constructed from 16 bytes of data. Let P', M' and K' be single bytes with the same index from P, M and K. Also, let M'_j be the jth bit of M', where $0 \leq j \leq 7$. Using P' and K', the corresponding byte of output from the substitute byte function at the start of the next round is

$$M' = S(P' \oplus K').$$ (21.1)

So the intermediate value to be analysed is the direct output of the substitute byte function S from the first round.

The attacker runs the algorithm N times with N different known (plaintext) input blocks P and records the N power traces. As the non-linear substitute byte function S is a byte operation, it must be performed 16 times, once on each of the 16 bytes of the data block. The attacker looks at the first round (strictly speaking, round 0) and considers each byte position in turn, using the formula (21.1) in which P' is known. Since K' is unknown, the attacker XORs all the 256 possible values of K' (0x00 to 0xFF) with P' and generates all the 256 possible values of M'. This yields an $N \times 256$ byte array of intermediate values by considering each of the N plaintext blocks.

In order to pick the correct value of a key byte out of the 256 candidates, the attacker considers each possible value of K' in turn. Having selected one for investigation, he takes the jth bit of M' for some j where $0 \leq j \leq 7$ and splits the power traces into two sets: W_1 which is the set of power traces where $M'_j = 1$ should hold if the key byte guess K' is correct, and W_0 which is the complementary set of power traces where $M'_j = 0$ should hold. Then the *differential power* ΔW is calculated by subtracting the mean power trace computed over the members of W_0 from the mean power trace corresponding to W_1. A function of time t is obtained for the time interval during which the substitution is performed:

$$\Delta W(t) = |W_1|^{-1}(\Sigma_{tr \in W_1} tr(t)) - |W_0|^{-1}(\Sigma_{tr \in W_0} tr(t))$$ (21.2)

where $tr(t)$ is the power measured at time t for one of the N traces tr. In other words, for each time t at which power measurements were taken in the N traces, the attacker computes the difference of the average power consumptions over the two sets W_1 and W_2. A typical power trace is given in Fig. 21.8 and some differential power traces using the above formula are given in Fig. 21.9.

Since there are 256 columns in the $N \times 256$ matrix of intermediate values, the above calculation has to be repeated 256 times, once for each possible K', and the resulting differential power traces ΔW plotted in a graph. The plot of ΔW is always close to zero for almost the whole period except for a few noticeable peaks.

These peaks are generated at the moments where M'_j or related key data is processed by the microcontroller (μC). From the resulting 256 graphs the K' value that generates the largest peak is chosen as the correct key byte. This choice can be confirmed by repeating the calculation on other bit positions j.

The underlying intuition justifying this choice for K' is that in positions where key data is not used, the calculations are, on average, the same in both sets of data. Hence, the (average) power used in both cases should be almost the same, with random noise being cancelled. This leads to no peaks. However, where the calculation is key-dependent and the correct key byte has been chosen, there will be points during the calculation where the processed bits are strongly biased in one way for W_1 (because the bit value "1" is processed) and in another way for W_0 (because the opposite bit value "0" is processed). Typically, this may happen if and when M' is placed on the bus for storing in memory. This causes differential power usage, and hence a peak because of the different way in which electronic components, such as transistors, behave with different bit inputs. In between, if the wrong value is chosen for the key byte one can expect some, but less, bias at points where the byte is used, and the extent of the bias to depend on how close the wrong value is to the correct value. This is because not so many traces in W_1 are processing a "1", but the number is still more than for the traces in W_0. It generally results in smaller peaks, although it requires a more detailed analysis of the circuitry to justify formally. All the bytes of the secret key can be extracted in this way.

Other functions in a round of the AES can also be attacked using the same technique to determine the key providing the data input or data output is known.

Since the introduction of DPA, several researchers have proposed countermeasures that disturb the dependency of the power consumption on the intermediate data or function. Examples include masking [8, 15], hiding [22], bus encryption [3, 12] and constant power consuming differential logic circuits [34]. However, the resulting implementations may still be vulnerable to second or higher order DPA [23].

21.5.2 Practical Implementation of DPA

The process of recovering a cipher key using DPA involves several steps. These are: measuring the power consumption traces, calculating the intermediate values, analysing the traces, plotting the graphs and finally choosing and verifying the correct keys. In this section we discuss the practicalities of performing these steps to recover an AES cipher key from the PIC16F84A [25] and PIC16F876 [26] microcontrollers.

21.5.2.1 Triggering the Oscilloscope

When measuring the power consumption of the target device a *trigger signal* is needed to enable the oscilloscope to capture the power trace of a particular function within the algorithm at the right time. The oscilloscope can be configured to be triggered by the edge or event of this trigger signal. Once the signal is received, the oscilloscope updates its memory and screen display. In our case the oscilloscope was configured to update its display every time the trigger signal rose from 0 to 5 V. This is known as a *positive edge* trigger.

The trigger can be delivered in different ways depending on whether one has control over the internal implementation of the algorithm or not. In the former case, it is possible to insert a couple of instructions at the point of interest in the implemented algorithm in order to trigger the oscilloscope from the microcontroller. This is what has been done for the results presented here. However, it may not be possible for a commercial device. Then the attacker has to find another way of triggering the oscilloscope. For example, if the device were a smart card the attacker could design a custom reader to trigger the oscilloscope at a precise time delay after a message is sent to the card for encrypting.

The oscilloscope used in our attack was a LeCroy WaveRunner LT264M [33] which is capable of measuring at a rate of up to 1 billion samples per second (1 GS/s). The samples have 8-bit accuracy within a pre-selected range. The oscilloscope has to be connected to the microcontroller using a specialised probe which, in our case, was the Pomona 6069 A [28], a 1.2 m co-axial cable with 250 MHz bandwidth, 10 MΩ input resistance and 10 pf input capacitance. In general, the kit required for a successful attack is not expensive if the target device contains no counter-measures.

subsubsectionMeasuring the Power Traces

The power consumption of the microcontroller is measured by the oscilloscope as a voltage drop across a resistor inserted between the ground pin of the microcontroller and the ground voltage source. Using a serial RS-232 cable, the microcontroller is also connected to a computer which initiates each measurement. The overall configuration of the microcontroller, computer and the oscilloscope is presented in Fig. 21.6. In general, cables should be kept as short as possible to reduce capacitance in the circuit and shielding should be introduced if there is a danger of inductance. In particular, a high-specification short co-axial cable, a *probe*, is required between the oscilloscope and the microcontroller. Of course, shielding is essential when EMR is used as a side

Fig. 21.6 Experimental power measurement configuration for both microcontrollers

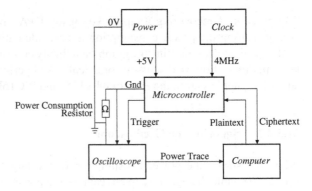

channel source. A detailed circuit diagram of the circuit board on which the PIC was mounted may be obtained by contacting the authors or through [29].

The current flow from the ground pin of the microcontroller to the ground pin of the voltage source is small. Hence the resistor value must be small enough in order not to disrupt the correct operation of the device, but large enough to enable a precise measurement of the voltage drop across it. Generally, this is closely related to the oscilloscope's input impedance. For our tests we used 1 Ω and 10 Ω resistors respectively for the PIC16F84A and the PIC16F876. The voltage drop is captured, sampled and saved by the digital oscilloscope connected in parallel across the resistor.

Before saving any measurements for analysis, it is important to have the equipment properly warmed up beforehand so that it is all running at a uniform temperature throughout the data collection phase. Otherwise the results may be inconsistent because electrical properties change with temperature. This requires the ambient temperature of the laboratory to be maintained at a constant level for several hours beforehand, and for the first few measurements in any run to be discarded. The measurements should then all be taken in a single session.

At the start of the measurement process a plaintext block is sent from the computer to the microcontroller and the power consumption is recorded while the microcontroller runs the algorithm. In our attacks the recorded power traces were sampled at a rate of 100 MHz for both microcontrollers when they were run with a 5 V power supply and a clock speed of 4 MHz. A sufficiently high sampling rate is required in order to isolate the point in the clock cycle when the logic gates process the key bits from the other points which will contribute unwanted noise to the measurements. The sampling rate is increased until the desired peaks show in the differential averages (see Fig. 21.9). In practice the rate is bounded by the limits of the oscilloscope and the quantity of data that can be stored. Typically, the sample rate is in the order of 2^6 times the clock speed of the device being investigated but, in our case, the slightly lower sampling rate of 25 per clock cycle proved adequate.

For most microcontrollers and a reasonable implementation such as those used on our two PICs, samples from the first 720 to 1200 or so clock cycles after the trigger should cover the first add round key and substitute byte functions, which is what is needed. This allows 45–75 cycles for executing each instance of (21.1), depending on the available instruction set. With 20,000 sample measurements for the PIC16F876 and 50,000 for the PIC16F84A, each recorded with 8-bit accuracy, the individual files of a trace were not enormous. Around 2^{15} traces had to be obtained but, after the averaging described in the next paragraph, only 2^{10} averaged traces needed to be stored for analysis. They were saved in Matlab* .dat files [20] as lists mostly of 10-character decimal numbers (11 characters when a sign was necessary), leading to 232 and 570 KB per trace file respectively. Had space been at a premium, two bytes per sample for a binary representation could easily have been used, thereby cutting storage by a factor of nearly 6. However, overall, there were totals of around 284 and 700MB of uncompressed data ready to be fed into the analysis stage.

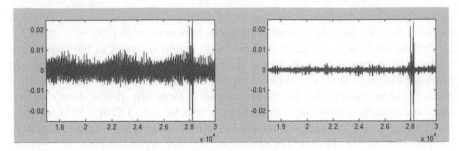

Fig. 21.7 Differential power traces for the PIC16F84A calculated without and with averaging

21.5.2.2 Noise Reduction

The digital oscilloscope is connected to the microcontroller on two of the oscillo-scope's channels, one for measuring power consumption and the other for receiv-ing the trigger signal. Every time the oscilloscope receives a new trigger signal it takes new measurements of the voltage drop across the resistor. However, the power consumption measured by the oscilloscope also includes noise introduced by the measurement environment. This noise can disturb the dependency of the measured power on the intermediate data. So, in order to reduce the noise, each plaintext was sent to the microcontroller 32 times and the average of the 32 power traces was saved as the final power consumption of the device for the particular plaintext. This cuts the standard deviation of the random noise by a factor of $\sqrt{32}$. Figure 21.7 illustrates the effect of this when calculating the differential power ΔW defined in (21.2): 1,000 single traces were used for the left hand differential power trace, but 1,000 averaged traces for the right hand trace. The difference is quite visible: the peaks which do not represent noise now stand out much more clearly from the underlying noise. In com-mon with all our diagrams, the horizontal axis is the time axis with sample number as the unit of measurement and the vertical axis gives voltage which is measured in volts and calibrated relative to its average.

21.5.2.3 Pre-Processing of Power Traces

After the power traces are captured and saved, pre-processing them before calculat-ing the differential power is important for maximising the efficiency of the attack. This pre-processing of samples includes but is not limited to alignment of traces, resampling of captured traces and removal of unwanted traces.

The captured power traces might contain the power consumption of the device while running unwanted functions before and after that being studied for the DPA attack. In that case the attacker needs to remove the unwanted samples from each end of the traces and perhaps from within them. For example, in our attack we needed the power consumption of the device for the period when the trigger signal

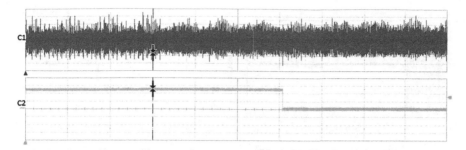

Fig. 21.8 Oscilloscope display of averaged power trace and trigger for the PIC16F84A

is high (5 V) since that is when the microcontroller runs the add round key and substitute byte functions of the first round. However, our measurement also contains the consumption of the device while executing some other subsequent functions of AES. That is because the oscilloscope monitors the power consumption continuously, and stores the samples from the trigger high signal for a set period of time within its memory capacity rather than only as far as the trigger signal falling to low. Since we could position the trigger high signal precisely, there were no unwanted samples at the start of the trace, only at the end. So, unwanted samples outside the period of interest are removed by aligning the power traces with the trigger signal. In our case, the rising edge of the trigger signal was clean enough for this to be done accurately, with aligned samples corresponding to the same point in the microcontroller clock cycle with an accuracy of less than the sampling period.

In Fig. 21.8, graph C1 shows an averaged power consumption trace for the PIC16F84A. The scale is 100 mV per vertical division, with the upper and lower edges giving the range to which the 8-bit sample measurements apply. Graph C2 is the trace of the trigger signal with 2 V per division. The horizontal scale is 50 μs per division. With a sampling frequency of 100 MHz, the 500 μs width means the trace contains 50 k samples. The power consumption while the trigger is high (5 V) is the power consumption for the add round and substitute byte functions. Consequently, as the two graphs are synchronised, we can deduce that only the first 30395 from the total of 50001 samples belong to the add round key and substitute byte functions of the AES algorithm. We removed the other samples from each recorded power trace as they were redundant.

As defined in Sect. 21.5.1, Eq. (21.2), the differential power is calculated by splitting the power traces into two sets and taking the difference in their mean values at each sample position. For this it is necessary to make sure that corresponding samples lie at the same point in the time axis. This is done using the trigger signal as a reference. *Alignment* is the process of moving the trace samples along the time axis to make sure that the corresponding samples of all the power traces lie at the same point.

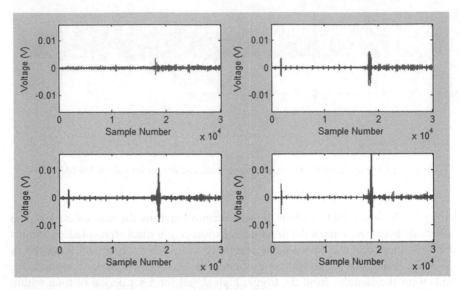

Fig. 21.9 Plot of ΔW for one correct and three incorrect key byte guesses for the PIC16F84A

21.5.2.4 DPA Calculation and Results

The usual objective of differential power analysis is to reveal secret information used by a cryptographic algorithm in a device. For our DPA experiment, 1,000 sets of 32 power traces were recorded from each target microcontroller while they were processing different plaintexts. Each set had the same plaintext input and its members were averaged to yield a trace with reduced noise as described earlier (see Fig. 21.7).

As discussed previously, the differential power trace ΔW is calculated by using one bit of the intermediate value M' to split the traces into two sets and taking the difference of their mean values. (Matlab code for this is available at [29].) Once the differential power is calculated for each possible key byte value, the resulting 256 ΔW traces are compared. The graph of ΔW with the largest spike is assumed to be that generated by the correct key byte guess. Figure 21.9 shows the plots of ΔW for four guesses for the PIC16F84A microcontroller. Out of the four graphs, the last generates the largest spike and indeed corresponds to the correct guess for K'. The average absolute noise in ΔW is about 120 µV for the PIC16F84A set-up, which is about three thousandth of that in an averaged trace where it was around 40 mV. This is why 1,000 or so trace sets are needed for DPA. More ΔW plots for the PIC16F84A are available at [29].

However, when the differential powers were computed for each splitting bit, it turned out that some correct key bytes generated smaller spikes than some wrong ones. This means that it is unreliable to deduce the whole key by using only one bit of the intermediate byte. To eliminate such errors, the average was calculated for all the differential powers over all the 8 bits of the intermediate byte, and the

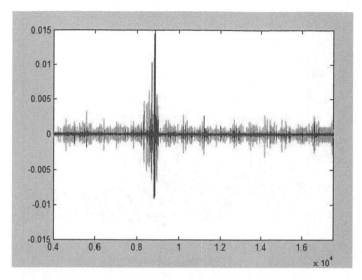

Fig. 21.10   The PIC16F876: 256 ΔWs for all possible values of a key byte overlaid on one graph

key byte yielding the maximum was chosen. This turned out to be a much more reliable indicator of the correct key bytes. Moreover, the number of bits for which the chosen key byte gives the maximum differential value can be used as an indicator of the reliability of the choice. (More sophisticated analysis is possible, of course.) After the analysis, the whole 128 bit AES key was correctly recovered from both microcontrollers. The choice of key is easily verified by comparing outputs generated from the same input using the unknown secret key and using the deduced key.

Figure 21.10 shows the plots of all 256 averaged differential powers for one typical key byte for the microcontroller PIC16F876, superimposed in one graph. Each trace has already been averaged over 32 instances to reduce the noise, and then averaged over the 8 bit positions. All differential powers are plotted in gray except the one generating the highest peak, which is plotted in black. The trace in black is the most likely to correspond to the correct key and that was always the case for us.

To make this clearer, Figs. 21.11 and 21.12 show just the maximum and minimum points of the 256 ΔW traces used in Fig. 21.10. The first graph is the plot of all maximum values and the second one is the plot of all minimum values. From the graphs we can see that there is one point standing out clearly above and one point just below the rest in the maximum and minimum graphs respectively. These two points both belong to the ΔW generated by the correct key byte guess. (The physical properties of logic gate switching mean that local maxima and minima are paired in the power trace. So the minima display similar properties to the maxima, with the lowest minimum of the 256 traces identifying the correct key byte. However, they are not known to contain any information which is not already present in the maxima. Indeed, interchanging W_0 and W_1 in (21.2) simply changes maxima to minima and *vice versa*.)

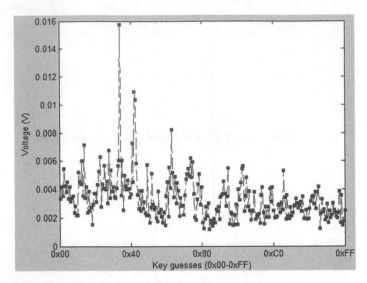

Fig. 21.11 The PIC16F876: maximum points of all the 256 ΔW

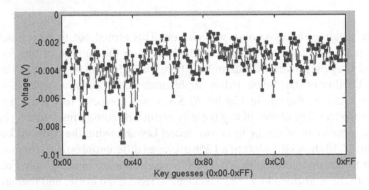

Fig. 21.12 The PIC16F876: minimum points of all the 256 ΔW

At the point where the highest black peak value is plotted in Fig. 21.10, there are other smaller peak values in gray. That is because certain incorrect byte values are sufficiently closely related to the correct values that they perform similar calculations at the critical points during the calculation of (21.1) when side channel information is released. We have to hope that the calculations are still sufficiently dissimilar not to generate side channel leakage which could be confused with the true value. Fortunately, in Fig. 21.11 the maximum points for the correct key byte are about 25 % larger than the largest maximum points given by any of the other key byte choices. This is typical for each byte of the secret key K in this implementation of AES, but it is possible for some correct byte peaks to be indistinguishable from the peaks of incorrect choices in other circumstances and contexts. Where the correct value is unclear, the attacker can try all 256 cases to see if any generates the desired AES output. Unlike a brute force task on the whole key, this is feasible if only one or two key bytes are guessed incorrectly.

The side channel information leaked by both microcontrollers is similar. So, although the PIC16F876 is superior in terms of performance and storage capacity, both chips still leak a significant level of important side channel information which is enough to reveal the key of mathematically strong algorithms such as AES. Better performance may not mean better resistance to side channel attack.

21.6 Conclusion

This chapter started with an introduction to microcontrollers in general and PIC microcontrollers in particular. We then discussed the different modules of a microcontroller, and their internal architecture, and briefly described the tools needed to develop an application for them. Finally, the principal theme was looking at how side channel information can be used to recover secret keys from them. To that end, we provided details of the experimental set-up and calculations involved, and illustrated these in the case of an implementation of AES.

PIC microcontrollers are popular among prototype designers. They are not the only option available on the market but clearly they are not ideal solutions for use in the field if side channel leakage could be an issue. In this chapter we have seen how secret keys can easily be recovered from them using differential power analysis. This is a powerful method that can be used on any general purpose microcontroller to retrieve sensitive information. Readers are therefore recommended to use only microcontrollers with built-in security countermeasures for security applications that will be used in hostile environments.

References

1. L. E. Bassham, III. The Advanced Encryption Standard Algorithm Validation Suite (AESAVS). NIST Cryptographic Algorithm Validation Program (CAVP), February 2002. http://csrc.nist.gov/groups/STM/cavp/documents/aes/AESAVS.pdf
2. M. Bates. Interfacing PIC Microcontrollers: Embedded Design by Interactive Simulation. Elsevier, 2006.
3. R. M. Best. Preventing Software Piracy with Crypto-Microprocessors. In Proc. IEEE Spring COMPCON '80, pages 466–469. IEEE Computer Society, 25–28 Feb, 1980.
4. E. Biham and A. Shamir. Differential Cryptanalysis of DES-like Cryptosystems. In Advances in Cryptology - CRYPTO '90, volume 537 of Lecture Notes in Computer Science, pp. 2–21. Springer, 1991. http://dl.acm.org/citation.cfm?id=646755.705229
5. E. Biham and A. Shamir. Differential Fault Analysis of Secret Key Cryptosystems. In Advances in Cryptology - CRYPTO '97, volume 1294 of Lecture Notes in Computer Science, pp. 513–525. Springer, 1997. http://dl.acm.org/citation.cfm?id=646762.706179
6. D. Boneh, R. DeMillo, R. Lipton. On the Importance of Checking Cryptographic Protocols for Faults. In Advances in Cryptology - EUROCRYPT '97, volume 1233 of Lecture Notes

in Computer Science, pp. 37–51. Springer, 1997. http://dl.acm.org/citation.cfm?id=1754542.
1754548

7. Xi Xi Chen. Simple Power Analysis a Threat in Embedded Devices. Master's thesis, University
   of Waterloo, Ontario, Canada, 2004.
8. J.-S. Coron and L. Goubin. On Boolean and Arithmetic Masking against Differential Power
   Analysis. In Cryptographic Hardware and Embedded Systems - CHES '00, volume 1965 of
   Lecture Notes in Computer Science, pages 1–14. Springer, 2000.
9. J. Daemen and V. Rijmen. The Design of Rijndael—Information Security and Cryptography.
   Springer, Heidelberg, 2002.
10. J.-F. Dhem, F. Koeune, P. Leroux, P. Mestré, J.-J. Quisquater, and J.-L. Willems. A Practical
    Implementation of the Timing Attack. In Smart Card Research and Applications—CARDIS
    '98, volume 1820 of Lecture Notes in Computer Science, pp. 167–182. Springer, 2000.
    http://dl.acm.org/citation.cfm?id=646692.703439
11. M. Dworkin. Computer Security: Recommendation for Block Cipher Modes of Operation.
    NIST Special Publication 800-38A. NIST, 2001. http://csrc.nist.gov/publications/nistpubs/
    800-38a/sp800-38a.pdf
12. R. Elbaz, L. Torres, G. Sassatelli, P. Guillemin, C. Anguille, M. Bardouillet, C. Buatois, and
    J. B. Rigaud. Hardware Engines for Bus Encryption: A Survey of Existing Techniques. In
    Proc. Design, Automation and Test in Europe (DATE '05), Vol. 3, pp. 40–45. IEEE Computer
    Society, 2005. http://dx.doi.org/10.1109/DATE.2005.170
13. Federal Information Processing Standards. FIPS PUB 197 - Announcing the Advanced Encryp-
    tion System (AES), November 2001. http://csrc.nist.gov/publications/fips/fips197/fips-197.
    pdf
14. M. Godfrey and D. Hendry. The Computer as von Neumann Planned It. Annals of the History
    of Computing, IEEE, 15(1):11–21, 1993.
15. J. Golic and C. Tymen. Multiplicative Masking and Power Analysis of AES. In Cryptographic
    Hardware and Embedded Systems—CHES 2002, volume 2523 of Lecture Notes in Computer
    Science, pages 198–212. Springer, 2003.
16. H. Handschuh and P. Paillier. Smart Card Crypto-Coprocessor for Public-Key Cryptography.
    In Smart Card Research and Applications—CARDIS 1998, volume 1820 of Lecture Notes for
    Computer Science, pages 386–394. Springer, 2000.
17. J. Iovine. PIC Microcontroller Project Book: For PIC Basic and PIC Pro Compilers. McGraw-
    Hill, second edition, 2004.
18. P. Kocher. Timing Attacks on Implementations of Diffie-Hellman, RSA, DSS, and other Sys-
    tems. In Advances in Cryptology—CRYPTO '96, volume 1109 of Lecture Notes in Computer
    Science, pp. 104–113. Springer, 1996. http://dl.acm.org/citation.cfm?id=646761.706156
19. P. Kocher, J. Jaffe, B. Jun. Differential Power Analysis. In Advances in Cryptology—CRYPTO
    '99, volume 1666 of Lecture Notes in Computer Science, pp. 388–397. Springer, 1999.
    http://dx.doi.org/10.1007/3-540-48405-1_25
20. MathWorks™. MATLAB and SIMULINK, MathWorks™ website visited October 2012.
    http://www.mathworks.com
21. M. Matsui. Linear Cryptanalysis Method for DES Cipher. In Advances in Cryptology—
    EUROCRYPT '93, volume 765 of Lecture Notes in Computer Science, pp. 386–397. Springer,
    1994. http://dl.acm.org/citation.cfm?id=188307.188366
22. R. McEvoy, C. Murphy, W. Marnane, M. Tunstall. Isolated WDDL: A Hiding Countermea-
    sure for Differential Power Analysis on FPGAs. ACM Trans. Reconfigurable Technol. Syst.,
    2(3):1–23, March 2009. http://doi.acm.org/10.1145/1502781.1502784
23. T. Messerges. Using Second-Order Power Analysis to Attack DPA Resistant Software. In
    Cryptographic Hardware and Embedded Systems—CHES '00, volume 1965 of Lecture Notes
    in Computer Science, pp. 238–251. Springer, 2000. http://dl.acm.org/citation.cfm?id=648253.
    752407
24. Microchip Technology Inc. MPLAB Integrated Development Environment, Microchip, web-
    site visited October 2012. http://www.microchip.com/stellent/idcplg?IdcService=SS_GET_
    PAGE&nodeId=1406&dDocName=en019469&part=SW007002

25. Microchip Technology Inc. PIC16F84A Data Sheet, Microchip website visited October 2012. http://ww1.microchip.com/downloads/en/devicedoc/35007b.pdf
26. Microchip Technology Inc. PIC16F876 Data Sheet, Microchip website visited October 2012. http://ww1.microchip.com/downloads/en/devicedoc/30292c.pdf
27. Microchip Technology Inc. Website visited October, 2012. http://www.microchip.com
28. Pomona Electronics. 6069A Scope Probe, website visited October 2012. www.pomonaelectronics.com/pdf/d4550b-sp150b_6_01.pdf.
29. Royal Holloway, University of London. Smart Card Centre website. http://www.scc.rhul.ac.uk/books/ssed/embedded/chapter_21
30. J. Sanchez and M. Canton. Microcontroller Programming: The Microchip PIC. CRC Press, 2007. http://www.crcpress.com
31. D. Smith. PIC in Practice: A Project Based Approach. Newnes, 2006.
32. V. Soso. PIC Simulator IDE, Oshon Software Project, website visited October 2012. http://www.oshonsoft.com/pic.html
33. Teledyne LeCroy Corporation, website visited October 2012. http://teledynelecroy.com/oscilloscope/
34. K. Tiri and I. Verbauwhede. Design Method for Constant Power Consumption of Differential Logic Circuits. In Design, Automation and Test in Europe, 2005. Proceedings, volume 1, pages 628–633, March 2005.
35. W. Tuchman. A Brief History of the Data Encryption Standard. In Dorothy E. Denning and Peter J. Denning, editors, Internet Besieged, pp. 275–280. ACM Press, Addison-Wesley Publishing Co., 1998. http://dl.acm.org/citation.cfm?id=275737.275754
36. W. van Eck. Electromagnetic Radiation From Video Display Units: An Eavesdropping Risk? Computer Security, 4(4):269–286, December 1985. http://dx.doi.org/10.1016/0167-4048(85)90046-X
37. J. von Neumann. First Draft of a Report on the EDVAC. Technical report, Moore School of Electrical Engineering, University of Pennsylvania, June 1945.

Chapter 22
An Introduction to Java Card Programming

Raja Naeem Akram, Konstantinos Markantonakis and Keith Mayes

Abstract Java Cards support a Java virtual machine that interprets code written in a subset of Java language. This may help programmers with prior knowledge of Java language to program smart cards. However, the programming paradigm of Java Card can be articulated as somewhat different than traditional Java programming. In this chapter, we will provide an introduction to smart card programming using Java Card and the subtleties of a restricted environment on application design.

Keywords Java Card · Java · Terminal · Programming · Tools · Testing

22.1 Introduction

Smart cards have evolved from a limited, purpose-built fixed function to a dynamic and multi-application environment. The technology has progressed and we can now program smart cards in high-level languages like Java and C (e.g. Java Card [5] and Multos [1]). Nevertheless, the restrictive nature of the smart card environment in terms of storage, processing, and physical attributes (i.e. size, and power consumption) still play a vital role in application development.

The aim of this chapter is to provide an appreciation of the challenges encountered by the smart card developers, which have to balance the application requirement and

R. N. Akram (✉)
Department of Computer Science, University of Waikato, Hamilton, New Zealand
e-mail: rnakram@waikato.ac.nz

K. Markantonakis · K. Mayes
Information Security Group, Smart Card Centre, Royal Holloway,
University of London, Egham, United Kingdom
e-mail: k.markantonakis@rhul.ac.uk

K. Mayes
e-mail: keith.mayes@rhul.ac.uk

K. Markantonakis and K. Mayes (eds.), *Secure Smart Embedded Devices,*
Platforms and Applications, DOI: 10.1007/978-1-4614-7915-4_22,
© Springer Science+Business Media New York 2014

497

resource consumption. In most cases, such considerations are not weighed when programming for personal computers or powerful servers. Therefore, even if a programmer knows a smart card supported language (e.g. Java and C), it is still a different programming style that he or she has to adopt. During this chapter, we take the Java Card as a running example for smart card programming and assume that readers do not have any prior knowledge of it. We start with a brief introduction of smart card and Java Card architectures and their role in application design, programming, and testing. Furthermore, we extend this discussion into a practical example of a Java Card application referred to as "My First Applet".

22.2 Smart Card Programming

In this section, we will look into the Java Card 3 architecture along with a generic smart card hardware architecture; subsequently discussing how these affect the programming model.

22.2.1 Smart Card Architecture

Figure 22.1, illustrates a smart card architecture based on the Java Card 3 specification [8]. The smart card hardware layer hosts the smart card operating system (SCOS) and native code. The SCOS is usually implemented in the native language for the given hardware; however, for performance reasons, cryptographic algorithms are implemented in hardware and related APIs are in native code.

The SCOS layer will include libraries that are specifically designed for the underlying hardware and in the case of Java Card they are accessible only to the Java Card

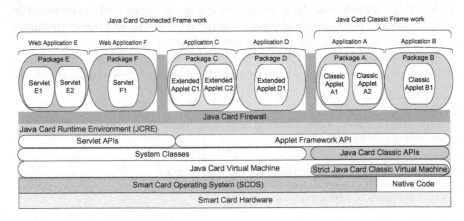

Fig. 22.1 Smart card architecture (Java Card 3 [8])

virtual machine (JCVM). For Multos, and Basic Card [9] an application can directly access SCOS services via SCOS application programming interfaces (APIs). Therefore, Java Card is not an SCOS, but a smart card platform; where the Multos and Basic Card are examples of SCOS. If we have Multos or Basic Card as an SCOS then there will be no Java Card runtime environment (JCRE) as most of the functionality provided by the JCRE is already part of the Multos and Basic Card. The native code is basically libraries that provide cryptographic services. The rationale behind having the native code layer is to provide better performance for resource intensive and time critical services. Above the runtime environment is the application layer, which is isolated by the platform's firewall mechanism (e.g. Java Card and Multos firewalls [10]).

At the application layer, the Java Card 3 supports three different application models: classic applets, extended applets and Web applications. In this chapter, we will focus on the classic applet architecture. It is backward compatible with previous Java Card specifications and to date the most deployed Java Card architecture.

We will discuss the Java Card Connected and Classic frameworks later in this chapter, but in subsequent sections we describe a generic smart card hardware and communication framework.

22.2.2 Smart Card Hardware

In a simplistic manner we can describe a smart card hardware as a single silicon chip that consists of a central processing unit (CPU), random access memory (RAM), read only memory (ROM) and electrically erasable programmable read only memory (EEPROM). When programming a smart card, the capabilities of these components should be considered as they influence the application design. A typical smart card has a 16 to 32bit CPU, RAM in the range of 8 kB, ROM around 200 kB and perhaps (max) 64 kB of EEPROM. If a smart card has a Flash memory then it may not require to have ROM or/and EEPROM.

The components illustrated on the right side of Fig. 22.2 are optional, but most of the high-end smart cards include them. These components provide dedicated hardware for cryptographic processing (i.e. crypto-copocessor) and also can include internal clock multiplier or clock circuitry [12]. The crypto co-processor might sup-

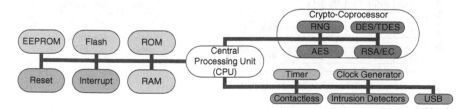

Fig. 22.2 Block diagram of a generic smart card hardware

port random number generator (RNG), advanced encryption standard (AES), data encryption standard (DES), triple DES (TDES) along with public key crypto systems like Rivest-Shamir-Adleman (RSA). Finally, a modern smart card may support a USB interface for faster communication with a terminal.

In addition to these storage and processing restrictions, an EEPROM memory only support limited number of write cycles (approx. 500,000 [13]) which may effect design considerations [12]. If an application extensively uses an EEPROM location, it will reach the EEPROM write cycle limit; thereby, disabling the respective memory cells of the EEPROM, which will render the application code/data inaccessible. One way to avoid this is to adopt an application design that only writes to the EEPROM when it is absolutely necessary; otherwise, for all remaining tasks it uses the RAM.

An additional restriction imposed on the application design is based on the non-existence of a user interface and the application environment in which a smart card is deployed. In most instances, a smart card acts as a security token with restrictions on the execution time such as in a transport application. Therefore, combining all these restrictions ranging from limited hardware support to restricted performance criteria makes programming for smart card's a resource-based design. A developer has to take into consideration all the contributing factors while programming the smart card including the communication architecture of smart cards. As there is no user interface, smart cards communicate with a terminal using a (restricted) application data unit (APDU) framework which is discussed in the next section.

22.2.3 Communication Architecture

To facilitate the communication between applications on a terminal and a smart card, the ISO/IEC 7816-4 [3] defines an APDU. The structure of the APDU is defined by the ISO/IEC 7816-4 standard and it is illustrated in Fig. 22.3.

The APDU message sent by a terminal is referred to as the "command APDU" and the response from a smart card is termed as "response APDU". Different elements of command and response APDU are described in Table 22.1. The command header is

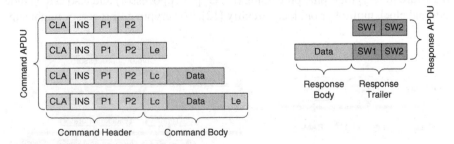

Fig. 22.3 Command and response APDUs

Table 22.1 APDU components and their explanations

Command APDU		
Field name	Byte length	Description
CLA	1 byte	Instruction class—this indicates the class of command: proprietary or interindustry
INS	1 byte	Instruction code—this indicates the instruction to be executed
P1	1 byte	Parameter bytes
P2	1 byte	
Lc	0, 1, or 3 byte	Number of data bytes in the command APDU
Data	1–65, 535 bytes	Data bytes
Le	0, 1, 2, or 3 byte	Number of data bytes expected in response from a smart card
Response APDU		
SW1	1 byte	Indicates the command processing status: successful or error
SW2	1 byte	
Data	1 - 65, 535 bytes	Data bytes

compulsory when using the APDUs to communicate with a smart card. The command body is optional depending upon the application design and requirements. From a programming perspective, a developer has to define adequate APDU structure, data length, both for the terminal and smart card applications. Therefore, we can articulate that a smart card application developer should also understand the terminal application's functionality and requirements.

From a Java Card's perspective, there are two types of APDUs that a smart card application can handle: normal and extended APDUs [5]. The normal APDUs has the Lc length as one byte thus only support 0–256 bytes for the data component of the command/response APDU. However, Java Cards (version 2.2.2 and 3) also support an extended APDU that provide the Lc length as two bytes supporting 0–32,767 bytes of data in the command APDU. The reason for this is the Java Card platform APIs, use a positive short intended to store the size of the APDU's data component, which has a range of 0–32,767. Therefore, unlike the ISO/IEC 7816-4 [5] standard that proposed the Lc of up-to three bytes giving the range of 65,536 (the first byte is 00, where the size of data is represented by byte two and three), Java Cards do not support them.

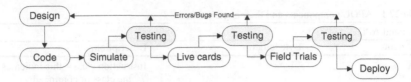

Fig. 22.4 Smart card application development lifecycle

22.2.4 Application Development Lifecycle

Figure 22.4 illustrates a smart card application development lifecycle. The first step is to gather the requirements and design the application. This stage should consider the target application and (smart card) hardware platform, as this will effect the design considerations. After the design of the application is completed, the next step is to program the application. Subsequently, it should be simulated as directly testing on a live hardware will be costly and in some cases difficult to debug. One of the most important aspect of the application development lifecycle is testing, the more you test the least time it will take to role out the service. During the testing process, if there are any bugs found then the development goes back to the application design and coding. More time spent on this process means saving time and money later, as going back to design and testing from later stages is more taxing.

Once the testing is completed, the application is then deployed on live smart cards that are selected for the actual deployment. The loaded application is tested again to find any possible bugs or design oversights. Further to the testing, the application should be put into a limited field trial to analyse the overall system and to find any unforeseeable issues that were not considered at the design stage. To keep the development of an application lifecycle simple we skip a few optional stages. For example, if an application has be to evaluated for its security and functional properties (e.g. EMV application [6]) then a developer should also include the security evaluation stage in the lifecycle.

Before we dive into the discussion on Java Card API framework and using them to program an application, we list some basic points that a developer should keep in his or her mind at the time of application development.

1. Understand the application requirements and the capability of the targeted smart card.
2. Base the application on a fault-tolerant robust design. During the lifecycle of an application, it may encounter unforeseeable commands or combinations of inputs. Therefore, the application design should have a fail-safe mode, to fall back if an undefined condition or command is encountered during the execution.
3. Design a clear and concise communication interface (the commands that an application will respond to if issued by a terminal).

4. Assign only essential data structures for storage in the EEPROM. Unnecessary storage in the EEPROM can reduce the life of the application and also the target smart card.
5. Consider the life of individual data structures. Modern smart cards may provide garbage collection. However, reducing the use of garbage collection during the execution of an application will increase its performance.
6. Reuse memory space (variables) within the application, especially data arrays if they do not create any security vulnerability.
7. Design a thorough and comprehensive testing plan for the application.

22.3 Java Card

In this section, we discuss the recent Java Card specification 3 that has two distinct editions: Java Card Classic and Connected.

22.3.1 Java Card Classic

The Java Card Classic edition is an incremental evolution to the Java Card 2.2.2 [5]. It maintains backward compatibility and is designed for resource restricted devices. The Java Card Classic virtual machine is similar to the previous JCVMs that had a split architecture. This architecture was divided between on-card and off-card processing of an application. Before loading an application on to a Java Card, the application code is analysed by an off-card virtual machine to verify its integrity and compliance with the Java Card specification. This analysed (pre-processed) application is then loaded onto a Java Card. The Java Card 3 Classic fixes some bugs, provide clarifications to the Java Card 2.2.2, along with providing support for security algorithms like 4096-bit RSA and NSA Suite B [2] and minor improvements on how contactless transactions are processed.

A typical Java Card 3 Classic edition and Java Card 2.2.2 supports the following features:

1. Boolean, byte and short data types.
2. Single thread execution.
3. Garbage collection (best effort).
4. Single dimension arrays.
5. Exceptions.
6. Java Card remote method invocation (RMI), and ISO/IEC 7816-4 (e.g. APDU) for off-card communication.
7. Dedicated API to support biometric data.

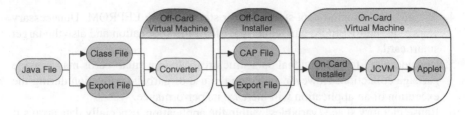

Fig. 22.5 Java Card classic edition application development cycle

22.3.1.1 Java Card Classic Application Development Cycle

The traditional Java Card application development cycle is illustrated in Fig. 22.5. After coding an application, a developer will compile the corresponding Java file to produce class and export files.

The class file contains the application related bytecode and export file has any associated configurations. These two files are then converted by the off-card virtual machine to a CAP[1] file and associated export file. The off-card virtual machine also performs the bytecode verification to check whether the application conforms with the Java Card specification.

After the generation of the CAP and export file by the off-card virtual machine, these files are input to the off-card installer that communicates them to an on-card installer. The on-card installer downloads the application code on to the respective smart card. Once the application is downloaded it registers itself with the JCVM. Now the application can execute and communicate with on-card or off-card entities. In most of the latest smart cards that support the Java Card specification, the GlobalPlatform specification[2] [4] is used for off-card and on-card installers, along with post issuance application management.

22.3.2 *Java Card Connected*

The Java Card Connected edition is being proposed for high-end smart cards. It contains a scaled down version of Java's connected limited device configuration (CLDC) virtual machine that is used in the mobile phones. Therefore, the Java Card Connected can support the feature rich programming environment provided by the CLDC virtual machines. It also enables a smart card to communicate concurrently on a variety of interfaces: using internet protocol (IP), ISO/IEC 7816-4 protocols, and contactless interfaces (ISO/IEC 14443 [7]).

---

[1] CAP: The Java Card Converted APplet (CAP) is an Java Card interoperable file format used to deploy an application on smart cards.

[2] GlobalPlatform: The GlobalPlatform specification provides a secure, reliable and interoperable application management framework for a multi-application smart cards.

The Java Card Connected edition moves the smart card into the role of a network node that is capable of acting as an independent platform providing security and privacy services. Therefore, the Java Card Connected edition may be the future for the smart card technology. The features provided by it are listed as below:

1. Support for all common data types except for float and double.
2. Multithreading.
3. Rich APIs like generic connection framework (GCF), servlet, sockets, threads and transactions, etc.
4. On-card bytecode verification.
5. Automatic garbage collection.
6. Multidimensional arrays.
7. Primitive wrapper classes (e.g. Boolean and Integer), string manipulation classes (e.g. StringBuffer, StringBuilder, etc.), input/output (I/O) classes (e.g. Reader, Writer, and Stream), and collection classes (e.g. Vector, Hashtable, Stack, and Iterator, etc.).
8. Java SE features like generics, metadata, assertions, enhanced for loop, varargs and static inputs, etc.

In addition to these features, the Java Card Connected edition supports the loading of a Java application from a Java Archive (JAR) file.

22.3.2.1 Java Card Connected Application Development Cycle

The Java Card Connected application development cycle is illustrated in Fig. 22.6. The application code is compiled to produce class and resource files along with supporting libraries. The Java application does not have to be in the CAP file format for the Java Card Connected edition, as it supports JAR format. Therefore, a packager will take the output of the compiler (e.g. supporting libraries, class and resource files) and packages them into a module (it can be either CAP or JAR depending upon the smart card's supported format) that is ready to be deployed. This module is then input to the off-card installer that loads the application to the respective smart card. As discussed before the Java Card Connected has an on-card bytecode verifier; therefore, there is no particular need to perform off-card byteode verification.

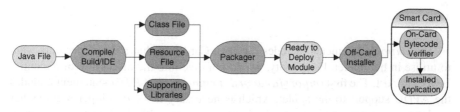

Fig. 22.6 Java Card connected application development cycle

Although, the application loading is different for the Java Card Classic and Connected editions, most of the modern integrated development environments (IDEs) can hide these details from the developer. Examples of the IDE for Java Cards are Net Beans and Eclipse. A developer can use other IDEs, but we choose these two for this chapter. The configuration details of these IDEs for Java Card development may be obtained by contacting the authors or through [11].

22.4 Java Card Programming

In this section, we begin with a short description of the traditional Java Card applet supported by Java Card version 2.2.2 and 3.1 Classic Edition. This will give an understanding of basic structure of a Java Card applet.

22.4.1 Java Card Applet Architecture

The basic required structure of a Java Card applet is shown in Listing 22.1. Every applet in the Java Card has to be associated with a package as shown in Fig. 22.1. A package can have multiple applets and it can be considered as a container on a smart card that has several components of the respective application.

Listing 22.1 Basic Architecture of a Java Card Applet

```
1     package myFirstApplet;
2
3     import javacard.framework.APDU;
4     import javacard.framework.Applet;
5     import javacard.framework.ISOException;
6
7     public class FirstApplet extends Applet {
8
9         private FirstApplet() {
10        }
11
12        public static void install(byte bArray[], short bOffset,byte bLength)
13            throws ISOException {
14            new FirstApplet().register();
15        }
16
17        public void process(APDU apduHandle) throws ISOException {
18
19        }
20    }
```

The *import* statements includes the Java Card API that a developer may want to utilise in her applet. For brevity, we do not dive into the discussion of the entire Java Card API. The first *import javacard.framework.APDU* statement includes the APDU support to the applet, which is necessary if the developer requires her applet to communicate with off-card entities using APDUs. Similarly, developers also have to import the *javacard.framework.Applet* from which all Java

Card applets have to be derived. The *install* and *process* methods listed at lines 12 and 17 are extended from the *javacard.framework.Applet*. The reasons for implementing these methods in the class is described later in this section. Finally, *javacard.framework.ISO Exception* is included to enable the applet to handle and throw exceptions. This is a useful feature for the robust design of an applet that if used properly can manage any possible errors during the application execution.

The two most important methods that a Java Card applet has to implement are: *install* and *process*. The *install* method is called by the JCRE to create an instance of the Applet subclass. In a simplistic manner, it can be explained as registering an applet with the JCRE. Once an applet is registered, it becomes selectable to an off-card entity enabling it to communicate with the applet. The *install* method is invoked at the end of the application loading, and usually it is requested by the off-card installer. If JCRE does not execute the *install* method, then the application is loaded on the smart card, but it would not be able to communicate with other entities (on-card or/and off-card). The *install* method accepts a byte array as an input parameter that can be sent by the off-card installer to personalise an application. The data items that a developer wants to execute only once in the lifetime of an applet should be declared in this method, examples of such data structures can be cryptographic keys associated with the application and any application personalisation information.

The *process* method is invoked by the JCRE, when an off-card entity sends an APDU for the applet. The input parameter of the process method is the APDU object which contains the byte array containing the APDU sent by the off-card entity. This method can be considered as the main function of the applet as it acts an interface between an off-card entity and the functionality implemented by the respective applet. Depending upon the design of the application, the applet will perform the required function.

22.5 My First Applet

In this section, we extend the basic structure of the applet to provide an example of the Java Card application (e.g. loyalty application).

22.5.1 Application Design

As an example, we will develop a simple counter application that can be considered as a loyalty application. The application stores a value, which can be increased, decreased or requested by an (off-card) terminal application. Note, in the sample implementation we do not include any authentication or cryptographic functionality. The reason behind this is to keep the code short and simple for easy explanation. A complete loyalty application with additional features is available—from authors.

Our loyalty application on a smart card has a data counter and support four commands as listed:

Table 22.2 Instruction bytes for individual commands

INS byte	Name	Description
0xE0	INS_RESET	Reset the counter
0xE1	INS_READ	Read the counter
0xE2	INS_INCREASE	Increase the counter with a value X
0xE3	INS_DECREASE	Decrease the counter with a value X

1. Increase value: This command will increase the counter by a value X, where X is provided as the input to the command.
2. Decrease value: This command will decrease the counter by a value X, where X is provided as the input to the command.
3. Read value: The terminal application requests the smart card application to provide the current value of the counter.
4. Reset counter: It resets the counter to the base value, which for our example is zero.

As per requirements, the only persistent data structure for our application is a counter value. If the counter value is declared as a byte then it will reduce the APDU manipulation as APDU data component is also a byte array. However, the size of a byte is limited to 255 and for this example we consider that a counter value should be declared as a "short" giving it the value of 33,767. By doing so, we have to convert the input bytes to short and vice versa.

Each package and its associated applets on a Java Card has a unique identifier referred as the application identifier (AID). The AID is a 5–16 byte identifier that consists of two components: registered identifier (RID) and Proprietary Application Identifier Extension (PIX). The RID is 5 bytes long and compulsory, on the other hand PIX can be 0–9 bytes long and is optional. If you are developing an application that would be used either nationally or internationally, you need to get a RID from designated authorities. However, for the example we do not require this, and you can use any AID for your example application. Therefore, the defined AID in hexadecimal format for the example package ($my First Applet$) is $0 \times D0\,0 \times 00\,0 \times 00\,0 \times 00\,0 \times 62\,0 \times 02\,0 \times 01\,0 \times 0C\,0 \times 0E$, and for the applet is $0 \times D0\,0 \times 00\,0 \times 00\,0 \times 00\,0 \times 62\,0 \times 02\,0 \times 01\,0 \times 0C\,0 \times 0E\,0 \times 0A$.

The loyalty application class byte is set to $0 \times C0$ and instruction bytes for individual commands are listed in Table 22.2.

In the next step, we define error messages that the loyalty application will throw if it encounters an unrecognisable input. The list of error messages with their descriptions are listed in Table 22.3. All values shown as the error messages are in hexadecimal format.

After documenting the basic structure, commands, and error messages, we can proceed with the development. Any changes to the basic structure, commands or error messages should be recorded back in the design document to give us a accurate view of the application's architecture.

Table 22.3 Loyalty application error messages and their description

Error	Name	Description
0xF0C0	SW_CLANOTSUPPORTED	CLA byte in the command APDU is invalid
0xF0C1	SW_INSNOTSUPPORTED	INS byte in the command APDU is invalid
0xF0C2	SW_LCNOTSUPPORTED	Length of input data is invalid
0xF0C3	SW_INCEXCEEDS	The counter value exceeds its limit ($input + counter > limit$)
0xF0C4	SW_DECEXCEEDS	The counter value become negative ($counter - input < 0$)
0xF0C5	SW_INVALID	Command structure is invalid
0xF0C6	SW_UNDEFINED	Undefined error occurred during execution

22.5.2 Coding

For brevity we will not list the complete code of the loyalty application here, which can be obtained by contacting the authors or through [11]; nevertheless, we continue on the code from Listing 22.1. First we declare the required data structures, which are illustrated in Listing 22.2.

Listing 22.2 Declarations of Data Structures used in FirstApplet

```
1   // Data variable to store the loyalty points
2       private short CounterValue = 0;
3
4   // FirstApplet specific constants
5       final static byte CLA = (byte)0xC0;
6       final static byte INS_RESET = (byte)0xE0;
7       final static byte INS_READ = (byte)0xE1;
8       final static byte INS_INCREASE = (byte)0xE2;
9       final static byte INS_DECREASE = (byte)0xE3;
10
11  // Applet's ERROR Status Words
12      final static short SW_CLANOTSUPPORTED = (byte)0xF0C0;
13      final static short SW_INSNOTSUPPORTED = (byte)0xF0C1;
14      final static short SW_LCNOTSUPPORTED = (byte)0xF0C2;
15      final static short SW_DECEXCEEDS = (byte)0xF0C4;
16      final static short SW_INCEXCEEDS = (byte)0xF0C3;
17      final static short SW_UNDEFINED = (byte)0xF0C6;
18      final static short SW_INVALID = (byte)0xF0C5;
```

Next, we program the `process` method that starts with handling the APDUs (both command and response APDUs), which are illustrated in Listing 22.3. The statement on line 3 declares a byte array that takes the command APDU from the apduHandle object. The byte array `apduBuffer` is used for both command and response APDUs. The first step in the command APDU handling is to check whether the command corresponds to the correct format. Therefore, on line 12 we verify whether the CLA byte of the command APDU is the one supported by the loyalty application. Next we use the "if" statements that define four cases for the INS byte of the command APDU, each associated with the relevant operation of the application. If the INS byte does not match any of the defined commands then the `FirstApplet` throws an `SW_INSNOTSUPPORTED` exception.

Listing 22.3 Main Functionality of the Loyalty Application

```
1    public void process(APDU apduHandle) throws ISOException {
2
3        byte[] apduBuffer = apduHandle.getBuffer();
4        short temporaryShortValue = 0;
5
6        // Return if the APDU is selection APDU
7        if(selectingApplet()){
8            return;
9        }
10
11       // For Command APDUs check their CLA value
12       if(apduBuffer[ISO7816.OFFSET_CLA]!= CLA){
13           ISOException.throwIt(SW_CLANOTSUPPORTED);}
14
15       // For Command APDUs check their INS value
16       if (apduBuffer[ISO7816.OFFSET_INS] = =INS_RESET){
17           CounterValue = 0;
18           return;
19       }
20
21       if (apduBuffer[ISO7816.OFFSET_INS] = =INS_READ){
22           converterShortToByte(apduBuffer);
23           apduHandle.setOutgoingAndSend((short)0, (short)2);
24           return;
25       }
26
27       if (apduBuffer[ISO7816.OFFSET_INS] = =INS_INCREASE){
28           if(apduBuffer[ISO7816.OFFSET_LC]= =(byte)0x02){
29               temporaryShortValue = converterByteToShort(apduBuffer,ISO7816.OFFSET_CDATA);
30               if(temporaryShortValue <= (short)(CounterValueLimit-CounterValue)){
31                   CounterValue = (short)(CounterValue+temporaryShortValue);
32                   return;
33               } else{
34                   ISOException.throwIt(SW_INCEXCEEDS);}
35           }else{
36               ISOException.throwIt(SW_LCNOTSUPPORTED);}
37       }
38
39       if (apduBuffer[ISO7816.OFFSET_INS] = =INS_DECREASE){
40           if(apduBuffer[ISO7816.OFFSET_LC]= =(byte)0x02){
41               temporaryShortValue = converterByteToShort(apduBuffer, ISO7816.OFFSET_CDATA);
42               if(temporaryShortValue <= CounterValue){
43                   CounterValue = (short)(CounterValue - temporaryShortValue);
44                   return;
45               }else{
46                   ISOException.throwIt(SW_DECEXCEEDS);}
47           }else{
48               ISOException.throwIt(SW_LCNOTSUPPORTED);}
49       }else{
50           ISOException.throwIt(SW_INSNOTSUPPORTED);}
51       return;
52   }
```

In case of the first command (`INS_RESET`), the `CounterValue` is simply set to zero. The counter read command (`INS_READ`) requests the short to byte converter method and returns the value to the terminal application. The conversion method copies the byte value of the `CounterValue` in the first two locations of the byte array (e.g. `apduBuffer`), which is then sent as part of the response APDU.

If the terminal application sends an `INS_INCREASE` command, the loyalty application first checks the length of data component of the APDU. If this value is not 2, it will throw `SW_LCNOTSUPPORTED` exception. After verifying the length of the input data, it will check whether adding the input value to the `CounterValue` would exceed the limit of short data types in Java Card. If it does not, it will add the value. As the input value is sent as an byte array, so loyalty application has implemented a helper method (`converterByteToShort`) to convert an input byte array to a short value. The `INS_DECREASE` command is handled in the similar manner by the loyalty application.

Listing 22.4 shows the helper methods that provide the short to byte conversion and vice versa. For brevity, we do not explain the rationale behind the statements in this listing as they are based on bit manipulation and how data is stored in a computer.

Once the coding of the application is completed we compile it using the Java compiler that will produce a "class" file, which is later converted by the capgen[3] to a CAP file. The compiling and CAP generation process can automatically be handled by the IDE. After the application is compiled and packaged, it can then be hosted on to the Java Card simulator.

Listing 22.4 Helper Methods of the Loyalty Application

```
 1    private void converterShortToByte(byte[] inputArray){
 2                  inputArray[0] = (byte) ((short)(CounterValue & (short)0xFF00)
 3                  >> (short)0x0008);
 4                  inputArray[1] = (byte) (CounterValue & (short)0x00FF);
 5        }
 6
 7    private short converterByteToShort(byte[] inputArray, short arrayOffset){
 8                  return (short) (((inputArray[arrayOffset] << 8)) |
 9                  ((inputArray[(short)(arrayOffset+(short)1)] & 0xff)));
10        }
```

22.5.3 Simulating and Testing

A Java Card 2.2.2 offers two flavors of Java Card simulators: CREF and JCWDE. Both of these simulators offer different features and it would be a good idea to check which supports the required functionality to simulate a particular application.

To test an application, there are two possible ways: scripting and developing a terminal application. In this chapter, we only discuss how to use scripts to test the loyalty application.

A typical terminal application will communicate with the loyalty application by first selecting the application and then requesting a particular process (i.e. commands listed in Table 22.2). A simplistic test script in hexdecimal format is shown in Listing 22.5.

---

[3] Capgen: It is part of the Java Card development kit and used to produce CAP files from class files.

Listing 22.5 A Simple Test Script for Loyalty Application

```
1    powerup;
2    // select FirstApplet applet
3    0x00 0xA4 0x04 0x00 0xA0 0xD0 0x00 0x00 0x00 0x62 0x02 0x01 0x0C 0x0E 0x0A 0x7F;
4    // send counter reset command
5    0xB0 0xE0 0x00 0x00 0x00 0x7F;
6    // send counter read command
7    0xB0 0xE1 0x00 0x00 0x00 0x7F;
8    // send counter increase command, increasing the value by 2
9    0xB0 0xE2 0x00 0x00 0x02 0x00 0x02 0x7f;
10   // send counter decrease command, decreasing the value by 1
11   0xB0 0xE3 0x00 0x00 0x02 0x00 0x01 0x7f;
12   // send counter read command
13   0xB0 0xE1 0x00 0x00 0x02 0x7f;
14   powerdown;
```

The first command APDU signals to the JCRE to select the loyalty application. Next, it resets the counter and then tries to read it. At this point the loyalty application should send the value zero. The script then requests the loyalty application to increase the counter value by two and subsequently decrease it by one. Finally, the script tries to read the counter value that should be returned as one. We can extend the script in many ways to check the application. The aim during the testing on a smart card application should be to evaluate all possible commands and scenarios, which enable the execution of most (preferably all) of the application code. For example, the script should have some error commands to test the robustness and fault-tolerance of the loyalty application. The more extensive the testing, the better the end product will be.

22.6 Conclusion

In this chapter, we described peculiarities of the smart card architecture that a developer has to consider. Although, the smart card technology has seen rapid development both in terms of hardware and inclusion of rich features; it is still a resource restricted device. We then moved to provide a simplistic example of a loyalty application. The complete code for the loyalty application along with additional examples and tutorial, readers are advised to check the developer's section of the book's official website.

Acknowledgments The authors want to thank the reviewers for their constructive comments which were helpful to improve this chapter.

References

1. Multos: The Multos Specification. http://www.multos.com/
2. NSA Suite B Cryptography. http://www.nsa.gov/ia/programs/suiteb_cryptography/index.shtml

3. ISO/IEC 7816–4, Identification cards - Integrated circuit cards - Part 4: Organization, security adn commands for interchange, (2005). http://www.iso.org/iso/iso_catalogue/catalogue_tc/catalogue_detail.htm?csnumber=36134
4. GlobalPlatform: GlobalPlatform Card Specification, Version 2.2, (2006). http://www.globalplatform.org/specificationscard.asp
5. Java Card Platform Specification; Application Programming Interface, Runtime Environment Specification, Virtual Machine Specification, (2006). http://java.sun.com/javacard/specs.html
6. EMV 4.2: Book 1 - Application Independent ICC to Terminal Interface Requirements, Book 2 - Security and Key Management, Book 3 - Application Specification, Book 4 - Cardholder, Attendant, and Acquirer Interface Requirements, (2008). http://www.emvco.com/specifications.aspx?id=155
7. ISO/IEC 14443: Identification Cards - Contactless Integrated Circuit(s) Cards - Proximity Cards, Part1: Physical Characteristics, Part 2: Radio Frequency Power and Signal Interface, Part3: Initialization and Anticollision, Part 4: Transmission Protocol, (2008). http://www.iso.org/iso/iso_catalogue/catalogue_tc/catalogue_detail.htm?csnumber=28728
8. Java Card Platform Specification: Classic Edition; Application Programming Interface, Runtime Environment Specification, Virtual Machine Specification, Connected Edition; Runtime Environment Specification, Java Servlet Specification, Application Programming Interface, Virtual Machine Specification, Sample Structure of Application Modules, (2009). http://java.sun.com/javacard/3.0.1/specs.jsp
9. BasicCard (Visited June, 2010). http://basiccard.com/
10. Akram, R.N., Markantonakis, K., Mayes, K.: Firewall Mechanism in a User Centric Smart Card Ownership Model. In: D. Gollmann, J.L. Lanet, J. Iguchi-Cartigny (eds.) Smart Card Research and Advanced Application, 9th IFIP WG 8.8/11.2 International Conference, CARDIS 2010, vol. 6035/2010, pp. 118–132. Springer, Passau, Germany (2010). DOI http://dx.doi.org/10.1007/978-3-642-12510-2
11. Royal Holloway, University of London. Smart Card Centre website. http://www.scc.rhul.ac.uk/books/ssed/embedded/chapter_22.
12. Rankl, W.: Smart Card Applications: Design Models for Using and Programming Smart Cards. Wiley (2007)
13. Rankl, W., Effing, W.: Smart Card Handbook. John Wiley & Sons, Inc., New York, NY, USA (2003)

Chapter 23
A Practical Example of Mobile Phone Application Using SATSA (JSR 177) API

Lishoy Francis

Abstract SIM as a security token is increasingly being used to secure mobile phone applications. Sensitive information such as PIN, security keys, etc are stored on the SIM card. To utilise the SIM functionalities, it is imperative that mobile phone applications interact with applets available on the SIM. The security features for mobile applications operating within the J2ME ecosystem are provisioned by SATSA API Framework. It allows support for cryptography, digital signatures, user credential management, communication with a smart card, and remote method invocation. The SATSA APDU Communication API provides support for mobile phone applications to interact with Java Card applets residing on a smart card, over the ISO7816 interface. This chapter provides a practical example of a mobile phone application implementing SATSA API. A MIDP 2.0 application or MIDlet that utilises the APDU package within SATSA API and a Java Card applet were developed. The MIDlet and applet were tested to work with each other on a PC-based development environment. The MIDlet was tested on Wireless Toolkit Emulator and the Java Card applet was tested on Java Card Platform Simulator. Freely available tools were used to create the above mentioned practical demonstrators.

23.1 Introduction

In this chapter, we show how to develop a Java 2 Platform, Micro Edition (J2ME) [1] Mobile information device profile (MIDP) [2] application (commonly known as MIDlet). This MIDlet implements the Security and trust services API (SATSA) [3] Application protocol data unit (APDU) Communication package. The SATSA APDU communication application programming interface (API) enables the MIDlet

L. Francis (✉)
Information Security Group, Smart Card Centre, Royal Holloway,
University of London, London, UK
e-mail: Lishoy.Francis.2005@live.rhul.ac.uk

K. Markantonakis and K. Mayes (eds.), *Secure Smart Embedded Devices,*
Platforms and Applications, DOI: 10.1007/978-1-4614-7915-4_23,
© Springer Science+Business Media New York 2014

to communicate with a Java Card applet residing on a smart card such as SIM [1][4].
We also show how to develop a Java Card applet that is based on Java Card Framework
v2.2.1 [5]. This applet is developed in order to work with or rather test the MIDlet.
The applet is made capable of processing APDU command messages received from
MIDlet over SATSA interface, and also to respond with the necessary APDU response
messages. Figure 23.1 illustrates the MIDlet, the Java Card applet, and their interaction over the SATSA APDU Communication interface.

Fig. 23.1 Design view

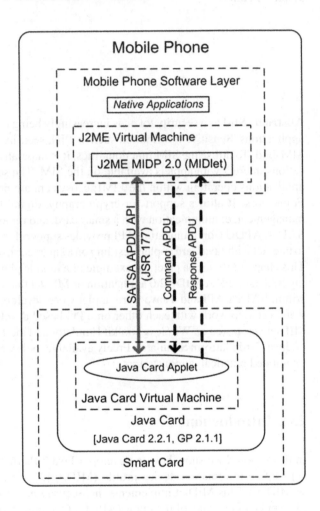

---

[1] SIM stands for Subscriber Identity Module [4] widely used to provision GSM network authentication security. But, here we refer to SIM as any smart card used in the mobile telecommunications
environment.

23.1.1 A Brief Overview of SATSA Framework

The Security and Trust Services API (previously known as Java Specification Request 177 or JSR 177) is an optional package within the J2ME framework that provides security features. This specification supports cryptographic operations, digital signature services, user credential management, communication with a smart card using the APDU interface, and Java card remote method invocation (JCRMI). SATSA connects J2ME and smart card ecosystems such as Java Card framework. SATSA is defined in four optional packages as described below.

- SATSA-CRYPTO: This optional package defines the cryptographic API. It consists of the following packages,

  - *java.security* package is a subset of J2SE *java.security*. It defines classes and interfaces required for the security framework. These include, *Key* and *PublicKey* interfaces, and classes such as *MessageDigest*, *KeyFactory*, and *Signature*
  - *java.security.spec* package is a subset of J2SE *java.security.spec*. It defines classes and interfaces for key specifications and algorithmic parameter specifications. These include, *X509EncodedKeySpec*, *KeySpec*, *AlgorithmParameterSpec*, and *EncodedKeySpec*
  - *javax.crypto* package is a subset of J2SE *javax.crypto*. It defines the *Cipher* class for cryptographic operations such as encryption and decryption. These include, cryptographic exceptions
  - *javax.crypto.spec* package is a subset of J2SE *javax.crypto.spec*. It defines classes for cryptographic key specifications.

- SATSA-PKI: This optional package mainly supports user credential management and signature services. It consists of the following two packages.

  - *javax.microedition.securityservice* package defines class *CMSMessageSignatureService*, and *CMSMessageSignatureServiceException*
  - *javax.microedition.pki* package defines class *UserCredentialManager*, and *UserCredentialManagerException*.

- SATSA-JCRMI: This optional package defines Remote Method Interface (RMI) interfaces. This package utilises Generic Connection Framework (GCF) and uses the package, *javax.microedition.io* which defines GCF Connector that supports Java Card Remote Method Invocation (JCRMI). SATSA-JCRMI package consists of the following packages.

  - *javacard.framework* package defines a subset of Java Card API. It includes Java Card API framework exceptions
  - *javacard.framework.service* package defines a subset of Java Card API. It includes Java Card API service exception
  - *javacard.security* package defines a subset of Java Card API. It includes a Java Card API crypto exception

- *java.rmi* package defines a subset of J2SE *java.rmi*. It includes *Remote Interface* and *RemoteException*
- *javax.microedition.jcrmi* package defines *JavaCardRMIConnection* interface, *RemoteStub* class and *RemoteRef* interface.

• SATSA-APDU: This optional package, *javax.microedition.apdu*, defines APDU-Connection interface supporting ISO7816-4 [6] APDU-based communications. This package utilises GCF and uses the package, *javax.microedition.io* which defines GCF Connector that supports APDU connections. We implement this API in our demonstrator.

23.1.2 A Brief Overview of Java Card Framework

Java Card technology has evolved since its inception in 1996. However, the initial design goals of portability and security have remained as the core part of its design. Java Card technology enables Java-based applications (commonly known as "Applets") to be run securely on resource-constrained devices such as smart cards. The Java Card Platform specifications were developed and managed by Sun Microsystems, which is now a subsidiary of Oracle Corporation. The latest Java Card Platform specification that has been released is version 3.0. Java Card utilises GlobalPlatform [7] specifications for securely managing applets on a smart card. Although Java Card is a subset of Java, it differs with Java in many aspects. For example, it has a limited virtual machine called Java card virtual machine (JCVM), and has limited programming language constructs. The JCVM differs from any standard Java virtual machine (JVM). Distinguishing between Java Card and Java is out of scope of this chapter. The reader is encouraged to explore Java Card technology further by referring to [5], [8].

23.2 Practical Example

In this section, we show how to develop a MIDP application implementing SATSA APDU Communication API and a Java Card applet based on Java Card Framework. We then test that both applications interact with each other, by using Java Card Platform Simulator and Wireless Toolkit Emulator.

23.2.1 Developing a MIDP Application (MIDlet) Implementing SATSA APDU Communication API

What You Will Need

• A beginner's level understanding of J2ME development process/life cycle.

- A beginner's level understanding of Eclipse IDE [9].
- Eclipse Integrated Development Environment (IDE) or any similar IDE.
- Oracle/Sun Java Wireless Toolkit for $CLDC^2$ and CDC^3 (WTK^4 v2.5.2.+), or Java ME Software Development Kit.
- J2SE (Java Standard Edition) environment installed on development PC.
- Optional:

  - A mobile phone supporting J2ME environment (MIDP v2.0+) and SATSA API, in order to install and test the developed MIDlet.
  - A code-signing [10] certificate issued from a trusted certificate authority in order to implement, sign and access restricted SATSA API.

In This Example We Used

- Eclipse IDE v3.4.1 (Ganymede) running on a PC.
- Oracle/Sun Wireless Toolkit v2.5.2_01.

We design a simple mobile phone application based on MIDP v2.0 [2], i.e. a MIDlet that would demonstrate a SATSA API. We chose to implement SATSA APDU Communication API in the MIDlet as this is the basic requirement for any interaction between the mobile phone and the SIM. This MIDlet would interact with a Java Card applet, residing on a SIM, that is capable of processing APDU messages.

We setup Oracle/Sun Java wireless toolkit (WTK) on Eclipse IDE v3.4.1 (Ganymede) installed in a PC based operating environment. One may use alternate IDEs, or command-line tools to build MIDlets. WTK is an advanced toolkit for developing wireless applications that are based on J2ME's MIDP and Connected limited device configuration (CLDC). This toolkit is designed to run on mobile phones, personal digital assistants, and other resource-constrained small mobile devices. The toolkit includes device emulation environments, performance tuning features, MIDlet signing, certificate management, integrated Over-the-air (OTA) emulation, push registry emulation, documentation, etc. WTK implements various device capabilities using standard APIs. These APIs are defined through Java Community Process (JCP). The supported APIs in WTK v2.5.2 includes the following as given in Table 23.1. In J2ME ecosystems, the configuration (e.g. CLDC) consists of a lightweight version of the JVM, and associated base class libraries. The profile (Mobile information device profile - MIDP) is built on top of base class libraries. Other APIs exist on top of the configuration and the profile layers. The configuration and profile are normally embedded within mobile devices.

We use the following APIs to implement our MIDlet example.

- Security and Trust Services API for J2ME (JSR 177).
- Mobile Information Device Profile (MIDP) 2.0 (JSR 118).
- Connected Limited Device Configuration (CLDC) 1.1 (JSR 139).

---

[2] CLDC stands for Connected Limited Device Configuration.

[3] CDC stands for Connected Device Configuration.

[4] WTK stands for Wireless Toolkit.

Table 23.1 APIs supported in Oracle/Sun WTK v2.5.2

API	Description
JSR 248	Mobile service architecture
JSR 185	Java technology for the wireless industry (JTWI)
JSR 139	Connected limited device configuration (CLDC) 1.1
JSR 118	Mobile information device profile (MIDP) 2.0
JSR 75	PDA optional packages for the J2ME platform
JSR 82	Java APIs for bluetooth
JSR 135	Mobile media API (MMAPI)
JSR 172	J2ME web services specification
JSR 177	Security and trust services API for J2ME
JSR 179	Location API for J2ME
JSR 180	SIP API for J2ME
JSR 184	Mobile 3D graphics API for J2ME
JSR 205	Wireless messaging API (WMA) 2.0
JSR 211	Content handler API
JSR 226	Scalable 2D vector graphics API for J2ME
JSR 229	Payment API
JSR 234	Advanced multimedia supplements
JSR 238	Mobile internationalization API
JSR 239	Java binding for the openGL(R) ES API

The design requirements of our simple MIDlet can be summarised as follows,

- MIDlet requires only basic user interactions and no user input.
- MIDlet should display messages on a simple graphical user interface on the screen.
- MIDlet should be able to open SATSA APDU Connection with a smart card.
- MIDlet should exchange and display APDU command/response messages with the smart card.
- MIDlet should handle and display any error-conditions/exceptions that may be encountered.

23.2.1.1 Compiling and Building MIDlet Source Code

The full source code of our MIDlet is made available in the Appendix, available at the end of this chapter. The reader is now encouraged to go through the source code before continuing to read any further.

Every MIDlet must extend the abstract class *MIDlet* which is found within *javax.microedition.midlet* package. The MIDlet must also override three methods of this class, such as *startApp()*, *pauseApp()*, and *destroyApp()*. Our MIDlet implements the command listener found in *javax.microedition.lcdui.commandListener* package.

The *startApp* method is incorporated with functions such as opening a SATSA connection, and exchanging APDU messages over the opened connection. It also displays the APDU command and response messages. In order to interact with the Java Card applet, we need to use a connection based on Generic connection framework (GCF). GCF handles application selection command and logical communication channel between the MIDlet and a Java Card applet residing on the smart card.

```
javax.microedition.apdu.APDUConnection
```

The above interface in GCF defines how to exchange APDUs with a Java Card applet installed on a smart card. It also defines other functions such as read Answer-to-reset (ATR), manage Personal Identification Number (PIN), reset card, etc.

A GCF connection is created by calling *Connector.open()* method. For establishing a SATSA channel, the Application identifier (AID) of the target Java Card applet and slot number of the smart card are required to be supplied within the connection Uniform resource locator (URL). For simplicity, from this point onwards let us call the GCF connection a SATSA connection. The format of connection URL is shown as below.

```
Protocol:[Slot-Identifier];Target=<AID>
```

where, *Protocol* parameter is set as *apdu* or *jcrmi*. *Slot-Identifier* is the number that indicates slot where the smart card is available. By default, this value is set to *0*. The *Slot-Identifier* field is optional. In case the mobile phone supports more than one reader/smart card, the *Slot-Identifier* can be discovered by using a special property called *microedition.smartcardslots* that is available by calling the method System.getProperty(). *Target* field is either an AID of the Java Card applet or the SIM application toolkit (SAT).

In our example, the connection URL string is as follows,

```
String urlSC =
"apdu:0;target=59.92.51.6E.83.6D.71.44.25.64.39.34.90";
```

With the SATSA connection open, the Java Card applet gets selected by default and is ready to process any APDUs received from the MIDlet. In order to exchange any APDU commands, the MIDlet must use *exchangeAPDU()* method. The SATSA connection must be closed after use so that the logical channel acquired to communicate with Java Card applet on the smart card is released. This can be performed by calling the method *Connector.close()*. The reader is encouraged to refer to the SATSA specification available at [3], for more detailed information such as exceptions raised for the methods discussed.

The following code snippet shows the SATSA connection being opened, with a Java Card applet with AID *59.92.51.6E.83.6D.71.44.25.64.39.34.90* that resides on the smart card available in slot number *0*, APDU messages being exchanged, and finally the SATSA connection is closed.

Listing 23.1 SATSA connection being opened by a Java Card applet.
```
1  byte[] apdu =
2  {
3  //command APDU
```

```
 4  //00 F2 00 00 14
 5  (byte) 0x00, (byte) 0xF2, (byte) 0x00, (byte) 0x00,
 6  (byte) 0x14
 7  };
 8
 9  String urlSC =
10  "apdu:0;target=59.92.51.6E.83.6D.71.44.25.64.39.34.90";
11
12  //Create and open a connection object
13  ISO_SM_Conn = (APDUConnection)Connector.open(urlSC);
14
15  //send APDU via exchangeAPDU method
16  response = ISO_SM_Conn.exchangeAPDU(apdu);
17
18  //Close connection
19  try{
20      ISO_SM_Conn.close();
21      }//try
22  catch(Exception ex)
23      {
24      ex.printStackTrace();
25      }//catch
```

Now that we have understood the usage of SATSA APDU Communication API and after incorporating it in the source code, we shall compile and build the Java sources. To compile and build the Java sources available in the Eclipse project, one must choose the *Build Project* option from the *Project* menu of Eclipse IDE as shown in Fig. 23.2. This step would compile the Java source file into Java class file.

The MIDlet requires mandatory security permission, *javax.microedition.apdu.aid*, in order to open a SATSA APDU connection. This type of permission is only granted to MIDlets in the operator, manufacturer, and third-party trusted domains. So it is important that we add *javax.microedition.apdu.aid* as the required MIDlet permission within the *application descriptor* or *manifest* file of the project.

```
MIDlet-Permissions: javax.microedition.apdu.aid
```

For more information on MIDP security permissions and code-signing, the reader is advised to refer to [10].

23.2.1.2 Preverification and Packaging the MIDlet

The next step is to preverify the compiled MIDlet class. This step is required to be performed by JVM to ensure the correctness of the class file in accordance to the JVM specification. For more detailed information on steps involved in preverification, the reader is encouraged to refer to the JVM specification [11]. In a nutshell, during the preverification process, information is added to the class file, such as variable types and operand stack items used. The inlines of all subroutines are added and jump operations are removed. The whole process is aimed to make verification on the device more efficient.

The final step is to package the MIDlet to make it ready for testing and deployment. The *application descriptor* or the *manifest* files can be personalised by adding addi-

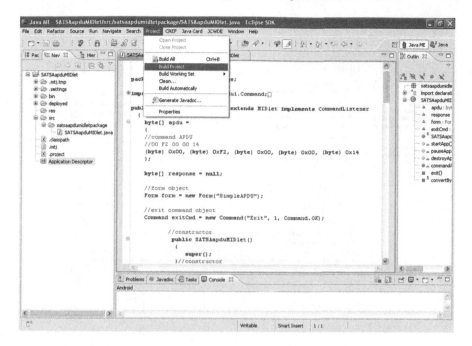

Fig. 23.2 MIDlet compile and build

tional information such as MIDlet Name, MIDlet vendor, MIDlet JAR URL, MIDlet version, etc. A screenshot of *application descriptor* open in the Eclipse workspace is shown in Fig. 23.3.

By clicking on the *create package* option within *application descriptor*, Eclipse IDE performs the preverification and the packaging of MIDlet. The result would be JAR (Java Archive) file and a Java application descriptor (JAD) file. The JAR file is the Java executable that needs to be installed and run on the target mobile phone. The JAD file contains all information from *application descriptor* and informs the mobile device what to expect within the JAR file. With *create package* successful run, you can see that the JAR and JAD files are created in the *deployed* folder of the project. The *application descriptor* or the JAD file of our MIDlet is as follows.

```
MIDlet-Version: 1.0.0
MIDlet-Vendor: MIDlet Suite Vendor
MIDlet-Jar-URL: SATSAapduMIDlet.jar
MicroEdition-Configuration: CLDC-1.1
MIDlet-1: SATSAapduMIDlet,
satsaapdumidletpackage.SATSAapduMIDlet
MicroEdition-Profile: MIDP-2.0
MIDlet-Permissions: javax.microedition.apdu.aid
MIDlet-Name: SATSAapduMIDlet MIDlet Suite
```

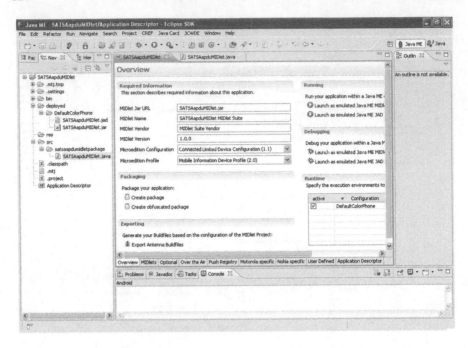

Fig. 23.3 MIDlet application descriptor overview

23.2.1.3 Running the MIDlet on WTK Emulator

Before deploying the MIDlet on the real device, it is important that it is tested on a mobile phone (WTK) emulator. In order to deploy the MIDlet on the WTK emulator, one should create a *Run Configuration* in Eclipse. The target project and emulator needs to be specified. A screenshot of *Run Configuration* is shown in Fig. 23.4.

The MIDlet when run on the WTK emulator would ask the user to grant permission to access the smart card as shown in Fig. 23.5. The permission should be granted to continue making the SATSA connection.

23.2.1.4 Deploying the MIDlet on a Mobile Phone

In order to test the midlet on a mobile phone supporting SATSA APDU API, one would require to sign the MIDlet [10]. This would enable MIDlet with required security privileges to access the smart card available on the SIM reader interface of the mobile phone. After a successful run of the MIDlet on the emulator and rigorous testing, it is ready to be installed on the mobile phone. This step can be performed by using many different methods. Some of these methods are listed as follows. The

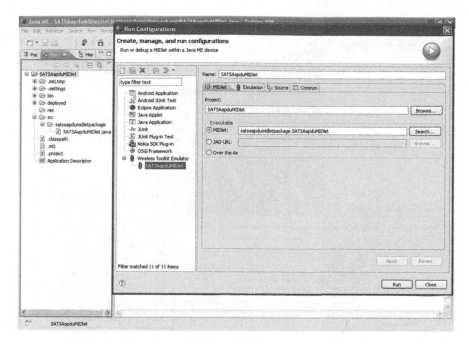

Fig. 23.4 MIDlet run configuration

reader is encouraged to find more details on following steps and adapt to the most suitable and convenient method.

- Installing the MIDlet over Bluetooth connection setup between mobile phone and PC.
- Installing the MIDlet over USB interface setup between mobile phone and PC.
- Installing the MIDlet from a memory card inserted on the mobile phone.
- Installing the MIDlet hosted on a webserver via Internet or Over-The-Air.

23.2.2 Developing a Java Card Applet

What You Will Need

- A beginner's level understanding of Java Card development process/life cycle.
- A beginner's level understanding of setting Java development environment.
- Oracle/Sun Java Card Development Kit (JCDK/JDK v2.2.+) setup to develop Java Card applets.
- Optional:

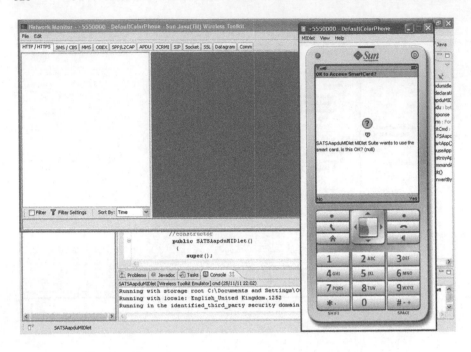

Fig. 23.5 MIDlet permission

Table 23.2 AIDs of applet and package

Category	RID	Length	PIX	Length	Total length
Package	0x59, 0x92, 0x51, 0x6E, 0x83	5	0x6D, 0x71, 0x44, 0x25, 0x64, 0x39, 0x34, 0x90, 0xA0, 0x01	10	15
Applet	0x59, 0x92, 0x51, 0x6E, 0x83	5	0x6D, 0x71, 0x44, 0x25, 0x64, 0x39, 0x34, 0x90	8	13

- A smart card supporting Java Card framework (v2.2+), in order to load, install and test the developed Java Card applet.

In this Example We Used

- Oracle/Sun Java Card Development Kit v2.2.1 setup on a PC.

For demonstration purposes of SATSA-APDU, a Java Card applet is required to work in conjunction with mobile phone application, i.e. MIDlet. In this section, we design a simple Java Card applet that can process APDUs and respond with a pre-determined message.

Table 23.3 Command interface - APDU commands

Command name	CLA	INS	P1	P2	Lc	Data	Le
SELECT	0x00	0xA4	0x04	0x00	0x0C	0x59, 0x92, 0x51, 0x6E, 0x83, 0x6D, 0x71, 0x44, 0x25, 0x64, 0x39, 0x34, 0x90	0x00
TEST	0x00	0xF2	0x00	0x00	NA	NA	0x14

Table 23.4 Command interface - APDU responses

Command name	Response	SW1	SW2
SELECT	0x90, 0x00	0x90	0x00
TEST	0x00, 0xF1, 0x00, 0x00, 0x0D, 0x74, 0x72, 0x61, 0x6E, 0x73, 0x6C, 0x61, 0x74, 0x65, 0x63, 0x32, 0x63, 0x78, 0x90, 0x00	0x90	0x00

23.2.2.1 Application Identifiers (AIDs) of Java Card Applet and its Package

According to Java convention, a Java Card applet class is defined within a package. Both package and applet requires to be assigned with an Application identifier (AID). AID is defined in ISO 7816-5 [6]. In our case, we assign the following AIDs as given in the Table 23.2.

23.2.2.2 Applet Command Interface

We assign a simple command interface for our Java Card applet. It consists of *SELECT* and *TEST* command APDUs. The APDU command message formats are given in Table 23.3, and APDU response message formats are given in Table 23.4.

23.2.2.3 Compile, Convert, and Run Applet on CREF Simulator

The Java Card Development Kit includes a Java Card Platform Simulator called C-language Java Card RE (CREF). This is the reference implementation of Java Card Runtime Environment and is written in C programming language [12]. CREF simulates persistent memory (EEPROM[5]) and allows applets to be installed on it. It allows us to save and restore EEPROM contents to disk files of the host computer. It performs Input/Output (I/O) over socket interface which simulates all interactions

---

[5] EEPROM stands for Electrically erasable programmable read-only memory. This is a type of non-volatile memory that is used to store data that must be saved when power is removed, in computers, smart cards and other electronic devices.

with a card reader by supporting communication protocols such as T=1, T=CL, or T=0. More detailed information of CREF and on how to use, can be found in user's guide, within documentation of any Java Card Development Kit [5]. We used Oracle/Sun's CREF Java Card Platform Simulator to test the developed applet.

Firstly, we describe the steps to develop the applet. Then *run* the applet on the simulator. For more detailed information of development process, reader is required to refer to Java Card documentation available in any Java Card Development Kit [5].

The *source code* used in this example is available in the Appendix, at the end of this chapter. The reader is encouraged to read through and understand the source code before proceeding to read this section any further. It is recommended to set the paths of *JAVA_HOME* and *JC_HOME*, logically mapped to Java Development Kit (J2SE) and Java Card Development Kit install directories respectively. In our case, these variables were set on a PC environment as follows.

```
set JAVA_HOME=C:\Program Files\Java\j2sdk1.4.2_05
set JC_HOME=C:\java_card_kit-2_2_1
```

The *source code* is available in the following path.

```
C:\java_card_kit-2_2_1\appcodes\testjsrpackage
\Testjsrapplet.java
```

The following command-line instructions were used to *compile* the source code of the applet.

```
"%Java_Home%"\bin\javac -classpath
C:\java_card_kit-2_2_1\lib\api.jar -g
C:\java_card_kit-2_2_1\appcodes\testjsrpackage
\Testjsrapplet.java
```

After successful compilation of Java source file with no errors and no warnings, the resulting class(es) are required to be converted to "java card executable" known as Java card converted applet (CAP). We convert the class file into CAP by using the command-line instructions as shown below.

```
%JC_Home%\bin\converter -exportpath
%JC_Home%\api_export_files -d C:\java_card_kit-2_2_1
\appcodes -applet 0x59:0x92:0x51:0x6E:0x83:0x6D:0x71
:0x44:0x25:0x64:0x39:0x34:0x90 Testjsrapplet -classdir
C:\java_card_kit-2_2_1\appcodes testjsrpackage 0x59:0x92
:0x51:0x6E:0x83:0x6D:0x71:0x44:0x25:0x64:0x39:0x34:0x90
:0xA0:0x01 1.0
```

Now that we have our CAP file ready, we need to generate the applet download-script using *scriptgen* tool available in the Java Card Development Kit. The resulting script file would contain all the necessary APDUs for downloading the CAP file onto the virtual-card. For this, we invoke the following command-line instruction.

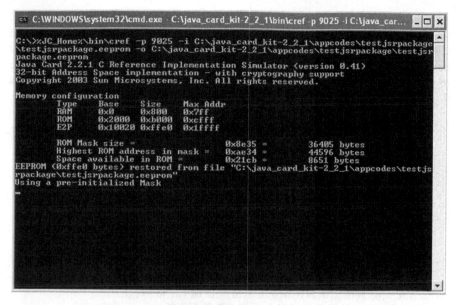

Fig. 23.6   Java Card applet running on CREF simulator

```
%JC_Home%\bin\scriptgen -o C:\java_card_kit-2_2_1
\appcodes\testjsrpackage\testjsrpackage.script
C:\java_card_kit-2_2_1\appcodes\testjsrpackage\javacard
\testjsrpackage.cap
```

The generated script file would be similar to the one shown in the Appendix 23.4.3 at the end of this chapter. It is to be noted that first 2 lines (powerup and selecting the installer applet), and the last 3 lines (creating our applet, selecting installer applet and powerdown) have been added in generated script file.

The next step is to setup the virtual-card image (EEPROM image) that contains the required applet. For this, run CREF with any suitable name that virtual-card image should be saved as on the disk. We created the image with the name *testjsr-package.eeprom* by invoking the follow command-line instruction.

```
start %JC_Home%\bin\cref -o C:\java_card_kit-2_2_1
\appcodes\testjsrpackage\testjsrpackage.eeprom
```

You should now get the confirmation from CREF that the virtual-card image has been saved, as shown in Fig. 23.6.

With CREF running we load the applet onto the virtual-card image (*testjsrpack-age.eeprom*) by using apdutool with script-file (*testjsrpackage.script*) via the following command-line instruction.

```
%JC_Home%\bin\apdutool C:\java_card_kit-2_2_1
\appcodes\testjsrpackage\testjsrpackage.script
```

If the applet has been successfully loaded and installed onto the virtual-card image, window running CREF should shut automatically without any error messages. The CREF exits and saves the virtual-card image as a file on the disk with the name you had specified. This step needs to be performed only once unless required otherwise to clear the state of the applet or the virtual-card image. The message-run of script download using apdutool would look similar to the following.

Listing 23.2 Message-run of script download using apdutool.

```
1  Java Card 2.2.1 APDU Tool, Version 1.3
2  Copyright 2003 Sun Microsystems, Inc. All rights reserved.
3  Use is subject to license terms.
4  Opening connection to localhost on port 9025.
5  Connected.
6  Received ATR = 0x3b 0xf0 0x11 0x00 0xff 0x00
7  CLA: 00, INS: a4, P1: 04, P2: 00, Lc: 09, a0, 00, 00, 00, 62, 03, 01, 08, 01, Le: 00, SW1: 90, SW2: 00
8  CLA: 80, INS: b0, P1: 00, P2: 00, Lc: 00, Le: 00, SW1: 90, SW2: 00
9  CLA: 80, INS: b2, P1: 01, P2: 00, Lc: 00, Le: 00, SW1: 90, SW2: 00
10 CLA: 80, INS: b4, P1: 01, P2: 00, Lc: 1c, 01, 00, 19, de, ca, ff, ed, 01, 02, 04, 00, 01, 0f, 59,
11 92, 51, 6e, 83, 6d, 71, 44, 25, 64, 39, 34, 90, a0, 01, Le: 00, SW1: 90, SW2: 00
12 CLA: 80, INS: bc, P1: 01, P2: 00, Lc: 00, Le: 00, SW1: 90, SW2: 00
13 CLA: 80, INS: b2, P1: 02, P2: 00, Lc: 00, Le: 00, SW1: 90, SW2: 00
14 CLA: 80, INS: b4, P1: 02, P2: 00, Lc: 20, 02, 00, 1f, 00, 19, 00, 1f, 00, 11, 00, 0b, 00, 2e, 00, 0c,
15 00, eb, 00, 0a, 00, 18, 00, 00, 00, 66, 00, 00, 00,00, 00, 00, 01, Le: 00, SW1: 90, SW2: 00
16 CLA: 80, INS: b4, P1: 02, P2: 00, Lc: 02, 01, 00, Le: 00, SW1: 90, SW2: 00
17 CLA: 80, INS: bc, P1: 02, P2: 00, Lc: 00, Le: 00, SW1: 90, SW2: 00
18 CLA: 80, INS: b2, P1: 04, P2: 00, Lc: 00, Le: 00, SW1: 90, SW2: 00
19 CLA: 80, INS: b4, P1: 04, P2: 00, Lc: 0e, 04, 00, 0b, 01, 02, 01, 07, a0, 00, 00, 00, 62, 01, 01, Le:
20 00, SW1: 90, SW2: 00
21 CLA: 80, INS: bc, P1: 04, P2: 00, Lc: 00, Le: 00, SW1: 90, SW2: 00
22 CLA: 80, INS: b2, P1: 03, P2: 00, Lc: 00, Le: 00, SW1: 90, SW2: 00
23 CLA: 80, INS: b4, P1: 03, P2: 00, Lc: 14, 03, 00, 11, 01, 0d, 59, 92, 51, 6e, 83, 6d, 71, 44, 25, 64,
24 39, 34, 90, 00, 8c, Le: 00, SW1: 90, SW2: 00
25 CLA: 80, INS: bc, P1: 03, P2: 00, Lc: 00, Le: 00, SW1: 90, SW2: 00
26 CLA: 80, INS: b2, P1: 06, P2: 00, Lc: 00, Le: 00, SW1: 90, SW2: 00
27 CLA: 80, INS: b4, P1: 06, P2: 00, Lc: 0f, 06, 00, 0c, 00, 80, 03, 02, 00, 02, 07, 01, 00, 00, 00, 99,
28 Le: 00, SW1: 90, SW2: 00
29 CLA: 80, INS: bc, P1: 06, P2: 00, Lc: 00, Le: 00, SW1: 90, SW2: 00
30 CLA: 80, INS: b2, P1: 07, P2: 00, Lc: 00, Le: 00, SW1: 90, SW2: 00
31 CLA: 80, INS: b4, P1: 07, P2: 00, Lc: 20, 07, 00, eb, 00, 05, 10, 18, 8c, 00, 02, 18, 10, 14, 90, 0b,
32 3d, 03, 03, 38, 3d, 04, 10, f1, 38, 3d, 05, 03, 38, 3d, 06, 03, 38, Le: 00, SW1: 90, SW2: 00
33 CLA: 80, INS: b4, P1: 07, P2: 00, Lc: 20, 3d, 07, 10, 0f, 38, 3d, 08, 10, 74, 38, 3d, 10, 06, 10, 72,
34 38, 3d, 10, 07, 10, 61, 38, 3d, 10, 08, 10, 6e, 38, 3d, 10, 09, 10, Le: 00, SW1: 90, SW2: 00
35 CLA: 80, INS: b4, P1: 07, P2: 00, Lc: 20, 73, 38, 3d, 10, 0a, 10, 6c, 38, 3d,
36 10, 0b, 10, 61, 38, 3d, 10, 0c, 10, 74, 38, 3d, 10, 0d, 10, 65, 38, 3d, 10, 0e, 10, 63, 38, Le: 00,
37 SW1: 90, SW2: 00
38 CLA: 80, INS: b4, P1: 07, P2: 00, Lc: 20, 3d, 10, 0f, 10, 32, 38, 3d, 10, 10, 10, 63, 38, 3d, 10,
39 11,10, 78, 38, 3d, 10, 12, 10, 90, 38, 3d, 10, 13, 03, 38, 87, 00, 18, Le: 00, SW1: 90, SW2: 00
40 CLA: 80, INS: b4, P1: 07, P2: 00, Lc: 20, 05, 90, 0b, 3d, 03, 10, 90, 38, 3d, 04, 03, 38, 87, 01, 7a,
41 02, 30, 8f, 00, 03, 3d, 8c, 00, 04, 8b, 00, 05, 7a, 04, 21, 19, 8b, Le: 00, SW1: 90, SW2: 00
42 CLA: 80, INS: b4, P1: 07, P2: 00, Lc: 20, 00, 06, 2d, 1a, 04, 25, 75, 00, 41, 00, 02, ff, a4, 00, 0d,
43 ff, f2, 00, 27, 19, 8b, 00, 07, 3b, 19, ad, 01, 92, 5b, 8b, 00, 08, Le: 00, SW1: 90, SW2: 00
44 CLA: 80, INS: b4, P1: 07, P2: 00, Lc: 20, 19, ad, 01, 03, ad, 01, 92, 5b, 8b, 00, 09, 70, 22, 19, 8b,
45 00, 07, 3b, 19, ad, 00, 92, 5b, 8b, 00, 08, 19, ad, 00, 03, ad, 00, Le: 00, SW1: 90, SW2: 00
46 CLA: 80, INS: b4, P1: 07, P2: 00, Lc: 0e, 92, 5b, 8b, 00, 09, 70, 08, 11, 6d, 00, 8d, 00, 0a, 7a, Le:
47 00, SW1: 90, SW2: 00
48 CLA: 80, INS: bc, P1: 07, P2: 00, Lc: 00, Le: 00, SW1: 90, SW2: 00
49 CLA: 80, INS: b2, P1: 08, P2: 00, Lc: 00, Le: 00, SW1: 90, SW2: 00
50 CLA: 80, INS: b4, P1: 08, P2: 00, Lc: 0d, 08, 00, 0a, 00, 00, 00, 00, 00, 00, 00, 00, 00, 00, Le: 00,
51 SW1: 90, SW2: 00 CLA: 80, INS: bc, P1: 08, P2: 00, Lc: 00, Le: 00, SW1: 90, SW2: 00
52 CLA: 80, INS: b2, P1: 05, P2: 00, Lc: 00, Le: 00, SW1: 90, SW2: 00
```

53 CLA: 80, INS: b4, P1: 05, P2: 00, Lc: 20, 05, 00, 2e, 00, 0b, 02, 00, 00, 00, 02, 00, 00, 01, 06, 80,
54 03, 00, 01, 00, 00, 00, 06, 00, 00, 01, 03, 80, 03, 01, 03, 80, 0a, Le: 00, SW1: 90, SW2: 00
55 CLA: 80, INS: b4, P1: 05, P2: 00, Lc: 11, 01, 03, 80, 0a, 07, 03, 80, 0a, 09, 03, 80, 0a, 05, 06, 80,
56 07, 01, Le: 00, SW1: 90, SW2: 00
57 CLA: 80, INS: bc, P1: 05, P2: 00, Lc: 00, Le: 00, SW1: 90, SW2: 00
58 CLA: 80, INS: b2, P1: 09, P2: 00, Lc: 00, Le: 00, SW1: 90, SW2: 00
59 CLA: 80, INS: b4, P1: 09, P2: 00, Lc: 1b, 09, 00, 18, 00, 08, 7b, 0f, 2d, 08, 03, 0f, 08, 03, 00, 0c,
60 05, 8a, 04, 03, 07, 15, 09, 0b, 06, 09, 0b, 08, Le: 00, SW1: 90, SW2: 00
61 CLA: 80, INS: bc, P1: 09, P2: 00, Lc: 00, Le: 00, SW1: 90, SW2: 00
62 CLA: 80, INS: ba, P1: 00, P2: 00, Lc: 00, Le: 00, SW1: 90, SW2: 00
63 CLA: 80, INS: b8, P1: 00, P2: 00, Lc: 0f, 0d, 59, 92, 51, 6e, 83, 6d, 71, 44, 25, 64, 39, 34, 90, 00,
64 Le: 0d, 59, 92, 51, 6e, 83, 6d, 71, 44, 25, 64, 39, 34, 90, SW1: 90, SW2: 00
65 CLA: 00, INS: a4, P1: 04, P2: 00, Lc: 09, a0, 00, 00, 00, 62, 03, 01, 08, 01, Le: 00, SW1: 90, SW2: 00

In order to make the virtual-card image work with our SATSA MIDlet, we can run the Java Card applet by using the command-line instruction shown below when needed. CREF would listen to the port *9025* for any APDU commands (for instance, those received from Wireless Toolkit Emulator).

```
%JC_Home%\bin\cref -p 9025 -i C:\java_card_kit-2_2_1
\appcodes\testjsrpackage\testjsrpackage.eeprom -o
C:\java_card_kit-2_2_1\appcodes\testjsrpackage
\testjsrpackage.eeprom
```

The state of the virtual-card image is saved and available for subsequent use, unless a new image is created and over-written on the existing one. Now that the virtual-card image is loaded with our applet it is time to test MIDlet's SATSA APDU Communication interface.

23.2.3 Results: Testing MIDlet and Java Card Applet

Now, let us test the SATSA MIDlet and Java Card applet together. The Java Card applet saved on the virtual-card image needs to be running on CREF before we fire-up the MIDlet to run on WTK Emulator.

```
%JC_Home%\bin\cref -p 9025 -i C:\java_card_kit-2_2_1
\appcodes\testjsrpackage\testjsrpackage.eeprom -o
C:\java_card_kit-2_2_1\appcodes\testjsrpackage
\testjsrpackage.eeprom
```

Figure 23.6 shows the Java Card applet running on CREF simulator.

Now let us run the SATSA MIDlet on WTK Emulator. For this, invoke the *Run Configuration* of the project as shown in Fig. 23.4 and allow smart card access permission when requested, as shown in Fig. 23.5.

With the MIDlet successfully opening the SATSA connection, it would send the command APDU-*SELECT* implicitly, and followed by the command APDU -*TEST*. If there is no error in any of the previous operations, the APDU response from Java Card applet would be displayed on the MIDlet as shown in Fig. 23.7. After sufficient

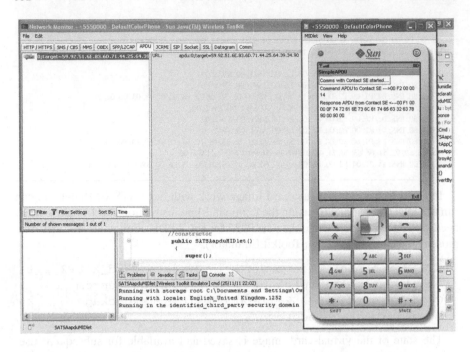

Fig. 23.7 SATSA MIDlet interacting with Java Card applet

rest, it is recommended that reader try and implement more complex features in the MIDlet using SATSA APIs and Java Card applets. Please enjoy coding!

23.3 Conclusion

Smart cards implemented with SIM applications are increasingly employed to provision security in mobile telecommunication systems. The SIM is used to provide authentication to services, and also for storing sensitive information such as secret keys and PIN on a mobile phone. In order to utilise the security services based in the SIM and to interact with the applications installed on it, mobile phone applications would need to implement an advanced API called SATSA API (JSR 177). The core part of SATSA is the APDU Communication API. This API provides support for mobile phone applications to interact with Java Card applets residing on the SIM card over ISO7816 contact interface. In this chapter, we have presented a practical example of developing a mobile phone application (MIDlet) that implemented SATSA APDU Communication API and a corresponding Java Card applet. The MIDlet and the applet were enabled to interact with each other on a PC-based development environment. Both applications were developed and tested by using freely available software tools. Additional resources may be obtained by contacting the authors or through [13].

23.3.1 Source Code of MIDP Application (MIDlet)

Listing 23.3 Source code for SATSAapduMIDlet.java.

```
1   package satsaapdumidletpackage;
2
3   import javax.microedition.lcdui.Command;
4   import javax.microedition.lcdui.CommandListener;
5   import javax.microedition.lcdui.Displayable;
6   import javax.microedition.midlet.MIDlet;
7   import javax.microedition.midlet.MIDletStateChangeException;
8   import javax.microedition.lcdui.Display;
9   import javax.microedition.lcdui.Form;
10  import java.io.IOException;
11  import javax.microedition.io.Connector;
12  import javax.microedition.apdu.APDUConnection; //SATSA APDU API
13
14  public class SATSAapduMIDlet extends MIDlet
15  implements CommandListener
16  {
17  byte[] apdu =
18  {
19  //command APDU
20  //00 F2 00 00 14
21  (byte) 0x00, (byte) 0xF2, (byte) 0x00, (byte) 0x00, (byte) 0x14
22  };
23
24  byte[] response = null;
25
26  //form object
27  Form form = new Form("SimpleAPDU");
28
29  //exit command object
30  Command exitCmd = new Command("Exit", 1, Command.OK);
31
32  //constructor
33  public SATSAapduMIDlet()
34    {
35  super();
36    }//constructor
37
38  protected void startApp() throws MIDletStateChangeException
39  {
40  //add command to form object
41  form.addCommand(exitCmd);
42
43  //add command listener
44  form.setCommandListener(this);
45
46  //set the display to form object
47  Display.getDisplay(this).setCurrent(form);
48
49  //initialise connection object
50  APDUConnection ISO_SM_Conn = null;
51
```

```
52  try
53  {
54  form.append("Comms with Contact SE started....\n");
55
56  //construct URL string with Java Card Applet AID
57  //AID is 0x59, 0x92, 0x51x 0x6E, 0x83, 0x6D, 0x71, 0x44, 0x25, 0x64, 0x39, 0x34,
58  0x90.
59  String urlSC = "apdu:0;target=59.92.51.6E.83.6D.71.44.25.64.39.34.90";
60
61  //Create and open a connection object
62  ISO_SM_Conn = (APDUConnection)Connector.open(urlSC);
63
64  //print the apdu send to form
65  form.append("Command APDU to Contact SE −−−>"+ convertByteToHexString(apdu) +"\n");
66  response = null;
67
68  //send APDU via exchangeAPDU method
69  response = ISO_SM_Conn.exchangeAPDU(apdu);
70
71  if(response !=null)
72  {
73  //Last two bytes define the status of the response
74  if (response.length > 2)
75  {
76  //expecting a response of more than 2 bytes long (20 bytes) print response to form
77  form.append("Response APDU from Contact SE <−−−"+ convertByteToHexString(response)
78  +"\n");
79  }//end of if
80  else
81  {
82  //expecting an error code Status Word print response to form
83  form.append("Response to cmd:\n" + convertByteToHexString(response));
84  }//end of else
85
86  response = null;
87  }
88  }//end of try
89  catch(IOException e)
90  {
91  form.append(e.getMessage());
92  e.printStackTrace();
93  }//end of catch
94  finally
95  {
96  if (ISO_SM_Conn != null)
97  {
98  //Close connection
99  try{
100 ISO_SM_Conn.close();
101 }//try
102 catch(Exception ex)
103 {
104 form.append(ex.getMessage());
105 ex.printStackTrace();
106 }//catch
107 }//if
108 }//finally
109
110 }//end of startApp
111
112 protected void pauseApp()
113 {
114 }//end of pauseApp()
```

```
115
116  protected void destroyApp(boolean arg0)
117  throws MIDletStateChangeException
118  {
119  notifyDestroyed();
120  }//end of destroyApp
121
122  //handle Exit Command
123  public void commandAction(Command arg0, Displayable arg1)
124  {
125  if(arg0 == exitCmd)
126  {
127   exit();
128  }//if
129  }//commandAction
130
131  private void exit()
132  {
133   try
134    {
135   destroyApp(false);
136   notifyDestroyed();
137    }//try
138   catch(MIDletStateChangeException ex)
139    {
140    }//catch
141  }//exit()
142
143  //convert a byte array into hexadecimal string
144  private static String convertByteToHexString(byte[] data)
145  {
146  StringBuffer sbuff = new StringBuffer();
147
148  for (int i = 0; i < data.length; i++)
149  {
150  String bufftemp = Integer.toHexString(data[i] & 0xFF);
151       bufftemp = bufftemp.toUpperCase();
152
153  if (bufftemp.length() == 1)
154    {
155   sbuff.append(0);
156    }//end of if
157   sbuff.append(bufftemp + " ");
158  }//end of For
159
160  return sbuff.toString();
161  }//end of convertByteToHexString
162
163  }//end of class SATSAapduMIDlet
```

23.3.2 Source Code of Java Card Applet

Listing 23.4 Source code for testjsrpackage.java.

```
1  package testjsrpackage;
2  import javacard.framework.*;
3
4  public class Testjsrapplet extends Applet
5  {
6  //initialiase command variables
7  private final static byte
8  INS_ONE = (byte) 0xA4; //command SELECT
```

```
 9  private final static byte
10  INS_TWO = (byte) 0xF2; //command TEST
11
12  //response apdu for command TEST
13  byte[] output_buffer2 =
14  {
15  //00 F1 00 00 0F 74 72 61 6E 73 6C 61 74 65 63 32
16    63 78 90 00
17  (byte) 0x00, (byte) 0xF1, (byte) 0x00, (byte) 0x00,
18  (byte) 0x0F, (byte) 0x74, (byte) 0x72, (byte) 0x61,
19  (byte) 0x6E, (byte) 0x73, (byte) 0x6C, (byte) 0x61,
20  (byte) 0x74, (byte) 0x65, (byte) 0x63, (byte) 0x32,
21  (byte) 0x63, (byte) 0x78, (byte) 0x90, (byte) 0x00
22  };
23
24  byte[] output_buffer3 =
25  {
26  //90 00
27  (byte) 0x90, (byte) 0x00
28  };
29
30  //constructor
31  public Testjsrapplet()
32  {
33  }//endo of Testjsrapplet()
34
35  //install method
36  public static void install(byte[] buffer, short offset,
37  byte length) throws ISOException
38  {
39  //register
40  new Testjsrapplet().register();
41  }
42
43  //process method
44  public void process(APDU apdu) throws ISOException
45  {
46  //buffer variable to store apdu buffer
47    byte[] buffer = apdu.getBuffer();
48
49  switch (buffer[ISO7816.OFFSET_INS])
50  {
51  //if SELECT return SW: 9000
52  case (byte) INS_ONE:
53    apdu.setOutgoing();
54    apdu.setOutgoingLength((byte) output_buffer3.length);
55    apdu.sendBytesLong(output_buffer3, (short) 0,
56  (byte) output_buffer3.length);
57      break;
58
59  //if TEST return 00F100000D7472616E736C617465633263789000
60  case (byte) INS_TWO:
61    apdu.setOutgoing();
62    apdu.setOutgoingLength((byte) output_buffer2.length);
63    apdu.sendBytesLong(output_buffer2, (short) 0,
64  (byte) output_buffer2.length);
65      break;
66
67    default:
68    ISOException.throwIt(ISO7816.SW_INS_NOT_SUPPORTED);
69
70  }//end of switch−case construct
71  }//end of process method
```

```
72  }//end of class Testjsrapplet
```

23.3.3 Java Card Applet Download-Script

Listing 23.5   Source code for testjsrpackage.script.

```
 1  powerup;
 2
 3  // Select the installer applet
 4  0x00 0xA4 0x04 0x00 0x09 0xa0 0x00 0x00 0x00 0x62
 5  0x03 0x01 0x08 0x01 0x7F;
 6
 7  0x80 0xB0 0x00 0x00 0x00 0x7F;
 8
 9  // testjsrpackage/javacard/Header.cap
10  0x80 0xB2 0x01 0x00 0x00 0x7F;
11  0x80 0xB4 0x01 0x00 0x1C 0x01 0x00 0x19 0xDE 0xCA
12  0xFF 0xED 0x01 0x02 0x04 0x00 0x01 0x0F 0x59 0x92
13  0x51 0x6E 0x83 0x6D 0x71 0x44 0x25 0x64 0x39 0x34
14  0x90 0xA0 0x01 0x7F;
15
16  0x80 0xBC 0x01 0x00 0x00 0x7F;
17
18  // testjsrpackage/javacard/Directory.cap
19  0x80 0xB2 0x02 0x00 0x00 0x7F;
20  0x80 0xB4 0x02 0x00 0x20 0x02 0x00 0x1F 0x00 0x19
21  0x00 0x1F 0x00 0x11 0x00 0x0B 0x00 0x2E 0x00 0x0C
22  0x00 0xEB 0x00 0x0A 0x00 0x18 0x00 0x00 0x00 0x66
23  0x00 0x00 0x00 0x00 0x00 0x00 0x01 0x7F;
24
25  0x80 0xB4 0x02 0x00 0x02 0x01 0x00 0x7F;
26  0x80 0xBC 0x02 0x00 0x00 0x7F;
27
28  // testjsrpackage/javacard/Import.cap
29  0x80 0xB2 0x04 0x00 0x00 0x7F;
30  0x80 0xB4 0x04 0x00 0x0E 0x04 0x00 0x0B 0x01 0x02
31  0x01 0x07 0xA0 0x00 0x00 0x00 0x62 0x01 0x01 0x7F;
32
33  0x80 0xBC 0x04 0x00 0x00 0x7F;
34
35  // testjsrpackage/javacard/Applet.cap
36  0x80 0xB2 0x03 0x00 0x00 0x7F;
37  0x80 0xB4 0x03 0x00 0x14 0x03 0x00 0x11 0x01 0x0D
38  0x59 0x92 0x51 0x6E 0x83 0x6D 0x71 0x44 0x25 0x64
39  0x39 0x34 0x90 0x00 0x8C 0x7F;
40
41  0x80 0xBC 0x03 0x00 0x00 0x7F;
42
43  // testjsrpackage/javacard/Class.cap
44  0x80 0xB2 0x06 0x00 0x00 0x7F;
45  0x80 0xB4 0x06 0x00 0x0F 0x06 0x00 0x0C 0x00 0x80
46  0x03 0x02 0x00 0x02 0x07 0x01 0x00 0x00 0x00 0x99
47  0x7F;
48
49  0x80 0xBC 0x06 0x00 0x00 0x7F;
50
51  // testjsrpackage/javacard/Method.cap
52  0x80 0xB2 0x07 0x00 0x00 0x7F;
53  0x80 0xB4 0x07 0x00 0x20 0x07 0x00 0xEB 0x00 0x05
54  0x10 0x18 0x8C 0x00 0x02 0x18 0x10 0x14 0x90 0x0B
```

```
55  0x3D 0x03 0x03 0x38 0x3D 0x04 0x10 0xF1 0x38 0x3D
56  0x05 0x03 0x38 0x3D 0x06 0x03 0x38 0x7F;
57  0x80 0xB4 0x07 0x00 0x20 0x3D 0x07 0x10 0x0F 0x38
58  0x3D 0x08 0x10 0x74 0x38 0x3D 0x10 0x06 0x10 0x72
59  0x38 0x3D 0x10 0x07 0x10 0x61 0x38 0x3D 0x10 0x08
60  0x10 0x6E 0x38 0x3D 0x10 0x09 0x10 0x7F;
61  0x80 0xB4 0x07 0x00 0x20 0x73 0x38 0x3D 0x10 0x0A
62  0x10 0x6C 0x38 0x3D 0x10 0x0B 0x10 0x61 0x38 0x3D
63  0x10 0x0C 0x10 0x74 0x38 0x3D 0x10 0x0D 0x10 0x65
64  0x38 0x3D 0x10 0x0E 0x10 0x63 0x38 0x7F;
65  0x80 0xB4 0x07 0x00 0x20 0x3D 0x10 0x0F 0x10 0x32
66  0x38 0x3D 0x10 0x10 0x10 0x63 0x38 0x3D 0x10 0x11
67  0x10 0x78 0x38 0x3D 0x10 0x12 0x10 0x90 0x38 0x3D
68  0x10 0x13 0x03 0x38 0x87 0x00 0x18 0x7F;
69  0x80 0xB4 0x07 0x00 0x20 0x05 0x90 0x0B 0x3D 0x03
70  0x10 0x90 0x38 0x3D 0x04 0x03 0x38 0x87 0x01 0x7A
71  0x02 0x30 0x8F 0x00 0x03 0x3D 0x8C 0x00 0x04 0x8B
72  0x00 0x05 0x7A 0x04 0x21 0x19 0x8B 0x7F;
73  0x80 0xB4 0x07 0x00 0x20 0x00 0x06 0x2D 0x1A 0x04
74  0x25 0x75 0x00 0x41 0x00 0x02 0xFF 0xA4 0x00 0x0D
75  0xFF 0xF2 0x00 0x27 0x19 0x8B 0x00 0x07 0x3B 0x19
76  0xAD 0x01 0x92 0x5B 0x8B 0x00 0x08 0x7F;
77  0x80 0xB4 0x07 0x00 0x00 0x20 0x19 0xAD 0x01 0x03 0xAD
78  0x01 0x92 0x5B 0x8B 0x00 0x09 0x70 0x22 0x19 0x8B
79  0x00 0x07 0x3B 0x19 0xAD 0x00 0x92 0x5B 0x8B 0x00
80  0x08 0x19 0xAD 0x00 0x03 0xAD 0x00 0x7F;
81  0x80 0xB4 0x07 0x00 0x0E 0x92 0x5B 0x8B 0x00 0x09
82  0x70 0x08 0x11 0x6D 0x00 0x8D 0x00 0x0A 0x7A 0x7F;
83  0x80 0xBC 0x07 0x00 0x00 0x7F;
84
85  // testjsrpackage/javacard/StaticField.cap
86  0x80 0xB2 0x08 0x00 0x00 0x7F;
87  0x80 0xB4 0x08 0x00 0x0D 0x08 0x00 0x0A 0x00 0x00
88  0x00 0x00 0x00 0x00 0x00 0x00 0x00 0x00 0x00 0x7F;
89  0x80 0xBC 0x08 0x00 0x00 0x7F;
90
91  // testjsrpackage/javacard/ConstantPool.cap
92  0x80 0xB2 0x05 0x00 0x00 0x7F;
93  0x80 0xB4 0x05 0x00 0x20 0x05 0x00 0x2E 0x00 0x0B
94  0x02 0x00 0x00 0x00 0x02 0x00 0x00 0x01 0x06 0x80
95  0x03 0x00 0x01 0x00 0x00 0x00 0x06 0x00 0x00 0x01
96  0x03 0x80 0x03 0x01 0x03 0x80 0x0A 0x7F;
97  0x80 0xB4 0x05 0x00 0x11 0x01 0x03 0x80 0x0A 0x07
98  0x03 0x80 0x0A 0x09 0x03 0x80 0x0A 0x05 0x06 0x80
99  0x07 0x01 0x7F;
100 0x80 0xBC 0x05 0x00 0x00 0x7F;
101
102 // testjsrpackage/javacard/RefLocation.cap
103 0x80 0xB2 0x09 0x00 0x00 0x7F;
104 0x80 0xB4 0x09 0x00 0x1B 0x09 0x00 0x18 0x00 0x08
105 0x7B 0x0F 0x2D 0x08 0x03 0x0F 0x08 0x03 0x00 0x0C
106 0x05 0x8A 0x04 0x03 0x07 0x15 0x09 0x0B 0x06 0x09
107 0x0B 0x08 0x7F;
108 0x80 0xBC 0x09 0x00 0x00 0x7F;
109
110 0x80 0xBA 0x00 0x00 0x00 0x7F;
111
112 // create Testjsrapplet applet
113 0x80 0xB8 0x00 0x00 0xf 0xd 0x59 0x92 0x51 0x6E
114 0x83 0x6D 0x71 0x44 0x25 0x64 0x39 0x34 0x90 0x00
115 0x7F;
116
117 // Select the installer applet
```

```
118  0x00 0xA4 0x04 0x00 0x09 0xa0 0x00 0x00 0x00 0x62
119  0x03 0x01 0x08 0x01 0x7F;
120
121  powerdown;
```

References

1. Java Platform Micro Edition (Java ME). http://www.oracle.com/technetwork/java/javame/.
2. JSR-000118 Mobile Information Device Profile 2.0. http://jcp.org/aboutJava/communityprocess/final/jsr118/.
3. JSR 177 Experts Group. Security and Trust Services API (SATSA) v2.1 for J2ME. http://jcp.org/aboutJava/communityprocess/final/jsr177/index.html.
4. Third Generation Partnership Project (3GPP). Specification of the Subscriber Identity Module-Mobile Equipment (SIM-ME) interface (Release 1999). TS 11.11 V8.14.0 (2007–06). http://www.3gpp.org/.
5. Oracle/Sun Microsystems. Java Card Platform Specification v2.2.1. http://www.oracle.com/technetwork/java/javacard/downloads/index.html.
6. International Organization for Standardization. ISO/IEC 7816 parts 1–15. 2005. http://www.iso.org/.
7. GlobalPlatform. Card Specification v2.2. http://www.globalplatform.org/.
8. Z. Chen. Java Card Technology for Smart Cards: Architecture and Programmer's Guide. Addison-Wesley Longman Publishing Co., Inc., Boston, MA, USA., 2000.
9. Eclipse Open Source Community. http://www.eclipse.org/.
10. Java Code Signing for J2ME. http://www.oracle.com/technetwork/java/index.html.
11. T. Lindholm and F. Yellin. The Java Virtual Machine Specification, Second Edition. http://java.sun.com/docs/books/jvms/.
12. B. W. Kernighan and D. M. Ritchie. The C programming Language. Prentice Hall, 1988.
13. Royal Holloway, University of London. Smart Card Centre website. http://www.scc.rhul.ac.uk/books/ssed/embedded/chapter_23.

Chapter 24
Wireless Sensors (Languages/Programming/ Developments Tools/Examples)

Jérémie Albert, Lionel Barrère, Serge Chaumette and Damien Sauveron

Abstract This chapter focuses on three major wireless sensor node technologies (Sun SPOTS, Arduino and TinyOS) to help the reader choose what would best fit his/her applications. Our goal is to provide the basic useful information required to quickly start working (or just playing) with them in less than a few hours.

24.1 Introduction

As can be seen from the previous chapters (and by doing some basic search on the Web), there is a large number of different wireless sensors and communicating objects that are available to be used in real world applications. It is of course impossible to know all of them and furthermore to learn/explain how to program or use all of them. Therefore we have chosen a number of architectures/systems/platforms which we believe are representative both in terms of hardware and of software layers: Sun SPOT, Arduino, TinyOS. These are also the most widely used in real world applications. For each of these architectures, we briefly describe it, the software environment that must be installed to develop for this target and we give basic code examples that

J. Albert (✉)
Solutions Architect, Ezakus, Bruges, Belgium
e-mail: jeremie.albert@ezakus.com

L. Barrère
H5 Audits, Paris, France
e-mail: lionel.barrere@h5audits.com

S. Chaumette
LaBRI, University of Bordeaux, Bordeaux, France
e-mail: serge.chaumette@labri.fr

D. Sauveron
University of Limoges, Limoges, France
e-mail: damien.sauveron@unilim.fr

K. Markantonakis and K. Mayes (eds.), *Secure Smart Embedded Devices,*
Platforms and Applications, DOI: 10.1007/978-1-4614-7915-4_24,
© Springer Science+Business Media New York 2014

Fig. 24.1 A Sun SPOT node

we believe will make it easier for the reader to get started. It should also be noted
that NFC and RFIDs will be major components of the Cyber Physical Space. They
are now combined with sensing capabilities and thus would have a natural space in
this chapter. For example, WIMA USA, which is the leading Near Field Communication (NFC) event, recently decided on talking about sensor technology. Products
are also becoming available in the medical sector. Nevertheless, these technologies
are covered in a another chapter of this book.

24.2 Sun SPOTs (Sun Small Programmable Object Technology)

24.2.1 Introduction

Sun SPOTs [12] (see Fig. 24.1) are small nodes that host a wireless connection and
that are powered by a battery which is rechargeable by USB. They embed a number of
sensors and actuators. Dedicated sensor boards can be bought/designed and plugged
on top of the other boards to achieve some specific operation or acquire some specific
measures. Thanks to the embedded radio board (IEEE 802.15.4 [5]), Sun SPOTs can
also be used as nodes of a global sensor network. It should be noted that a software
layer supporting mesh networking and including a Link Quality Routing Protocol
(LQRP) is provided, should it be useful. From a user perspective Sun SPOTs and
the associated software environment, provide for quick and easy hardware/software
design. The associated development language is Java, a Sun SPOT application being
a MIDlet.

To summarize, as described by Oracle on sunspotworld.com, "The Sun SPOT is
a Java programmable embedded device designed for flexibility."

Fig. 24.2 The different layers of a Sun SPOT

24.2.2 History

The history of Sun SPOTs started in 2004, when it was decided in the Sun Labs to create an in-house sensor that would be powerful and easy to program. After Sun was bought by Oracle, the distribution of Sun SPOTs was suspended for a while. Nevertheless, the project was still active; for instance, a University Session has been run at JavaOne 2010 in San Francisco. SUN SPOTs are now developed in the Oracle labs and have been rebranded and are available to buy on the Wcb [3].

24.2.3 Hardware Overview

A Sun SPOT is a physically stack of several layers: a battery, a processor board and a number of sensor boards (see Fig. 24.2). On top of that lies what Oracle calls a removable Sunroof that both protects the sensors and closes the stack of hardware boards.

The sensor boards that come with the SPOTs included in the development kit (see Fig. 24.1) contain:

- Sensing components: a tricolor light sensor, an accelerometer and an IR receiver. It should be noted that a temperature sensor is embedded on the main (processor) board.
- Output components: an audio speaker, tricolor LEDs, etc.
- A number of I/O interfaces: 4 digital GPIO pins, 4 analog in lines, an I2C interface, and a serial line coming from the processor board.

For a complete list of features the reader is referred to the official specifications [13].

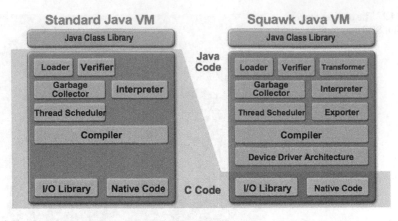

Fig. 24.3 The Squawk VM compared to the standard Java VM. (Image courtesy of Oracle)

24.2.4 Software Overview

A Sun SPOT runs a small Java Virtual Machine called Squawk [9].[1]

This virtual machine has been designed for small devices and is thus perfectly suited for Sun SPOTs. A Software Development Kit is available for download (see below) that contains the tools that make programming Sun SPOTs, debugging and testing applications relatively easy. We will be using the latest version (as of writing) of this kit, i.e. rev8 of v6.0 which is referred to as the yellow release. Squawk obeys the Java CLDC 1.1 specification and one of its major differences with a standard Java stack is that most of it is written in Java (see Figs. 24.2 and 24.3).

24.2.5 How to Start with a Sun SPOT

24.2.5.1 Buying a Simple Test Platform

It is possible to buy a basic development kit from Oracle that contains:

- two Sun SPOTs;
- a base station, i.e. a Sun SPOT that does not support stacking of other boards and is most of the time connected to a PC via a USB cable and used to communicate with the other SPOTs (referred to as Free Range SPOTs) and to bring data back to the PC (where it can be processed);
- the required software tools.

---

[1] Oracle and Java are registered trademarks of Oracle and/or its affiliates. Other names may be trademarks of their respective owners.

Fig. 24.4 The Sun SPOT
manager tool

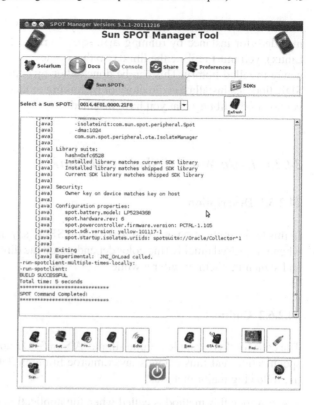

24.2.5.2 Installation of the Development Environment

In the tutorials given by Oracle, they use the Netbeans IDE environment. In order
to be consistent, we have decided on using the same environment. We use the 7.1
release, which is freely available from the Netbeans web site [6]. Squawk being a
Java virtual machine, the Java environment is required. In our tests we use a Java 6
JRE installed on a Linux machine.

It is also necessary to install a piece of software called the Sun SPOT Manager (see
Fig. 24.4), which is the heart of the development platform. It is available from the Sun
SPOT web site. The installation process is initiated by going to a web page [10]. The
installation script then checks the configuration. It checks in particular the availability
of a Java Development Kit and of the Ant system [15]. It offers to install Netbeans,
but it is recommended to do so beforehand, since this installation sometimes causes
problems if your station is not in a proper configuration (it may be the case that this
results in an old version of NetBeans being installed). It also installs the Netbeans
Sun SPOT modules. The Sun SPOT development kit uses the RXTX library [8] to
talk with the SPOTs. RXTX provides (among other features) a serial communication
layer for Java applications. If you are using a 64 bit machine, the library that comes
with the Sun SPOT Manager is not the proper one (it is for a 32 bit platform).

You thus need to install the 64 bit version yourself (if not already installed). Once installed (for instance by running `apt-get install librxtx-java` under Linux), you must replace the librxtx.so that comes with the Sun SPOT SDK by the newly installed library. This can be done either by copying your new library in the SDK libraries installation directory or by linking in that same directory to the 64 bit version of the library that you have installed.

24.2.6 Hello World ("Shake and Blink")

24.2.6.1 Description

In this basic application, we will make a Sun SPOT say hello by switching its LEDs when it is moved quickly (the color depending on the direction in which it is moved) and switching them off after a while.

24.2.6.2 Coding

The code for this example is quite simple. A Sun SPOT application obeys the MIDlet specification and thus extends javax.microedition.midlet.MIDlet by implementing the following major methods:

- startApp(): this method is called when the application is started.
- destroyApp(): this method is called when the application id destroyed.
- pauseApp(): this method is called when the application is paused.

To exit a MIDlet it is required to call the notifyDestroy() method. A MIDlet interacts with the embedded sensors and the radio communication board via a set of predefined APIs which are well documented and easy to understand. From a practical point of view the simplest way to proceed is to copy one of the projects that come with the SDK, so as to automatically inherit the configuration/compilation/deployment xml files used by Ant. You simply need to change the name of the project, the names of the source files (if required) and to update the MANIFEST file, which can be accessed through the Files tab associated to the project in Netbeans. Listing 24.1 gives the code of our HelloWorld application, which is straightforward to understand.

Listing 24.1 HelloWorld.java

```
1    package org.sunspotworld.demo;
2
3    import com.sun.spot.resources.Resources;
4    import com.sun.spot.resources.transducers.IAccelerometer3D;
5    import com.sun.spot.resources.transducers.ITriColorLEDArray;
6    import com.sun.spot.service.BootloaderListenerService;
7    import javax.microedition.midlet.MIDletStateChangeException;
8
9    public class HelloWorld extends javax.microedition.midlet.MIDlet {
10
```

```
11        private ITriColorLEDArray LED; /* get a ref. to the array of LEDs */
12        private IAccelerometer3D accelerometer; /* get a ref. to the accelerometer */
13
14        protected void startApp() throws MIDletStateChangeException {
15
16            // Listen to USB input and pass control to the bootloader
17            BootloaderListenerService.getInstance().start();
18            LED = (ITriColorLEDArray)Resources.lookup(ITriColorLEDArray.class);
19            accelerometer = (IAccelerometer3D) Resources.lookup(IAccelerometer3D.class);
20
21                    while (true){
22                      try {
23                          if (accelerometer.getAccelX()>1.0) {
24                              LED.setOn(); LED.setRGB(255,0,0);
25                              Thread.sleep(500);
26                          }
27                          else if (accelerometer.getAccelY()>1.0) {
28                              LED.setOn(); LED.setRGB(0,255,0);
29                              Thread.sleep(500);
30                          }
31                          else if (accelerometer.getAccelZ()>1.0) {
32                              LED.setOn(); LED.setRGB(0,0,255);
33                              Thread.sleep(500);
34                          }
35                          else
36                              LED.setOff();
37                      } catch (Exception ex) {
38                          ex.printStackTrace();
39                      }
40                    }
41        }
42
43        protected void pauseApp() { }
44        protected void destroyApp(boolean unconditional)
45            throws MIDletStateChangeException { }
46    }
```

24.2.6.3 Deployment

Deployment on the SPOT connected to the station is easily achieved in Netbeans
by selecting Build and then Deploy to SPOT from the project menu. It may be the
case that you need to upgrade your Sun SPOT, which can be achieved by using the
Sun Spot Manager. The Manager also enables to check that the SPOT is properly
connected (USB cable) to the station and thus taken into account by the software
environment.

24.2.6.4 Execution

The MIDlet can be launched either by:

1. Resetting the Sun SPOT (by pressing the Control Button at the bottom of the
   SPOT).
2. Selecting Run in the menu of the project in Netbeans. By doing so (which of
   course requires the SPOT to remain connected to the PC) it makes it possible

to see the output (println) of the application in Netbeans, what is useful for debugging purposes.

24.2.7 Networked Sun SPOTs Applications

24.2.7.1 Description

The application that we are going to show now will use two SPOTs, a free range SPOT collecting light measurements at regular intervals and sending them to a base station SPOT that will print them to its standard output so that it can be seen in the Netbeans interface.

24.2.7.2 Communicating Between SPOTs

Communication between SPOTs is supported by several APIs. Here, for the sake of simplicity, we will only describe Stream Connections, but packet oriented communications are also available. A connection between two spots is setup using the following piece of code.

```
1   StreamConnection conn = (StreamConnection)
2   Connector.open(''radiostream://nnnn.nnnn.nnnn.nnnn:xxx'');
```

where nnnn.nnnn.nnnn.nnnn stands for the address of the SPOT to connect to and xxx is a port number (optional). The other SPOT must do the same so that the connection is established. Thereafter input and output streams can be retrieved in a classical manner.

```
1   DataInputStream dis = conn.openDataInputStream();
2   DataOutputStream dos = conn.openDataOutputStream();
```

24.2.7.3 Coding

The application is composed of two MIDlets. A first MIDlet called Collector.java is run on the base station SPOT to collect and print the samples that are sent by the second MIDlet (Sampler.java) running on the free range SPOT.

Listing 24.2 Collector.java
```
1   package org.sunspotworld.demo;
2   [...]
3
4   public class Collector extends javax.microedition.midlet.MIDlet {
5
6
7       protected void startApp() throws MIDletStateChangeException {
8       try {
9           // Listen USB input and pass control to the bootloader
10          BootloaderListenerService.getInstance().start();
11          StreamConnection con =
```

```
12        (StreamConnection) Connector.open("radiostream://0014.4F01.0000.3817");
13        DataInputStream in = new DataInputStream(con.openInputStream());
14
15        while (true) {
16          System.out.println("Collecting next sample");
17          System.out.println("Value = " + in.readDouble());
18        }
19      } catch (Exception ex) {
20            ex.printStackTrace();
21      }
22
23    }
24
25    [...]
26  }
```

Listing 24.3 Sampler.java

```
1   package org.sunspotworld.demo;
2
3   [...]
4
5   public class Sampler extends javax.microedition.midlet.MIDlet {
6
7       private ITriColorLEDArray LED; // get a ref. to the LEDs
8       private IAccelerometer3D accelerometer; // get a ref. to the accelerometer
9
10      protected void startApp() throws MIDletStateChangeException {
11
12      try {
13          // Listen to USB input and pass control to the bootloader
14          BootloaderListenerService.getInstance().start();
15          StreamConnection con=
16            (StreamConnection) Connector.open("radiostream://0014.4F01.0000.21F8");
17          DataOutputStream out = new DataOutputStream(con.openOutputStream());
18
19          LED = (ITriColorLEDArray) Resources.lookup(ITriColorLEDArray.class);
20          ILightSensor lightSensor =
21            (ILightSensor) Resources.lookup(ILightSensor.class);
22          LED.setOn();
23
24          while (true) {
25            LED.setOn(); LED.setRGB(255, 0, 0);
26            Thread.sleep(500);
27            LED.setOff();
28
29            out.writeDouble(lightSensor.getAverageValue());
30            out.flush();
31            Thread.sleep(500);
32
33            LED.setOn(); LED.setRGB(0, 255, 0);
34          }
35      } catch (Exception ex) {
36            ex.printStackTrace();
37      }
38    }
39
40    [...]
41  }
```

24.2.7.4 Deployment and Execution

Deployment and execution is achieved as for the HelloWorld application. It should be noted that, given the configuration, the Sampler code on the free range SPOT is run by resetting it and that the Collector on the base station is run using the Netbeans menu Run feature. Netbeans then displays the output of the SPOT in one of its windows.

Communication between the base station and a standalone application running on the PC it is connected to will not be described here. An API is provided to achieve this and is documented in the Sun SPOT Programmers Manual [11].

24.3 Arduino

24.3.1 Introduction and History

Arduino [18, 20] is an open hardware platform and an open-source software stack [1]. It has been designed for diverse research field projects. It can be used for prototyping innovative ideas as well as teaching classical issues related to embedded systems programming. The hardware is based on a set of independent boards (named shields), sensors, lights, motors, etc. that can be connected to each other depending on the intended objective. Arduino nodes are always made up of a main board and of some (may be zero) optional devices. The main board is an Atmel AVR processor with I/O access. Examples of additional shields, sensors, etc. are presented in the next section. The Arduino project began in Ivrea, Italy in 2005. Massimo Banzi and David Cuartielles, who founded this project, named it after Arduin of Ivrea (955–1015). He is the main historical character of this Italian town and was king of Italy from 1012 to 1014. This project is a fork of the open-source Wiring Platform.

An example of a real world application based on Arduino is MPGuino [16]. This system allows to get a real-time measurement of the miles per gallon consumption for 90s (even late 80s) vehicles that use injectors that do not provide this kind of information. The consumption is given in real-time to the driver using a LCD screen. This application relies on an open platform called *opengauge* [7].

24.3.2 Hardware Overview

As explained above, Arduino is a set of independent open hardware and open-source software. This makes possible the design of non official Arduino-compatible shields. Figure 24.5 presents the Arduino Duemilanove board which is one of the official Arduino main boards and the Arduino WiFi shield which is one of the optional Arduino shields.

Fig. 24.5 An Arduino main board and a WiFi shield

The Arduino Duemilanove board is based on the ATmega168 (Flash memory 16 KB, SRAM 1 KB and EEPROM 512 Bytes) or ATmega328 (Flash memory 32 KB, SRAM 2 KB and EEPROM 1 KB) 16 MHz microcontroller. It operates using 5 Volts derived from 6 to 20 Volts input. It has 14 digital I/O pins and 6 analog input pins. The ATmega328 board version is very similar to the Arduino Uno board. The Arduino WiFi shield allows an Arduino node to communicate using a WiFi connection. Among the other related shields are the Arduino Bluetooth shield whose name speaks for itself.

Arduino hardware and software are available under a copyleft license. However the name Arduino can only refer to official boards. The non-official Arduino-compatible products are named using duino name variants (e.g. CraftDuino [4] or SunDuino [14] to name just two).

Arduino has been designed to test and implement solutions which rely on unusual electronic hardware combinations. The low cost of classical boards and shields is also a factor that encourages prototyping with this platform.

24.3.3 Software Overview

The most efficient way to start coding Arduino software is to download the IDE (Integrated Development Toolkit) from the official website. This IDE is available for several environments such as Microsoft Windows, Mac OS X and Linux (32 bits and 64 bits) operating systems. It comes with a C/C++ library that is called Wiring (as explained above). Source code of this IDE (which can also be downloaded from the official website) is under the GNU Public General Licence v2. Once installed, the

Fig. 24.6 Overview of the
Arduino hardware and software stacks

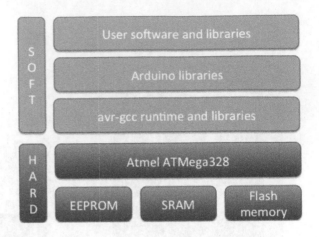

IDE enables writing and cross-compiling C/C++ source code for Arduino nodes. It
also allows to easily flash a USB-connected Arduino node.

Figure 24.6 presents a global view of the Arduino hardware and software stacks.
The software stack relies on the avr-gcc (available from the gcc.gnu.org web site)
compiler and the corresponding libc library. Arduino nodes execute code compiled
from C-like source code. The programming paradigm is thus imperative.

24.3.4 How to Start with a Arduino

24.3.4.1 Buying a Simple Test Platform

The best way to learn who the Arduino hardware distributors are is to refer to the
Arduino official website. Arduino Uno boards can be bought for some 30 USD. This
board only provides computing capabilities. As previously explained, Arduino is a
very flexible solution for prototyping, therefore each additional component can be
bought independently from the others.

ZigBee, WiFi and Bluetooth shields are commonly used to provide Arduino nodes
with a wireless communication means. For instance Ethernet shields can be used for
wired Ethernet-based communication. Starter kits often contain numerous devices
that can sense temperature, power, light, torsion, noise, etc. LEDs, push-buttons and
small electric engines are also often provided in these starter kits.

There are also LCD screens (as illustrated in Fig. 24.8) that can be plugged into
Arduino nodes. They are the good illustration of the numerous possibilities given by
this technology.

Fig. 24.7 Red and yellow LEDs, push-buttons and an electric engine

Fig. 24.8 A LCD screen for Arduino nodes

24.3.4.2 Installation of the Development Environment

Arduino is a very well documented technology. Its official website provides an Integrated Development Environment (IDE) for Linux, Mac OS X and Microsoft Windows operating systems. It simplifies the development and the compilation of suitable C/C++ source code and the installation of the corresponding binaries on Arduino nodes.

24.3.5 Hello World ("Blinking SOS")

24.3.5.1 Description

In a classical configuration, Arduino main boards do not have any screen. Therefore the following source code does not print any "Hello World" character string, but it blinks the Morse code for "SOS". It is however a good starting point to understand the structure of a Arduino program. It is composed of the setup and loop functions which are both mandatory. The setup function is run once at the start of the program and the loop function is then ceaselessly called. We make the node say "SOS" thanks to its led which blinks every second (by means of a timed loop).

24.3.5.2 Coding and Deployment

Listing 24.4 HelloWorld.cc

```
1    int led = 13; // configure with the pin number for the led on your board
2    int time = 400;
3    int timeShort = time / 4;
4    int timeLong = 3 * time / 4;
5    int delayBetweenSignals = time / 4;
6
7    void setup(){
8      pinMode(led, OUTPUT);
9      digitalWrite(led, LOW);
10   }
11
12   void shortBlink(){
13     digitalWrite(led, HIGH);
14     delay(timeShort);
15     digitalWrite(led, LOW);
16     delay(delayBetweenSignals);
17   }
18
19   void longBlink(){
20     digitalWrite(led, HIGH);
21     delay(timeLong);
22     digitalWrite(led, LOW);
23     delay(delayBetweenSignals);
24   }
25
26   void morseCodeS(){
27     longBlink();
28     longBlink();
29     longBlink();
30   }
31
32   void morseCodeO(){
33     shortBlink();
34     shortBlink();
35     shortBlink();
36   }
37
38   void loop(){
39     morseCodeS();
40     morseCodeO();
41     morseCodeS();
42   }
```

Source code of Listing 24.4 corresponds to our Hello World program and is quite straightforward to understand. The pinMode function (line 8) configures the pin to which the led is connected to be an output pin. The digitalWrite function (line 9) sets the voltage of the pin. The voltage is set to 5V (or 3.3V on 3.3V boards) if the second parameter is HIGH and is set to 0V (ground) if the second parameter is LOW. However, if the pin had been configured to be an input pin (using the pinMode function), writing a HIGH value with digitalWrite would have enabled a 20K pullup resistor and writing a LOW value would have disabled this pullup. The Morse code for "SOS" is output using the led of the board.

Deployment on the node connected to the station is achieved by using the IDE. It is really straightforward (it can be achieved in one click).

24.3.6 Networked Arduino Application

24.3.6.1 Communication Between Arduino Nodes

Arduino nodes can communicate with each other using short-range communication technologies such as WiFi, Bluetooth or ZigBee (to do so the main boards have to be equipped with the required shields).

The following program is very similar to the previous one. For space reasons, its source code is not completely given here, but it is sufficient to understand communication principles between Arduino nodes. We assume here that the nodes are equipped with ZigBee enabled shields plugged into their serial I/O interface.

The sender program sets the serial interface speed to 9600 baud. Then it repeats what follows. First, it sends the value 1 to its serial interface, turns its led on and waits for 500 ms. Second, it sends the value 0 to its serial port, turns its led off and waits for 500 ms.

Listing 24.5 Sender.cc

```
1   int led = 8; // configure with the pin number for the led on your board (8 here)
2
3   void setup(){
4     // Start up our serial port
5     // We configured our XBEE devices at 9600 bps
6     Serial.begin(9600);
7   }
8
9   void loop(){
10    Serial.print(1);
11    digitalWrite(led, HIGH);
12    delay(500);
13    Serial.print(0);
14    digitalWrite(led, LOW);
15    delay(500);
16  }
```

The receiver program also sets the speed of its serial interface to 9600 baud. It is of course fundamental that the sender and the receiver choose the same speed. Then, the node reads incoming data from its serial interface. When data is available, it turns its led off if the received value is 0 and it turns it on if the received value is 1.

Listing 24.6 Receiver.cc

```
1   int incomingByte = 0;
2   int led = 8;  // configure with the pin number for the led on your board (8 here)
3
4   void setup(){
5     // Start up our serial port
6     // We configured our XBEE devices at 9600 bps
7     Serial.begin(9600);
8   }
9
10  void loop(){
11    if (Serial.available() > 0) {
```

Fig. 24.9 Overview of the TinyOS hardware and software stacks

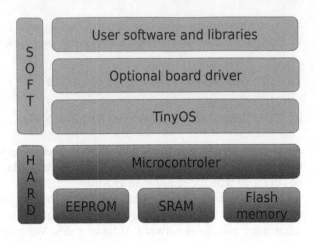

```
12        // read the incoming byte:
13        incomingByte = Serial.read();
14    }
15    if(incomingByte ==  '0'){
16        digitalWrite(led, LOW);
17    }
18    else if (incomingByte == '1'){
19        digitalWrite(led, HIGH);
20    }
21    }
```

As a result, when the receiver is within range of the sender, the two nodes blink synchronously.

24.4 TinyOS

24.4.1 Introduction

TinyOS [17, 22] is an open-source operating system that relies on the event based programming paradigm; it has been especially designed for sensor networks. It provides application developers with an API (see Fig. 24.9) and the underlying stack to collect data from sensors and to communicate between nodes or with a base station.

The programming language is the NesC, which is an extension of the C language that makes it possible to define components and their relationships (for instance how a given component uses the output of another component). Every NesC source code is composed of two files:

- the first file describes the interfaces that are used / defined by the application (like a *.h* header file);

- the second file contains the implementation (event handlers) of the application features (like a .c file).

The application and the kernel are compiled and linked together using a cross compiling approach to get one single executable that can then be uploaded to the target chip. The goal of this section is to show how TinyOS is different from the other platforms by giving its main characteristics, therefore we will not enter in any configuration or compilation details.

24.4.2 Hardware Overview

There are many vendors who have developed hardware supporting TinyOS and the associated drivers (see Table 24.1).

Each TinyOS hardware configuration is composed of a main board (usually named a mote) embedding a microcontroler, a wireless communication unit, a power supply, some control leds and a connector to plug in sensors shields. Figure 24.10a shows a Crossbow Mica2 mote that is historically the most popular WSN hardware platform. It also shows the two shields, Crossbow MTS300 and Crossbow MTS420. The Crossbow MTS300 board is equipped with a temperature sensor, a light sensor, a microphone and a 4 kHz buzzer (Fig. 24.10b). The Crossbow MTS420 board, a high end one, is composed of a humidity and temperature sensor, a light sensor, a barometer, a two axes accelerometer and a GPS chip (Sirf Star II) (Fig. 24.10c).

Platforms embedding ZigBee are becoming the standard, and the most popular configurations, as of writing, are MicaZ, TelosB and IRIS.

Table 24.1 Available TinyOS compatible hardware platforms

Model	Networking	Architecture	Specifications
Crossbow Mica2	433/915 Mhz	Atmel ATMEGA 128L	8-bit, 16 Mhz
Crossbow MicaZ	2.4 Ghz, 802.15.4 (ZigBee Baseband)	Atmel ATMEGA 128L	8-bit, 16 Mhz
IRIS	2.4 Ghz, 802.15.4	Atmel ATmega1281	8-bit, 16 Mhz
Mulle	Bluetooth	Renesas M16C/62 MCU	16-bit, 10 Mhz
TelosB	2.4 Ghz, 802.15.4	Texas Instruments MSP430	16-bit, 8/16 Mhz
BTnode	433/915 Mhz et Bluetooth	Atmel ATMEGA 128L	8-bit, 16 Mhz
Imote (Intel Mote)	2.4 Ghz, 802.15.4	Intel XScale	32-bit, 416 Mhz

Fig. 24.10 Crossbow Mica2 mote and two sensor shields

24.4.3 How to Start with TinyOS

24.4.3.1 Buying a Simple Test Platform

The first step is to choose a hardware platform. There are many parameters to take into account: one can decide on a specific configuration because of its cost, its radio communication baseband, the available sensors, etc. The reader should know that some motes such as Mica2, MicaZ or IRIS need a base station to be able to connect to a computer whereas TelosB comes with a built-in USB interface. Internet is the best way to get a TinyOS platform, each vendor (see Table 24.1) having a sales area on the web, should you require a quotation.

24.4.3.2 Installing TinyOS

The TinyOS development kit is available as a Debian package. One just has to add the TinyOS repositories to the Debian */etc/apt/sources.list* file. TinyOS can then be installed as follows:

```
1   $ apt–get update
2   $ apt–get install tinyos−2.1.1
```

The TinyOS install directory is */opt/tinyos-2.1.1* and contains all what is needed to use it with Mica or TelosB motes: examples of applications, drivers for most used sensorboards, etc. The install directory contains two major subdirectories:

- *tos*, that contains the TinyOS core and the hardware drivers (*platforms* and *sensorboards*);
- *apps*, that contains some example applications.

More configuration details are available in the TinyOS documentation wiki [17].

24.4.4 Hello World ("Sense and Blink")

24.4.4.1 Description

In this section we present a simple example that consists in using the temperature sensor and then showing the sampled value using the embedded leds.

TinyOS compatible hardware platforms do not provide (at least in the basic configurations) any screen to get feedback from the running application. There are thus two ways to give information to the user:

- the first, the simplest, consists in using the embedded leds;
- the second, the best, is to communicate via a serial interface or over the network to eventually display information on a desktop.

For this simple example, we have chosen the first approach.

24.4.4.2 Coding

The first file (Listing 24.7) defines which components will be used in the application and how the events will be handled. The second file (Listing 24.8) contains the application code, i.e. the functions that handle the events coming from the different components.

Listing 24.7 DemoSenseApp.nc

```
1    configuration DemoSenseAppC
2    {
3    }
4    implementation {
5      // define which implementations will be used
6      components Sense, MainC, LedsC, new TimerMilliC(), new TempC() as Sensor;
7
8      Sense.Boot -> MainC;
9      Sense.Leds -> LedsC;          // Sense.Leds is implemented by LedsC
10     Sense.Timer -> TimerMilliC;
11     Sense.Read -> Sensor;
12   }
```

Listing 24.8 DemoSense.nc

```
1    module DemoSense
2    {
3      uses {// define the interfaces that are used
4        interface Boot;
5        interface Leds;
6        interface Timer<TMilli>;
7        interface Read<uint16_t>; // We use the interface Sense.Read with unsigned
8                                   // short result
9      }
10   }
11   implementation // application source code (handlers)
12   {
13     // sampling frequency in milliseconds
14     #define SAMPLING_FREQUENCY 100
```

```
15
16      event void Boot.booted() {
17        call Timer.startPeriodic(SAMPLING_FREQUENCY);
18      }
19
20      event void Timer.fired()
21      {
22        call Leds.led0Toggle();
23        call Read.read();
24      }
25
26      event void Read.readDone(error_t result, uint16_t data)
27      {
28        call Leds.led0Toggle();
29        if (result == SUCCESS){
30          call Leds.led1On();
31          call Leds.led2Off();
32        }else{
33          call Leds.led1On();
34          call Leds.led2Off();
35        }
36      }
37    }
38  }
```

24.4.4.3 Deployment and Execution

Once the source code of the application is written, it has to be compiled to run on the target platform. The configurations for the most popular platforms are already defined in the distribution of TinyOS, therefore compiling for instance for Mica2 motes just requires to run the make mica2 command. Of course avr-gcc is needed for cross compiling, but it is automatically installed with TinyOS. Once the application is compiled, it can be uploaded to the mote. If the mote has an interface to connect to a host computer, as is the case for the TelosB platform, it can be done through this interface. In the other cases a base station with a serial or USB interface is needed, where the target mote must be plugged in. For example in the case of a Mica2 mote connected through a MIB510 serial base station on the /dev/ttyS0 serial port, the application is uploaded using the following command:

```
1   $ make mica2 reinstall mib510,/dev/ttyS0
```

24.4.5 Networking with TinyOS

Networking is needed by most if not all real world TinyOS applications because the most popular application domain is the collection of data related to the evolution of some natural phenomenon at different locations. For space reasons, we will not give here a TinyOS application that would use the network. We just want to emphasise on the fact that TinyOS is event based, and networking is no exception. Thus, receiving a message is done by implementing a function that handles an event:

```
1    event message_t* Receive.receive(message_t* bufPtr,void* payload, uint8_t len){
2                      [....]
3    }
```

It is also possible to use high level network interfaces for instance by using an IP stack such as BLIP [2] (Berkeley Low-power IP stack). An IP stack protocol can be interesting for routing or security reasons, but it depends on the target application and on the acceptable battery consumption that its use would imply.

24.5 Sensor Network Deployment: An Example

24.5.1 Introduction

In this section, an example of a Sensor Network deployment is presented to illustrate some common issues which the reader may face during a real deployment. The goal of the application that we consider is to track a physical phenomenon. Dynamic data are collected by several sensor nodes which communicate them to their neighbours using broadcast based communication (routing would not be suitable here because we need to save energy and to be tolerant to the failure or disconnection of nodes). A detailed description of the application can be found in [19]. We focus here on the the main problems we faced when deploying it. Such an application can be used in different contexts (battlefields, crisis management, etc.) and thus the reader can imagine any adaptation that would fit his/her own needs. The application does not only observe the considered phenomenon, but it performs some additional actions to react to the appearance of specific patterns in the collected data that can mean potential threats or emerging trends. For example, if some data samples are missing (there can be several reasons for that which will be discussed later), the application should use interpolation or extrapolation techniques to provide a continuous view of the distributed phenomenon. The physical world value that is collected by our application in this particular example is the temperature.

24.5.2 Hardware Architecture

The Wireless Sensors network which we consider here is composed of Crossbow Mica2 nodes on top of which are sensor boards measuring temperature (e.g. Crossbow MTS420—see Sect. 24.4.2). In this configuration, all the nodes do not play the same role. We can distinguish two categories (see Fig. 24.11):

1. Basic sensor nodes whose role is to take a measurement, to broadcast it and to forward the values received from the other nodes.
2. Client nodes which are more powerful: they are composed of a station (laptop, desktop, or any device with a reasonable amount of computing capabilities) that embeds a Mica2 mote so as to be able to communicate with the basic sensor

Fig. 24.11 A sensor network
configuration to sample and
collect a physical value

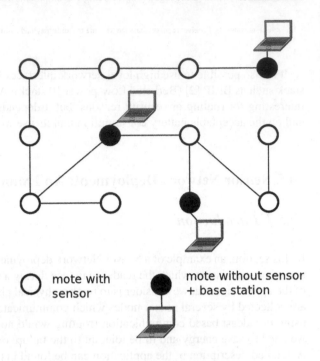

nodes. These client nodes communicate with other motes to collect the measurement samples in order to build a global view of the overall phenomenon in the area covered by the network (a global view is thus constructed from local measurements).

24.5.3 The Time Synchronization Issue

Synchronizing the motes is required to be able to achieve a coherent temporal fusion of the measurements taken by the different sensors. The synchronization algorithm cannot only rely on the broadcast and propagation of the time from a single node as the many delays that can appear during the transmission of a message have to be taken into account and must be compensated. These delays are due to the time needed for the sending mote to access the transmission channel, pack and effectively send the message. For the receiver, these delays are due to the time needed to receive and process a message. To overcome this issue, we implemented the Flooding Time Synchronization Protocol [21] by using the Time Stamping library of TinyOS. The goal is to get an accurate estimation of the time required to receive a message, taking into account the transmission delays and the local clock drifts (one cannot afford to embed expensive clocks that would provide small drifts in sensor nodes). The readers interested in this feature for their applications are encouraged to read [21].

24.5.4 Data Collection, Location and Network Load Issues

The goal of our algorithm is to collect the temperature still keeping track of where and when each measurement was taken. These measurements are collected by the client nodes after they have been sent in broadcast mode and propagated over the network by the basic sensors nodes. Our algorithm is quite simple. Once a basic sensor node is synchronized, each time its timer is fired it measures the temperature and broadcasts it. To avoid overloading the network with useless messages (already dealt with or too old), a time to live (TTL) field is added to the messages. This is a well-known method: the TTL field is decremented at each hop, and its initial value can be adapted depending on the radius of the network and on the number of nodes that are deployed. However when the density of motes in the broadcast range is high, using a TTL is not sufficient to properly regulate the number of messages in the network. Therefore in our algorithm, we added a table in which we store the identifiers of the most recently received messages. When receiving a message, each mote thus checks if it has already forwarded it, and if this is the case it just discards it.

It should be noted that a mote will not send its own measurements before being synchronized, even though it will forward the messages that it receives. This is typically what happens when we introduce a new mote in a network, the components of which are already synchronized.

We assume in our application that each mote knows its own location. It should be noted that in a configuration where only some motes know their location, a triangulation algorithm based on the coordinates of these motes could be used to compute the location of all the other sensors.

24.5.5 The Problem of Missing Information

When taking measurements in an adverse environment (e.g. a battlefield), which is often the case, data can be lost for many reasons:

- communication can be jammed because of interference coming from other radio equipments (or generated by the enemy);
- a sensor can run out of battery power;
- a sensor can be damaged or destroyed.

The question is then to provide the user with a continuous view of the discontinuous collected information.

In our application, we addressed this issue in a module in charge of collecting the individual measurements. For these missing values, it can provide replacements that can either be interpolated or extrapolated.

24.5.6 Conclusion

The goal of this section was to help the reader understand some of the challenges he/she will face when deploying a wireless sensor node application. We do not pretend that we covered all the problems, because they depend on the application at hand. For instance, in many cases it will be necessary to take issues related to energy and security into account. Nevertheless, most of these issues are addressed in some form or another throughout this book and the reader is thus encouraged to explore it thoroughly.

Acknowledgments The authors want to thank the reviewers for their constructive comments which were helpful to improve this chapter.

References

1. Arduino web page, http://www.arduino.cc/
2. BLIP (Berkeley Low-power IP stack). http://docs.tinyos.net/tinywiki/index.php/BLIP_Tutorial
3. Buying Sun SPOTs on the Web. http://www.sunspotworld.com/products/index.html
4. CraftDuino board. http://robocraft.ru/
5. IEEE 802.15.4. http://www.ieee802.org/15/pub/TG4.html
6. Netbeans. http://netbeans.org/
7. Open Source Fuel Efficiency Instrumentation utilities (opengauge). http://opengauge.googlecode.com
8. RXTX library. http://rxtx.qbang.org/wiki/index.php/Main_Page
9. Squawk Development Wiki. http://java.net/projects/squawk/pages/SquawkDevelopment
10. Sun SPOT Manager installation/launch web page. http://www.sunspotworld.com/SPOTManager/SPOTManager.jnlp
11. Sun SPOT Programmers Manual. http://www.sunspotworld.com/docs/index.html
12. Sun SPOT world. http://www.sunspotworld.com/
13. Sun SPOTs specification. http://sunspotworld.com/products/
14. SunDuino board. http://www.sunduino.neth.pl/
15. The Apacahe Ant project. http://ant.apache.org/
16. The MPGuino (Miles per Gallon) wiki page. http://ecomodder.com/wiki/index.php/MPGuino
17. TinyOS wiki. http://docs.tinyos.net/. Last seen 03/2012
18. Banzi, M.: Getting Started with Arduino. O'Reilly, Media (2011).
19. Barrère, L., Chaumette, S., De Peretti, C.: Delay tolerant dynamic data collection over a sensor network. In: Proceedings of the 26th IEEE Military Communications Conference (MILCOM 2007), pp. 1–7. Orlando, Fl., États-Unis (2007). DGA.
20. Margolis, M.: Arduino Cookbook. O'Reilly Series. O'Reilly, Media (2011).
21. Maróti, M., Kusy, B., Simon, G., Lédeczi, Á.: The flooding time synchronization protocol. In: SenSys '04: Proceedings of the 2nd international conference on Embedded networked sensor systems, pp. 39–49. ACM Press, New York, NY, USA (2004).
22. P. Levis, S.M., Polastre, J., Szewczyk, R., Whitehouse, K., Woo, A., Gay, D., Hill, J., Welsh, M., Brewer, E., Culler., D.: Ambient Intelligence, chap. TinyOS: An Operating System for Sensor Networks, pp. 115–148. Springer (2005).

Errata to: Secure Smart Embedded Devices, Platforms and Applications

Konstantinos Markantonakis and Keith Mayes

Errata to:
K. Markantonakis and K. Mayes (eds.), *Secure Smart Embedded Devices, Platforms and Applications*, DOI 10.1007/978-1-4614-7915-4

Biography for Dr. Keith Mayes in book front matter should read as below:

Keith Mayes B.Sc. Ph.D. (Bath) CEng FIET received his BSc (Hons) in Electronic Engineering in 1983 from the University of Bath and his PhD degree in Digital Image Processing in 1987. He is an active researcher/author with over a 100 publications in conferences, books and journals. His interests include the design of secure protocols, mobile/fixed communications systems and security tokens/NFC/RFID as well as associated attacks/countermeasures. During his first degree he was employed by Pye TVT (Philips) which designed and produced TV broadcast and studio equipment. His PhD was sponsored by Honeywell Aerospace and Defence and on completion he accepted their offer of a job. In 1988 he started work for Racal Research where he worked on a wide range of research and advanced development products and was accepted as a Chartered Engineer. In 1995 he joined Racal Messenger to continue work on a Vehicle Licence plate recognition system (Talon) and an early packet radio system (Widanet/Paknet). In 1996 Keith joined Vodafone as a Senior Manager working within the Communication Security and Advanced Development group. Early work

---

The online version of the original book can be found under DOI 10.1007/978-1-4614-7915-4

K. Markantonakis · K. Mayes
Information Security Group, Smart Card Centre, Royal Holloway, University of London, Egham, Surrey, UK
e-mail: k.markantonakis@rhul.ac.uk

K. Mayes
e-mail: keith.mayes@rhul.ac.uk

K. Markantonakis and K. Mayes (eds.), *Secure Smart Embedded Devices,*
Platforms and Applications, DOI: 10.1007/978-1-4614-7915-4_25,
© Springer Science+Business Media New York 2014

concerned advanced radio relaying systems and involved participation in international standardisation (ETSI SMG2). Later he led the Maths and Modelling team and eventually took charge of the Fraud and Security group. During this time he was training in intellectual property and licensing, culminating in membership of the Licensing Executives Society and the added responsibility for patent issues in Vodafone UK. In 2000, following some work on m-commerce and an increasing interest in Smart Cards he joined the Vodafone International organisation as the Vodafone Global SIM Card Manager, responsible for SIM card harmonisation and strategy for the Vodafone Group. In 2002, Keith left Vodafone to set up Crisp Telecom (www.crisptele.com) and in November 2002 he was also appointed as the Director of the Smart Card Centre (www.scc.rhul.ac.uk) at Royal Holloway University of London, reporting to Professor Fred Piper in the world renowned Information Security Group (www.isg.rhul.ac.uk). Keith is a Fellow of the Institution of Engineering and Technology, a Founder Associate Member of the Institute of Information Security Professionals and a member of the editorial board of the Journal of Theoretical and Applied Electronic Commerce Research (JTAER). He has also had director experience within a London stock market listed company and a subsidiary of an American communications company. Recent high profile activity included leading the expert team that carried out counter-expertise work on the Ov-Chipkaart for the Dutch transport ministry, following published attacks on the MIFARE Classic chip card; and acting as General Chair for ESORICS 2013.

In Chap. 10, following corrections need to be corrected:

Page 235, Section 10.3.2, first paragraph after bullet points, Line 3 should refer Fig. 10.3.
Page 249, Section 10.6.2, Paragraph 1, Line 6, "specialisthardware", typo - still need to add a space between the words, i.e. it should read as 'specialist hardware'.
Page 255, Section 10.7.1 , Paragraph 2, Line 7, "RTM [42]", non-existent reference—just delete [42] after RTM.
Page 256, Section 10.7.2, Paragraph 2, Line 2, "resistance", typo - still need to change resistance to resistant.

In Chap. 12, following corrections need to be corrected:

In Section 12.3.3, sentence "They offer a number of pre-implemented cryptographic services such as DES/3DES, AES, hash functions, long number arithmetic for public key operations, RSA, ECC, secure generation of random numbers,." should not have a comma before the period.

Errata to: Secure Smart Embedded Devices, Platforms and Applications

Konstantinos Markantonakis and Keith Mayes

Errata to:
K. Markantonakis and K. Mayes (eds.), *Secure Smart Embedded Devices, Platforms and Applications*, DOI 10.1007/978-1-4614-7915-4

Page vii, Foreword, Paragraph 3 Line 2: RFID NFC should read "Radio Frequency Identity (RFID), Near Field Communication (NFC)"
Page x, Preface, Paragraph 2 Line 10: FPGA should read "Field Programmable Gate Array (FPGA)"
Page x, Preface, Paragraph 3 Line 16: WSN should read "Wireless Sensor Nodes (WSN)"
Page xi, Preface, Paragraph 1 Line 11: SCADA should read "Supervisory Control And Data Acquisition (SCADA)"
Page xv, Acknowledgements, Paragraph 1 Line 8: Naem should read "Naeem"
Page xxxviii, Contributors, Paragraph 4 Line 7: Saarbrcken should be "Saarbrucken"
Page 55, Section 3.3.1, Paragraph 1 Line 7: Acorn RISC machine should read "Advanced Risk Machine"
Page 100, Section 5.3, Paragraph 2 Lines 7 and 12: GE should read "Gate Equivalent (GE)"
Page 134, Table 6.1, Heading: unit should be "μs" not "us"

Capitalization of acronyms was incorrect in the published volume. The corrected list of acronyms is provided below:

The online version of the original book can be found under DOI 10.1007/978-1-4614-7915-4

K. Markantonakis (✉) · K. Mayes
Information Security Group, Smart Card Centre, Royal Holloway, University of London, Egham, Surrey, UK
e-mail: k.markantonakis@rhul.ac.uk

K. Mayes
e-mail: keith.mayes@rhul.ac.uk

K. Markantonakis and K. Mayes (eds.), *Secure Smart Embedded Devices,*
Platforms and Applications, DOI: 10.1007/978-1-4614-7915-4_26,
© Springer Science+Business Media New York 2014

Page	Section	Para	Expansion (Abbreviation)	Correct form
3		Abstract	Radio Frequency Identification (RFID)	radio frequency identification (RFID)
3	1	1	Radio Frequency Identity (RFID)	radio frequency identity (RFID)
4	1, list 1	1	Identification Friend or Foe (IFF)	identification friend or foe (IFF)
5	1.2.1	4	Subscriber Identity Module (SIM)	subscriber identity module (SIM)
5	1.2.1	4	Universal Integrated Circuit Card (UICC)	universal integrated circuit card (UICC)
6	1.2.1	3	Identities (IDs)	identities (IDs)
7	1.2.2	1	Automatic Teller Machines (ATM)	automatic teller machines (ATM)
7	1.2.2	1	Point of Sale terminals (POS)	point of sale terminals (POS)
7	1.2.2, list 5	1	Personal Identification Numbers (PIN codes)	personal identification numbers (PIN codes)
9	1.2.4	1	Set Top Boxes (STB)	set top boxes (STB)
12	1.2.7	2	High Medium or Low (H:M:L)	high medium or low (H:M:L)
13	1.3.1	2	Security Element (SE)	security element (SE)
13	1.3.1	2	Near Field Communication (NFC)	near field communication (NFC)
15	1.3.2	4	Application protocol Data Unit (APDU)	application protocol data unit (APDU)
17	1.4.2	1	Message Authentication Code (MAC)	message authentication code (MAC)
17	1.4.2	1	Key Management System (KMS)	key management system (KMS)
17	1.4.2	1	Near Field Communication (NFC)	near field communication (NFC)
20	1.6	1	Read Only Memory (ROM)	read only memory (ROM)
21	1.7	2	Operating System (OS)	operating system (OS)
21	1.7	2	Trusted Service Manager (TSM)	trusted service manager (TSM)
23	1.7	5	2-D Discrete cosine transform (DCT)	2-D discrete cosine transform (DCT)
31	Table 2.1		Level 1 (L1)	level 1 (L1)
38	2.3.1, list 2	2	Level 2 (L2)	level 2 (L2)
38	2.3.1, list 2	2	Level 3 (L3)	level 3 (L3)
38	2.3.1, list 2	2	Instruction set architecture (ISA)	instruction set architecture (ISA)
40	2.3.2	2	Advanced RISC Machine (ARM)	advanced RISC machine (ARM)
41	2.3.2	7	Voice-over-IP (VoIP)	voice-over-IP (VoIP)
46	2.3.4, list 3	3		

(continued)

(continued)

Page	Section	Para	Expansion (Abbreviation)	Correct form
49	Abstract	1	Central Processing Units (CPU)	central processing units (CPU)
50	3.1	3	Joint Test Action Group (JTAG)	joint test action group (JTAG)
51	3.2	1	Very Long Instruction Word (VLIW)	very long Instruction Word (VLIW)
51	3.2	1	Digital Signal Processors (DSPs)	digital signal processors (DSPs)
51	3.2	3	Integrated Circuit (IC)	integrated circuit (IC)
51	3.2	3	Bit Slice Processors (BSP)	bit slice processors (BSP)
51	3.2	3	Programmable Read Only Memories (PROM)	programmable read only memories (PROM)
51	3.2	3	Digital Signal Processor (DSP)	digital signal processor (DSP)
52	3.3	4	Integrated Circuit (IC)	integrated circuit (IC)
53	3.3	1	Complementary Metal Oxide Semiconductor (CMOS)	complementary metal oxide semiconductor (CMOS)
54	3.3.2	1	Field Programmable Gate Array (FPGA)	field programmable gate array (FPGA)
55	3.3.3	1	Acorn RISC Machine (ARM)	acorn RISC machine (ARM)
55	3.3.3	1	Differential Power Analysis (DPA)	differential power analysis (DPA)
56	3.4	2	Serial Peripheral Bus (SPI)	serial peripheral bus (SPI)
56	3.4	2	Control Area Network (CAN)	control area network (CAN)
56	3.4	2	Universal Serial Bus (USB)	universal serial bus (USB)
57	3.4	2	Application-Specific Integrated Circuits (ASICs)	application-specific integrated circuits (ASICs)
58	3.4	1	Electrically Erasable Programmable Read Only Memory (EEPROM)	electrically erasable programmable read only memory (EEPROM)
58	3.4	2	Peripheral Interface Controller (PIC)	peripheral interface controller (PIC)
59	3.4	1	One Time-Programmable (OTP)	one time-programmable (OTP)
61	3.4.1	1	Static Random Access Memory (SRAM)	static random access memory (SRAM)
61	3.4.1	1	Read Only Memory (ROM)	read only memory (ROM)
62	3.4.1	2	One-Time Programming (OTP)	one-time programming (OTP)
62	3.4.1	3	Electrically Erasable PROM (EEPROM)	electrically erasable PROM (EEPROM)
65	3.4.2	2	In-Circuit-Emulator (ICE)	in-circuit-emulator (ICE)

(continued)

(continued)

Page	Section	Para	Expansion (Abbreviation)	Correct form
67	3.5	1	Differential Electromagnetic Analysis (DEMA)	differential electromagnetic analysis (DEMA)
68	3.6	1	Trusted Platform Modules (TPM)	trusted platform modules (TPM)
72	4.1	1	Trusted Computing Group (TCG)	trusted computing group (TCG)
72	4.1	1	Trusted Multi-tenant Infrastructure and trusted network connect	trusted multi-tenant infrastructure and trusted network connect
76	4.3.1	4	TPM Entity (TPME)	TPM entity (TPME)
77	4.3.2	Table 4.1	Platform Configuration Registers (PCRs)	platform configuration registers (PCRs)
81	4.3.4.2	2	Initial Program Loader: IPL	initial program loader: IPL
108	5.5	4	Reduced Instruction Set Computing (RISC)	reduced instruction set computing (RISC)
125	6.3.1	2	Data Encryption Standard (DES)	data encryption standard (DES)
128	6.3.1.3	6	Advanced Encryption Standard (AES)	advanced encryption standard (AES)
129	6.3.1.4	4	Cipher Block Chaining (CBC)	cipher block chaining (CBC)
131	6.3.1.5	1	Cyclic Redundancy Check (CRC)	cyclic redundancy check (CRC)
134	6.3.2	1	Optimal Asymmetric Encryption Padding (OAEP)	optimal asymmetric encryption padding (OAEP)
134	6.3.3	2	Cyclic Redundancy Check (CRC)	Cyclic redundancy check (CRC)
135	6.3.3	1	Message Authentication Code (MAC)	Message authentication code (MAC)
136	6.4.1	5	Public Key Infrastructure (PKI)	public key infrastructure (PKI)
136	6.4.1	6	Certification Authority (CA)	certification authority (CA)
137	6.4.1	1	Registration Authority (RA)	registration authority (RA)
137	6.4.1	1	Point of Sale (POS)	point of sale (POS)
137	6.4.1	2	Certificate Revocation List (CRL)	certificate revocation list (CRL)
137	6.4.2	4	Near Field Communication (NFC)	near field communication (NFC)
139	6.5	1	Security Elements (SE)	security elements (SE)
147	7.2	4	Data Encryption Standard (DES)	data encryption standard (DES)
147	7.2.1	5	Advanced Encryption Standard (AES)	advanced encryption standard (AES)
148	7.2.1	2	Permuted Choice 1 (PC1)	permuted choice 1 (PC1)

(continued)

(continued)

Page	Section	Para	Expansion (Abbreviation)	Correct form
148	7.2.1	2	Permuted Choice 2 (PC2)	permuted choice 2 (PC2)
168	7.5.3.1	5	Chinese Remainder Theorem (CRT)	Chinese remainder theorem (CRT)
179	Chapter 8	2	Graphics Processing Units (GPUs)	Graphics processing units (GPUs)
	8.1.1	4	Compute Unified Device Architecture (CUDA)	compute unified device achitecture (CUDA)
	8.1.3	1	Close-to-Metal (CTM)	close-to-metal (CTM)
	8.2.1	3	Data Encryption Standard (DES)	data encryption standard (DES)
212	9.2.3	1	Side-channel Attack Standard Evaluation BOard (SASEBO)	Side-Channel Attack Standard Evaluation Board (SASEBO)
218	9.3.2.4	4	PROM (Programmable Read-Only Memory)	PROM (programmable read-only memory)
227	10 (Abstract)	1	Global System for Mobile Communications (GSM)	global system for mobile communications (GSM)
227	10 (Abstract)	1	Universal Mobile Telecommunications System (UMTS)	Universal mobile telecommunications system (UMTS)
227	10 (Abstract)	1	Subscriber Identity Module (SIM)	subscriber identity module (SIM)
228	10.1	1	Near Field Communication (NFC)	near field communication (NFC)
228	10.1	2	Subscriber Identity Module (SIM)	subscriber identity module (SIM)
229	10.1	1	Mobile Equipment (ME)	mobile equipment (ME)
229	10.2	2	Total Access Communications (TACS)	total access communications (TACS)
230	10.2.1	2	Authentication Centre (AuC)	authentication centre (AuC)
230	10.2.1	3	International Mobile Subscriber Identity—IMSI	international mobile subscriber identity—IMSI
230	10.2.1	3	Secret Key (Ki)	secret key (Ki)
230	10.2.1	4	secret key (Ki)	secret key (Ki)
230	10.2.1	4	session key (Kc)	session key (Kc)
232	10.2.1	1	IntegrityKey (IK)	integritykey (IK)
234	10.3.2	6	Mobile Equipment (ME)	mobile equipment (ME)
238	10.4.2	1	Over The Air (OTA)	over the air (OTA)
240	10.4.7	5	Near Field Communication (NFC)	near field communication (NFC)
240	10.4.7	5	Security Element (SE)	security element (SE)

(continued)

(continued)

Page	Section	Para	Expansion (Abbreviation)	Correct form
243	10.5.2	3	Focussed Ion Beam (FIB)	focussed ion beam (FIB)
246	10.5.4	2	Chinese Remainder Theorem (CRT)	Chinese remainder theorem (CRT)
247	10.6.1	2	Pure Software SIM (PSSIM)	pure software SIM (PSSIM)
247	10.6.1.1	3	Telemetry/Machine-to-Machine (T/M2M)	telemetry/machine-to-machine (T/M2M)
249	10.6.2		Hardware Shared Security Software SIM Solution (HS-SSIM)	hardware shared security software SIM solution (HS-SSIM)
249	10.6.2	2	Advanced RISC Machines Limited (ARM)	advanced RISC machines limited (ARM)
249	10.6.2	6	Data Rights Management (DRM)	data rights management (DRM)
250	10.6.2	3	Small Terminal Interoperability Platform (STIP)	small terminal interoperability platform (STIP)
251	10.6.3	4	Standalone Hardware Security SIM (SH-SIM)	standalone hardware security SIM (SH-SIM)
252	10.6.3.2	3	Electrically Erasable Programmable Read-Only Memory (EEPROM)	electrically erasable Programmable Read-Only Memory (EEPROM)
252	10.6.3.2	3	Hardware Security Module (HSM)	hardware security module (HSM)
254	10.7	2	Trusted Platform Module (TPM)	trusted platform module (TPM)
254	10.7	2	Trusted Computing Platform Alliance (TCPA)	trusted computing platform alliance (TCPA)
254	10.7	2	Trusted Computing Group (TCG)	trusted computing group (TCG)
255	10.7.1	5	Root of Trust for Measurement (RTM)	Root of trust for measurement (RTM)
255	10.7.1	5	Root of Trust for Storage (RTS)	Root of trust for storage (RTS)
255	10.7.1	5	Root of Trust for Reporting (RTR)	Root of trust for reporting (RTR)
255	10.7.1	6	Core Root of Trust for Measurement' (CRTM))	core root of trust for measurement (CRTM)
255	10.7.1	6	BIOS Boot Block (BBB)	BIOS boot block (BBB)
255	10.7.1	6	Platform Configuration Registers (PCRs)	platform configuration registers (PCRs)
256	10.7.2	1	Platform Configuration Registers (PCRs)	platform configuration registers (PCRs)
257	10.7.3	2	Mobile Trusted Platform (MTM)	mobile trusted Platform (MTP)
257	10.7.4	Section title	Mobile Trusted Platform (MTP)	Mobile trusted platform (MTP)
258	10.7.4	1	Mobile Local-Owner Trusted Module (MLTM)	mobile local-owner trusted module (MLTM)
258	10.7.4	1	Mobile Remote-Owner Trusted Module (MRTM)	mobile remote-owner trusted module (MRTM)

(continued)

(continued)

Page	Section	Para	Expansion (Abbreviation)	Correct form
258	10.7.4	3	Protection Profile (PP)	protection profile (PP)
261	10.8	1	Pure Software SIM (PSSIM)	pure software SIM (PSSIM)
261	10.8	2	Hardware Shared Software SIM (HS-SSIM)	hardware shared software SIM (HS-SSIM)
264	10.8.1	2	Security Element (SE)	security element (SE)
268	11.2	4	Real-Rime Locating System (RTLS)	real-time locating system (RTLS)
275	11.3.1.1	1	Angle of Arrival (AoA)	Angle of arrival (AoA)
276	11.3.1.3	2	Time of arrival (ToA)	time of arrival (ToA)
276	11.3.1.3	2	Time difference of arrival (TDoA)	time difference of arrival (TDoA)
276	11.3.1.3	2	Round trip time (RTT)	round trip time (RTT)
278	11.4	2	Round-trip-time (RTT)	round-trip-time (RTT)
279	11.4	1	Ultra-WideBand (UWB)	ultra-wideBand (UWB)
287	12.1	2	Electronic control units (ECU)	electronic control units (ECU)
287	12.1	2	Controller Area Network (CAN)	controller area network (CAN)
288	12.1		Local interconnect network (LIN)	local interconnect network (LIN)
288	12.1	2	Dedicated short range communication (DSRC)	dedicated short range communication (DSRC)
288	12.1	2	Vehicle-to-Vehicles communication - V2V	vehicle-to-vehicles communication - V2V
288	12.1	2	Vehicle-to-Infrastructure communication =- V2I	vehicle-to-Infrastructure communication =- V2I
288	12.1	2	Universal Mobile Telecommunications System (UMTS)	Universal Mobile Telecommunications System (UMTS)
288	12.1	2	Global Positioning System (GPS)	Global positioning system (GPS)
295	12.2.2.2	6	Radio Data System (RDS)	radio data system (RDS)
295	12.2.2.2	6	Global Positioning System (GPS)	global positioning system (GPS)
296	12.2.4	3	Anti-lock braking system (ABS)	anti-lock braking system (ABS)
296	12.2.4	3	Electronic stability program (ESP)	electronic stability program (ESP)
298	12.2.6	6	Original Equipment Manufacturer (OEM)	original equipment manufacturer (OEM)
299	12.2.8	7	Engine control unit (ECU)	engine control unit (ECU)

(continued)

(continued)

Page	Section	Para	Expansion (Abbreviation)	Correct form
300	12.2.8	1	Center for Automotive embedded systems security (CAESS)	Center for Automotive Embedded Systems Security (CAESS)
300	12.2.8	1	Tyre pressure monitoring system (TPMS)	tyre pressure monitoring system (TPMS)
302	12.3	3	Open vehicular secure platform (OVERSEE)	open vehicular secure platform (OVERSEE)
302	12.3	3	Electronic Control Unit (ECU)	electronic control unit (ECU)
303	12.3.1	2	Operating System (OS)	operating system (OS)
303	12.3.1	2	Real Time Operating System (RTOS)	real time operating system (RTOS)
303	12.3.1	3	Electronic code unit (ECU)	electronic code unit (ECU)
306	12.3.2	3	Lightweight directory access protocol (LDAP)	lightweight directory access protocol (LDAP)
306	12.3.2	3	NSS (Name Switch Service)	NSS (name switch service)
306	12.3.2	3	PAMs (Pluggable Authentication Modules)	PAMs (pluggable authentication modules)
307	12.3.3	1	Electrically Erasable Programmable Read-Only Memory (EEPROM)	electrically erasable programmable read-only memory (EEPROM)
307	12.3.3	2	Protection Profiles (PP)	protection profiles (PP)
307	12.3.3	8	Secure Hardware Extension (SHE)	secure hardware extension (SHE)
308	12.3.3	1	E-safety Vehicle Intrusion Protected Applications (EVITA)	E-safety vehicle intrusion protected applications (EVITA)
336	14.2	2	national oceanographic and atmospheric administration (NOAA)	National Oceanographic and Atmospheric Administration (NOAA)
336	14.2	2	defense advanced research projects agency (DARPA)	Defense Advanced Research Projects Agency (DARPA)
340	14.4	1	Inter-Integrated Circuit (I2C)	inter-integrated circuit (I2C)
345	14.5.2.1	2	Denial-of-Service (DoS)	denial-of-service (DoS)
345	14.5.2.1	4	Frequency-Hopping Spread Spectrum (FHSS)	frequency-hopping spread spectrum (FHSS)
357	15.3.2	3	Universal Integrated Circuit Card (UICC)	universal integrated circuit card (UICC)
357	15.3.2	4	SigIn-SigOut-Communication (S2C)	SigIn-SigOut-communication (S2C)
359	15.4	6	Portable Operating System Interface (POSIX)	Portable operating system interface (POSIX)

(continued)

(continued)

Page	Section	Para	Expansion (Abbreviation)	Correct form
362	15.5.3	3	GSM association (GSMA	GSM Association (GSMA)
362	15.5.3	4	Digital Rights Management (DRM)	digital rights management (DRM)
364	15.6.1	4	Japanese industrial standard (JIS)	Japanese Industrial Standard (JIS)
364	15.6.2		NFC Data Exchange Format (NDEF)	NFC data exchange format (NDEF)
365	15.6.2	1	Multipurpose Internet Mail Extensions (MIME)	multipurpose internet mail extensions (MIME)
369	Abstract	1	Basic Input Output System (BIOS)	basic input output system (BIOS)
369	Abstract	1	Trusted Platform Module (TPM)	trusted platform module (TPM)
370	16.1.1	1	Random Access Memory (RAM)	random access memory (RAM)
370	16.1.1	2	Operating System (OS)	operating system (OS)
370	16.1.1	3	User Interface (UI)	user interface (UI)
383	17 (Abstract)	1	Hardware Security Modules (HSMs)	Hardware security modules (HSMs)
383	17 (Abstract)	1	Tamper Resistant Security Modules (TRSMs)	tamper resistant security modules (TRSMs)
383	17 (Abstract)	1	Personal Identification Numbers (PINs)	personal identification numbers (PINs)
383	17 (Abstract)	1	Federal Information Processing Standard (FIPS)	federal information processing standard (FIPS)
383	17.1	2	Hardware Security Module (HSM)	hardware security module (HSM)
384	17.1	1	Tamper Resistant Security Modules (TRSM)	tamper resistant security modules (TRSM)
384	17.1	1	Secure Cryptographic Devices (SCD)	secure cryptographic devices (SCD)
384	17.1	1	Certification Authority (CA)	certification authority (CA)
384	17.1	1	Point of Sales terminals (POS terminals)	point of sales terminals (POS terminals)
384	17.2	3	Payment card industry (PCI)	payment card industry (PCI)
385	17.2	2	Card Verification Value (CVV)	card verification value (CVV)
385	17.2	2	Card Verification Key (CVK)	card verification key (CVK)
385	17.2	2	Personal Identification Number (PIN)	personal identification number (PIN)
385	17.2	3	PIN Verification Key (PNK)	PIN verification key (PNK)
385	17.2	4	Automatic Teller Machine (ATM)	automatic teller machine (ATM)
386	17.2	4	Message Authentication Code (MAC)	message authentication Code (MAC)

(continued)

(continued)

Page	Section	Para	Expansion (Abbreviation)	Correct form
388	17.3	3	Host Master Key (HMK)	host master key (HMK)
390	17.4	1	Federal Information Processing Standard (FIPS)	federal information processing standard (FIPS)
392	17.4	Table 17.2	Code of Federal Regulations (CFR)	Code of federal regulations (CFR)
395	17.5	3	PIN Entry Device (PED)	PIN entry device (PED)
396	17.6	1	Key Encrypting Key (KEK)	key encrypting key (KEK)
397	17.6	4	Electronic Codebook (ECB)	electronic codebook (ECB)
398	17.6	4	Cipher Block Chaining (CBC)	cipher block chaining (CBC)
398	17.6	4	Initial Vector (IV)	initial vector (IV)
401	17.7	4	Account Related Data (ARD)	account related data (ARD)
410	18.2	3	Protection Profile (PP)	protection profile (PP)
413	18.2.1	2	(PIN Transaction Security) PTS	(PIN transaction security) PTS
413	18.2.2	4	information technology Security Evaluation Facility (ITSEF)	information technology security evaluation facility (ITSEF)
413	18.2.2	4	CommerciaL Evaluation Facility (CLEF)	commerciaL evaluation facility (CLEF)
414	18.2.2	3	common criteria management committee (CCMC)	Common Criteria Management Committee (CCMC)
414	18.2.2	3	common criteria development board (CCDB)	Common Criteria Development Board (CCDB)
415	18.2.3	4	Common Evaluation Methodology (CEM)	common evaluation methodology (CEM)
421	18.4.1	3	Derived Test Requirements (DTRs)	derived test requirements (DTRs)
433	19.2.2.2	2	Static Random Access Memory (SRAM)	static random access memory (SRAM)
452	20.2	2	Programmable Logic Controller (PLC)	programmable logic controller (PLC)
452	20.2	2	Remote Terminal Units (RTU)	remote terminal units (RTU)
453	20.2	3	Human Machine Interface (HMI)	Human machine interface (HMI)
453	20.2	4	Application programming interface (API)	application programming interface (API)
457	20.3.2	5	American petroleum institute (API)	American Petroleum Institute (API)
458	20.3.2	6	electric reliability organisation (ERO)	Electric Reliability Organisation (ERO)
464	20.4.3	3	Network Intrusion Detection System (NIDS)	network intrusion detection system (NIDS)

(continued)

(continued)

Page	Section	Para	Expansion (Abbreviation)	Correct form
475	21.1	1	Central Processing Unit (CPU)	central processing unit (CPU)
477	21.3.1	1	Peripheral Interface Controller (PIC)	Peripheral interface controller (PIC)
477	21.3.1	1	Microprocessor unit (MPU)	microprocessor unit (MPU)
479	21.3.4	1	Integrated Development Environment (IDE)	integrated development environment (IDE)
481	21.4.1	1	National Institute of Standards and Technology (NIST)	National Institute of Standards and Technology (NIST)
498	22.2.1	1	Java Card virtual machine (JCVM)	Java card virtual machine (JCVM)
499	22.2.1	1	Java Card runtime environment (JCRE)	Java card runtime environment (JCRE)
508	22.5.1	2	Proprietary Application Identifier Extension (PIX)	proprietary application identifier extension (PIX)
515	23.1	1	Mobile information device profile (MIDP)	mobile information device profile (MIDP)
515	23.1	1	Application protocol data unit (APDU)	application protocol data unit (APDU)
517	23.1.1	1	Java card remote method invocation (JCRMI)	Java card remote method invocation (JCRMI)
517	23.1.1	8	Remote Method Interface (RMI)	remote method interface (RMI)
517	23.1.1	8	Generic Connection Framework (GCF)	generic connection framework (GCF)
517	23.1.1	8	Java Card Remote Method Invocation (JCRMI)	Java card remote method invocation (JCRMI)
518	23.1.2	4	Java card virtual machine (JCVM)	Java card virtual machine (JCVM)
518	23.1.2	4	Java virtual machine (JVM)	Java virtual machine (JVM)
519	23.2.1	2	Integrated Development Environment (IDE)	integrated development environment (IDE)
519	23.2.1	12	Connected limited device configuration (CLDC)	connected limited device configuration (CLDC)
519	23.2.1	12	Over-the-air (OTA)	over-the-air (OTA)
519	23.2.1	12	Java Community Process (JCP)	Java community process (JCP)
519	23.2.1	14	Mobile Information Device Profile (MIDP)	Mobile information device profile (MIDP)
519	23.2.1	15	Connected Limited Device Configuration (CLDC)	Connected limited device configuration (CLDC)
521	23.2.1	1	Generic connection framework (GCF)	generic connection framework (GCF)
521	23.2.1	3	Answer-to-reset (ATR)	answer-to-reset (ATR)
521	23.2.1	3	Personal Identification Number (PIN)	personal identification number (PIN)

(continued)

(continued)

Page	Section	Para	Expansion (Abbreviation)	Correct form
521	23.2.1	4	Application identifier (AID)	application identifier (AID)
521	23.2.1	4	Uniform resource locator (URL)	uniform resource locator (URL)
523	23.2.1.2	1	Java application descriptor (JAD)	Java application descriptor (JAD)
527	23.2.2.1	1	Application identifier (AID)	application identifier (AID)
542	24.1	1	Near Field Communication	near field communication
542	24.2	Heading	Sun Small Programmable Object Technology (SPOTs)	Sun small programmable object technology (SPOTs)
542	24.2.1	1	Link Quality Routing Protocol	link quality routing protocol
553	24.3.4.2	1	Integrated Development Environment (IDE)	integrated development environment (IDE)

Index

K. Markantonakis and K. Mayes (eds.), *Secure Smart Embedded Devices,*
Platforms and Applications, DOI: 10.1007/978-1-4614-7915-4,
© Springer Science+Business Media New York 2014

Printed in the United States
By Bookmasters